The Theory and Practice of Innovation Policy

PRIME SERIES ON RESEARCH AND INNOVATION POLICY IN EUROPE

Series Editor: Philippe Larédo, *ENPC (France) and University of Manchester, UK*

The last decade has seen dramatic transformations in the configuration of national systems of innovation in Europe and in the way in which knowledge is produced. This important new series will provide a forum for the publication of high quality work analysing these changes and proposing new frameworks for the future.

In particular it will address the changing dynamics of knowledge production within the NBIC (Nano, Bio, Information, Cognitive) sciences and within different industries and services. It will also examine the changing relationship between science and society and the growing importance of both regional and European public authorities.

The series will include some of the best empirical and theoretical work in the field with contributions from leading and emerging scholars.

Titles in the series include:

Universities and Strategic Knowledge Creation
Specialization and Performance in Europe
Edited by Andrea Bonaccorsi and Cinzia Daraio

The Handbook of Technology Foresight
Concepts and Practice
Edited by Luke Georghiou, Jennifer Cassingena Harper, Michael Keenan, Ian Miles and Rafael Popper

The Sociology of Scientific Work
The Fundamental Relationship between Science and Society
Dominique Vinck

The Theory and Practice of Innovation Policy
An International Research Handbook
Edited by Ruud E. Smits, Stefan Kuhlmann and Philip Shapira

The Theory and Practice of Innovation Policy

An International Research Handbook

Edited by

Ruud E. Smits
Utrecht University, The Netherlands

Stefan Kuhlmann
University of Twente, The Netherlands

Philip Shapira
University of Manchester, UK; Georgia Institute of Technology, USA

PRIME SERIES ON RESEARCH AND INNOVATION POLICY IN EUROPE

Edward Elgar
Cheltenham, UK • Northampton, MA, USA

Published by
Edward Elgar Publishing Limited
The Lypiatts
15 Lansdown Road
Cheltenham
Glos GL50 2JA
UK

Edward Elgar Publishing, Inc.
William Pratt House
9 Dewey Court
Northampton
Massachusetts 01060
USA

A catalogue record for this book
is available from the British Library

Library of Congress Control Number: 2010925936

ISBN 978 1 84542 848 8 (cased)

Printed and bound by MPG Books Group, UK

Contents

List of Contributors

Anna Bergek is an Associate Professor in the Department of Management and Engineering, Linköping University, Sweden. She received a PhD in technology management from Chalmers University of Technology, Göteborg, Sweden in 2002, served on a government commission on energy research 2002-03 and was an assistant professor at Linköping University 2003-08. Dr Bergek's research interests include innovation and industry dynamics, especially the emergence of new industries and the renewal of mature industries following technological discontinuities, with implications for firm strategy and innovation policy. Particular issues of concern are external knowledge sourcing in R&D and innovation, the evolution of technological innovation systems, renewable energy technology and complex product systems.

Patries Boekholt is Director of the Technopolis Group (Amsterdam office). She holds an MA in political science (1987) from the Catholic University of Nijmegen, and a PhD (1995) from Aston Business School, UK, with a thesis on innovation policy in the European Community: reorienting towards diffusion. She has worked on science, technology and innovation policy for over 15 years. Activities include international benchmark and policy studies, strategic advice and evaluations of research programmes and institutions. Dr Boekholt has worked in many European countries and for several international organizations such as the European Commission, OECD and UNIDO. Before joining Technopolis she worked as senior consultant at the TNO Centre for Science and Technology Policy Studies, The Netherlands.

Knut Blind is Professor of Innovation Economics, Berlin University of Technology, Germany. He also heads the Competence Center for Regulation and Innovation of the Fraunhofer Institute for Systems and Innovation Research and is Professor of Standardisation at the Rotterdam School of Management, Erasmus University, The Netherlands. He studied economics at the University of Freiburg, Germany, and Brock University, Canada. Prof. Blind has published widely on the interface between regulation and innovation, with a focus on IPR regimes and standardization. He is author of *The Economics of Standards: Theory, Evidence, Policy* (Edward Elgar, 2004).

Bo Carlsson is Frank Tracy Carlton Professor of Economics at the Weatherhead School of Management at Case Western Reserve University in Cleveland, Ohio, where he also directs the Executive Doctor of Management Program. He received his BA in economics from Harvard College and his PhD in economics from Stanford University. From 1972 to 1984 he was a Research Associate and for five years

Associate Director of the Industrial Institute for Economic and Social Research (IUI) in Stockholm, Sweden. Prof. Carlsson has published over 20 books and numerous articles on industrial economics, small business and entrepreneurship, technological change and industrial policy. His current research interests include the digital economy, entrepreneurship, technology transfer, intellectual property management, and the nature and role of technological innovation systems in economic growth.

Cristina Chaminade is an Associate Professor in innovation studies at the Centre for Innovation, Research and Competence in the Learning Economy (CIRCLE), Lund University, Sweden. She received a PhD in economics from the Autonomous University of Madrid, Spain. Dr Chaminade is also affiliated with LEAP4D (Learning Economy Analysis for Development) at the Research Policy Institute, Sweden, and is a visiting professor at the HHL Leipzig Graduate School of Management, Germany. Her research focuses on understanding innovation and knowledge creation in the globalized learning economy and on innovation systems policy. She is coordinating several projects on globalization of innovation in China, India, South Africa and Central America. Dr Chaminade has authored numerous articles, book chapters and books in the fields of innovation systems and innovation system policy, particularly for developing countries.

Susan E. Cozzens is Professor of Public Policy and Associate Dean for Research, Ivan Allen College, Georgia Institute of Technology, Atlanta, USA. She holds a PhD from Columbia University and a BS from Michigan State University. Prof. Cozzens has been active in science, technology, and innovation policy for 30 years, as researcher and practitioner. She has been Director of the Office of Policy Support at the National Science Foundation and edited two major journals in the field. Her research is on STI and inequalities, including projects on international collaboration in biofuels; distributional consequences of emerging technologies; nanotechnology, equity, and development; and water and sanitation in developing countries.

Ben Dankbaar is Professor of Business Administration at the Nijmegen School of Management of the Radboud University Nijmegen, The Netherlands, where he also holds a Chair in Management and Technology in the Faculty of Science. He studied economics and social sciences at the University of Amsterdam and has a PhD in economics from the University of Maastricht. Earlier he worked with the University of Amsterdam, the Science Centre Berlin (Germany) and the Maastricht Economic Research Institute on Innovation and Technology (MERIT), University of Maastricht. Prof. Dankbaar is an expert on the automobile industry and has been involved in studies of strategic issues in the industry since the early 1980s. He has published extensively on issues of technology and work organization, organization design, innovation management and innovation policy. His current work is concerned with radical innovation and industry-university collaboration in a range of industries.

Jakob Edler is Professor of Innovation Policy and Strategy, and Research Director at the Manchester Institute of Innovation Research, Manchester Business School, University of Manchester, UK. He holds a PhD (1999) from the University of Mannheim, Germany, and Masters degrees from the University of Mannheim and Dartmouth College, USA. Until 2007, he headed the Department for Innovation Systems and Policy at the Fraunhofer Institute for Systems and Innovation Research, Germany. Prof. Edler has written widely on public and private research and innovation strategies and policies, and on the analysis of dynamics in innovation systems. His publications include *Software Patents: Economic Impacts and Policy Implications* (Edward Elgar, 2005, with M. Friedewald and K. Blind) and *Changing Governance of Research and Technology Policy: The European Research Area* (Edward Elgar, 2003, with S. Kuhlmann and M. Behrens).

Charles Edquist holds the Ruben Rausing Chair in Innovation (since 2003) and is Director, Centre for Innovation, Research and Competence in the Learning Economy (CIRCLE), at Lund University, Sweden. He received his Docent PhD (1981) in economic history from the University of Lund, Sweden and his Habilitation (1985) in economic history from Umeå University, Sweden. He also holds an MA in economics (1974) from the University of California, Berkeley, USA. From 1986 to 2003, he was Professor in the Department of Technology and Social Change, Linköping University, Sweden. Prof. Edquist has published extensively on the comparison of innovation systems, innovation policy and systems of innovation, including *Systems of Innovation: Technologies, Institutions and Organizations* (Pinter Publishers/Cassell Academic, 1997).

Lennart Elg is an analyst with the Strategy Development Division of the Swedish Agency for Innovation Systems (VINNOVA). He is also a member of the Swedish delegation to the OECD/CSTP and the Swedish representative in the OECD/CSTP/TIP group. Mr Elg's fields of expertise include innovation and technology policy analysis, research on innovation and technology policy, and technology foresight.

Stuart J.H. Graham is an Assistant Professor of Strategic Management, College of Management, Georgia Institute of Technology and Kauffman Foundation Fellow, Berkeley Center for Law and Technology, University of California, Berkeley School of Law. He received his PhD in business economics from the University of California, Berkeley, and holds advanced degrees in law (JD) and business (MBA). Dr Graham's fields of interest include firm intellectual property strategies, intellectual property transactions and markets for technology, technology entrepreneurship and the legal environment. A selection of his recent publications includes 'Why do Start-ups Patent?' in the *Berkeley Technology Law Journal* (with T. Sichelman); 'Beyond Patents: The Roles of Copyrights, Trademarks, and Trade Secrets in Technology Commercialization' in *Advances in the Study of Entrepreneurship, Innovation, and*

Economic Growth; and 'Software Patents: Good News or Bad News?' in the American Enterprise Institute/Brookings Institution Joint Center, *Intellectual Property Rights in Frontier Industries: Software and Biotechnology* (with D.C. Mowery).

David H. Guston is Professor of Political Science and Co-Director of the Consortium for Science, Policy and Outcomes at Arizona State University, where he also directs the Center for Nanotechnology in Society at ASU. He has published widely on R&D policy, technology assessment and public participation in science and technology; his *Between Politics and Science* (Cambridge University. Press, 2000) was awarded the 2002 Don K. Price Prize by the American Political Science Association for best book in science and technology policy. Prof. Guston is the series editor of the *Yearbook of Nanotechnology in Society* (Springer) and the general editor of a two-volume *Encyclopedia of Nanoscience and Society* (Sage, 2010). He is the North American editor of *Science and Public Policy*. He holds a BA from Yale and a PhD from MIT.

Marko Hekkert is Professor of Dynamics of Innovation Systems and Research Director of the Department of Innovation and Environmental Sciences, Copernicus Institute, Utrecht University, The Netherlands. He has a background in chemistry (MSc 1995) and energy and material systems modelling (PhD 2000). Prof. Hekkert's research focus on the transition to a sustainable energy system, uncertainties in energy transitions and energy innovation systems.

Pim den Hertog is a founding partner of Dialogic Innovation & Interaction, a research-based consultancy with specializations in innovation (measurement, strategic advice and policy studies), broadband/content industries, E-government and demand steering/user involvement in innovation. He is currently seconded to the Amsterdam Academic Centre for Service Innovation (AMSI) at the Amsterdam Business School. Mr den Hertog graduated in economic geography from Utrecht University. While at TNO Strategy, Technology and Policy (1990-98), he began working on service innovation, which has developed into one of his major research interests. His interests also include research and advice on innovation systems and their governance, cluster studies, innovation policies and evaluation of policy schemes.

Staffan Jacobsson is Professor in Science and Technology Policy at Chalmers University of Technology, Sweden. He holds a BA from the University of Lund, Sweden (1977) and a DPhil in economics from the University of Sussex, UK (1985). He was appointed as an Assistant Professor at Chalmers in 1987 and received his professorial appointment in 1998. Prof. Jacobsson's research initially focused on industrialization issues in newly industrializing countries where he studied the development of a capital goods sector. His current research focuses on developing a conceptual framework for studying large-scale industrial transformation processes, policy issues related to the transformation to a sustainable energy sector, and science policy.

Stefan Kuhlmann is Professor and Chair of the Department of Science, Technology, and Policy Studies (STePS) at the University of Twente, The Netherlands. He received his PhD (Dr. rer.pol., 1986) and Habilitation (1998) from University of Kassel, Germany. He has worked at the University of Kassel (1979-88), the Fraunhofer Institute for Systems and Innovation Research, Karlsruhe (1988-2006), and the University of Utrecht (2001-06). Prof. Kuhlmann's research interests include research and technological innovation as social and political processes, with a focus on the governance of science, research and innovation systems and policies. He has published extensively in this field. Prof. Kuhlmannn is a co-editor of *Research Policy*, and an editor (with P. Shapira) of *Learning from Science and Technology Policy Evaluation: Experiences from the United States and Europe* (Edward Elgar, 2003) and (with J. Edler and M. Behrens) *Changing Governance of Research and Technology Policy – the European Research Area* (Edward Elgar, 2003).

Ben R. Martin is Professor of Science and Technology Policy Studies at SPRU (Science and Technology Policy Research) at the University of Sussex, UK, and former Director of SPRU (1997 to 2004). He has carried out research for 30 years in the field of science policy, helping to establish techniques for the evaluation of research, and pioneering the notion of foresight as a tool for looking into the longer-term future of science and technology. Prof. Martin has published seven books and 50 journal articles. He is an editor of *Research Policy*, and the 1997 winner of the Derek de Solla Price Medal for Science Studies.

Rutger van Merkerk works at the University Medical Centre in Utrecht, The Netherlands, where his current activities focus on translating innovative ideas into commercially available products and services. He holds a PhD from the Department of Innovation and Environmental Sciences at Utrecht University, The Netherlands (2007). His research was on constructive technology assessment of emerging technologies and nanotechnology in particular. Dr van Merkerk's interests also include characterizing emerging technologies and anticipating prospective technological developments.

Jordi Molas-Gallart is Research Professor at the Spanish Council for Scientific Research (CSIC). He works at INGENIO, a joint research centre of CSIC and the Universitat Politècnica de València. He holds a Licenciatura in economics from the Universitat Autònoma de Bacelona, the MA in international relations from the School of Advanced International Studies of the Johns Hopkins University, and a DPhil from SPRU at the University of Sussex. Prof. Molas Gallart was a Fellow in the Department of Politics of the University of Glasgow (1989-91) and worked at SPRU between 1991 and 2005, first as a Research Fellow, and afterwards as a Senior

Fellow and a Senior Lecturer. His research focuses on the on the study of innovation in the defence and aerospace industries and the evaluation of science, technology and innovation policies.

Daniel Sarewitz is Professor of Science and Society, and Co-Director of the Consortium for Science, Policy, and Outcomes, at Arizona State University, USA. He holds a PhD in geological sciences from Cornell University (1986). From 1989 to 1993 he worked on Capitol Hill, first as a Congressional Science Fellow, and then as science consultant to the House of Representatives Committee on Science, Space, and Technology. Prof. Sarewitz's research focuses on the societal outcomes of science and technology policies. He is co-editor (with A. Lightman and C. Desser) of *Living with the Genie: Essays on Technology and the Quest for Human Mastery* (Island Press, 2003). He is also the co-editor of *Prediction: Science, Decision-Making, and the Future of Nature* (Island Press, 2000) and the author of *Frontiers of Illusion: Science, Technology, and the Politics of Progress* (Temple University Press, 1996).

Philip Shapira is Professor of Innovation, Management and Policy at the Manchester Institute of Innovation Research, Manchester Business School, University of Manchester, UK; and Professor of Public Policy at Georgia Institute of Technology, Atlanta, USA. He holds a PhD in city and regional planning and an MA in economics from the University of California, Berkeley, USA, as well as an MCP from the Massachusetts Institute of Technology, and a DipTP from Gloucestershire College of Art and Design. He served as a Congressional Fellow and analyst with the Office of Technology Assessment, United States Congress (1986-88). Prof. Shapira's research interests include science, technology and innovation management and policy; industry analysis; regional innovation; R&D and knowledge measurement; and policy evaluation. He has authored or co-authored numerous articles on science, technology, innovation and development topics, including in *Research Policy*, the *Journal of Technology Transfer*, the *International Journal of Technology Management*, *Scientometrics*, and other leading journals. Prof. Shapira is a Fellow of the Royal Society of Arts.

Keith Smith is Head of Science and Innovation Analysis for the UK Department of Innovation, Universities and Skills, and Professor of Innovation, Australian Innovation Research Centre. Previously, he was Professor at the Institute for New Technologies, United Nations University, The Netherlands (2001-03); Professor of Economics of Technology at the Norwegian University of Science and Technology (1995-2001); and Director of the Group for Studies in Technology, Innovation and Economic Policy (STEP) in Oslo (1991-2000). He has also worked with the European Commission at the Joint Research Centre in Seville, Spain, on policy issues related to the European Research Area. Prof. Smith has widely published on the nature and dynamics of innovation processes, on innovation systems theories, on innovation in low-technology industries, innovation and growth in resource-based economies, the development and analysis of innovation indicators, and on the foundations of public policies supporting innovation.

Ruud Smits is Professor of Governance of Innovation Systems and Chair of the Innovation Studies Group of the University of Utrecht, The Netherlands (since 1999). He studied physics at the University of Eindhoven (1972) and trained in system dynamics at the Massachusetts Institute of Technology and Dartmouth College. He received his PhD (1991) in Technology Assessment and Technology Policy from Amsterdam Free University. He spent the earlier part of his career with the TNO-Centre for Technology and Policy Studies, serving as its director (1988-96). Prof. Smits focuses his work on the management of innovation processes, the development of user-oriented types of innovation policy and management, technology assessment and the development of instruments to facilitate participation of relevant actors in decision-making processes on innovation, and on the dynamics of innovation systems.

Morris Teubal is Professor of Economics, Hebrew University, Jerusalem, Israel. His PhD in economics (1967) is from the University of Chicago on inflation and economic growth (1967). His research interests include economic growth processes, technological change, technology infrastructure policy, and the evolution of national systems of innovation (particularly Israel's). Prof. Teubal has written on a range of technology and innovation topics, including market-building policies: a framework for research, technology and innovation policies for Europe and for NICs; the formulation of an R&D policy for Israel; and high-tech start-ups and venture capital: evolutionary and systemic perspectives.

Geert Vissers is a Research Fellow in ISIS, the Institute for Science, Innovation, and Society of Radboud University Nijmegen, Faculty of Science, The Netherlands. He studied sociology at Nijmegen University, and obtained a PhD in business science from Erasmus University Rotterdam. Dr. Vissers is involved in Micord, a research program focusing on university-industry research collaboration in three sectors (agrofood, chemicals, and machinery and equipment); and in Eurodite, a European research project investigating the contribution of firm and territorial knowledge dynamics to regional development. He has published on simulation gaming in the contexts of organizational strategy development and policy evaluation, creativity in multidisciplinary teams, the knowledge economy, and virtual collaboration.

Acknowledgments

The development of this book received support from the PRIME Network of Excellence, a network sponsored by the 6th Framework Programme of the European Commission to advance long-term research and shared infrastructures on policies for research and innovation in the move towards the European Research Area (ERA). PRIME-sponsored editor-author workshops in Utrecht (2005), Karlsruhe (2006) and Enschede (2007) helped to prepare the book. Additional support was received from the Dutch Knowledge network on System Innovations and Transitions (KSI). We gratefully acknowledge these sponsors. We also warmly acknowledge the Innovation Studies Group at the Faculty of Geosciences, Utrecht University, for institutional support. We are particularly grateful to Ineke Bakker, Camiel Smits, Jenny Canters and Hilde Meijer for their assistance in bringing this book to publication and to Annie Bidgood and Stephen Carley for their help in proofreading. We must add that all opinions, findings and recommendations expressed in this volume are those of the authors and do not necessarily reflect the views of our sponsors.

Ruud Smits, Utrecht, The Netherlands
Stefan Kuhlmann, Enschede, The Netherlands
Philip Shapira, Manchester, UK, and Atlanta, USA

1. Introduction. A Systemic Perspective: The Innovation Policy Dance

Stefan Kuhlmann, Philip Shapira and Ruud Smits

This book is about the interaction between innovation practice, innovation policy and innovation theory. The book aims to increase insight into this interactive process, searching for options to improve the effectiveness and efficiency of policy and of innovative practice. The book also identifies conceptual or empirical lacunae and questions that can guide future research. We seek to inform and support policymakers, innovators and innovation scientists – not so much by providing a state-of-the-art, but by presenting a theory-based vision on the interrelated dynamics of innovation policy, practice and research.

In this introductory chapter we set out a framework for the book by taking three steps. First we introduce a systemic perspective on innovation as a social phenomenon as well as a subject of public policy and of socio-economic analysis. Second we offer the reader a guiding metaphor for the book's underlying approach: the 'dance of innovation practice, policy, and theory'. Finally we present an overview of the book's chapters.

INNOVATION IN A SYSTEMIC PERSPECTIVE

However named and valued, 'innovation' – the development and adoption of new and improved ways of addressing social and economic needs and wants – has occurred in society and economy since the early days of mankind. The stirring power of innovation changed the social structure of medieval societies as the art of printing unleashed the spreading of ideas and aided the birth of modern science and technology (e.g. Zilsel et al. 2003; Roberts et al. 2007). In the mid 19th century, the train, telegraph and other communication technologies transformed countries from loosely connected cities and villages to coherent nations, while the conveyer belt – together with Tayloristic organization principles – resulted in a tremendous increase in the productivity of factories in the first half of the 20th century (e.g. Geels 2006; Hughes 1989; Mumford 1964). Innovation is of a social, economic and technological character. It emerges sometimes spontaneously, sometimes as result of actors' strategic endeavor. Innovations vary according to subject, location, actors, speed and numerous other characteristics. One may be in favor of an innovation or dislike it, facilitate or try to impede it; yet innovation occurs, in one way or another. Joseph von Schumpeter (1934) defined innovation as 'creative destruction': innovators afford 'new combinations' of hitherto disconnected ideas, knowledge domains, technologies, or markets. In this perspective, successful innovation requires transgressing limits. Also sociological analyses tell us that it is often border-crossers who dare to move around in different worlds with diverging rationalities leading them to discover new combinations and facilitating innovation (e.g. Burt 2003). Crossing boundaries is not easy: facing complex, fuzzy environments, we are inclined to simplify, to reduce our cognitive horizon ('bounded rationality') and to solidify borders. While we might sense undiscovered insights and options behind borders, we tend to be impeded by vested interests and the inertia of existing institutional, political and organizational structures from exploring such options and becoming exposed to unknown consequences.

Empirical innovation research has revealed the high level of 'path dependence' in the evolution of knowledge and institutions (Nelson and Winter 1977; Ziman 2000). Streamlined 'technological regimes' (Nelson and Winter 1982) are established which are characterized by specific patterns of technical change and mostly incremental innovations molded by the frameworks of engineers in that industry. Van den Ende and Kemp (1999) defined a technological regime 'as the complex of scientific knowledge, engineering practices, production process technologies, product characteristics, user practices, skills and procedures, and institutions and infrastructures that make up the totality of a technology'.

Rip and Kemp (1998) explicitly added to the 'grammar' of a regime the policies and actions of other technology actors including public authorities. Industrialists, scientists, policymakers and other actors in policy arenas typically follow bounded rationalities: experienced in a given arena, with limited scope, actors in one domain often lack insight into other arenas. Policy instruments and public regulation normally develop in an incremental and only rarely radical way; sometimes, though, actors might see a need for strategic change. In short, both innovation practice and innovation policy have to cope with a tense polarity between path dependence and creative destruction.

If one wants to discern the origins and dynamics of innovation, observing one company, one area of knowledge, one field of technology, or one policy normally would not suffice – the scope of inquiry has to be widened. Since the mid-1980s, both innovation research and policymakers, in search of the determinants of innovation as a means to strengthen economies, have adopted a systems perspective (e.g. Gaudin 1985; OECD 1991), leading to the development of the concept of innovation systems. With the growing share of knowledge-intensive products and services in international trade, explanations for the differing degrees of competitiveness of economies, especially of their ability to innovate, were sought. It was recognized that the variety of national, regional and sectoral patterns of technological specialization and diffusion, each rooted in historical origins, characteristic and unique industrial, scientific, state and politico-administrative institutions and inter-institutional networks, crucially affected the ability of research and economic actors to produce knowledge and innovation as well as of policymakers to invest and regulate.

Innovation systems encompass, according to a widely accepted understanding (Freeman 1987; Lundvall 1992; 2007; Nelson 1993; Metcalfe 1995; Edquist 1997; Kuhlmann 2001), the 'biotopes' of all those institutions which are engaged in scientific research and the accumulation and diffusion of knowledge, which educate and train the working population, develop technology, produce innovative products and processes, and distribute them. Hereto belong the relevant regulatory bodies (standards, norms, laws), as well as the state investments in appropriate infrastructures. Innovation systems extend over schools, universities, research institutions (education, private sector and science systems), industrial enterprises (economic system), the politico-administrative and intermediary authorities (political system) as well as the formal and informal networks of the actors of these institutions. As complex and heterogeneous systems, they represent a section of the society that carries far over into other societal areas, e.g. through education, or through entrepreneurial innovation activities and their socio-economic effects. As a result, innovation systems are believed to have a key influence on the modernization processes of societies (OECD 1999).

Successful innovation systems develop their special competitive scientific, educational, technological profiles and strengths rather slowly, in the course of decades, or even centuries, and change is also often slow to occur. Leading innovation systems are based on well-established exchange relationships among the institutions of science and technology, industry and the political system. They make possible the formation of a characteristic, system-specific spectrum of diverse role definitions of the actors actively involved, develop their own negotiation arenas, and stabilize mutual expectations of behavior. Finally, they bear particular intermediary fora and bodies which facilitate the transactions of the actors of innovation systems. Chris Freeman classically detailed these processes of system stabilization (and destabilization) when he compared the British and the German innovation systems during the first and the second industrial revolutions. During the first, the British system was superior due to excellent links between the science, political, cultural and industrial sub-systems. During the second, the British system eroded because of a widening gap between the science system and the other three components. At the same time the German system improved by building bridges among (industrial-) research, production and the political and cultural sub-systems (Freeman 1997).

In sum, the innovation system perspective helps us to understand the dynamics of innovation processes by pointing at path dependency and structural sclerosis as well as the potential for new combinations, related chances and options, and opportunities for innovation policy.

Historically, the institutional infrastructures and networks of research and innovation systems did not come into existence spontaneously and without control:

in the past 150 years, this area of society was shaped by state political interventions. National political systems, themselves increasingly differentiated, developed science, technology and innovation policy activities in which they acted as catalysts, promoters and regulators of the innovation bodies which were emerging within their countries. The establishment and growing economic significance of colleges of engineering and technical universities in France, Germany or in the USA document this clearly. The innovation systems of the industrialized countries that developed in the course of the twentieth century in co-evolution with their national political systems assumed a country-specific character. It is because of this close interweaving with the political systems that one speaks of 'national innovation systems'. Some analysts would include also regional (Braczyk et al. 1998) and sectoral innovation systems (Malerba 2002), with public innovation policies increasingly developed at these levels too.

As Boekholt in her chapter in this book demonstrates, state interventions shaping innovation systems cover a large and (since the 1960s) ever growing scope of instruments (working with financial, regulatory or information incentives) and areas of application (see also Rothwell and Zegveld 1981; 1985; Rothwell and Dodgson 1992; Dodgson and Bessant 1996; Branscomb and Keller 1998; Archibugi et al. 1999; Feldman and Link 2001; Biegelbauer and Borrás 2003). The European 'Trendchart' intelligence service distinguishes 25 categories of innovation policies, under five major headings: improve innovation governance and strategic intelligence for policymaking; foster an innovation-friendly environment; encourage technology and knowledge transfer to enterprises and development of innovation poles and clusters; promote and sustain the creation and growth of innovative enterprises; and strengthen entrepreneurial innovation including the protection and commercialization of intellectual property (European Commission 2006; see also Nauwelaers and Wintjes 2008). An OECD taxonomy suggests an even broader perspective which covers innovation policies, targeted at, as well as cutting across, sectors. These include policies directed at innovating industries and economic growth, and innovation policies in a wider sense aiming also to improve quality of life (see Table 1.1).

Policy in general can be defined as 'a purposive course of action followed by an actor or a set of actors in dealing with a matter of concern. Public policies are those developed by governmental bodies and officials' (Anderson 1990). In our case, the matter of concern is 'innovation'. If we accept the broad definition of innovation policy adopted by the OECD report (above) it appears wise to 'emphasize a holistic view of policy-making, a belief that the whole is greater than the sum of its parts, that individuals, institutions, interactions, and ideology all matter, even if there is notable disagreement about the proportional importance of each' (Nelson 1996). This holistic view of policymaking fits with the perspective adopted in this book: not surprisingly we will apply a quite generic notion of innovation policy which could be defined as 'a set of policy actions to raise the quantity and efficiency of innovative activities, whereby "innovative activities" refers to the creation, adaptation and adoption of new or improved products, processes or services' (Cowan and van de

Table 1.1 A taxonomy of innovation policy

Goals	Sectoral innovation policy	Multi-sectoral innovation policy
Innovation policy, i.e. aimed primarily at innovating industries and economic growth	Innovation policy in a limited sense (basically technology and industrial policies)	Integrated STI policies
Innovation policy in a wider sense, i.e. aimed at economic growth and quality of life	Innovation policies in other sectoral domains, e.g. innovation policies in health, innovation policies in the environment	Horizontal/comprehensive/ integrated or coherent/ systematic innovation policies

Source: OECD (2005, 22). STI = science, technology and innovation.

Paal 2000). Obviously, this broad definition covers a large number of heterogeneous policy actors, instruments and measures – as will be shown in several chapters of this book. Thereby innovation policy can be characterized as 'systemic' in a double sense: as a system-wide distribution of varieties of innovation-related policies across domains (e.g. manufacturing, services), or as policies designed to work on system characteristics, such as demand-oriented policies (see Chapter 13 by Edler in this book) or strategic policies (see Chapter 17 by Smits, Kuhlmann and Teubal).

At this point we can conclude the following: (1) Innovation practice and innovation policy normally do not start from a systemic/holistic perspective; nevertheless – as several contributions to this book will show – such a perspective is receiving increasing attention. (2) In this book when considering innovation processes we take such a perspective, in the double sense of domain variety and system-level action. (3) In analyzing innovation policies we start from a holistic concept aiming at improvement of policy and practice. In doing so, we will now introduce the guiding metaphor of this book – the three dancers and their potential for interactive learning.

THE INNOVATION POLICY DANCE: INNOVATION PRACTICE AND INNOVATION THEORY AS PARTNERS

It has been suggested that there is no coherent theory on innovation and on innovation systems (Lundvall 2007). Rather, related research is rooted in heterogeneous theoretical strands: evolutionary economics, in particular growth theory; micro-economics; innovation economics; sociology, in particular sociological system theory; action theories; neo-institutional approaches; sociology of organization; sociology of science; political science, in particular policy analysis; neo-corporatist analysis; international relations; science policy studies. For the time being we use the term 'innovation system' just as a helpful heuristic aide (Kuhlmann 2001; Lundvall 2007) speaks of it as an 'analytical focusing device'.

The theoretical perception of 'innovation' and 'innovation policy' has undergone considerable change in recent decades. While since the 1950s in economics and sociology research, development and innovation processes were seen as subsequent activities of institutionally and organizationally distinct units (linear approach) this changed in the course of the 1980s and 1990s. Today, most concepts and theoretical constructs agree largely on the interactive character of idea generation, scientific research, development, product and process innovation (in a narrow sense), and introduction into markets or other areas of use. This has been simplified with the tag of 'Mode 2' innovation as suggested by Gibbons et al. (1994). In a way, the Mode 2 perspective on knowledge production and innovation builds on a long strand of studies into the relation of science and technology (e.g. De Solla Price 1965; Rip 1992; Weingart 1997) and, at least implicitly, rediscovering older, more systemic concepts from the late eighteenth through to the early twentieth centuries (in the work of Smith, Marx, List and Schumpeter, among others; see Lundvall 2007). Meanwhile the innovation system approach has been made more dynamic, inter alia by introducing the concept of 'functions' or 'critical success factors' of innovation systems (see Chapter 6 by Bergek, Jacobsson, Hekkert and Smith; also Hekkert et al. 2006) which can be related to innovation policy instruments. Still, the question remains how actual policymaking refers to functions, and how functions translate into policymaking and policy effects.

Three chapters of this book explore the emergence and conceptual potential of the innovation system notion in some detail (Carlsson, Elg and Jacobsson; Chaminade and Edquist; and Bergek, Jacobsson, Hekkert and Smith). Other chapters address several of the key weaknesses to date of this influential concept. The innovation system approach is most dominant in Europe. In the United States, innovation scholars refer to innovation systems, but there is not much use of the concept in the policymaking and business communities. The innovation systems approach has been communicated widely by OECD and others, and has received considerable interest in other part of the world – although the fit is less clear with the needs of 'less developed economies' (see Chapter 15 by Cozzens). Other weaknesses of the innovation system approach are addressed in Chapter 17 by Smits, Kuhlmann and Teubal. These include the tendency to conceive systems as stable if not static instead of also conceptually modelling them as permanently evolving social structures; the fact that policymakers only occasionally design interventions in light of a systems perspective; and weak attention of policy designs for actors at the micro level.

A Learning Perspective

The relationship between innovation practice, innovation policy and innovation theory could be considered as an expression of 'co-evolution' between various actors interested in influencing technological change and innovation in terms of their own goals:

> Firms will think in terms of market success and strategic advantages. NGOs have their issues to pursue, for example health improvement or a clean environment. National governments and other government agencies have overall goals like security, quality of life, sustainability, under which a variety of actions are formulated and implemented. Their assessments of the situation, actions and interactions actually contribute to the co-evolution. (Rip 2008)

In this book, we do not over-emphasize the notion of co-evolution with its allusion to biological processes. Instead we place as central to system development the conditions and modes of interactive learning of organized actors in innovation policy arenas.

We start from the basic assumption that the ideas, rationales and instruments of innovation policy emerge as a result of interactive learning among actors involved in innovation practice (I), innovation-related public intervention strategies (P), and innovation research and theory (T). One may conceive the interactive learning space between I, P and T as a 'dance floor'.[1] For the purposes of this book we suppose that the three dancers while moving observe each other, and react to the others' movements: they copy, comment, complement, counteract, neglect, and thereby learn. Mutually learning, they constantly create and change IPT configurations. Sometimes innovation practice is the driving force in a configuration, sometimes theory, sometimes public or private policy. The dancers may happen to bump into each other or may enjoy phases of pure harmony. Table 1.2 makes an attempt to characterize the dance of the three groups in a systematic way.[2]

Learning on the innovation policy dance floor may occur as first-order or as second-order learning (Argyris and Schön 1978). First-order learning links outcomes of action to organizational strategies and assumptions that are modified so as to keep organizational performance within the range set by accepted organizational norms. The norms themselves remain unchanged. Second-order learning concerns sorts of inquiry that resolve incompatible organizational norms by setting new priorities and relevance of norms, or by restructuring the norms themselves together with associated strategies and assumptions, hence escaping tunnel vision and crossing borders. In our perspective not only is innovation itself a matter of new combinations but also the 'innovation policy dance' between innovation practice, theory and policymaking.

Through the chapters of this book we hope to demonstrate how practice, theory and policymaking interact – across a number of aspects: how they 'dance' with each other and thereby learn, both first-order, i.e. reacting to observed changes in a conservative manner, and second-order, i.e. adopting or developing new assumptions,

1 Drawing on Kuhlmann (2007) and Smits and Kuhlmann (2004). The dancing metaphor has earlier been used by Arie Rip (1992) with respect to the relation of science and technology, inspired by Derek de Solla Price's discussion of this relation (1965).

2 This table is a variation of the PIT Matrix presented by Smits and Kuhlmann (2004).

Table 1.2 The innovation practice, policy and theory dance floor (IPT matrix)

	Innovation Practice	Theory	Policy
Innovation Practice (e.g. sustainable housing)		I→T **Learning by searching** (e.g. researchers learn on user-producer relations from real life experiments with sustainable housing) **Learning by interacting** (e.g. researchers use experiences of actors as empirical input for user-producer research)	I→P **Learning by using** (e.g. policymakers learn form the impact of their policies by evaluations) **Learning by interacting** (e.g. policymakers learn from the impact of their policies by talking to actors in the field)
Policy (e.g. environmental innovation policies)	P→I **Learning by using** (e.g. entrepreneurs learn by using policy measures)	P→T **Learning by searching** (e.g. researchers learn on user-producer relations from (the impact of) policies focusing on sustainable housing) **Learning by interacting** (e.g. researchers use experiences of policymakers as empirical input for user-producer research)	
Theory (e.g. user-producer interaction)	T→I **Formal learning** (e.g. entrepreneurs learn from theories on user-producer innovation and change their mental frame, conceptual use) **Learning by interacting** (e.g. researchers act as consultants for entrepreneurs)		T→P **Formal learning** (e.g. policymakers learn from theories on user-producer innovation and change their mental frame, conceptual use) **Learning by interacting** (e.g. researchers act as consultants for policymakers)

targets and measures. While moving on the dance floor, learning and finally even changing perspectives, bounded rationality will remain the actors' prevailing mode of guidance. Even second-order learning could be subject to trends, fashions, or waves in innovation practice, theory or policy – or in the surrounding 'music'. In other words, a note of caution is required, asking for reflexivity, in particular on 'our' side, the side of 'theory' as a dancing partner (Kuhlmann 2007).

External changes (new 'music fashions') could imply new roles of dancers on the floor, or even the appearance or (temporary) farewell of an actor. Has, for example, theory always been a relevant dancing partner from the perspectives of innovation practice or policy? Also, obviously there is not just one theory: innovation practice might prefer other dances with theory than public policy would like (e.g. different choices of theories, with different policy targets and designs as a consequence). Furthermore we know that different national political systems and related political elites revolve around different intellectual traditions and styles (Galtung 1981), hence expose marked preference for specific innovation theories. In the US, for example, there is a strong interest among actors in innovation policy and practice for (quantitative) economics-based analytical concepts (since 2005 re-emphasized in the 'Science of Science and Innovation Policy' initiative of the National Science Foundation). This can be compared with the continental European tradition of sociology and evolutionary economics-based analysis of public and private institutions as the shaping force of 'innovation systems'. These variations in conceptual emphasis have influenced the design of evaluation regimes for science and technology policy in the US and Europe (Shapira and Kuhlmann 2003).

Innovation Theory and Innovation Practice as Dancing Partners

Practical innovations, i.e. technology-based processes or products and social and organizational innovations in private and public services, are at the heart of the working of capitalist economies. The social and economic centre of gravity of practical innovation locates in the (mostly private) 'firm'. Dankbaar and Vissers state Chapter 3 in this book that the 'firm will continue to play the central entrepreneurial role in processes of innovation and technological change in capitalist economies – and even in economies that are not formally capitalist, decision making on innovation is decentralizing towards the level of the firm'.

Modern innovation theory is largely based on empirical analyses of social and economic change in the 'practice' of firms and public organizations, starting not only with Joseph Schumpeter (1883-1950), but even with the research and writings of Adam Smith (1723-90) and Karl Marx (1818-83) (see Antonelli 2008). Over decades, a large number of economic and sociological studies into the patterns of (in particular technological) change in firms and in sectors have helped to accumulate knowledge on the dynamics of innovation as a core phenomenon of modern economies (e.g. Freeman and Louça 2001), a phenomenon with generic and systemic characteristics: searching in and interacting with innovation 'practice' has been a key resource of theory-building.

At the same time innovation 'practice', though socially located quite distant from academic theory, has been influenced by innovation studies in many ways. This certainly holds for the broad scope of 'innovation management' literature which has had a considerable (and sometimes fashionable) impact on firms' understanding of innovation and on their strategies (see Dankbaar and Vissers, Chapter 3 this book). By and large, the systemic perspective on innovation as a socio-economic and technological process has advanced – to some degree also in firms – that any 'innovation success' in product creation processes should rather be viewed as by-products along 'innovation journeys' than as end results (van de Ven et al. 1989). Such 'journeys' are characterized by numerous setbacks along the road; when one wants to deal with contingencies during a journey, retrospective attributions of success to certain approaches or persons is risky. Following this perspective the understanding of innovation management as fundamentally a control problem unpacks as a myth: rather it should be seen as one of orchestrating a highly complex, uncertain and probabilistic process of collective action in a systemic context (van de Ven et al. 1989).

Innovation Theory and Innovation Policy as Dancing Partner

The evolutionary approach of Nelson and Winter (1977), the innovation system tradition as inspired by Freeman (1987) and developed further by Lundvall (1992), Nelson (1993) and Edquist (1997) as well as the cluster approach advocated by Porter (1990) heavily influenced the development of innovation policy concepts and instruments. These assume that with globalization, dynamic clusters of innovative firms and knowledge-generating public organizations are becoming key factors in a country's capacity to attract the international investment that generates new technological expertise, to interest investors in innovation (venture capital, etc.) and to benefit from the international mobility of skilled personnel (OECD 1999; 2001). The Dutch cluster policies of the 1990s represent a prominent example which applies these theories (Jacobs and de Man 1996; Jacobs 1998). Also bodies of knowledge dealing with the broader embedding of innovation processes, such as for instance the social construction of technology (Bijker et al. 1987), technology assessment (Smits et al. 1995) and research after the role of users in innovation processes (von Hippel 1988; Lundvall 1992; Moors et al. 2003), had a considerable impact on the policymaking processes and their design. The underlying constructivist perspective has, for example, facilitated the development of policy-related specialized Technology Assessment organizations (see the detailed discussion in Chapter 16 by Smits, van Merkerk, Guston and Sarewitz), or the rise and high visibility in innovation policymaking of foresight exercises (see e.g. Georghiou et al. 2008).

In addition to economics- and sociology-based research and theory, innovation policy has benefited from contributions from the academic domain of policy analysis. The political science notion of governance offers a heuristic, denoting the dynamic interrelation of involved (mostly organized) actors, their resources, interests and power, fora for debate and arenas for negotiation between actors, rules of the game, and policy instruments applied (Schön and Rein 1994; Kuhlmann 2001; Benz 2007). The governance heuristic helps to understand the context, options and limitations of public policy; it refers to analytically distinguishable forms of institutional coordination of autonomous but interdependent actors.[3] Hierarchy, competition, network, association and community are such ideal types of governance capturing the rules of a game at a highly generalized level (Hollingsworth and Boyer 1997). In reality, these governance forms are often interconnected, thus forming governance regimes. Actors have to find out how to cooperate with competitors or to compete with partners in networks, to negotiate an agreement under tight organizational constraints or to find approval for the outcome in external arenas in their own organization or group (Benz 2007). In order to also explore the dynamics of emerging new actor constellations and institutional settings we consider it necessary to also analyse 'soft' forms of social rules, not yet frozen into codified regulations. Following Scott (1995) one can conceptualize institutions as sets of rules of a regulatory, normative or cognitive character providing stability and meaning to social behavior, transported by various carriers like cultural patterns or routines. Institutions guide human behavior by (1) utility-oriented rules which may be enforced by coercion ('regulatory'), (2) norm-based obligations ('normative') and (3) immediate participation in taken-for-granted models of reality ('cognitive'); institutions store historical experience in idiosyncrasies guiding the day-to-day action of their members.[4] This broadened definition of 'institution' implies a notion of governance exceeding the classical definition of political control and steering. Mayntz (1998) identified two even more extensive definitions: (1) 'a more cooperative mode where state and non-state actors participate in mixed public/private networks' and (2) 'modes of coordinating individual actions, or basic forms of social order'. The second definition includes both the classical steering idea as well as network-oriented 'softer' aspects. The identification and reflection of the dynamics and impacts of experiences inscribed in institutions may help to better understand the actual orientations and strategies of corporatist and other organized actors in the actual governance of innovation systems, thereby interpreting governance as an evolutionary social order.

The governance perspective has gained relevance with the increasing inter- and transnationalization (but also regionalization) of both innovation practice and innovation policy initiatives. Increasingly, we see innovation issues being negotiated under conditions of multi-level, multi-actor agency and governance (e.g. Grande 2001; Kuhlmann 2001). Take as an example the various attempts of national or regional governments to launch overarching 'innovation policy strategies' (LEG 2008) In the United Kingdom, a new ministerial department for Innovation, Universities and Skills was established in 2007, bringing together policies on skills, higher education and innovation for the first time, aiming at an integrated approach.

3 'The term designates a certain analytical perspective determined to make apparently opaque and over-complicated structures and processes of collective action in the state, economy and society comprehensible. As an analytical tool, the term can be used to describe or evaluate reality' (Benz 2007).

4 I.e. through 'scripts', 'taken-for-granted-rules', or embedded in 'central themes' (Powell and DiMaggio 1991; Lepsius 1995; Scott 1995).

In Denmark, the government adopted a horizontal 'Globalization Strategy', including 350 measures pertaining to the areas of education, research, innovation and entrepreneurship. This strategy has been built up with the involvement of several ministries and a wide range of stakeholders. In 2006 the German Federal Government announced an overarching 'High-tech Strategy', designed as a holistic innovation policy concept. Although the effects of the suggestions and measures of this initiative can be evaluated only after years, the strategy can be characterized as a major attempt of strategic co-ordination at the national level. The initiative consists of diverse measures, including defining areas of competence for the Federal Ministry for Education and Research and the Federal Ministry for Economics and Technology and fostering public procurement for innovation, a 'better' IPR regime, thematic programming, public-private partnership models, venture capital, spin-off activities, cluster financing and increased spending in education.

In short, the holistic concept appealed to policymakers, not least because the systemic perspective provided an argument for a broadened scope and reach of public STI policy (Smits and Kuhlmann 2004). Some have used it as a branding device. For example, the Swedish state office for innovation policy calls itself the 'Governmental Agency for Innovation Systems'. In other Scandinavian countries, and also in the Netherlands and Germany, relevant policy initiatives and agencies draw justification from this notion. Indeed, it turns out that the very concept of innovation systems, while being designed by innovation researchers, was also inspired and strongly supported by Scandinavian policymakers (see Carlsson, Elg and Jacobsson, Chapter 7 in this book) and by the OECD (Lundvall 2007). The concept became 'theory in action'. Of course, our Swedish academic colleagues could have tried to maintain academic distance to the borrowing of their concepts and findings by policymakers and practitioners in innovation – but they chose to 'pro-actively' offer information, heuristics, analysis and theory to advance the policymakers' own theories. In other words, they danced with practice and policy and even jointly composed new melodies, while the innovation system concept served as a 'boundary object' ('fuzzy boundaries distinguishing academic and policy circles allowed for cross-fertilization by theoretical and practical considerations, and this accelerated the spread of the [this] concept', Sharif 2006).

Innovation Practice and Innovation Policy as Dancing Partners

Since the early days of innovation-driven capitalist economic development, 'innovation practice' in companies and private or public R&D labs has evolved in an interwoven manner with public policy. Innovators have asked for and received public financial support, pushed for changes in regulation facilitating the adoption of innovative solutions, and exploited state-guaranteed Intellectual Property Rights (IPR). From the perspective of innovation practitioners, public policy has to be responsive and to adapt to new, innovation-driven needs. This often gives rise to conflicts of interest among economic actors or with organizations representing societal interests; think of the widespread concerns about innovation based on genetically modified organisms (GMO) and the inconsistent strategies of industries vis-à-vis 'functional food' innovations. Thereby, innovation practitioners may also borrow arguments from innovation research and theory – as societal organizations and public policy also often do. Also, the debates around 'open innovation' and patenting of software products provide a telling example: policy actors have made extensive use of theory-based empirical studies of patenting practice in order to develop strategic policies coping with contradicting industry positions (large software producers versus small open-source software developers; see Chapter 10 by Blind).

At the same time, innovation policy – though with different underlying rationales, institutions, and means across various cultural and national contexts (Hollingsworth and Boyer 1997; Hall and Soskice 2001) – has claimed to 'shape' supportive conditions for innovation-based economic growth and welfare. The above-mentioned adoption of the innovation system concept by many policy-making bodies is a case in point. The normative ambition of innovation system-based policymaking vis-à-vis innovation practice has occasionally been criticized as camouflaging deliberate political decisions into 'natural' and historically 'inevitable' measures (for the case of Finland, see Miettinen 2002).

Following the global financial and economic crisis that started in late 2008, the interaction between innovation practice and innovation policy as dancing partners is likely to become even more complex: practitioners are calling for more supportive innovation conditions, and there are broader pressures to foster sustainable economies through innovation. Policymakers may be preoccupied with immediate pressures to rescue financial institutions, housing markets and industrial companies, to address joblessness and to postpone tighter environmental regulations, even as they recognize the importance of innovation as a driver of renewed economic growth. It remains to be seen how innovation theory can adapt to this changed situation. Just as there has been a resort to Keynesian theory to justify deeper public borrowing and spending to address the macro-economic slump, there may well be a revival of interest in Schumpeterian arguments about the role of innovation in overcoming downturns in business cycles. But these are old ideas. A challenge for innovation theorists will be to critically contribute new ideas and analyses to the formulation of strategies and practices for innovation in and beyond the present period of economic duress.

Summing up, one can ask how distinct the identities of the three dancers actually are. Although at first sight industry, academia and the politico-administrative system appear clearly different in terms of membership, constituency, relevance criteria and reward mechanisms, taking a closer view one would face a more blurred picture. Innovation practice, for example, extends far beyond classical industrial innovation and includes socio-technical or organizational innovation in public services (such as health), thus overlapping with the sphere at public policy (though not the playground of classical innovation policy, hence hinting to domain and governance issues inherent to the public sector). Also, the spheres of innovation theory are in relevant instances closely interwoven with the realm of public policymaking, with key figures developing their careers across both fields. Examples of this include the functioning of OECD as a social organization, or the development of the innovation system approach in Sweden (see Chapter 7 by Carlsson, Elg and Jacobsson), both providing exciting cases of blurred borders between dancers. Hence, while the distinction of three dancers is useful, their roles, identities and relations may change and overlap, and the intensity and nature of interaction varies over time. The chapters in this book provide more detailed evidence about the dynamics of these relationships.

A THEMATIC INNOVATION POLICY HANDBOOK

This book combines attributes of a classical handbook and a thematic book. The book does not seek to provide an all-encompassing review of the state of the art of innovation policy, practice and theory. Instead we have tried to present a coherent, theory-based vision of the interrelated dynamics of innovation policy, practice and research. The handbook elements provide the reader with an updated overview of dimensions, concepts and challenges of innovation policy from a systemic perspective. The thematic discussion revolves around the mutual learning between three 'dancers': innovation practice, innovation theory and innovation policy, the last being the focal aspect of interest.

The first part of the book examines some basic driving forces of innovation practice, theory and policy – in a way the background music of the innovation policy dance. In his chapter, Martin addresses open questions in the context of changes in the modes of knowledge production and the emergence of new policies and institutional configurations for facilitating border-spanning research collaboration and increasingly science-based innovation. What are the consequences for the 'innovation systems' concept and for the role of public policy? And do we have the necessary 'holistic' policy instruments required for effective, integrated science and innovation policy, and for overcoming the main types of policy failures? Dankbaar and Vissers consider the changing role of the 'firm' in innovation processes and the way this has been reflected in academic literature. Processes of differentiation and specialization, requiring new organizational capabilities, have affected innovation, like so many other activities. Thereby interactions between theory and practice are not simply processes of mutual reinforcement: experiences made in one company, one sector or one country lead to concepts and understandings, which are then applied to completely different companies, sectors and countries. Theory then becomes a force for change and innovation of its own right. Smith explores data and recent trends in globalization, and the background to these trends in policy decisions over the past three decades. He argues that the combinations of liberalizing economic policies and new international institutions have created an environment in which far greater economic interdependence now exists. In many areas, including innovation, it is claimed by many that the new economic environment constrains or prohibits independent policy action by government. Smith argues that this is not so. The key point about innovation policy is that, properly conceived, it affects the structure and operations of the innovation system – that is, the system of institutions and organizations that shape the behaviour of firms.

The second part of the book focuses on the conceptual evolution of the systems perspective as a result of various innovation policy dance configurations. In so doing three major flaws currently associated with this approach (too static, no policy consequences, the role of actors on the micro level) are highlighted. In 1987 the Swedish National Board for Technical Development (STU) invited a group of Swedish scholars in economics, engineering, management and sociology to do a joint study of 'Sweden's Technological System and Future Competitiveness'. In their chapter Carlsson, Elg and Jacobsson trace how that group – in interaction with policymakers – improved the understanding of the structural components of an innovation system, how such systems are formed, and how the development processes vary between specific systems. Chaminade and Edquist explore the implications of the adoption of the innovation system approach for public policy (what to do, when and how to do it), more in particular they discuss the rationales of innovation policies when this approach is adopted. They compare the basic assumptions of the neoclassical and evolutionary-systemic theories and the implications of the adoption of one or another for the rationales for public intervention. The next chapter builds on the preceding: Bergek, Jacobsson, Hekkert and Smith investigate the 'functionality' of innovation systems as a rationale and guide in innovation policy. They offer taxonomy of system functions, as a step towards an understanding of how systems affect innovation. Furthermore they suggest that functions of a system can form a framework for understanding and shaping both the foundations and content of policies in support of innovation.

As innovation policy covers in our understanding a very broad range of targets, instruments and areas of application the third part of the book deals with an inevitably heterogeneous variety of ongoing and new issues. In his chapter, Shapira addresses the relationship between innovation and small and midsize enterprises (SMEs) as a longstanding and central issue of innovation theory and policy. In particular he looks at the relationships between innovation system dynamics and policy strategies for promoting innovation in SMEs as an important mediating mechanism that policymakers use to understand innovation dynamics.

The next two chapters deal with the options various regulatory policies provide for the support of innovation. Blind offers an overview of types of regulations and their ambivalent impacts on innovation, and a new taxonomy of product market regulations. The role of product market regulations with respect to various functions of innovation systems is also discussed. Graham's chapter examines the role of regulatory policies for innovation, focussing on a specific case of the co-evolution of intellectual property (IP) protections and innovation policy within the context of a particular IP institution: the 'continuing' patent application procedure in the United

States. The continuation procedure allows a US patent applicant to postpone the issue of a patent, affording inventors several strategic opportunities, among which are delay and secrecy.

There has long been a relationship between civil and military innovation. Yet, the separation between military and civilian innovation systems has not always been so clear-cut. Different technological, economic and political environments have co-evolved with changing policy perceptions and interventions. Molas-Gallart explores the evolving relationship in his chapter, focusing on the experience of the US and some Western European countries since the Second World War and the systemic shifts now occurring since the start of the 21st century. He argues that the sharp distinction between military and civilian innovation systems is starting to fall apart. Mutually interdependent changes in technology, the political and security context, and innovation policies are leading to a new, more complex and integrated, innovation system. Edler addresses an area of innovation policy that in recent years has (again) attracted the attention of policymakers: demand-oriented innovation policies, defined as public measures to induce innovations or speed up the diffusion of innovations through increasing the demand for novel applications, defining new functional requirement for products and services or better articulating demand. Edler offers a policy typology and gives an overview of recent empirical developments. However, he also highlights severe governance challenges that set limits to demand orientation in innovation policymaking.

Den Hertog draws the reader's attention to the growing relevance of innovation in the service sector: he describes a slow but steady process in which new views on the transformation to a service economy and its implications for state-of-the-art innovation practices, innovation research and statistics and appropriate (innovation) policy interventions were exchanged. Service innovations were analyzed in more depth, new theories and typologies introduced and new statistical indicators introduced. An increasing number of firms realized that managing technological innovation needs to be combined with managing service innovation and some of these were looking for more formalized knowledge on how to accomplish this. In some countries innovation policymakers started to explore new, more service-friendly innovation schemes.

So the range of targets, instruments and areas of application of innovation policy has become very broad. Boekholt shows in her chapter the evolution of innovation-related policy instruments. In the 1950s until the 1970s there were two separate systems (the science policy system and the industrial policy system) that each followed different dynamics. The concepts of technology policy and later innovation policy emerged only in the late 1970 to mid 1980s. The first generation instruments clearly showed the characteristics of 'linear model thinking' and focused for instance on technology transfer, e.g. bringing inventions and technologies from the shelves of the laboratories to the manufacturing plants of industry to be transformed into innovations. In the 1980s, influenced by conceptual thinking about the interactive chain link models and the many feedback loops between science, strategic research, applied research and the market, gradually more sophisticated bridging and networking policies were developed. The innovation systems approach, finally, became accepted as a useful analytical model for policymaking around the turn of the century, although first only in a small number of countries. Today it is more widely accepted but at the same time its practical use as a guidance for now well policy instruments can be designed is limited. In recent years the debate has shifted towards the definition of 'holistic' and 'horizontal' policy approaches where different policy domains coordinate the strategic approach to tackle specific issues.

The final part of the book looks ahead, identifying new dynamics in the innovation policy dance partly as a consequence of mutual learning, partly due to changes in the external environment ('new music'). By focusing on economic development, innovation systems research has contributed to the debate on how to reduce inequality between nations and also regions. Innovation systems research, however, has not given much attention to another dimension of development, namely, inequalities between individuals, households, or groups, as Cozzens shows in her chapter. A lack of a perspective on the role of individual actors in innovation systems is one of the major flaws of the innovation systems approach. This chapter attempts to fill that gap. Econometric studies have shown that income inequality within countries hinders economic development, and ethnic fragmentation has also been demonstrated to do the same. Such studies reflect the human reality that inequalities place barriers in the path of using innovation for development and keep it from achieving everything it could. We cannot have effective theory, policy, or practice in this area without addressing these issues.

The systemic character of innovation in general and the requirements of the 'innovation policy dance' in particular increase the actors' need for information enabling them to engage in innovative activities in an adequate and effective way. In this handbook this information is conceptualized as Strategic Intelligence (SI); actors involved in innovation require it to develop their visions, strategies and plans of action. Apart from this 'instrumental' role, Smits, van Merkerk, Guston and Sarewitz argue in their chapter that SI helps to reflect on the development, interaction and effectiveness of innovation theory, practice and intervention. By this, SI provides an important input in the further development of these three concepts. The chapter concentrates on one particular strand: technology assessment (TA).

All the chapters of this book show that although innovation studies have made advances over the last two decades, the prevailing innovation systems approach has some major flaws; in particular it is still conceptualized much in too static a manner. The dynamics of the 'innovation policy dance floor' are hardly reflected by the concept. Smits, Kuhlmann and Teubal argue in their chapter that one should differentiate 'strategic policies' dealing with radical changes in policies, setting up new interfaces and arenas requiring broad processes of consultation and strategic intelligence to support this, against on the other hand 'systemic instruments' that are only meaningful in a systemic and co-evolutionary perspective in relatively steady development stages of innovation systems.

Finally, the editors conclude the book by offering an overall summary and outlook (Chapter 18). This chapter notes that the concept of innovation policy as a systems-oriented activity is gaining ground and that the importance of one of the three dancers – theory – vis-à-vis the other two is growing. This requires researchers to take another role: they can contribute to reinforcing learning processes between the three dancers from a systems perspective. Here, encouraging more 'hybrid actors' – people and organizations who can serve as bridges and intermediaries between sectors – is likely to be useful. There is also a case for related changes in our educational systems related to science, technology, management, organizational and public policy, developed in the perspective of a long-term innovation agenda. This should assist both policymakers and practitioners in pursuing such agendas and in customizing them to best suit their own innovation system circumstances. Enhancements in strategic intelligence infrastructures and greater support for reflexivity and deliberation about innovation policy and practice are also important. There is a relationship between means and ends: we suggest that through such investments in the processes and tools of innovation policy, practice and theory – and by strengthening interactions and learning relationships among these three domains – then tangible steps forward can be made along the journey of building knowledge-intensive, equitable and sustainable economies.

REFERENCES

Anderson, J. (1990), *Public Policymaking*, Boston: Houghton Mifflin.

Antonelli, C. (ed.) (2008), *The Economics of Innovation*, Vol. 1, London, UK and New York, USA: Routledge.

Archibugi, D., Howells, J. and Michie, J. (eds) (1999), *Innovation Policy in a Global Economy*, Cambridge, UK: Cambridge University Press.

Argyris, C. and Schön, D. (1978), *Organizational Learning: A Theory of Action Perspective*, Reading, MA: Addison-Wesley Publishing Company.

Benz, A. (2007), 'Governance in connected arenas: political science analysis of coordination and control in complex control systems', in D. Jansen (ed.), *New Forms of Governance in Research Organizations: From Disciplinary Theories towards Interfaces and Integration*, Berlin, Germany, Heidelberg, Germany and New York, USA: Springer-Verlag, pp. 3-22.

Biegelbauer, P. and Borrás, S. (eds) (2003), *Innovation Policies in Europe and the US: The New Agenda*, Aldershot: Ashgate Publishing.

Bijker, W., Hughes, T. and Pinch, T. (eds) (1987), *The Social Construction of Technological Systems: New Directions in the Sociology and History of Technology*, Cambridge, MA, USA and London, UK: MIT Press.

Bozeman, B. (2000), 'Technology transfer and public policy: a review of research and theory', *Research Policy*, **29** (4-5), 627-655.

Braczyk, H.-J., Cooke, P. and Heidenreich, M. (eds) (1998), *Regional Innovation Systems*, London: University College London Press.

Branscomb, L. and Florida, R. (1999), *Challenges to Technology Policy in a Changing World*, Cambridge, MA, USA and London, UK: MIT Press.

Branscomb. L. and Keller, J. (eds) (1998), *Investing in Innovation: A Research and Innovation Policy that Works*, Cambridge, MA, USA and London, UK: MIT Press.

Burt, R. (2003), 'Social origins of good ideas', paper, University of Chicago.

Cowan R. and van de Paal, G. (2000), 'Innovation policy in a knowledge-based economy', European Commission Directorate General, DG Enterprise, Brussels.

De Solla Price, D. (1965), 'Is technology historically independent of science? A study in statistical historiography', *Technology and Culture*, **6** (4), 553-568.

Dodgson, M. and Bessant, J. (1996), *Effective Innovation Policy: a New Approach*, London: International Thomson Business Press.

Edquist, C. (ed.) (1997), *Systems of Innovation: Technologies, Institutions and Organisations*, London, UK and New York, USA: Pinter.

Ende, J. van den, and Kemp, R. (1999), 'Technological transformations in history: how the computer regime grew out of existing computing regimes', *Research Policy*, **28** (8), 833-851.

European Commission (2006), 'European Innovation Progress Report 2006', Trendchart, Luxembourg: Office for Official Publications of the European Communities.

Feldman, M. and Link, A. (2001), *Innovation Policy in the Knowledge-Based Economy*, Berlin, Germany, Heidelberg, Germany and New York: Springer-Verlag.

Freeman, C. (1997), 'The diversity of national research systems', in R. Barré, M. Gibbons, J. Maddox, B. Martin and P. Papon (1997), *Science in Tomorrow's Europe*, Paris: Economica International, pp. 183-194.

Freeman, C. (1987), *Technology Policy and Economic Performance: Lessons from Japan*, London, UK and New York, USA: Pinter

Freeman, C. and Louça, F. (2001), *As Time Goes By: From the Industrial Revolutions to the Information Revolution*, Oxford, UK and New York, USA: Oxford University Press.

Galtung, J. (1981), 'Structure, culture, and intellectual style: an essay comparing saxonic, teutonic, gallic and nipponic approaches', *Social Science Information*, **20** (6), 817-856.

Gaudin, T. (1985), 'Definition of innovation policies', in G. Sweeney (ed.), *Innovation Policies: An International Perspective*, London, UK and New York, USA: Pinter, pp. 11-47.

Geels, F. (2006), 'Major system change through stepwise reconfiguration: a multi-level analysis of the transformation of American factory production (1850-1930)', *Technology in Society*, **28** (4), 445-476.

Georghiou, L., Cassingena Harper, J., Keenan, M., Miles, I. and Popper, R. (eds) (2008), *The Handbook of Technology Foresight: Concepts and Practice*, Cheltenham, UK and Northampton, MA, USA: Edward Elgar.

Gibbons, M., Limoges, C., Nowotony, H., Schwartzman, et al. (1994), *The New Production of Knowledge: The Dynamics of Science and Research in Contemporary Societies*, London: Sage Publishing.

Grande, E. (2001), 'The erosion of state capacity and the European innovation policy dilemma: a comparison of German and EU information technology policies', *Research Policy*, 30 (6), 905-921.

Hall, P. and Soskice, D. (eds) (2001), *Varieties of Capitalism: The Institutional Foundations of Comparative Advantage*, Oxford, UK and New York, USA: Oxford University Press.

Hekkert, M., Suurs, R., Negro, S., Kuhlmann, S. and Smits, R. (2006), 'Functions of innovation systems: a new approach for analysing technological change', *Technological Forecasting and Social Change*, 74 (4), 413-432.

Hippel, E. von (1988), *The Sources of Innovation*, Cambridge, MA, USA and London, UK: MIT Press.

Hollingsworth, R. and Boyer, R. (eds) (1997), *Contemporary Capitalism: The Embeddedness of Institutions*, Cambridge, UK: Cambridge University Press.

Hughes, T. (1989), *American Genesis: A Century of Invention and Technological Enthusiasm, 1870-1970*, New York: Viking Press.

Jacobs, D. (1998), 'Innovation policies within the framework of internationalization', *Research Policy*, 27 (7), 711-714.

Jacobs, D. and de Man, A-P. (1996), 'Clusters, industrial policy and firm strategy: a menu approach', *Technology Analysis and Strategic Management*, 8 (4), 425-437.

Kuhlmann, S. (2007), 'Governance of innovation: practice, policy, and theory as dancing partners', address delivered upon the acceptance of the Chair Foundations of Science, Technology and Society Faculty Management and Governance, University of Twente, University of Twente, Enschede.

Kuhlmann, S. (2001), 'Governance of innovation policy in Europe: three scenarios', in H. Klein, S. Kuhlmann and P. Shapira (eds), 'Innovation Policy in Europe and the US: New Policies in New Institutions', *Research Policy*, Special Issue, 30 (6), 953-976.

Lepsius, M. (1995), 'Institutionenanalyse und Institutionenpolitik', in B. Nedelmann (ed.), *Politische Institutionen im Wandel*, Opladen: Westdeutscher-Verlag, pp. 392-403.

Lisbon Expert Group (LEG) (2008), 'Lisbon strategy: between revolution and illusion – the governance challenge for knowledge policies', synthesis report of the Lisbon Expert Group, Brussels.

Lundvall, B.-A. (2007), 'Post Script: innovation system research. Where it came from and where it might go', mimeo.

Lundvall, B.-A. (1992), *National Systems of Innovation: Towards a Theory of Innovation and Interactive Learning*, London, UK and New York, USA: Pinter.

Malerba, F. (2002), 'Sectorial system of innovation and production', *Research Policy*, 31 (2), 247-264.

Mayntz, R. (1998), 'New challenges to governance theory', European University Institute, Jean Monnet Chair Paper RSC No. 98/50 (http://www.iue.it/RSC/Mayntz.htm).

Metcalfe, J. (1995), 'The economic foundations of technology policy: equilibrium and evoltionary perspectives', in P. Stoneman (ed.), *Handbook of Economics of Innovation and Technolology Change*, Oxford: Blackwell, pp. 409-512.

Miettinen, R. (2002), *National Innovation System: Scientific Concept or Political Rhetoric*, Helsinki: Edita Prima Ltd.

Moors, E., Enzing, C., Van der Giessen, A. and Smits, R. (2003), 'User-producer interactions in functional genomics innovations', *Innovation: Management, Policy and Practice*, 5 (2-3), 120-143.

Mumford, L. (1964), *The Myth of the Machine: The Pentagon of Power*, New York, USA and San Diego, USA: Harcourt Brace Jovanovich.

Nauwelaers, C. and Wintjes, R. (2008), *Innovation Policy in Europe: Measurement and Strategy*, Cheltenham, UK and Northampton, MA, USA: Edward Elgar.

Nelson, B. (1996), 'Public policy and administration: an overview', in R. Goodin and H.-D. Klingemann (eds), *A New Handbook of Political Science*, Oxford, UK and New York, USA: Oxford University Press, pp. 551-592.

Nelson, R. (1993), *National Innovation Systems*, Oxford, UK and New York, USA: Oxford University Press.

Nelson, R. and Winter, S. (1982), *An Evolutionary Theory of Economic Change*, Cambridge, MA, USA: Harvard University Press.

Nelson, R. and Winter, S. (1977), 'In search of a useful theory of innovation', *Research Policy*, **6** (1), 36-76.

OECD (2005), *Governance of Innovation Systems*, Vol. 1, Paris: OECD Publishing.

OECD (2001), *Innovative Clusters: Drivers of National Innovation Systems*, Paris: OECD Publishing.

OECD (1999), *Boosting Innovation: The Cluster Approach*, Paris: OECD Publishing.

OECD (1991), 'Technology and productivity: the challenge for economic policy', TEP-programme, Paris.

Porter, M. (1990), *The Competitive Adantage of Nations*, New York: The Free Press.

Powell, W. and DiMaggio, P. (eds) (1991), *The New Institutionalism in Organizational Analysis*, Chicago, USA and London, UK: University of Chicago Press.

Rip, A. (2008), 'Processes of technological innovation in context – and their modulation', in B. van Looy (ed.), *Festschrift in honour of Rene Bouwen (Louvain)*, forthcoming.

Rip, A. (1992), 'Science and technology as dancing partners', in P. Kroes and M. Bakker (eds), *Technological Development and Science in the Industrial Age*, Dordrecht, The Netherlands, Amsterdam, The Netherlands, Boston, USA and London, UK: Kluwer, pp. 231-270.

Rip, A. and Kemp, R. (1998), 'Technological change', in S. Rayner and L. Malone (eds), *Human Choice and Climate Change*, Vol. 2, *Resources and Technology*, Washington, DC: Batelle Press, pp. 327-400.

Roberts, L., Schaffer, S. and Dear, P. (eds) (2007), *The Mindful Hand: Inquiry and Invention from the Late Renaissance to Early Industrialization*, Chicago, USA and London, UK: University of Chicago Press

Rothwell, R. and Dodgson, M. (1992), 'European technology policy evolution: convergence towards SMEs and regional technology transfer', *Technovation*, **12** (4), 223-238.

Rothwell, R. and Zegveld, W. (1985), *Reindustrialization and Technology*, Harlow: Longman.

Rothwell, R. and Zegveld, W. (1981), *Industrial Innovation and Public Policy: Preparing for the 1980s and the 1990s*, Portsmouth: Greenwood Publishing Group.

Schön, D. and Rein, M. (1994), *Frame Reflection: Toward the Resolution of Intractable Policy Controversies*, New York: Basic Books.

Schumpeter, J. (1934), *The Theory of Economic Development*, Cambridge, UK: Polity.

Scott, R. (1995), *Institutions and Organizations*, London: Sage Publishing.

Shapira, P. and Kuhlmann, S. (2003), 'Learning from science and technology policy evaluation', in P. Shapira and S. Kuhlmann (eds), *Learning from Science and Technology Policy Evaluation: Experiences from the United States and Europe*, Cheltenham, UK and Northampton, MA, USA: Edward Elgar, pp. 1-17.

Sharif, N. (2006), 'Emergence and development of the National Innovation Systems concept', *Research Policy*, **35** (5), 745-766.

Smits, R. and Kuhlmann, S. (2004), 'The rise of systemic instruments in innovation policy', *International Journal of Foresight and Innovation Policy*, **1** (1-2), 4-32.

Smits R., Leyten, A. and Den Hertog, P. (1995), 'Technology assessment and technology policy in Europe: new concepts, new goals, new infrastructures', *Policy Sciences*, **28** (3), 271-299.

Ven, A. van de, Polley, H., Garud, D., Schroeder, R. and Venkataraman, S. (1989), *Research on the Management of Innovation: The Minnesota Studies*, New York: Harper and Row.

Weingart, P. (1997), 'From "Finalization" to "Mode 2": old wine in new bottles?' *Social Science Information*, **36** (4), 591-613.

Zilsel, E., Raven, D., Krohm, W. and Cohen, R. (eds) (2003), *The Social Origins of Modern Science*, Dordrecht, The Netherlands, Amsterdam, The Netherlands, Boston, USA and London, UK: Kluwer.

Ziman, J. (ed.) (2000), *Technological Innovation as an Evolutionary Process*, Cambridge, UK: Cambridge University Press.

I. Driving forces of innovation practice, theory and policy

2. Inside the Public Scientific System: Changing Modes of Knowledge Production

Ben R. Martin

INTRODUCTION

Over the last 20 years, we have witnessed significant changes in the ways (or 'modes') in which knowledge is produced (Gibbons et al. 1994; Nowotony et al. 2001). Amongst other things, these changes pose a fundamental challenge to the traditional distinction between 'basic' and 'applied' research. Those researchers who are involved in 'Mode 2' knowledge production or in research in 'Pasteur's Quadrant' (Stokes 1997), or who are pursuing 'frontier research' (Harris et al. 2005), may well be concerned both with seeking new understanding about the world and at the same time with generating useful knowledge that can be applied in developing a new technology. As a result, there appears to be a much closer and more intimate connection between the resulting science and technology than in previous decades, with few of the barriers that arise when basic research and applied research are carried out separately, as they were to a significant extent in previous decades. Those engaged in 'Mode 2' or in 'frontier research' pursue questions with little regard for established disciplinary boundaries. This may involve multi-, inter- or trans-disciplinary research that brings together researchers from different disciplinary or technological backgrounds, with different theoretical and conceptual approaches, techniques, methodologies and instrumentation, and perhaps even different goals and motivations (for example, advancing scientific understanding, developing technology, or creating an innovation).

In parallel with these changes in the modes of knowledge production, we have seen the emergence of new policies and institutional configurations for facilitating border-spanning research collaboration and for stimulating increasingly science-based innovation. This co-evolution of knowledge production practice and policy has been accompanied by a third set of developments – namely, theoretical and conceptual advances in the field of science, technology and innovation (STI) policy research. Indeed, the starting point for this book is the observation that the processes involved in the production of knowledge ('science'), the application of knowledge (in 'technology') and the successful exploitation of knowledge ('innovation') are undergoing fundamental change. The book focuses on the interactions between three components – (1) science, technology and innovation (STI), (2) STI policy, and (3) STI policy studies (or STI 'theory'). While other chapters (e.g. Chaminade & Edquist) focus more on the external forces driving these changes, the aim of this chapter is to analyse the internal driving forces and their consequences, in particular for policy.

The focus of the chapter is, first of all, on the production of scientific knowledge as opposed to technological knowledge or other knowledge inputs to innovation processes. Here, 'scientific' is defined quite broadly (in the Continental European rather than the Anglo-Saxon sense) to include not only the natural sciences and engineering but also social sciences and human sciences. Secondly, the emphasis is on changes in the modes of knowledge production, and in the causes and consequences of these. Thirdly, the primary focus is on the public scientific system, as opposed to the private or corporate scientific system.

In what follows, I look first at the predominant modes of knowledge production in the three or four decades after 1945. During this time, there was a strong emphasis on the distinction between 'basic' and 'applied' research along with an associated institutional division of labour. Next, I identify the main influences that have brought about changes in the modes of knowledge production from roughly the 1980s onwards. The chapter then examines the main conceptual frameworks that have been put forward for interpreting these changes – Mode 1/Mode 2, the Triple Helix, Pasteur's Quadrant, frontier research, and changing search regimes – and critically analyses their strengths and weaknesses and the extent to which they impact on policy. In the subsequent section, I attempt a synthesis of these various frameworks to identify the key characteristics of scientific research and knowledge production today. The chapter concludes with an examination of the main consequences of changing modes of knowledge production for public policy.

TRADITIONAL MODES OF KNOWLEDGE PRODUCTION – THE SIMPLIFIED MODEL

The work of historians of science and technology suggests that for much of human history up to the late 19th century, the processes of knowledge production, application and exploitation have generally been closely intertwined. In other words, science, technology and innovation were overlapping rather than distinct activities. From the time of Archimedes (who, in addition to working on mathematics, was in charge of developing military innovations to defend Syracuse against the Romans – see Molas-Gallart, in this book), most 'scientists' (as we would now term them) were involved not only in producing knowledge but also in developing technologies and innovations to address the military, social or economic needs of the day. In the British Industrial Revolution, for example, it was networks of industrialists, inventors, engineers and 'natural philosophers' (formalised, for example, in the

Lunar Society of Birmingham (Schofield 1957[1]) and other provincial 'philosophical' societies[2]) that were often central in acquiring and sharing the knowledge needed to develop important innovations.

However, all this began to change in the 19th century. Following the introduction of the Humboldtian university model in which research and teaching were seen as an essential unity, 'science' gradually became more of a profession carried out by trained scientists working primarily in university departments (often departments of 'philosophy' or 'natural philosophy'). Over time, those departments then divided and separated on the basis of the emerging scientific disciplines (e.g. biology, physics, chemistry etc.), and their research became more 'academic' in orientation. Later, the separation between science and innovation was further widened as large firms began to establish their own R&D laboratories towards the end of the 19th century. About the same time, the first government R&D laboratories were also created.

From 1945 through to around the 1980s, the following simplified model was an approximate representation of the knowledge production process. The primary contributor to knowledge production was the university or the higher education institution. Universities were predominantly discipline-based in terms of their organisation of teaching as well as research, and the subsequent professional careers of researchers depended primarily on their contribution to their discipline and how those contributions were judged by peers from within that discipline. University research was mainly funded through institutional core funding (except in the United States where virtually all university research was financed via individual project grants). In addition (for universities outside the US), there was some project funding for university research coming mainly from national research funding agencies (again often organised through discipline-based committees), who judged the research proposals submitted to them using 'peer review' (i.e. with little recourse to any 'external' criteria) in a 'responsive mode' of operation.

Another substantial contributor to knowledge production was the public sector laboratory. This took a variety of forms including (1) government laboratories or agencies (e.g. to carry out research and development in nuclear energy, space or defence), (2) research institutes such as 'big science' centres and academy institutes, and (3) research council laboratories or institutes. The other main type of contributor to knowledge production was 'high-tech' industry, primarily in the form of a few, large, science-based companies in sectors such as electronics (e.g. AT&T, IBM, Philips, Siemens), aerospace (e.g. Lockheed, Rolls Royce), chemicals and pharmaceuticals (e.g. BASF, Dupont, Hoechst, ICI) and oil (e.g. BP, Shell).

1 Schofield (1957) concluded that 'the Lunar Society represented an eighteenth century technological research organization'.

2 In its earliest years, the Royal Society of London also provided a means for bringing together natural philosophers and 'entrepreneurs' (as we would now term them), before it became more 'academic' in orientation.

CATEGORISATION OF DIFFERENT MODES OF KNOWLEDGE PRODUCTION

In the period after 1945, the distinction between 'basic' research (knowledge produced 'for its own sake') and 'applied' research (knowledge produced with a specific application in mind) became increasingly important. This was linked to the development of R&D statistics in the 1960s (and especially the *Frascati Manual* first produced by OECD in 1963), although the distinction predates that (Godin 2000; 2005). The basic/applied dichotomy emerged from earlier distinctions between theoretical and practical knowledge (linked back to the ancient Greeks who developed a hierarchy of knowledge in which 'theory' was ranked ahead of 'practice'), and between 'pure'[3] and applied research (cf. Calvert & Martin 2001). During the 20th century, basic research came to be very closely associated with 'academic research' (indeed Ziman (2000) regarded them as virtually equivalent), with most basic research being carried out in academic institutions, and most of the research performed by academic institutions being 'basic' in character rather than 'applied' (and with relatively little being 'development'). As we note below, this was perhaps rather less true in the case of the US, where mission agencies such as the Department of Defense, the Department of Energy, the US Department of Agriculture and the National Institutes of Health were the main funders of university research, investing far more than the National Science Foundation, the body responsible for supporting basic research. Yet even there (and notwithstanding the tradition of the 'land-grant universities' dedicated to supporting agriculture and mechanical engineering), universities continued to see basic research as their main mission rather than applied research or development.[4]

These categories of research and of knowledge production were closely linked with an institutional division of labour (although again the description provided here is highly idealised). In this, basic research was largely performed in universities and other academic institutions (e.g. Max Planck institutes in Germany or CNRS laboratories in France) to produce 'scientific' knowledge along with trained graduates; applied research was mostly performed in government laboratories and some industrial laboratories (but also in university engineering departments), creating 'technological' knowledge; and development was carried out mainly in industrial laboratories to produce 'inventions' which, if introduced commercially, then became 'innovations'.[5] This notional division of labour was closely linked with

---

3 As the term 'pure' research lost its allure, the notion of 'fundamental' research became more common, before it, in turn, was largely superseded in the second half of the 20th Century by the concept of 'basic' research.

4 According to Bozeman (2000), 70% of US university departments viewed basic research as a major mission, compared with 23% for technological development.

5 This was the original differentiation made by Schumpeter (1934) between 'invention' and 'innovation', but the latter term is often used rather loosely today, making this distinction less

Vannevar Bush (1945) and the 'linear model' (the 'science-push' one, not the later 'demand-pull' one). Hence, this provides an early example of the co-evolution of theory, policy and practice.

The linear model proved very useful from rhetorical, ideological and political points of view, being used to justify rapidly growing government investment in basic and academic research during the 1950s, '60s and '70s. However, it should be recognised that it was never more than a crude over-simplification. As noted above, much university research – in the United States in particular but also elsewhere – was funded by mission agencies to help address wider societal needs in sectors not easily served by private R&D (e.g. defence, health, space, energy, agriculture). Here, the other linear model of innovation, one based on 'demand-pull' rather than 'science-push', was more relevant, with societal needs 'calling forth' the research required to develop new technologies and innovations (Bozeman 2000).[6] In addition, there were significant non-university contributions to basic research, most notably by AT&T's Bell Labs, which produced a number of Nobel Prize winners.

This idealised picture of different types of research and knowledge production[7] was provided with more substantial underpinnings by the social sciences. From economics came the notion of 'market failure' used by Nelson (1959) and Arrow (1962) to develop a rationale for government funding of basic research based on certain simplified and idealised properties of scientific knowledge (i.e. that it is a non-rival, non-excludable, 'public good', the production of which is subject to considerable uncertainty, and the benefits from which cannot be fully appropriated by the producers – see Smith, 2000, and Chaminade & Edquist, in this book). In addition, sociologists, and in particular Robert Merton (1973), attempted to explain the production of scientific knowledge in terms of certain norms (the 'CUDOS norms') that influence the behaviour of (basic) researchers.

CHANGES IN KNOWLEDGE PRODUCTION FROM THE 1980S ONWARDS

Following the economic crises of the 1970s[8] when many governments found themselves struggling to cope with problems of inflation, unemployment and stagnation, research and the production of knowledge began to experience substantial changes from the 1980s onwards. These were the combined result of various driving

straightforward.

6 This represents another early example of the co-evolution of theory, policy and practice.

7 Moreover, it should be stressed that this highly idealised picture was far less true in earlier historical eras (i.e. before the second half of the 20th century).

8 In attempting to give a date, one must inevitably generalise and oversimplify. There was, of course, some variation in timing across countries – for example, Sweden's main economic crisis came later.

forces, which have significantly changed the institutional environment in which knowledge production takes place.

The first such driving force is globalisation and the opening up of numerous economies to market capitalism with a consequent escalation in economic competition. One consequence of this has been a growing emphasis on innovation in industrialised countries as a response to the increasing competition from low-cost economies. A second has been a trend on the part of many corporations that were previously quite active in research to reduce or outsource their research to universities or to specialised organisations. A third consequence has been the growing scale of multi-nationals buying foreign companies, again with major implications for STI policy and practice; governments, whether national or regional, need to ensure that the local knowledge base is strong and therefore attractive to the rest of the world.

A second key driver relates to the growing importance of scientific and technological knowledge and skills as we began to move towards a more knowledge-intensive society or the 'knowledge economy' (e.g. Drucker 1994; Stehr 1994; Cowan et al. 2000). As we shall see below, one aspect of this is that innovation is coming to depend increasingly on science and technology. Another is the rapid growth in the scale of high education (sometimes termed the 'massification' of HE), and with it a considerable increase in the number of academics expecting to be funded to do research in combination with their teaching. This, coupled with the rising costs of research, has placed a considerable strain on governments.

Thirdly, the rising costs and expectations with regard to research have unfortunately occurred at a time of increasing constraints on government spending, some external (such as the consequences of the economic crises referred to above, which led governments to seek to get more from less), others the result of explicit decisions by politicians such as Mrs Thatcher to downsize government and its spending. As a result, there have been demands for more explicit and wider public accountability for all areas of government spending, including that on research, stimulating the growth in evaluations and the use of performance-related indicators.[9] The earlier, rather 'loose' social contract between science and society that characterised the decades after the Second World War has been replaced by much tighter social contract with more direct expectations that public funding of research should yield benefits to the economy and society (Guston & Keniston 1994; Martin 2003).

Lastly, there has been a growing concern among the public with certain scientific and technological activities and with the perceived risks associated with them (e.g. nuclear energy, GM crops – see e.g. Wynne 2001). This has brought about a shift from the previous approach to involving the public (based on the 'public understanding of science' and the 'deficit model' – see e.g. Hilgartner 1990; Durant

---

9 There has also been increasing emphasis in government policy on international collaboration as a way to share the costs of expensive research (e.g. on climate change, or mapping the genome), another factor stimulating the increase in collaboration mentioned above.

et al. 1992) to a more interactive form of 'public engagement', preferably 'upstream' rather than at the 'end of the pipe' (Wilsdon & Willis 2004). This is part of the broader trend for a growing number of institutional actors to be involved in matters to do with STI, with the result that there is greater heterogeneity (Rip 2000), an increasing number of cross-linkages and increased complexity to be confronted in decision-making processes.

These driving forces and the resulting changes in the institutional environment in which knowledge production takes place have brought about a number of changes in the knowledge production process. One relates to changes in funding mechanisms. There are a number of aspects to this including the widespread shift from general university funding (GUF) to project-based funding, from responsive funding to a more directed mode of funding, and from peer review to 'merit review' along with greater involvement of users. These reflect the revised 'social contract' referred to above – the increasing political expectations that government funding for science should be linked to the delivery of benefits to 'users' whether in industry or society more widely. Consequently, there has been a growth of targeted research programmes, centres and other initiatives in selected priority areas. This, in turn, is linked to the development of Technology Foresight and other forms of strategic intelligence as a means to help overcome 'institutional myopia' and to identify the priority areas for selective funding (Irvine & Martin 1984; Martin & Irvine 1989; Martin 1995; Martin & Johnston 1999). The result has been to concentrate funding and hence research activities on areas of research seen as being closely linked to the development of new technologies and innovations.

Closely linked to the process of globalisation described above is the growing international nature of science, technology and innovation. In the case of science, for example, international collaboration has risen steadily from the 1960s, when inexpensive international air-travel based on wide-body jets first became available, a growth subsequently stimulated by easy, cheap, international phone calls, then by the development of fax machines, and later by email and other electronic communication (Katz et al. 1995). As a result, competition in science, technology and innovation has become ever fiercer as more 'players' join in and as those new entrants learn 'to play the game' more effectively.[10] The escalating competition and the growing costs of research have also contributed to an increasing concentration of research resources and the emergence of a more pronounced hierarchy of research universities (i.e. with a bigger gap between 'the best' and the rest).

A third set of changes are those relating to the previous institutional division of labour. As noted above, one feature of the last two decades has been a reduction in the level of research carried out internally by many firms, a trend often accompanied

10 In the case of science, this is reflected in the trends in shares of scientific publications on the ISI-Thomson Scientific database, where the US share has been falling as scientists in China, India and elsewhere succeed in getting their results published in the leading journals covered by ISI (Leydesdorff & Wagner 2006).

by greater reliance on outsourcing, whether to universities or to specialist R&D organisations. Partly as a result of this, innovation now often involves interactions between a number of firms and other organisations – i.e. it takes the form of a 'distributed' or 'networked' innovation process[11] (see Dankbaar & Vissers in this book).[12] There has also been a significant downsizing of the traditional government laboratory sector (e.g. through closures, mergers or privatisation). Where new publicly funded research centres have been established, these have often been of fixed duration. They also tend to be cross-disciplinary in nature, sometimes based on university-industry collaboration or other institutional 'hybrids'.

A fourth trend is that technology and innovation are apparently becoming more science-based. During the last part of the 20th century, science and technology seem to have become more closely linked, interacting and overlapping (as suggested, for example, by the growing number of patent citations to scientific publications – see Narin et al. 1997). As noted above, in some respects this represents a shift back to the situation pertaining in an earlier era when science and technology undoubtedly overlapped to a considerable extent. However, one difference is that now new technologies such as information technology, biotechnology and nanotechnology do seem to depend to a much greater extent on developments in science than was the case with new technologies and innovations in previous centuries.

These fundamental changes in knowledge production processes have been accompanied by significant shifts in terminology. In particular, the traditional basic/applied categorisation was first replaced by a three-fold distinction between basic, 'strategic' and applied research (Irvine & Martin 1984). Later, other new terminology was introduced (in particular the concepts of 'Pasteur's Quadrant' and 'frontier research', as we shall see below). The shift in terminology is part of a broader trend – the development of an ever more sophisticated understanding of the nature of innovation process, of the relationship of innovation to technology and to science, of 'systems of innovation' be they national, regional or sectoral (see Chaminade & Edquist in this book), and of how best to encourage or steer knowledge production and innovation through policy (e.g. through the use of more 'holistic' policy instruments).

At the same time, work by social scientists and others has steadily undermined the earlier intellectual underpinnings of the traditional scientific knowledge production model (such as the notion of 'knowledge for its own sake' and of science

---

11 According to Smith (2000), one benefit of this is that cooperating firms tend to be more innovative.

12 In this respect, it has certain similarities to the situation in the 19th Century before firms set up their own R&D laboratories, when they were likewise forced to draw upon inputs from a variety of sources in developing innovations – see the earlier discussion of the role of philosophical societies in providing entrepreneurs with a means to engage in 'networked' innovation during the 18th Century in Britain.

driven by its own internal logic, the linear model,[13] the 'market-failure' rationale and the CUDOS norms). These have been replaced by more complicated models and conceptual frameworks including the 'chain-link' interactive model of the innovation process (Kline & Rosenberg 1986), the 'social construction' of science and technology (e.g. Pinch & Bijker 1984; Bijker et al. 1987), evolutionary models of innovation and economics (Nelson & Winter 1982), the notion of 'national systems of innovation' (Freeman 1987; Lundvall 1992; Nelson 1993), and the 'system-failure' rationale (see Chaminade & Edquist in this book).

These developments have stemmed from several decades of efforts by science and innovation policy researchers, and illustrate the 'dance' between theory and practice that forms the focus of this book. As our understanding and models of the innovation process has become more sophisticated, and as the innovation process has itself apparently become more complex,[14] this has placed greater demands on policy-makers to make sense of this increasing complexity. The new, more sophisticated innovation models are undoubtedly much harder to explain and to justify to policy-makers, politicians and others. Here, we may well face a problem in the form of the limited 'absorptive capacity' of government policy-makers for understanding and acting on ideas developed by the STI research community, a problem often made worse by the comparatively rapid turnover of staff (e.g. following the introduction of 'new public management' approaches by government). In the concluding section, we return to consider the question of whether simpler models of the innovation process may sometimes be more useful to policy-makers.

CONCEPTUAL FRAMEWORKS FOR INTERPRETING THE CHANGES IN KNOWLEDGE PRODUCTION

Over the last 15 years, a number of conceptual frameworks have been proposed for analysing and understanding these changes in the knowledge production process. These include the analysis in terms of the 'Mode 1' and 'Mode 2' versions of the knowledge production process, the 'Triple Helix' model of the interactions between universities, government and industry, the notion of 'Pasteur's Quadrant' research and the closely related concept of 'frontier research' and shifts in 'search regimes'.

13 Curiously, only the 'science-push' linear model continues to be repeatedly slain today; the 'demand-pull' linear model is, for some mysterious reason, spared this ritual treatment. Where once it had equal prominence to the science-push model, the demand-pull model seems almost to have been 'airbrushed' out of the history of our field!

14 As with many phenomena studied by social scientists, it is difficult to be certain whether a given phenomenon is actually becoming more complex, or whether our ability to study and conceptualise it in a more sophisticated manner merely gives the impression that its complexity is increasing.

In the following sections, we critically analyse each of these frameworks in turn.

Mode 1 and Mode 2 Knowledge Production

According to Gibbons and his colleagues (1994), the world is witnessing a shift from the 'Mode 1' form of knowledge production to 'Mode 2'. Mode 1 knowledge production is largely academic or university-based, and is mostly mono-disciplinary in nature, pursuing research questions generated within the scientific discipline concerned. Researchers engaged in Mode 1 who receive public funding are accountable to their scientific peers only. Mode 2, in contrast, is carried out in a growing variety of institutions. It is often transdisciplinary, not least because it addresses 'external' research questions rather than those posed within a single discipline. As a result, it generates knowledge 'in the context of application'. Publicly funded Mode 2 researchers are accountable externally (to funders, politicians and the public more generally) as well as to other scientists. There is some evidence that the notion of Mode 1/Mode 2 has influenced the thinking of policy-makers (Rip 2000) and hence shaped (or re-enforced) policy – in other words, there has been an element of a 'dance' between practice, theory and policy.

However, as critics such as Pestre (1997, 2003), Weingart (1997), Godin (1998) and Shinn (2000) have pointed out, there are various problems with the Mode 1/ Mode 2 thesis. First, its claims appear to be based more on assertion than evidence. For example, what evidence is there that there is now more transdisciplinary research than in previous times?[15] Secondly, the claim that we are moving towards a *new* mode of knowledge production is historically suspect. Mode 2 knowledge production is certainly not new. It was prominent, for example, in the 19th century and the early part of the 20th century. From the time of the emergence of modern science, there has always been 'research in the context of application'. Indeed, historians of science have found few instances of scientific research where external influences were entirely absent. For example, even Sir Isaac Newton may have been motivated at least in part by very practical military and industrial considerations (Hessen 1931). Likewise, multi-, inter- and transdisciplinary research has always existed; indeed, this has often been the origin of what then in due course has become a separate 'discipline'. Similarly, for much of the history of science, external 'strings' have usually been attached to the funds provided to support scientific research with some form of external accountability. Even during the period from 1945 to the 1980s when Mode 1 knowledge production was most prominent (under the Vannevar Bush 'social contract'), much research in universities had at least some Mode 2 characteristics, even though this tends to be ignored in academic rhetoric.

In short, the claims of Gibbons et al. are perhaps better reformulated in terms of shifts in the balance of Mode 1 and Mode 2 over time. The proportion of Mode

15 There have been some attempts to investigate, for example by Grupp (1992), Katz et al. (1995) and Hicks and Katz (1996).

1 was probably higher for the period from 1945 to the 1980s than in the periods before and after. However, this raises the question of what is the most appropriate balance between the two modes of knowledge production for society as a whole.

The Triple Helix

According to the proponents of the Triple Helix (Etzkowitz & Leydesdorff 1997; 1998; 2000; Leydesdorff & Etzkowitz 1996; 1998), universities, government research organisations and industrial laboratories were in the past essentially separate in relation to knowledge production (see the earlier discussion of the institutional division of labour). However, over the last 15-20 years, we have seen the emergence of a Triple Helix relationship, with growing interaction and overlap of the three sectors. As part of this, there has been a gradual blurring of the boundaries between these sectors as universities have begun to assume some of the research responsibilities previously confined largely to industrial laboratories of exploiting as well as creating knowledge. Consequently, universities are now contributing more directly to meeting the knowledge needs of government, industry and society, in some cases formally assuming this as a 'third mission' along with the two traditional missions of research and teaching. According to Etzkowitz (1997; 2000), this corresponds to 'the second academic revolution',[16] from which are emerging 'entrepreneurial universities' embracing all three missions. Moreover, the Triple Helix has been picked up as a 'motif' for certain policies, for example in Latin America and Scandinavia,[17] a specific example of the co-evolution of STI theory and practice.

The problems with the Triple Helix model are similar to those for Mode 1/Mode 2 (for a critique, see Shinn 2000). First, the 'third mission' is certainly not new. It was particularly prominent, for example, in technical universities in Germany, in 'land-grant' universities in the United States, and in the engineering departments of many other universities in the latter part of the 19th century (Martin & Etzkowitz 2001). Likewise, strong relations between higher educational institutions and the state are not a new phenomenon; indeed, this was a distinctive characteristic of higher education institutions in 19th century France and Prussia. And in that same century, universities often had links with industry, with engineering professors and others helping firms to address their technological needs in an era when firms had yet to establish their own R&D laboratories (Rosenberg & Nelson 1994). Even in the period from 1945 to the 1980s, US universities were heavily funded by the Department of Defense and were closely integrated in the 'military-industrial complex' – i.e. they were part of a Triple Helix relationship.

Another problem is the 'Triple Helix' metaphor itself – how much does it add to the earlier formulation of 'Sábato's Triangle' – the three-sided interaction between

16 The first academic revolution occurred when universities formally took on research as a formal mission, to be combined with teaching.

17 For examples, see papers presented at the 'Triple Helix' conferences in Brazil and Denmark.

government, industry and university put forward in Latin America by Sábato some two decades earlier (Sábato 1975, cited in Etzkowitz & Leydesdorff 2000 and Etzkowitz et al. 2000)? It is not clear why the diagrammatic representation shows helical strands rather than three circles (either linked or overlapping); indeed, many papers supposedly espousing the 'Triple Helix' model actually contain figures based on the latter (i.e. 'Sábato's Triangle') rather than on helices.[18] But is it meaningful to regard 'government' as a single helical strand, as EU and regional policies grow in importance relative to national government policies? Similarly, to what extent can 'industry' be regarded as a single strand? What corresponds to the 'bonds' holding the three helices together? And would any configuration of three helices ever be stable? In short, while the 'Triple Helix' (with its resonance with DNA and evolution) may offer an attractive mental image or logo,[19] it is not clear whether it adds much as an intellectual model in terms of adding to our understanding of knowledge production and exploitation, let alone generating testable hypotheses.

Pasteur's Quadrant

As noted above, from the mid-20th century onwards, much discussion of science policy was predicated on the dichotomy between 'basic research' and 'applied research' – i.e. research could be either one or the other. However, in an influential book Stokes (1997) argued that one needed to replace this basic/applied dichotomy with a two-by-two matrix. Besides research aiming to generate new understanding but not to develop a new application (Bohr's Quadrant), and research aiming to develop a new application but not new understanding (Edison's Quadrant), one can also have research aiming both to produce new understanding and to develop a new application and thus meet some economic or societal need (Pasteur's Quadrant).[20] There is some similarity here to the earlier notion of 'strategic research', although the latter was generally seen more as occupying an intermediary position on a one-dimensional spectrum from pure to applied research (Irvine & Martin 1984).

The concept of 'Pasteur's Quadrant', and the related concept of 'basic technology'[21] put forward by Branscomb and Keller (1999) at about the same time,

---

18 For examples, again see papers presented at the various 'Triple Helix' conferences.

19 One undoubted success of the 'Triple Helix', however, has been in providing a focus for a series of regular international conferences, bringing together researchers in the field of STI policy research from all round the world.

20 The remaining quadrant has received much less attention. It corresponds to work where the aim is neither to create new knowledge nor to produce a new application. Possible examples include work on taxonomy and curation.

21 Branscomb's notion of 'basic technology' was picked up by the UK Government around 2000 and used to justify an increase in public funding for this 'new' type of activity. The author was centrally involved in this transfer, providing a neat example of the co-evolution of theory and policy.

proved very influential politically and economically in the United States. Prior to that, there had been a traditional partisan split between Democrats and Republicans as to whether or not it was appropriate for government to spend federal funds on 'applied research', with Republicans viewing government expenditure on much civil applied research as inappropriate in a free market system and opposing it on the grounds that it represented 'corporate welfare'.[22] The shift in terminology to 'Pasteur's Quadrant' effectively moved the boundary as to what forms of research the Federal Government could legitimately fund, with the Republican-dominated Congress often vying to outspend the Democratic Administration on medical research (the clearest example of Pasteur's Quadrant) over the second half of the 1990s.[23] This constitutes another example of the 'dance' between practice, theory and policy.

There are two main problems with Pasteur's Quadrant. First, it is a useful descriptive taxonomy rather than an analytical model. Secondly, the great majority of research conducted in universities and public research laboratories falls in Pasteur's Quadrant (especially if one adopts a long-term view) rather than Bohr's Quadrant. Scientists in even the most 'basic' of sciences tend to argue that ultimately society will benefit in some (yet to be defined) way from the results of their work, just as the purest research in the 19th and early 20th centuries resulted eventually in developments in electronics, nuclear power and so on. Indeed, a more specific problem is how to categorise research that in the short term fits in Bohr's Quadrant but in the longer term relates more to Pasteur's Quadrant.

Frontier Research

The concept of 'frontier research', which was developed in a recent EU report (Harris et al. 2005), is broadly similar to Pasteur's Quadrant. The term 'frontier' emphasises certain characteristics of much current knowledge production: (1) research that is at the very forefront of developing new knowledge and understanding; (2) research that is risky, unpredictable, and difficult or even impossible to 'plan'; (3) research that is concerned with developing both new understanding and useful knowledge, so there are closer and more intimate connections between 'science' and 'technology', with the research being at the frontiers of both; and (4) research that shows no respect for disciplinary boundaries – in other words, research that is intrinsically multi-, inter- or transdisciplinary (ibid.).

As with Pasteur's Quadrant', the shift in terminology to 'frontier research' has been influential politically in Europe, helping to justify the creation of the new European Research Council, which is to operate in an entirely different manner from

---

22 Although there were some exceptions, such as agriculture and health, where there was already bipartisan support for federal funding.

23 The ideas of Stokes and Branscomb first began to attract attention even before their respective books were published.

the European Commission as an independent research council more like DFG.[24] In other words, the adoption of the term 'frontier research' represents another example of the co-evolution of theory and practice.[25] (It is ironic to note in passing that the French example of Pasteur was influential in the United States, while the term 'frontier', which is far more redolent in the US, proved more influential in Europe!)

Search Regimes

A somewhat different approach to analysing changes in the modes of knowledge production has recently been proposed by Bonaccorsi (2008), who analyses those changes in terms of shifts in 'search regimes'. A 'regime' is 'a consistent set of dynamic properties of the search process' in a scientific field or discipline (ibid.). Bonaccorsi categorises different scientific domains in terms of three dimensions of search regimes: (1) the rate of growth (measured via the growth of new scientific words); (2) the degree of diversity in terms of the various directions of research (measured by the concentration of scientific words – i.e. whether the same words are used again and again); and (3) the level of complementarity – that is, 'the extent to which different human or material resources are needed, in addition to the intellectual resources of the scientist himself' (ibid.).

On the basis of this categorisation scheme, Bonaccorsi identifies four main categories of research or knowledge production:

1. Low diversity/low complementarity: e.g. traditional chemistry;
2. High diversity/low complementarity: e.g. mathematics;
3. High diversity/high complementarity: e.g. high-energy physics, space research;
4. High growth, high diversity and high complementarity: e.g. biomedical/biotechnology research, computer science, materials science and nanotechnology.

According to Bonaccorsi (ibid.), category (4) includes the 'new leading sciences' – sciences characterised by rapid growth, divergent dynamics of search with new hypotheses continually generating new research programmes, and new complementarities especially the need for wide-ranging cross-disciplinary competences. Bonaccorsi's analysis is more empirically based than the other conceptual models of changing modes of knowledge production discussed above, although one can debate how satisfactory are data based on scientific 'words' for

24 *Deutsche Forschungsgemeinschaft.*

25 Several members of the science policy research community including A. Bonaccorsi, S. Kuhlmann, P. Stephan and the author were members of the team that produced the report on the challenges of 'frontier research' and the role that a new ERC could play in meeting them, and were therefore at the heart of this co-evolutionary process.

measuring growth and diversity. It is, as yet, probably too soon to detect any influence on policy.[26]

Nevertheless, there are a number of questions with the Bonaccorsi framework that need to be addressed. First, while it provides a static snapshot of the situation today, it is not apparent how one might incorporate a dynamic element. For example, how does a scientific domain switch from one category to another? And what might bring about such a change? Secondly, what is the relationship of this framework to earlier historical periods? How well does this model fit knowledge production in the 19th or early 20th centuries? Thirdly, how much further forward does the Bonaccorsi framework take us compared with Kuhn's analysis (1962) in terms of 'pre-paradigmatic science', 'revolutionary science', and 'normal science'?

SYNTHESIS – KEY CHARACTERISTICS OF TODAY'S FRONTIER RESEARCH

Mode 2, the Triple Helix, Pasteur's Quadrant, 'frontier research' and 'search regimes' all represent attempts to characterise changes to research and the production of scientific knowledge in recent years.[27] Common to all of these are a number of broad features.

First, much research is neither 'purely' basic nor completely applied, but instead aims to produce both new understanding and useful knowledge. Linked to this, the traditional boundary between 'science' and 'technology' is becoming less clear or more blurred.[28] Some institutions and researchers are involved in both at the same time. This may generate certain benefits as there are fewer barriers to be overcome in exploiting scientific advances in new technology, but it may also generate problems (see below).

---

26 However, Bonaccorsi's ideas were certainly influential in the group that produced the report on 'frontier research' by Harris et al. (2005) (see above). Also, European Commission policy-makers concerned with fostering the 'European Research and Innovation Area' have shown interest in the concept since it offers a perspective on knowledge production beyond national constituencies.

27 However, as noted elsewhere, it is debatable how much of this is actually 'new' compared, for example, with 19th century research. There has arguably been some shift over the last 15 years compared to the previous 30 years, but the situation now with regard to knowledge production certainly has some similarities with that in the 19th century and the earlier part of the 20th century.

28 Again, it is difficult to disentangle whether it is actually becoming increasingly blurred (as e.g. Narin et al. (1997) suggest on the basis of growing numbers of patent citations to scientific articles), or whether we are merely coming to appreciate that our earlier conceptions of 'science' and 'technology' as essentially distinct activities were too simplistic.

Secondly, research and the production of new scientific knowledge is carried out in a great variety of institutions and contexts – not just in individual universities and public research laboratories, but also in research centres, networks, collaborations, and other hybrid organisations (temporary or permanent) cross-cutting academia, government and industry (as the Triple Helix proponents stress). In other words, there is a richer, denser set of links integrating the institutional actors that comprise the national system of innovation. As a result, there are arguably fewer and/or smaller institutional barriers to knowledge transfer and exploitation. However, this may also create problems. For example, with universities becoming involved in industrial or business activities, this may conflict with the two original missions of teaching and research, and result in a clash between the traditional CUDOS norms of 'open science' and the PLACE norms (Proprietary, Local, Authoritarian, Commissioned and Expert) of the knowledge production process in 'commercialised science' (Ziman 2000).

Thirdly, much of the 'frontier research' or 'leading science' of today draws on several traditional scientific disciplines – on different theoretical perspectives, methods, instruments etc. In other words, it is intrinsically inter-, multi- or trans-disciplinary. This results in various problems or challenges:

1. *Funding*: existing research funding agencies (and their internal divisions or committees) are often structured around traditional disciplines, while peer review is also generally organised on the basis of disciplines; as a result, research proposals that are not in the disciplinary mainstream often 'fall between two stools' in the form of two discipline-based committees for each of whom the research in question is not a top priority, or where one or more of the peer reviews is less enthusiastic about the proposal.
2. *Research-performing organisations*: universities in particular are still in most cases organised primarily around discipline-based departments, and these can therefore be slow to respond to new research opportunities. Some universities (for example MIT – see Etzkowitz 2000) have tried to overcome this by creating interdisciplinary research centres to pursue the emerging opportunities in new fields. However, these then run the risk of becoming separated from undergraduate teaching, most of which is conducted in discipline-based departments.
3. *Evaluations*: assessments, like the research assessment exercise (RAE) in the UK, are generally discipline-based and consequently tend to be biased in favour of mono-disciplinary research. This runs counter to policies encouraging research that addresses the needs of 'users', research that inevitably involves inter- or multi-disciplinary research. There is thus an inherent contradiction at the heart of science policy in the UK and to some extent also in other countries where policies aiming to encourage Mode 2 research co-exist with evaluation systems tailored to assessing Mode 1 research.

4. *Culture clashes between disciplines*: as Becher (1989) has argued, researchers in different disciplines to some extent resemble 'academic tribes' with their own norms, values, specialist language and so on. While much of today's research may require collaboration between different disciplines, significant effort may need to be devoted to establishing and managing such collaborations in order to minimise disputes and maximise the synergy. If the disciplinary differences are too great, the costs of managing such collaborations may be greater than the benefits (Katz & Martin 1997).

CONCLUSIONS – SOME POLICY CONSEQUENCES OF CHANGING MODES OF KNOWLEDGE PRODUCTION

In this final section, I examine some of the main implications of changing modes of knowledge production for public policy. I look in turn at the implications for research performers, for research-funding institutions, for national systems of innovation, and for the rationale for government intervention in research and innovation. I also consider the issue of whether a balance needs to be struck between complexity and utility with regard to the models of innovation developed by STI researchers. I conclude by raising a fundamental question, namely: 'What are science and knowledge production for?'

The first set of policy implications relate to research performers. In the case of universities, some may decide to pursue 'third mission' activities. Such institutions will need to develop suitable reward and incentive structures (e.g. criteria for promotion that reflect 'third mission' contributions as well as those to teaching, research and university administration). They will also have to build up new funding streams for 'third mission activities' and determine how best to allocate these funds, which in turn may involve deciding how to assess often quite diverse 'third mission' contributions. Universities that embrace 'third mission' activities should also consider the potential benefits of organisational restructuring away from traditional discipline-based departments. Another requirement for universities expanding their 'third mission' activities is for close monitoring to ensure that those activities do not result in significant damage to the two traditional university missions of teaching and research, as some academics fear (e.g. Elzinga 1997; Ziman 2000; Stephan 2001; David 2004; Nelson 2004).

The changes in modes of knowledge production also have policy implications for research centres, which are assuming a growing variety of forms (e.g. university-industry hybrids). Here, one should avoid creating permanent laboratories like those set up in the 1950s and 1960s. Instead, when new research centres are created, they should be regarded as having a fixed lifetime only (e.g. of 10 years),[29] after which they

29 For example, the interdisciplinary research centres in the UK, and the 'competence centres' in Austria.

are wound down, with the resources being reallocated to new growth areas. Where large, non-university public-sector research organisations have a more permanent existence (as in Germany, USA, France and the Netherlands, for example), they are likely to need organisational modes fostering the influx of new research themes and perspectives, creativity-supporting funding and evaluation criteria, and a regulatory regime facilitating inter-institutional collaboration with external public and private partners (e.g. Heinze & Kuhlmann 2007).

In the case of industry and business, a growing range of business organisations are involved in research and the generation of scientific knowledge. Examples include software companies and consultancies, although in many such cases their research activities are often not captured, or not captured fully, in official R&D surveys as a result of using 'old' definitions of research (see Dankbaar & Visser in this book, on the changing role of the firm). This points to the need for revised definitions and improved survey techniques. Lastly, an increasing proportion of knowledge production is carried out by collaborations, consortia and networks rather than by individual research organisations. Again, this should be recognised and reflected in public policy (cf. Bozeman (2000) on the 'cooperative technology policy' paradigm).

A second set of policy implications relate to research funding institutions. Given the changes to knowledge production processes, these need to adopt a more holistic approach to policy and to develop appropriate systemic policy instruments (cf. Smits & Kuhlmann 2004), such as Technology Foresight and strategic intelligence, in order to avoid the dangers of compartmentalisation as well as institutional myopia. In addition, they may have to engage in periodic restructuring to reduce the level of organisational inertia because they are 'locked in' to traditional disciplines (Rip 2000). Their internal organisational structure is likely to require periodic restructuring in order to reflect the shift from the traditional 19th century disciplines. In particular, funding agencies must ensure that initiatives from newly emerging areas of science are not held back by a peer review system that often tends to favour traditional disciplines. Given the changes noted above, research funders must also acquire the flexibility to fund a growing range of research performers, not just academic researchers. This also means finding an appropriate balance between peers and 'users' with regard to the bodies involved in strategy formulation as well as funding decisions.

Thirdly, there are implications of changing modes of knowledge production for national systems of innovation. As we have seen, these systems are becoming more complex, making it more difficult to intervene through policy measures. There is a much less clear distinction among the sectors of science, technology and innovation than previously, and consequently a weaker division of labour or responsibility between different types of institutional actors. At the same time, however, there are fewer and weaker 'barriers' impeding knowledge transfer and interaction between science, technology and innovation. Related to this is a shift to greater emphasis on demand-oriented innovation policy (Edler, in this book).

Another major change is the shift from purely 'national' systems of innovation to multiple levels – regional, national, European and global. During the period from

1945 to the 1980s, the only significant involvement by government in STI in most countries was at the national level. However, in many European countries, policies and initiatives by regional governments and the EU are becoming increasingly important, while the influences of globalisation (as manifest, for example, in the R&D activities of multinational corporations) are also important (see Smith, in this book). The 'national system of innovation' is consequently comprised to a certain extent of a number of 'regional systems of innovation' (with different 'sectoral systems of innovation' forming another component – see Carlsson et al., in this book), while it also forms part of an emerging 'European system of innovation'[30] and perhaps even of a nascent 'global system'. There is therefore a growing need to ensure that policy interventions at these different levels are coherent (perhaps based on some agreed division of labour) and effectively integrated so that they do not give rise to contradictory effects at different levels.

Fourthly, changes in knowledge production processes have a number of implications for policy. One involves the need to integrate what were often previously separate policies for science, for technology and for industry (and indeed for higher education) into a more holistic 'innovation policy'. This may often require restructuring of government ministerial responsibilities (where the separate policies were previously handled by different ministries).[31] In the case of the UK, for example, in recent years responsibility for funding the Research Councils has passed from the Department of Education and Science (as it then was) to the Office of Science and Technology (OST), which, after a brief period in the Cabinet Office, is now embedded in the Department of Trade and Industry. Indeed, in 2006, following a further re-organisation to bring together 'innovation' and 'science and technology', this was transformed into the Office of Science and Innovation (OSI). However, such re-organisations may not be immediately effective because inter-ministerial jealousies (and 'the territorial imperative' that seems to be exhibited by ministries in governments across the world) tend to linger on as a constraint on attempts to develop more holistic, government-wide policy.[32] It was arguably far simpler under the 'old' system (where a single ministry was responsible for a given area) to

30 It is not clear (indeed, this needs further investigation) whether we are witnessing an 'integration' of formerly national systems, accompanied even by some 'centralisation' of funding policies, or rather an open-ended 'trans-national' reshuffling of responsibilities and policies towards some new form of governance (e.g. Kuhlmann 2001; Edler & Kuhlmann 2008).

31 In countries such as Germany and the Netherlands, there was for several decades a formal division of labour with regard to institutional responsibilities for research that was based on the traditional 'basic/applied' dichotomy. As that distinction has broken down, so the 'boundaries' in terms of the respective missions and responsibilities of those institutions have become more blurred.

32 In Germany, the formal differentiation between the Helmholtz, Max Planck, Leibniz and Fraunhofer institutions has increasingly impeded inter-institutional collaboration (Heinze & Kuhlmann 2007).

formulate challenges and the appropriate policy response to them (e.g. in the form of developing nuclear energy, or waging the 'war on cancer') than now when cross-ministry agreement must be negotiated.

Another implication for policy is the requirement to involve a greater range of stakeholders including the public in major policy decisions. This is linked to the growing emphasis on different forms of strategic intelligence such as technology assessment (see Smits et al., in this book) and technology foresight.[33] All of this reflects a substantial shift from top-down approaches to policy to network steering or 'horizontal' policies, policies that involve not only a greater number but also a more heterogeneous range of institutional actors. This growing complexity obviously places greater demands on policy-makers.

As noted earlier, not only is the innovation process growing more complex, but so are our models of the innovation process. Yet policy-makers have a limited absorptive capacity. This raises an interesting, unresolved question: 'What is the optimum level of sophistication of models in the world of policy studies?'. Social scientists, in their efforts to establish their research reputation by making an 'original' contribution, are driven to a significant degree by attempts to find shortcomings in the earlier theoretical and conceptual models developed by their predecessors, and to replace these with their own, inevitably more sophisticated models. Yet that increased complexity probably comes at a price in terms of utility to policymakers. In the area of science and innovation policy, over the last 10-20 years, we may well have succeeded in developing models that provide a better representation of the complex realities they are meant to describe. But are they still as useful to policy-makers? And are the resulting policies that attempt to reflect this more sophisticated understanding of the innovation process more effective than those, say, of 20 or 30 years earlier?

Finally, the changes in the modes of knowledge production have implications for the rationale for government intervention. As we have seen, for several decades that rationale was predicated on the notion of market failure and the early work of authors such as Nelson and Arrow. However, advances in our understanding based on the last 40 years of research in the field of STI policy have brought about a shift from 'market failure' to 'system failures' as the primary rationale for government intervention (see Chaminade & Edquist in this book; also Smith 2000; Smits & Kuhlmann 2004). Whereas past policies largely focused on building up strong actors (firms, government labs, universities) with regard to knowledge production and innovation, now the emphasis has shifted to nurturing and strengthening the links between those actors so that the national (or regional) innovation system as a whole works as effectively as possible.

Let us end by pointing to the need to revisit the fundamental question of 'What is science (or knowledge production) for?' The traditional response for basic research

33 See, for example, the discussion of the German *Futur* programme in Smits and Kuhlmann (2004), another example of the co-evolution of theory and practice.

has been 'To satisfy human curiosity'. However, with this response, it is difficult to justify very large government funding on the scale common today (and, as we have seen, the concept of 'basic research' is increasingly inappropriate anyway). To justify larger government funding, one needs to extend the response in some way, for example, 'To generate useful scientific knowledge that in the longer-term will help to meet the needs of society'. But this then raises the question of who determines what are the most important needs of society and the relative importance of those that relate to economics and wealth creation as opposed to those that are more concerned with the quality of life and human well-being.[34] For much of the last two decades, science and technology have come to be seen as primarily for developing the innovations that will yield economic benefits and contribute to maximising GDP growth. However, given the scale of the challenges now posed by global warming (see e.g. Stern 2006), and given the intriguing finding that, above a certain level of GDP per capita, well-being does not appear to increase with further wealth (Layard 2005), we need to reconsider whether this should continue to be the dominant aim. Perhaps the time has come when, for the developed world at least, the goal should be widened, say, to 'addressing global challenges' such as sustainable development (and protecting the environment), and enhancing our quality of life and well-being.

ACKNOWLEDGEMENTS

This chapter has benefited appreciably from comments by other authors of chapters in this book, in particular, Bo Carlsson, Cristina Chaminade, Susan Cozzens, Ben Dankbaar, Staffan Jacobsson, Jordi Molas-Gallart, Philip Shapira, Keith Smith, Ruud Smits, Morris Teubal and Rutger van Merkerk, who discussed earlier drafts at workshops held in Utrecht in January 2006 and Karlsruhe in June 2006.

REFERENCES

Arrow, K. (1962), 'Economic welfare and the allocation of resources for invention', in R. Nelson (ed.), *The Rate and Direction of Inventive Activities*, Princeton: Princeton Universtity Press, pp. 609-625.

Becher, T. (1989), *Academic Tribes and Territories: Intellectual Enquiry and the Cultures of Disciplines*, Buckingham, UK and Philadelphia, USA: Open University Press.

Bijker, W., Hughes, T. and Pinch, T. (eds) (1987), *The Social Construction of Technological Systems: New Directions in the Sociology and History of Technology*, Cambridge, MA, USA and London, UK: MIT Press.

34 There is also the question of whether attaching such 'strings' to research funding will irreparably damage the scientific endeavour, as some such as Elzinga (1997), Ziman (2000) and Nelson (2004) fear.

Bonaccorsi, A. (2008), 'Search regimes and the industrial dynamics of science', *Minerva*, **46** (3), 285-315.

Bozeman, B. (2000), 'Technology transfer and public policy: a review of research and theory', *Research Policy*, **29** (4-5), 627-655.

Branscomb, L. and Keller, J. (eds) (1999), *Investing in Innovation: Creating a Research and Innovation Policy That Works*, Cambridge, MA, USA and London, UK: MIT Press.

Bush, V. (1945, reprinted in 1990), *Science: The Endless Frontier*, Washington, DC: National Science Foundation.

Calvert, J. and Martin, B. (2001), 'Changing Conceptions of basic research?', Background Report prepared for the OECD Workshop on Policy Relevance and Measurement of Basic Research, 29-30 October 2001, Oslo.

Cowan, R., et al. (2000), 'Innovation policy in a knowledge-based economy', DG Enterprise, European Commission, Luxembourg.

David, P. (2004), 'Understanding the emergence of "open science" institutions: functionalist economics in historical context', *Industrial and Corporate Change*, **13** (4), 571-589.

Drucker, P. (1994), *Post Capitalist Society*, Boston, USA and London, UK: Butterworth-Heinemann.

Durant, J., Evans, G. and Thomas, G. (1992), 'Public understanding of science in Britain: the role of medicine in the popular representation of science', *Public Understanding of Science*, 1 (2), 161-182.

Edler, J. and Kuhlmann, S. (2008), 'Integration of European research systems: a multi-dimensional phenomenon', in D. Jansen (ed.), *Towards a European Research Area*, Baden-Baden: Nomos Publishers.

Elzinga, A. (1997), 'The science-society contract in historical transformation: with special reference to "epistemic drift"', *Social Science Information*, **36** (3), 411-445.

Etzkowitz, H. (2000), *The Second Academic Revolution: MIT and the Rise of Entrepreneurial Science*, London: Gordon and Breach.

Etzkowitz, H. (1997), 'The entrepreneurial university and the emergence of democratic corporatism', in L. Leydesdorff and H. Etzkowitz (eds), *A Triple Helix of University-Industry-Government Relations*, London: Cassell Academic.

Etzkowitz, H., Webster, A., Gebhardt, C. and Terra, B. (2000), 'The future of the university and the university of the future: evolution of ivory tower to entrepreneurial paradigm', *Research Policy*, **29** (2), 313-330.

Etzkowitz, H. and Leydesdorff, L. (2000), 'The dynamics of innovation: from national systems and "Mode 2" to a Triple Helix of university-industry-government relations', *Research Policy*, **29** (2), 109-123.

Etzkowitz, H. and Leydesdorff, L. (1998), 'The endless transition: a "Triple Helix" of university-industry-government relations', *Minerva*, **36** (3), 203-208.

Etzkowitz, H. and Leydesdorff, L. (eds) (1997), *Universities and the Global Knowledge Economy: A Triple Helix of University-Industry-Government Relations*, London: Cassel Academic.

Freeman, C. (1987), *Technology Policy and Economic Performance: Lessons from Japan*, London, UK and New York, USA: Pinter.

Gibbons, M., Limoges, C., Nowotony, H., et al. (1994), *The New Production of Knowledge*, London: Sage Publishing.

Godin, B. (2005), 'The linear model of innovation: the historical construction of an analytical framework', project on the History and Sociology of S&T Statistics Paper No. 30, Canadian Science and Innovation Indicators Consortium, Montreal.

Godin, B. (2000), 'Measuring science: is there "basic research" without statistics?', project on the History and Sociology of S&T Statistics Paper No. 3, Observatoire des Sciences et des Techniques, Montreal.

Godin, B. (1998), 'Writing performative history: the new *New Atlantis*?', *Social Studies of Science*, **28** (3), 465-483.

Grupp, H. (1992), *Dynamics of Science-Based Innovation*, Berlin, Germany, Heidelberg, Germany and New York, USA: Springer-Verlag.

Guston, D. and Keniston, K. (1994), 'Updating the social contract for science', *Technology Review*, **97** (8), 60-68.

Harris, W., Martin, B., et al. (2005), 'Frontier research: the European challenge', report of the High-Level Expert Group, European Commission, Brussels.

Heinze, T. and Kuhlmann, S. (2007), 'Analysis of heterogeneous collaboration in the German research system with a focus on nanotechnology', in D. Jansen (ed.), *New Forms of Governance in Research Organizations. From Disciplinary Theories towards Interfaces and Integration*, Berlin, Germany, Heidelberg, Germany and New York, USA: Springer-Verlag.

Hessen, B. (1931), 'The social and economic roots of Newton's principia', in N. Bukharin, et al. (eds), *Science at the Crossroads: Papers from the Second International Congress of the History of Science and Technology*, London: Frank Cass – Co. Ltd. (1971, reprint), pp. 147-203.

Hicks, D. and Katz, J. (1996), 'Where is science going?', *Science, Technology & Human Values*, **21** (4), 379-406.

Hilgartner, S. (1990), 'The dominant view of popularization: conceptual problems, political uses', *Social Studies of Science*, **20** (3), 519-539.

Irvine, J. and Martin, B. (1984), *Foresight in Science: Picking the Winners*, London, UK and New York, USA: Pinter.

Jacob, M. and Hellström, T. (eds) (2000), *The Future of Knowledge Production in the Academy*, Buckingham, UK and Philadelphia, USA: Open University Press.

Katz, S., Hicks, D., Sharp, M. and Martin, B. (1995), 'The changing shape of British science', STEEP Special Report SSR 3, SPRU, University of Sussex, Brighton.

Katz, S. and Martin, B. (1997), 'What is research collaboration?', *Research Policy*, **26** (1), 1-18.

Kline, S. and Rosenberg, N. (1986), 'An overview of innovation', in R. Landau and N. Rosenberg (eds), *The Positive Sum Game*, Washington, DC: National Academy Press.

Kuhn, T. (1962), *The Structure of Scientific Revolutions*, Vol. 2, Number 2, International Encyclopaedia of Unified Science, Chicago, USA and London, UK: University of Chicago Press (1970).

Kuhlmann, S. (2001), 'Governance of innovation policy in Europe – three scenarios', in Special Issue of H. Klein, S. Kuhlmann and P. Shapira (eds), 'Innovation policy in Europe and the US: new policies in new institutions', *Research Policy*, 30 (6), 953-976.

Layard, R. (2005), *Happiness: Lessons from a New Science*, London, UK and New York, USA: Penguin.

Leydesdorff, L. and Wagner, C. (2006), 'Is the United States losing ground in science? A global perspective on the world science system in 2005', manuscript, available electronically from 1st author's website.

Leydesdorff, L. and Etzkowitz, H. (eds) (1998), *A Triple Helix of University-Industry-Government Relations: The Future Location of Research?*, New York: Science Policy Institute, State University of New York.

Leydesdorff, L. and Etzkowitz, H. (1996), 'Emergence of a Triple Helix of university-industry-government relations', *Science and Public Policy*, **23** (5), 279-286.

Lundvall, B-Å. (1992), *National Systems of Innovation: Towards a Theory of Innovation and Interactive Learning*, London, UK and New York, USA: Pinter.

Martin, B. (2003), 'The changing social contract for science and the evolution of the university', in A. Geuna, A. Salter and W. Steinmueller (eds), *Science and Innovation: Rethinking the Rationales for Funding and Governance*, Cheltenham, UK and Northampton, MA, USA: Edward Elgar, pp. 7-29.

Martin, B. (1995), 'Foresight in science and technology', *Technology Analysis and Strategic Management*, 7 (2), 139-168.

Martin, B. and Etzkowitz, H. (2001), 'The origin and evolution of the university species', *Journal for Science and Technology Studies (Tidskrift för Vetenskaps- och Teknikstudier, VEST)*, **13** (3-4), 9-34.

Martin, B. and Johnston, R. (1999), 'Technology foresight for wiring up the national innovation system: experiences in Britain, Australia and New Zealand', *Technological Forecasting and Social Change*, **60** (1), 37-54.

Martin, B. and Irvine, J. (1989), *Research Foresight: Priority-Setting in Science*, London, UK and New York, USA: Pinter.

Merton, R. (1973), *The Sociology of Science*, Chicago, USA and London, UK: University of Chicago Press.

Narin, F., Hamilton, K. and Olivastro, D. (1997), 'The increasing linkage between U.S. technology and public science', *Research Policy*, **26** (3), 317-330.

Nelson, R. (2004), 'The market economy, and the scientific commons', *Research Policy*, **33** (3), 455-71.

Nelson, R. (ed.) (1993), *National Innovation Systems: A Comparative Analysis*, Oxford, UK and New York, USA: Oxford University Press.

Nelson, R. (1959), 'The simple economics of basic scientific research', *Journal of Political Economy*, **67**, 297-306.

Nelson, R. and Winter, S. (1982), *An Evolutionary Theory of Economic Change*, Cambridge, MA, USA: Belknap Press.

Nowotony, H., Scott, P. and Gibbons, M. (2001), *Re-Thinking Science: Knowledge and the Public in an Age of Uncertainty*, Cambridge, UK: Polity.

OECD (1963), 'The measurement of scientific and technical activities: proposed standard practice for surveys of research and experimental development (Frascati Manual)', Organization for Economic Cooperation and Development, Paris.

Pestre, D. (2003), 'Regimes of knowledge production in society: towards a more political and social reading', *Minerva*, **41** (3), 245-261.

Pestre, D. (1997), 'La production des savoirs entre academiés et marché', *Revue d'Economie Industrielle*, **79** (1), 163-174.

Pinch, T. and Bijker, W. (1984), 'The social construction of facts and artefacts: or how the sociology of science and the sociology of technology might benefit each other', *Social Studies of Science*, **14** (3), 399-441.

Rip, A. (2000), 'Fashions, lock-ins and the heterogeneity of knowledge production', in M. Jacob and T. Hellström (eds), *The Future of Knowledge Production in the Academy*, Buckingham, UK and Philadelphia, USA: Open University Press, pp. 56-69.

Rosenberg, N. and Nelson, R. (1994), 'American universities and technical advance in industry', *Research Policy*, **23** (3), 323-348.

Schofield, R. (1957), 'The industrial orientation of science in the lunar society of Birmingham', *Isis*, **48** (4), 408-415.

Schumpeter, J. (1934), *The Theory of Economic Development*, Cambridge, MA, USA: Harvard University Press.

Shinn, T. (2000), 'The "Triple Helix" and "new production of knowledge" as socio-cognitive fields', paper presented at the Third Triple Helix Conference, 26-29 April 2000, Rio de Janeiro.

Smits, R. and Kuhlmann, S. (2004), 'The rise of systemic instruments in innovation policy', *International Journal of Foresight and Innovation Policy*, **1** (1-2), 4-32.

Smith, K. (2000), 'Innovation as a systemic phenomenon: rethinking the role of policy', *Enterprise & Innovation Management Studies*, **1** (1), 73-102.

Stehr, N. (1994), *Knowledge Societies*, London: Sage Publishing.

Stephan, P. (2001), 'Educational implications of university-industry technology transfer', *Journal of Technology Transfer* **26** (3), 199-205.

Stern, N. (2006), *The Economics of Climate Change*, London: H.M. Treasury.

Stokes, D. (1997), *Pasteur's Quadrant: Basic Science and Technological Innovation*, Washington, DC: Brookings Institution Press.

Weingart, P. (1997), 'From "Finalization" to "Mode 2": old wine in new bottles?', *Social Science Information*, **36** (4), 591-613.

Wilsdon, J. and Willis, R. (2004), *See-Through Science: Why Public Engagement Needs to Move Upstream*, London: Demos.

Wynne, B. (2001), 'Creating public alienation: expert cultures of risks and ethics on GMOs', *Science as Culture*, **10** (4), 445-481.

Ziman, J. (2000), *Real Science: What It Is, and What It Means*, Cambridge, UK: Cambridge University Press.

3. The Changing Role of the Firm

Ben Dankbaar and Geert Vissers

INTRODUCTION

Since the industrial revolution, capitalism has been characterized by technological change at a speed and breadth incomparable to any other period in the history of mankind. Inventions had been made in earlier periods too, but it was only in capitalist economies that inventions were actively sought and implemented by capitalists competing for markets and profits. It was the first time in history that human inventiveness was employed in a social order that allowed almost unlimited freedom for individuals to introduce new products and new methods without having to ask for permission from anyone. Therefore, the history of technological change in capitalism is not so much a story of scientists, inventors and laboratories; it is a story of firms and entrepreneurs. The firm is the place where decisions are made, initiatives are taken and inventiveness is turned into real products sold for profit. Karl Marx (Marx & Engels 1848; Marx 1867) was one of the first to point to capitalists, the bourgeoisie, as the driving force behind the development of the forces of production, as he called it. His insights were echoed half a century later in Joseph Schumpeter's doctoral thesis (1911) that is often quoted as the first major work on innovation in modern economics. Today, a century later, the firm is still the central actor and place of decision making in processes of innovation, but there have been important changes in the way this role is being played. Innovation, like so many other activities, has been affected by processes of differentiation and specialization. As a consequence, innovation is, more than in the past, the result of organized interactions between formally independent firms. This is truer for some sectors than for others, but as a result there is an overall increase in the complexity of innovation processes in the economy, requiring new organizational capabilities.

This chapter discusses the changing role of the firm in innovation processes and the way this has been reflected in academic and management literature. Obviously, our scientific understanding of the role of the firm in turn has an effect on the actual practice of innovation, because it influences the actions and decisions of managers, capitalists and government policy makers. Such interactions between theory and practice are not simply processes of mutual reinforcement. Experiences made in one company, one sector or one country may lead to new concepts and understandings, which are then applied to other, perhaps completely different companies, sectors and countries. Theory then becomes a force for change and innovation in its own right.

Insights derived from studies of the computer industry may be applied to processes of change in the automobile industry or they may form the basis for government policies directed at all sectors of the economy.

Practitioners also read theory and act upon it, so it is sometimes quite difficult to distinguish between *real* processes and the way they are reflected in theory. Is theory based on an adequate description of practice, or is practice based on practitioners' understanding of theory? For instance, the literature on the importance of concentrating on core competencies has no doubt influenced decision making on outsourcing in companies. It may even have caused a preference for outsourcing that did not exist before the theory was published. Changes in thinking are often based on changes in practice, but practice has always been more varied and more complex than theories could allow for. Generally speaking, major shifts in the focus of thinking, e.g. from individual enterprises to networks of enterprises, reflect changes in practice, but practice shows much more continuity than theory suggests. Theory also has its own internal dynamic, following the logic of its own concepts and arguments (Rynes et al. 2001). It should be noted, therefore, that changes in thinking about the role of the firm and changes in the role of the firm are not perfectly correlated, but part of a complex set of interactions between theory, practice and policy.

THE CLASSICAL VIEW

Schumpeter was an Austrian economist, who later moved to the United States and taught at Harvard. In his early work, Schumpeter criticized the view of capitalist economies arising from neoclassical equilibrium economics. In economic equilibrium, there is no such thing as profit, because profit results from the fact that markets are not in equilibrium. For Schumpeter, the essence of capitalism is disequilibrium. How can we understand capitalism in terms of equilibrium, he argued, if profit is the central motive for action in capitalism and profits are absent in equilibrium? Instead of the equilibrating market mechanism, he put the entrepreneur at the center of the capitalist economy. The entrepreneur, in his search for profitable opportunities, is constantly making 'new combinations': introducing new products, developing new methods, finding new markets. By doing so, the entrepreneur is causing disequilibrium in markets, creating an at least temporary monopoly position (until competitors move in), which gives rise to profits. These new combinations are called innovations. They have to be distinguished from inventions. Inventions are new ideas, technical or otherwise. Inventions come from inventors, scientists, tinkerers, practitioners. Inventions only become innovations if they are brought together with a market, i.e. with purchasing power. It is the entrepreneur who realizes this connection. Innovations are not necessarily based on new (technical) inventions: they may consist of 'new combinations' of hitherto well known techniques or product features applied for a novel purpose or in another setting.

Schumpeter's emphasis on the role of entrepreneurship was in line with the ideas of other Austrian thinkers (also referred to as the Austrian School, covering several generations of scholars like Menger, Von Mises, Hayek and Kirzner). It is also echoed in Peter Drucker's work (e.g. Drucker 1985); incidentally, this 'guru of all management gurus' was born in Vienna at about the time that Schumpeter was working on his dissertation. Schumpeter's ideas never fitted in very well with mainstream economics, which remained strongly focused on equilibrium thinking with a strong penchant for mathematical modeling. Innovation had no place in neoclassical economics. Schumpeter was much more empirical in his approach to economics. His work remained influential in heterodox approaches like institutional and evolutionary economics, in which innovation and technological change play a central role (Hodgson 1993).

THE RISE OF INDUSTRIAL R&D

It can be argued that the picture of the entrepreneur as the central actor in capitalism was already somewhat outdated by the time Schumpeter's book was published. The typical risk-taking entrepreneur was an icon of the 19th century, but at the beginning of the 20th century enormous enterprises had come into existence in many parts of the economy. These large enterprises often were at the forefront of innovation and technological change. In that sense, they played the entrepreneurial role that Schumpeter had identified, but that role was no longer embodied in a single person. Entrepreneurship was becoming a function of organizations to which large numbers of people, managers and employees contributed. One of the most important differences with 19th century companies was the presence of sometimes very large departments for research and development. In some respects, Thomas Edison could still be considered an old-fashioned inventor-entrepreneur, but the way he organized the research and development activities for his company in separate laboratories was clearly new (Hughes 2004). On the other side of the Atlantic, the German chemical industry was the first to set up large-scale corporate laboratories (Chandler 2005).

Reflecting on these developments, Schumpeter (1942) described them as the 'routinization of invention'. In his earlier work, inventions had basically come from outside the realm of the economy and were not necessarily produced with an economic motive. The coming of corporate laboratories had drawn invention at least partly into the economic sphere. Inventions were produced in laboratories much in the same way that other products were produced in factories. As a consequence, the entrepreneurial role had also changed. Professional managers could rely on a steady stream of inventions coming from the laboratory. There was still a need to realize new combinations, but the act became less singular and less heroic now that the supply of inventions was secure.

Obviously, the 'routinization of invention' was not perfect. Some research projects were unsuccessful; others achieved their goal, but did not result in useful commercial propositions. Investing in R&D was, in other words, a risky business that could only be undertaken by companies with deep pockets. In his early work, Schumpeter had argued that innovations were the origin of profits; now, he added that profits were a precondition for innovation. Only large companies dominating their markets were making enough surplus profits to invest in R&D. As a consequence, Schumpeter suggested an understanding of competition that sharply differed from established economic analysis. In neoclassical economics, competition is all about prices. In Schumpeter's view, competition is not so much related to prices as to innovation. Companies compete by doing things differently: advancing new products and services, achieving new levels of productivity, creating new markets, uncovering latent needs. Competition in this perspective is more violent because it does not end in equilibrium where everyone is the same. It may end in the complete demise of some of the players. Schumpeter called this 'creative destruction'. In Schumpeterian competition it pays to be big and to dominate your market. It gives you the wherewithal to fund R&D and to innovate.

Not long after Schumpeter published his observations, two nuclear attacks on Japan brought an almost immediate end to the Second World War. The successful development of the atomic bomb (the so-called Manhattan project) was widely seen as the final proof that invention could be organized in a systematic way. Moreover, the spectacular success of the atomic bomb reinforced already strong beliefs in the strategic importance of technology. As a result, the potential routinization of invention became an almost unquestioned article of faith among scientists, engineers, managers and politicians in East and West. Research and development became essential elements of economic competition throughout the economy after the Second World War, often closely intertwined with the political-military competition of the Cold War.

COMPETITION AND INNOVATION

The idea that large size and monopoly positions have a positive economic impact because they encourage innovation encountered both interest and skepticism among economists. In traditional economics, competition would always lead to the best result and the ideal form of competition consisted of a large number of companies competing on the same market. A considerable number of empirical studies were carried out in the 1950s and 1960s testing what came to be called the Schumpeterian hypotheses concerning the relationships between market structure (the level of concentration) and innovation, and between firm size and innovation. The results were rather inconclusive, especially regarding the relationship between market structure and innovation. In their survey of the literature as well as in their own investigations Kamien and Schwartz (1975; 1982) point out that the empirical work did not show a clear and positive relationship between firm size and R&D and innovation. In the relationship between market structure and innovation, intermediate structures appeared to be most supportive of innovation. Too much competition made it difficult for companies to invest in activities focused on the long term, but in the absence of competition companies had little incentive to do so. Oligopolistic markets in which a limited number of relatively large companies were competing seemed to offer an environment in which both the incentive and the resources for innovation were present. Thus, the empirical evidence appeared to support Schumpeter's idea that innovation would be more likely in an environment with relatively few and large competitors. In this context, Klein (1977) distinguished between static efficiency and dynamic efficiency. The traditional view of competition had always focused on the preconditions of achieving equilibrium with lowest cost and prices for a given set of products (static efficiency). These preconditions differ from those most supportive of achieving the largest number of new products and new technologies (dynamic efficiency). Under the latter conditions, prices may be higher in the short run, but there is more technological progress and therefore more welfare in the long run.

One way to overcome the relatively inconclusive results of all these efforts to test the 'Schumpeterian hypotheses' across sectors (see also van Cayseele 1998) is to investigate patterns of competition and innovation at a sectoral level. In an influential paper, Pavitt (1984) offered a typology of sectors based on empirical research, in which market structure and sources of technology were important differentiating elements. In sectors with a relatively large number of companies, patterns of competition and innovation will differ between sectors mainly producing their own technology (the specialist supplier sectors, like the machine tool sector) and sectors using technology produced elsewhere (the supplier dominated sectors, like the furniture industry). Competition and innovation in sectors with a relatively small number of players will also differ in relation to the sources of technology. In technology-producing, 'science-based' sectors like the chemical industry, large firms compete on the basis of R&D, whereas in technology-using, 'scale-intensive' sectors, firms compete on the basis of productivity and design. In a further elaboration of this typology, Dankbaar (1996) has pointed out that in reality not all firms in a sector behave in the same way. In every sector, there are firms of different size and these firms are following different innovation strategies, not necessarily related to their size. For instance, even in science-based sectors dominated by large corporations, small firms also survive. And some firms may follow offensive innovation strategies, investing heavily in R&D, while others survive in a defensive mode, following where the offensive firms lead them. In order to understand the problems of innovation management at the level of the individual firm, it is not enough to know the type of sector it is in. Figure 3.1 combines Pavitt's typology of sectors with a simple typology of firms. The 'typical' firms for each type of sector are in the four corners of the matrix. Problems and issues faced by the management of 'untypical' firms may be more similar to those typical of other sectors. For instance, innovation management in a small, offensive chemical firm may be more similar to that in a 'typical' specialist supplier firm than to that in a 'typical' large, science-based chemical company.

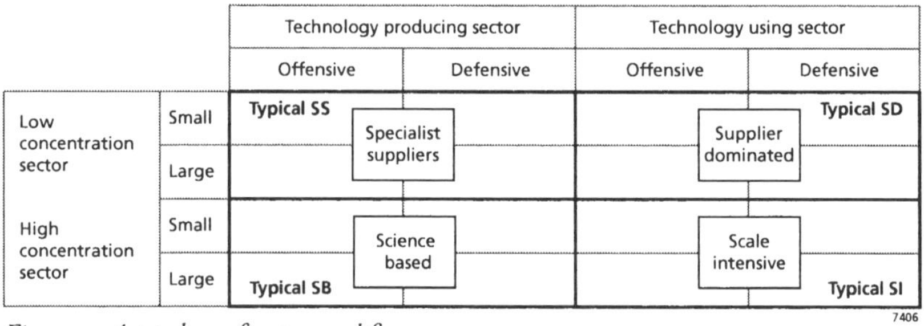

Figure 3.1 A typology of sectors and firms

In every sector, competition results from the strategies chosen by firms in view of the technological opportunities available to them. Market structure and technological regime are not just determinants of innovative action in the sector, but also the outcome (at a collective level) of choices made at the level of the firm. This kind of understanding of competition differs appreciably from that of the neoclassical tradition in economics (cf. McNulty 1968). Evolutionary economics has taken up the challenge of modeling competition in terms of individual companies making strategic choices in an environment offering technological opportunities that can only partly be known (Nelson & Winter 1982). Making choices involves taking risks, and success may depend on good fortune as much as on effort. Sometimes, patterns of technological development become clear and it is relatively safe to follow such 'technological trajectories', but firms searching far away from the well-known path may come up with 'disruptive' innovations that destroy the value of the competencies of all those who remained on well-known territory. It is obvious that Schumpeter has been an important source of inspiration for evolutionary economists.

EARLY INNOVATION POLICIES

Innovation policies in the 1950s and 1960s reflected the issues and concerns of the literature discussed above – and also the innovation practices found in companies. In fact, the expression 'innovation policy' was not used in those years. Nevertheless, innovation was a more or less explicit element in several areas of government policy. First of all, innovation became an issue and argument in competition policy, because arguments in favor of 'dynamic efficiency' led to conclusions different from those in favor of 'static efficiency'. More obvious, and maybe more important, was the considerable increase in government spending on science and technology, both at universities and at government laboratories. These 'science push' policies were complemented by government demand for new technologies in military applications, space and (nuclear) energy. Especially in the United States, important innovations in semiconductors, computing, aviation, satellites, imaging etc. were produced by complex interactions between private (sometimes competing) companies, government laboratories, universities and the military, fueled by Cold War military budgets of hitherto unknown proportions.

In spite of the importance of government demand in the dynamics of this 'Military Industrial Complex' (an expression used by President Eisenhower in his farewell address in 1961), the dominant perspective of that period is that of 'science push' (or 'technology push'). This perspective is characterized by a strong confidence in the effectiveness of research spending and little debate about possible choices or alternatives in technological development. Spending larger sums allows you to move faster along a path that seems to have few side tracks: research is about stronger rockets, more effective nuclear power plants, bigger bombs. And this perspective is not limited to the sphere of government. In companies there is a similar tendency to consider spending on research; a thing to do without much consideration for direction. Later, this was called 'first generation R&D spending': giving bright people a bundle of money and letting them decide what to do with it in full confidence that they will come up with something useful at some point in the future. It should be noted that these characterizations of the 'dominant perspective' during the first post-war decades (science push, first generation) have been produced afterwards, often by observers who advocated a change in perspective. Only if there is a change in practice, the specifics of past behavior become clearer.

There is hardly any literature on innovation management from the first postwar decades, probably because companies also considered innovation as a simple process of implementing whatever findings researchers produced. This was the age of mass production, an age without much of a marketing function in companies, in which companies proceeded on the realistic assumption that they would be able to sell whatever they were manufacturing.

THE DISCOVERY OF INNOVATION MANAGEMENT

In 1961, Burns and Stalker published *The Management of Innovation*. Yet, this book was not really concerned with innovation management in the sense of managing a specific process inside a company – the way we treat that subject today. Burns' and Stalker's study was primarily a contribution to organization theory. It focused on the limitations of the dominant Taylorist 'mass production' paradigm of organization, in which division of labor, standardization of tasks and separation between execution and control were the main elements of organization design. Burns and Stalker argued that innovation required an almost completely opposite design: not 'mechanistic', machine-like structures, but 'organic' structures. Innovation would thrive in organizational structures characterized by loosely defined tasks and responsibilities, horizontal rather than vertical communication, and considerable latitude for workers to guide and direct their own work. Obviously, these design principles were inspired by work organization in research laboratories, where the nature of the work makes it inherently difficult to apply Taylorist principles of work organization. Management in this context is more about encouraging communication, experimentation (with the chance of failure) and creativity than about standardization of tasks and hierarchical control. Burns and Stalker argued that firms in changing environments will need organizations that are more similar to research laboratories than to assembly lines. They need to be able to react quickly and to improvise. Management of innovation in Burns' and Stalker's study meant the ability to deal with rapid and unexpected changes in the environment.

The distinction between organic and mechanistic structures has remained important in organization theory, where it is still used to discuss choice in organizational design in relation to characteristics of the environment. The suggestion that innovative organizations necessarily have organic structures, however, has been questioned repeatedly. It was relatively clear that organic structures were appropriate in departments focused on *exploration* of new opportunities like research, development and maybe marketing, but would they also be appropriate in departments focused on *exploitation* of those opportunities like manufacturing? Even in highly innovative organizations, where new products regularly disrupt established organizational routines, one would expect production departments to be more mechanistic, focusing on productivity, standardization and automation. Was it necessary for the whole organization to be one or the other? Abernathy (1978), in his influential study of the automobile industry, suggested that firms indeed had to choose. They were faced by a real 'productivity dilemma': focusing on productivity (and therefore on Taylorist work structures as American car makers had done), they would inevitably become less innovative. European car manufacturers had focused much more on innovation, but as a consequence, they were less productive.

Schoonhoven and Jelinek (1990), however, investigated several innovative firms and found that their organizational structures were not really organic. Workers knew exactly what their jobs were, to whom they had to report, and how responsibilities had been allocated, all of which did not conform to the idea of organic structures sketched by Burns and Stalker. Schoonhoven and Jelinek pointed out, however, that the formal structure in these innovative firms was changing frequently and that it was also supplemented by a 'quasi-formal' structure of (often temporary) committees, teams, task forces and dotted-line relationships. A similar distinction was made several years later by Nonaka and Takeuchi (1995). In their concept of a 'hypertext organization', a firm is pictured in several layers: a business layer (focusing on the regular processes) and a project layer (focusing on processes of innovation and change), connected by a knowledge layer (in which knowledge of both types of processes is maintained). Workers move regularly between the layers.

Nowadays, this type of studies reflecting on the nature of innovative or flexible organizations is seldom catalogued under the heading of 'innovation management'. Next to – and to some extent in contrast with – the studies focusing on 'organic management', a voluminous body of literature and knowledge has come into existence that is taking a much more traditional 'control' perspective. Although it is clear that innovation processes cannot be controlled in the way that a chemical plant or an assembly line can be controlled, considerable efforts have been made to develop tools and practices to 'routinize' the innovation process. Around 1990, this type of innovation management had become a well-established discipline, based on academic studies as well as extensive consultancy work. Cooper (1988; 1993) introduced the highly influential 'stage-gate' model that sought to reduce risks by structuring the decision making process. Clark and Fujimoto (1991) explored Japanese methods of project management, emphasizing 'heavy' project management, multifunctional team work and simultaneous engineering. Roussel et al. (1991) argued that the time had come for 'Third generation R&D'. After the science push practices of the 1950s and 1960s (first generation R&D), companies had tried to link research much more closely to market demand by decentralizing research funding and project management (second generation R&D), resulting in a neglect of long-term research. Now it was time to think in terms of a portfolio of long-term as well as short-term R&D projects, linking R&D to the strategic choices of the corporation. Wheelwright and Clark (1992) produced a standard work on 'new product development' that presented the state of the art for this much more hands-on and control-oriented approach in innovation management.

CHANGING CONDITIONS AND CHANGING STRATEGIES

The development of modern innovation management, basically applying management knowledge to the work of 'knowledge workers' (Drucker 1969), had taken less than 20 years. Of course, some experiences and insights from managing corporate laboratories were available earlier on, but modern innovation management developed after the first oil crisis of 1973. For the first time since the Second World War all major capitalist economies were simultaneously experiencing an economic downturn. The whole idea of economic crisis had been absent from academic and management thinking for almost 30 years. Theories of crisis and economic development that had been popular in the 1930s were now rediscovered, among them also Schumpeter's work. In 1939, Schumpeter had published a two-volume work on business cycles in which he devoted considerable attention to the phenomenon of 'long waves' in economic development. He connected these waves of 50 to 60 years in length (which he named after the Russian economist Kondratieff, whose work on long waves had in turn been inspired by Dutch Marxian economists (cf. van Duijn 1979)) with a clustering of innovations in time, which caused an associated wave of investment and structural economic change. More specifically, each long wave seemed to be associated with a specific breakthrough technology that set the economy off on a new growth path (Freeman & Louçã 2001). The message was clear in the context of the 1970s: in order to get out of the economic slump, it was necessary to get a new long wave starting as soon as possible. And, coincidence or not, a new carrier technology was presenting itself: information technology.

In the second half of the 1970s, governments began to set up programs to support the 'micro-electronics revolution' and innovation in general (Rothwell & Zegveld 1981). These programs were mainly national in orientation. In some countries 'national champions' were supported and sometimes created; in other countries, this was impossible because there were no obvious candidates (or more than one) or because the principles of competition policy were held in higher esteem. There, however, governments often supported companies to collaborate in what came to be called 'pre-competitive' research. Government spending on science and applied research was maintained at high levels, but more money was flowing directly to companies as part of programs supporting not just research, but also diffusion of (micro-electronics) technology. As a result, the number of companies receiving subsidies was increasing. Not just large R&D-performing companies, but also small and medium-sized companies received attention from government agencies. Characteristic for these new innovation policies is the broader range of targets compared to earlier decades. Apart from the military and semi-military targets, the new innovation policies had targets that were inspired by an increasing awareness of the 'limits to growth' resulting from depletion of natural resources (Meadows et al. 1972) and of other negative impacts of the previous period of unprecedented economic expansion, especially environmental pollution. Targets therefore included the search for renewable energy sources, reduction of waste and pollution, creation of healthy work and living environments, etc.

Reaction to the economic downtown went beyond government. Firms also had to reassess their competitive positions. Reduced growth implied an intensification of competition. Competition was also increasingly global. Companies that used to control their home markets were confronted by powerful newcomers from other countries. Slowly but steadily the sellers' market of the era of mass production changed into a buyers' market. Consumers could ask for more features, more variety and better quality and companies had to comply in order to maintain their market share. The end of the long post-war boom was increasingly interpreted as the end of the era of mass production with wide ranging implications (Roobeek 1987). Competition would focus not just on price, but also on quality, speed, flexibility and increasingly on innovation. Consequently, scale advantages would become less important. Size and vertical integration could even become disadvantageous, because they were an impediment to flexibility. In other words, competition was not just changing in intensity, but also in character. It was becoming more 'Schumpeterian', with much more management attention needed for innovation, product variety and product differentiation (Piore & Sabel 1984). That was the reason why firms started to decentralize funding for R&D (second generation R&D). Putting the business units in charge of R&D implied a shift from long-term strategic research to much more short-term customer-oriented research. At the same time, firms were developing a marketing function in order to better understand the customers, who were no longer simply buying what they were offered. Inside the firm, managing innovation became a problem of organizing and managing the communication between marketing and research (introducing tools like Quality Function Deployment). Moreover, the intensification of competition implied that the time between product definition and market launch (time-to-market) was also becoming an issue (cf. Vesey 1992). Innovation management therefore also involved organizing communication between research, development and manufacturing (introducing tools like concurrent engineering and design for manufacturability).

This new combination of intensifying competition and increased innovation was not unproblematic. Competition ate into profits, but retained profits had been the main source of funding for corporate R&D. If only for that reason, companies had to look for other ways of funding R&D. Governments, as noted above, were sometimes willing to help. Government funding of research and development increased. Maybe more importantly, government encouraged firms to collaborate (Hemphill 2003). In the course of the 1980s, strategic alliances between competing firms became increasingly common. Many of these alliances were aimed at lowering costs and sharing risks of research and development. Nevertheless, corporate spending on long-term R&D was diminishing in many sectors of the economy. Third generation R&D, aiming at a healthy portfolio of long-term and short-term projects, was unable to undo that trend. Inside companies the long term seemed to become shorter, also under the influence of assertive groups of shareholders. For the long term, companies increasingly looked to universities, government-funded research institutes and (to a lesser extent) commercial laboratories. As a consequence, a new issue arose in the management of innovation: the problem of absorptive capacity (Cohen & Levinthal 1990). How can firms recognize, understand and absorb new insights from the world of long-term research if they no longer employ people who are engaged in similar research? In some countries and sectors, intermediary institutions have been created to facilitate communication between firms and universities. Universities themselves have also become more pro-active in the commercialization of ideas generated by their researchers. In electronics and especially in biotechnology, new patterns of knowledge transfer and acquisition have developed, in which universities generate ideas which form the basis for new high-tech start-ups. These start-ups bring the idea to fruition (or not), supported by a variable mix of public research funding and venture capital. If the company is successful, it is likely to be taken over by a large incumbent company. In that way, the incumbent company does not have to pay for all the start-ups that fail (but it does pay a hefty price to the venture capitalists who had it right) and its absorptive capacity does not have to be extremely well-developed. It just needs to be able to recognize successful start-ups.

SPECIALIZATION AND DIFFERENTIATION

Although it is difficult to measure, it was widely felt that the combined result of government policies and increased competition through innovation was a considerable increase in the speed of technological change in the 1990s. The number of scientists and engineers in the world was larger than ever before and continued to grow. More firms and more countries were active at the frontiers of technology. Communication about new findings was quicker than ever. Even the largest firms could no longer hope to be knowledgeable in all fields of technology that could be relevant for their competitive position. Alliances with companies with complementary capabilities were a logical result, but companies also consciously decided to give up certain capabilities and to concentrate on a limited number of 'core competencies' (Prahalad & Hamel 1990; cf. also Lei & Hitt (1995) and Patel & Pavitt (1999) for critical commentary). Outsourcing development of certain components to specialized suppliers and close collaboration with such suppliers in the elaboration of specifications for these components became a common practice in sectors with complex composite products like automobiles, computers and consumer electronics. Increased outsourcing was both a cause and a consequence of the growth of specialized supplier companies. It also was a cause as well as a consequence of increased modularization of product design in these sectors.

The end of mass production did not have all of its predicted consequences. The large firm did not disappear and some became even larger because of continuing concentration in the science-based and scale-intensive sectors. However, levels of vertical concentration have certainly gone down among the original equipment manufacturers and the number of specialized suppliers with in-house capabilities for product development has gone up – and some of these suppliers have become quite large organizations themselves. Processes of new product development that used to be taking place in one company are now distributed over a whole series of companies. Research, development, design and engineering used to be functions inside one company and sometimes they were hardly differentiated into separate departments. Nowadays, these functions are often performed by separate companies offering specialized services (knowledge-intensive business services or KIBS).

Due to alliances, outsourcing, specialization and differentiation, innovation management became increasingly concerned with the organization of processes across the boundaries of the firm. In the 1990s, networking became the new buzz-word. This incorporated networking not only within and between firm but also networking of computers locally and across the world (electronic data interchange). Information and communication technology (ICT), like generic (multi-purpose) breakthrough technologies of earlier 'long waves', was enabling new ways of manufacturing and doing business. The Internet became the ultimate network of all networks. But the role of ICT was not limited to the support of communication in innovation processes. Based on ICT, various tools and techniques have been developed, which Dodgson et al. (2005) have called 'innovation technologies' to distinguish them from the technologies used to automate production processes and to support processes of management and control. With the advent of rapid prototyping, simulation tools, virtual imaging and several other techniques, experimentation has become cheaper and easier and some parts of the innovation process can be automated or at least sped up appreciably. The cheapening and increasing user friendliness of some of these technologies make it easier for small companies to engage in innovative activities.

SYSTEMS OF INNOVATION

The management of innovation in networks has brought a number of new problems to the fore. Most prominent of these are issues of trust and the protection of intellectual property rights. Inside the boundaries of a single company, these issues are of much less importance. There had of course always been problems of industrial espionage and individual employees had sometimes been enticed into moving to a competitor firm, but in networks of collaborating companies knowledge has to flow between them and there has to be some level of trust that no-one will take undue advantage of this knowledge. Trust can be supported by formal agreements, but it is difficult (and costly) to make arrangements for every eventuality. Contrary to the situation in firms, no-one is formally in charge of a network. Some networks are quite hierarchical and the leading company may impose some rules, but other networks are more horizontal. In either case, decisions can not be enforced in a similar fashion as inside companies.

The building of trust takes time and it is often based on personal relationships. In spite of all the advances in communication technologies, most people prefer face-to-face contacts (at least once in a while) when they have to collaborate and trust each other. In innovation processes, there is the additional phenomenon that new ideas often arise when people with different disciplinary and functional backgrounds interact – not at a distance, but face-to-face. Creativity is probably more a function of interactions between people than the product of a single individual (Vissers & Dankbaar 2007). A lot of knowledge that is important for innovation turns out to be collective: it is available in a certain context where people have learnt to interact with each other in order to achieve certain results, but it is not known by any individual person (cf. Leiponen 2006).

These may all be reasons why the literature on innovation of the past 15 years has not just been concerned with (ICT-supported) networking, but also with geographical clustering of activities and with national and regional systems of innovation (Lundvall 1992; Nelson 1993; Braczyk et al. 1998). In a world of advanced communication technologies and global companies, proximity is still a factor of importance. In his study of the 'competitive advantage of nations' Porter (1990) emphasized the advantage of the clustering of related activities in a country or region. If all relevant activities for the production of a specific product are available in a limited space, competitive advantage will arise from competitive specialization, a wide range of suppliers, common training of workers, specialized research, appropriate infrastructure, specialized logistics, etc. Governments interested in promoting the competitiveness of their national (or regional) economies would therefore have to identify and foster such clustering. Well-known and frequently studied clusters are machinery and equipment in Baden-Württemberg (Germany) and the textile industry in Emilia Romagna (Italy), but the most famous cluster of all is the cluster of firms and institutions that developed around semiconductors, computers, computer peripherals and software in Silicon Valley, California. All over the world, governments, universities and companies have tried to imitate this example, if not always at the same scale. Proximity, the meeting of minds between firms and universities, between OEMs and their suppliers, between service providers and their customers, is organized and promoted in high-tech campuses, supplier parks, science parks, specialized industrial zones, etc.

Spatial proximity is also, albeit less explicit, an element in the voluminous literature on national and regional systems of innovation that appeared in the 1990s. Again, in a world full of debate about globalization, this literature emphasizes the continued importance of national and regional institutions and arrangements (Storper 1997). An important aspect of these systems is that they have developed over many decades if not centuries. In that sense, the literature about systems of innovation is more about the relevance of history than about the relevance of proximity. But the two obviously hang together. Institutions that are relevant for innovation – like the educational system, the universities, the industrial structure, industrial relations, the role of the state, culture, social legislation, traditions of entrepreneurship – have all co-evolved over time and are mutually dependent. An important implication is that successful policy instruments in one country may not be successful in another country. And successful constellations like Silicon Valley cannot be reproduced easily in another place (not even in the same country), as if history does not matter. Attention for systems of innovation has contributed to a more encompassing or 'holistic' approach to innovation policy, in which innovation policy is seen to co-evolve with the economy and institutions of a country. Firms too must realize that their efforts to compete and innovate are situated in specific institutional environments, and that their own history is also influencing the behavior of employees and managers. Innovation management cannot ignore the capabilities that were developed in the past, and the successes and failures of past innovative efforts will have an impact on attitudes and actions of today.

OPEN INNOVATION

Looking back over a century of entrepreneurship and innovation, a few important changes can be noted. First, compared to 100 years ago when Schumpeter first reflected on the matter, innovation has definitely become a business function. Although individual entrepreneurs still exist – some of them very wealthy and famous – innovation is now considered a regular responsibility in firms and many companies today have 'Chief Innovation Officers', who have overall responsibility for the innovation function. Second, the 'science push' practices of companies in the 1950s and 1960s have been replaced by much more focused and market-oriented approaches. Schumpeter looked at the new industrial laboratories of the early decades of the 20th century and called that development the 'routinization of invention'. The changes taking place especially after the First Oil Crisis (1973) may be called the 'routinization of innovation'. Not only was research (i.e. invention) being guided towards short-term targets, but the innovation process itself was increasingly managed and controlled. Third, over the past two decades innovation has become an increasingly distributed activity: distributed over different actors (companies, universities, research institutes) and over different locations (sometimes spread out across the entire world, sometimes inside a single region). This trend is more pronounced in the scale-intensive sectors (consumer electronics, automotive) than in others, but it reflects the growing importance of relatively small specialized supplier companies with significant development capabilities throughout the economy.

Henry Chesbrough (2003) has studied these distributed innovation activities and argues that there is more going on than a simple redistribution of activities. In his view, a new coherent set of principles has developed for innovation management, which he calls 'open innovation' in contrast to the 'closed innovation' of the past. The past, in this case, refers to the practices that prevailed in the 1970s and '80s and had become more or less codified around 1990. It has become common practice in management literature to set up such dichotomies to distinguish one's own, new approach from older and of course outdated ones. In reality, the new and old cannot be so easily distinguished, but the elements of Chesbrough's open innovation paradigm do point to developments in innovation management that appear to be a natural extension of the processes of specialization, differentiation and spatial distribution described above. Chesbrough's main point is that companies need to make use of internal as well as external ideas in a conscious and organized manner. External R&D can create significant value for a company, provided it has the capability (doing enough R&D of its own) to claim some portion of that value. It is possible to profit from ideas that have been generated elsewhere, because it is not necessary to be first on the market. It suffices to have the better business model.

In the background of these considerations is indeed a new development: a growing market for intellectual property rights. As noted above, if innovation becomes distributed among many actors, issues of intellectual property (IP) will inevitably arise. In recent years, companies have begun to learn how to deal with these issues and government legislation in the field has also been forthcoming. In addition, the rise of venture capital funds has encouraged the development of a market for IP and of methods for evaluating IP, because IP is often the main asset of the start-ups they are financing. As a result companies are now more prepared than in the past to consider the results of R&D in their own laboratories and those of others as products that can be bought and sold. They buy IP from others if they can use it and they sell IP generated in their own labs, if it does not fit in the current business. In the past unused IP simply stayed on the shelves, but now companies see possibilities of profiting from it.

In government policy, open innovation takes the form of increasing pressure on both universities and government laboratories to collaborate with the private sector. However, the institutional environment differs greatly between countries. There are major differences in the traditional role of universities, the organization of the (venture) capital market, the presence and size of government-funded research laboratories, attitudes towards intellectual property, etc. Much-needed insights into the complex linkages between all these elements of open innovation may be derived from the 'systems of innovation' literature. Interestingly, open innovation gives a more prominent role to university research compared to the period of big, 'closed' corporate R&D laboratories. This appears to give rise to a new type of 'science push' policies, in which governments provide substantial funding to fundamental research in selected areas of science and technology that promise to have significant economic applications in the future. Thus, part of the money available for fundamental research is now put into programmatic research. Several countries, among them also emerging economies like Singapore and South Korea, have developed large research programs in promising areas of biotechnology and nanotechnology.

The concept of open innovation is a good example of the complex interactions between practice, theory and policy. Chesbrough coined the concept, mainly based on his observations of developments in the computer industries, and offered open innovation as a general prescriptive model for all sectors and companies. Many firms and policy makers embraced the model as an accurate model of modern trends and a useful guide for further action. In their efforts to realize the trend, firms and policy makers encounter complex organizational issues (shaping the relationship between universities and firms, organizing trust in networks of firms and protecting intellectual property rights). The solutions found for these issues will no doubt give rise to new theories and concepts.

SERVICE INNOVATION

Although we are supposed to live in a post-industrial service economy, most of the literature on innovation and innovation management is concerned with the manufacturing sector. Why is that? Is innovation unimportant in services? Or is it so different that it requires separate treatment? (cf. Miles 2003)

The first point to make in answering these questions is that the expression 'post-industrial' is seriously misleading if it is taken to suggest that manufacturing industry will disappear. What is true is that manufacturing employment has been diminishing as a share of total employment in all developed economies. The same is true for agriculture, but that does not mean that agriculture has disappeared or has become unimportant. In both sectors, productivity has increased at a tremendous rate, while product demand has increased at a lower rate, resulting in lower levels of employment. Additionally, many activities that used to be carried out inside manufacturing firms are now carried out by separate service companies (in logistics, maintenance, engineering, research, financial administration, HRM). All these activities used to count as manufacturing in the past, but are now counted as services.

Another reason why services receive only limited attention in the innovation literature is that most of that literature is concerned with technological innovation, i.e. with the problem of generating new scientific and engineering knowledge and implementing it in new products meeting known or latent needs of customers (who may be consumers or businesses). Most technological innovation is taking place in the 'technology producing' sectors (like machinery and chemicals) and the products of these sectors are then often used in 'new combinations' in the 'technology using' sectors. Service sectors are typically technology using: they buy products from the manufacturing industry and apply them in the production of services. Innovation in the service sectors frequently involves the use of new technologies produced elsewhere. In that sense, service companies are not different from 'supplier-dominated' or 'scale-intensive' manufacturing sectors. Just like in manufacturing sectors, large 'scale-intensive' service companies like banks and insurance companies compete on the basis of scale advantages, but also in the innovative implementation of new technologies. In scale-intensive manufacturing sectors like consumer electronics and automobiles this may not just involve the use of components developed and manufactured by suppliers, but also internal knowledge-producing research and development activities (in that perspective, the distinction between science-based and scale-intensive may incorrectly suggest that the scale-intensive sectors have no base in science, where the distinction is rather one between chemistry and biology-based versus mechanics and physics-based sectors). In scale-intensive service sectors even the large companies have only very limited R&D activities, most of which are devoted to software development. Much of that is increasingly being outsourced. The financial sector for instance has certainly seen a lot of product innovation over the past decades (new types of mortgages, new types of investment funds, new instruments to finance mergers and acquisitions). The point is that thinking up these products requires only limited resources and manpower. Competitors can usually copy them easily. Competitive performance is based on the ability to sell and administer these products, which requires supportive software systems. Developing such systems and integrating them with other systems of the company is expensive and time consuming.

The software sector is a peculiar sector. On the one hand, it is similar to the machinery sector, when it develops one-of-a-kind customer-specific applications. On the other hand, it is similar to mass manufacturing, when it produces standard software packages. In the latter case, however, there is a major difference in that almost all production costs are made during product development, whereas production (and increasingly also distribution via Internet) of additional packages for sale is almost costless.

A major difference between manufacturing and services is that the production (delivery) of a service frequently involves close interaction with the customer. Innovation in services has frequently focused on the interaction with the customer.

The central concept here is self-service. Self-service became a household word with the introduction of supermarkets in retailing of groceries. With the advance of information and communication technologies, new possibilities arose for self-service. Internet banking is an example of self-service delivered by banks supported by ICT. The products of the ICT sector have therefore served a dual purpose in the service sectors, especially those sectors in which the transfer and processing of information plays an important role. ICT has enormously increased the productivity in information processing. Modern banking services would be impossible at the present scale without information technology. And ICT has created new possibilities for self-service. It has also allowed for a further individualization of services, e.g. by the construction of individual profiles of interest for the delivery of news or music (narrowcasting).

As a result of the characteristics of product development in services discussed above, government policies in direct support of service innovation are almost non-existent and service companies receive only limited innovation subsidies. However, government funding and policy making play an important role in the construction of infrastructure (roads, ports, telephone, cable television, Internet) that enables innovation in services and service delivery. Also, deregulation of the financial markets has enabled innovation in financial services in many countries. More generally, government regulation and deregulation strongly affect the possibilities for innovation in service sectors like communications, utilities, retailing and media.

It should be noted that products of the manufacturing industry have not just enabled innovation in services; they have also replaced services. In the 'self-service economy' (Gershuny 1978) households are using all kinds of products, e.g. washing machines, breadmaking machines and various tools, to carry out activities that used to be done by specialized service companies (laundry services, bakery shops, pipefitters).

SOCIAL INNOVATION

At the end of our observations concerning the changing role of the firm in innovation, it is useful to return briefly to the observations of Burns and Stalker (1961). They argued that innovation required 'organic' forms of organization. We noted that there was some debate about the extent to which this would be necessary for all of the organization. Some argued that organic forms of organization are especially appropriate for the departments directly concerned with innovation. In the end, however, all parts of the company are confronted with changes when new products are introduced. And if this is done frequently, the whole organization needs to be capable to deal with frequent changes, not just the R&D department.

In the course of the 1980s it was observed that the implementation of information technology also required a redesign of organizations. Too many projects failed because they simply involved 'the automation of bureaucracy' instead of using

IT to simplify the organization and enhance flexibility (Zuboff 1984). The principles for 'organic' organization design are well-known and were developed already in the 1950s and 1960s by the adherents of the so-called socio-technical systems approach (Morgan 2006). They include the introduction of team work, self regulation and product-flow-oriented structures (de Sitter et al. 1997). Notions like 'continuous improvement' (kaizen) were introduced later, but the underlying idea was the same, namely that the traditional Taylorist division of labor between execution and control has a negative impact on the flexibility and innovativeness of the organization. Apart from such considerations, it should be noted that production work in both manufacturing and service sectors has become less repetitive and more responsible because increasing levels of automation have eliminated most simple tasks. In that sense, work in production has become more similar to work in R&D – and because of the 'routinization of innovation' work in R&D has become more similar to production work.

Innovative organizations may have to redesign themselves repeatedly and a whole new discipline of 'change management' has come into existence in order to support such processes. While companies need to reorganize in order to deal with technological change and innovation, the same is true for society as a whole. The concept of 'social innovation' was coined in the 1980s to indicate that embarking on a new 'long wave' of economic development and the introduction of new breakthrough technologies like ICT involving major institutional changes (Freeman 1987). The economic crisis of the early 1970s also engendered debate about the welfare state, the system of industrial relations, the role of educational institutions, the organization of government, etc. In the following decades, many of these social institutions were subjected to reform and transformation. Social institutions are slow in changing, if only because the interests of many different people and organizations are involved. Slowly, all actors in the arena – politicians, trade unionists, employer associations, educators, all voters and citizens – are beginning to realize that social innovation too is not a one-off change, but a continuing process of adaptation to changing realities under the influence of innovation and technological change.

OPEN QUESTIONS

The firm will continue to play the central entrepreneurial role in processes of innovation and technological change in capitalist economies – and even in economies that are not formally capitalist, decision making on innovation is decentralizing towards the level of the firm. At the same time, 'the firm' has become a less homogeneous and unequivocal concept than it appeared to be in the past (Hodgson 2002). No doubt, the role of the firm will be changing in the future, just as it has changed over the past 100 years. These changes will result from continuing efforts to deal with unsolved problems of the present. Most of these have come forward in our overview. They are mainly related to the fact that, more than in the past, innovation has become a multi-level affair. Problems have surfaced at the level of the firm, but also at the level of networks, clusters, regions, nations and even at the global level. By way of summary, we will briefly recount them here.

First of all, there are questions at the level of the individual enterprise. How can the short term and the long term be balanced in innovation management in enterprises? Is there an antidote to the 'short-termism' imposed by hedge funds and other groups of impatient shareholders? How can companies realize more than incremental innovations without taking excessive risks? How and in what fields should a company maintain adequate 'absorptive capacity'? Then, there are questions at the level of the networks. What is the best way of organizing a network for innovation and what is the best way of operating as a member in such networks? Is it possible to distinguish different types of networks (for example hierarchical versus horizontal) and if so, will the optimal form be a function of the underlying technologies for the innovation at hand? Is there room for collective decision making in a network? How important is proximity in networks? Should a partner around the corner be preferred to one far away but with higher competencies?

Then, there are questions at the level of clusters and regions. As we move to this level, questions of government policy become more prominent. What are the circumstances that give rise to industrial clusters? Are they related to the presence of natural resources, advantages of location, the presence of human resources? How do these clusters of companies balance between destructive competition and destructive consolidation? How does competition policy deal with collaboration within a cluster of firms? How and to what extent should individual companies actively support the maintenance of a cluster? How do we recognize and support (potential) clusters that are located in different administrative regions or even different countries? At the level of national innovation systems, there are questions concerning the organization of communication between companies and educational institutions and between companies and research at universities and other research establishments. To what extent should companies influence curricula and research programs? More collaboration between companies, universities and public authorities seems highly desirable, but how should that be organized? How should intellectual property rights be protected in these networks?

Finally, at a national, but increasingly also at a global level, there are questions concerning the direction of innovation and technological change. How can we maintain and improve levels income and welfare without destroying the natural environment? How do we deal with the problem of global warming and climate change? Two hundred years of innovation in industrial capitalism have brought an enormous increase in income, welfare and security for large portions of mankind, but they have also caused enormous problems. There is no way back. Innovation management, innovation theory, innovation policies, entrepreneurship, creative theorizing and political leadership will have to come up with the answers for the future.

REFERENCES

Abernathy, W. (1978), *The Productivity Dilemma: Roadblock to Innovation in the Automobile Industry*, Baltimore: John Hopkins University Press.

Braczyk, H.-J., Cooke, P. and Heidenreich, M. (eds) (1998), *Regional Innovation Systems: The Role of Governances in a Globalized World*, London, UK and New York, USA: Routledge.

Burns, T. and Stalker, G. (1961), *The Management of Innovation*, London: Tavistock.

Cayseele, P. van (1998), 'Market structure and innovation: a survey of the last twenty years', *The Economist*, **146** (3), 391-417.

Chandler, A. Jr. (2005), *Shaping the Industrial Century: The Remarkable Story of the Evolution of the Modern Chemical and Pharmaceutical Industries*, Cambridge, MA, USA: Harvard University Press.

Chesbrough, H. (2003), *Open Innovation: The New Imperative for Creating and Profiting from Technology*, Boston: Harvard University School Press.

Clark, K. and Fujimoto, T. (1991), *Product Development Performance: Strategy, Organization and Management in the World Auto Industry*, Boston: Harvard Business School Press.

Cohen, W. and Levinthal, D. (1990), 'Absorptive capacity: a new perspective on learning and innovation', *Administrative Science Quarterly*, **35** (1), 128-152.

Cooper, R. (1993), 'Winning at new products. Accelerating the process from idea to launch', Reading: Addison Wesley.

Cooper, R. (1988), 'Predevelopment activities determine new product success', *Industrial Marketing Management*, **17** (3), 237-247.

Dankbaar, B. (1996), 'Patterns of technology management in European firms: an overview', in W. Cannell and B. Dankbaar (eds), *Technology Management and Public Policy in the European Union*, Oxford, UK and New York, USA: Oxford University Press, pp. 21-50.

Dodgson, M., Gann, D. and Salter, A. (2005), *Think, Play, Do: Innovation, Technology and Organization*, Oxford, UK and New York, USA: Oxford University Press.

Drucker, P. (1985), *Innovation and Entrepreneurship*, New York: Harper & Row.

Drucker, P. (1969), *The Age of Discontinuity. Guidelines to our Changing Society*, Boston, USA and London, UK: Butterworth-Heinemann.

Duijn, J. van (1979), *De lange golf in de economie: kan innovatie ons uit het dal helpen?*, Assen: Van Gorcum (English translation: *Long Waves in Economic Life*, London: Allen & Unwin.)

Freeman, C. (1987), *Technology Policy and Economic Performance: Lessons from Japan*, London, UK and New York, USA: Pinter.

Freeman, C. and Louçã, F. (2001), *As Time Goes By. From the Industrial Revolutions to the Information Revolution*, Oxford, UK and New York, USA: Oxford University Press.

Gershuny, J. (1978), *After Industrial Society? The Emerging Self-Service Economy*, London, UK and New York, USA: MacMillan.

Hemphill, T. (2003), 'Cooperative strategy, technology innovation and competition policy in the United States and the European Union', *Technology Analysis & Strategic Management,* **15** (1), 93-101.

Hodgson, G. (2002), 'The legal nature of the firm and the myth of the firm-market hybrid', *International Journal of the Economics of Business*, **9** (1), 37-60.

Hodgson, G. (1993), *Economics and Evolution. Bringing Life Back into Economics*, Cambridge, UK: Polity.

Hughes, T. (2004), *American Genesis: A Century of Invention and Technological Enthusiasm, 1870-1970*, Chicago, USA and London, UK: University of Chicago Press.

Kamien, M. and Schwartz, N. (1982), *Market Structure and Innovation*, Cambridge, UK: Cambridge University Press.

Kamien, M. and Schwartz, N. (1975), 'Market structure and innovation: a survey', *Journal of Economic Literature,* **13** (1), 1-37.

Klein, B. (1977), *Dynamic Economics*, Cambridge, MA, USA: Harvard University Press.

Lei, D. and Hitt, M. (1995), 'Strategic restructuring and outsourcing: the effect of mergers and acquisitions and LBOs on building firm skills and capabilities', *Journal of Management,* **21** (5), 835-859.

Leiponen, A. (2006), 'Managing knowledge for innovation: the case of business-to-business services', *Journal of Product Innovation Management,* **23** (3), 238-258.

Lundvall, B.-Å. (ed.) (1992), *National Systems of Innovation*, London, UK and New York, USA: Pinter.

Marx, K. (1867), *Das Kapital. Kritik der politischen Oekonomie*, Vol.1, Hamburg.

Marx, K. and Engels, F. (1848), *Manifest der kommunistischen Partei*, London: Imago.

McNulty, P. (1968), 'Economic theory and the meaning of competition', *The Quarterly Journal of Economics*, **82** (4), 639-656.

Meadows, D., Meadows, D., Randers, J. and Behrens, W. (1972), *The Limits to Growth*, New York: Universe Books.

Miles, I. (2003), 'Services innovation: coming of age in the knowledge-based economy', in B. Dankbaar (ed.), *Innovation Management in the Knowledge Economy*, London: Imperial College Press, pp. 59-81.

Morgan, G. (2006), *Images of Organization*, London: Sage Publishing.

Nelson, R. (ed.) (1993), *National Innovation Systems: A Comparative Analysis*, Oxford, UK and New York, USA: Oxford University Press.

Nelson, R. and Winter, S. (1982), *An Evolutionary Theory of Economic Change*, Cambridge, MA, USA: Harvard University Press.

Nonaka, I. and Takeuchi, H. (1995), *The Knowledge-Creating Company. How Japanese Companies Create the Dynamics of Innovation*, Oxford, UK and New York, USA: Oxford University Press.

Patel, P. and Pavitt, K. (1999), 'The wide (and increasing) spread of technological competencies in the world's largest firms: a challenge to conventional wisdom', in A. Chandler, P. Hagström and Ö. Sölvell (eds), *The Dynamic Firm. The Role of Technology, Strategy, Organization, and Regions*, Oxford, UK and New York, USA: Oxford University Press, pp. 192-213.

Pavitt, K. (1984), 'Sectoral patterns of technical change: towards a taxonomy and a theory', *Research Policy*, **13** (6), 343-373.

Piore, M. and Sabel, C. (1984), *The Second Industrial Divide: Possibilities for Prosperity*, New York: Basic Books.

Porter, M. (1990), *The Competitive Advantage of Nations*, New York: The Free Press.

Prahalad, C. and Hamel, G. (1990), 'The core competence of the corporation', *Harvard Business Review*, May-June, 79-91.

Roobeek, A. (1987), 'The crisis in fordism and the rise of a new technological paradigm', *Futures*, **19** (2), 129-154.

Rothwell, R. and Zegveld, W. (1981), *Industrial Innovation and Public Policy: Preparing for the 1980s and the 1990s*, London, UK and New York, USA: Pinter.

Roussel, P., Saad, K. and Erickson, T. (1991), *3rd Generation R&D: Managing the Link to Corporate Strategy*, Boston: Harvard Business School Press.

Rynes, S., Bartunek, J. and Daft, R. (2001), 'Across the great divide: knowledge creation and transfer between practitioners and academics', *Academy of Management Journal*, **44** (2), 340-355.

Schoonhoven, C. and Jelinek, M. (1990), 'Dynamic tension in innovative, high technology firms: managing rapid technological change through organizational structure', in M. von Glinow and S. Mohram (eds), *Managing Complexity in High Technology Organizations*, Oxford, UK and New York, USA: Oxford University Press, pp. 90-118.

Schumpeter, J. (1942), *Capitalism, Socialism and Democracy*, London: Unwin University Books, 1974.

Schumpeter, J. (1939), *Business Cycles: A Theoretical, Historical and Statistical Analysis of the Capitalist Process*, New York: McGraw-Hill.

Schumpeter, J. (1911), *Theorie der wirtschaftlichen Entwicklung*, Leipzig: Duncker & Humblot. (English translation: *Theory of Economic Development: An Inquiry into Profits, Capital, Credit, Interest, and the Business Cycle*, Cambridge, MA, USA: Harvard University Press.)

Sitter, L. de, Hertog, J. den and Dankbaar, B. (1997), 'From complex organizations with simple jobs to simple organizations with complex jobs', *Human Relations*, **50** (6), 497-534.

Storper, M. (1997), *The Regional World: Territorial Development in a Globalizing World*, New York: Guilford Press.

Vesey, J. (1992), 'Time-to-market: put speed in product development', *Industrial Marketing Management*, **21** (2), 151-158.

Vissers, G. and Dankbaar, B. (2007), 'A contribution to a multi-level theory of creativity', in E. Carayannis and J.-J. Chanaron (eds), *Leading and Managing Creators, Inventors, and Innovators: The Art, Science, and Craft of Fostering Creativity, Triggering Invention, and Catalyzing Innovation*, Westport: Praeger Publishers, pp. 171-192.

Wheelwright, S. and Clark, K. (1992), *Revolutionizing Product Development: Quantum Leaps in Speed, Efficiency and Quality*, New York: The Free Press.
Zuboff, S. (1984), *In the Age of the Smart Machine: The Future of Work and Power*, New York: Basic Books.

4. Globalisation and Innovation Systems: Policy Issues

Keith Smith

INTRODUCTION

Globalisation is an ill-defined and haphazardly used term that usually refers – with many differences of nuance and emphasis – to three related phenomena of the world economy at the present time. The first is the spread of neo-liberal economic doctrines and (to a lesser extent) practices. The second is increasing interdependence between economies in terms of demand patterns, productivity norms, production chains and innovation. The third is the spread of, or increased importance of, global economic institutions such as the International Monetary Fund (IMF), the World Trade Organization and specific agreements such as TRIPS or GATS.[1] Closely related to this institutional element is the spread of large-scale regional economic integration, of which by far the most important example is the European Union (EU). However, the North American Free Trade Area (NAFTA), the Asia-Pacific Economic Cooperation (APEC) forum and Mercosur (South American regional trade agreement) are all significant.

This chapter focuses on relations between innovation activity and these dimensions of globalisation, looking in particular at the challenges raised by globalisation for innovation support policies. A key argument here is that innovation issues cannot be addressed clearly without bearing in mind the doctrinal and institutional aspects of globalisation. Underlying globalisation in technology and innovation is greater interdependence in knowledge creation, in the forms of scientific collaboration, industrial R&D and the global mobility of highly skilled people. The chapter looks at definitional issues, then data and trends, and at the historical background to the new interdependence, before turning to the policy implications for national economies.

It is sometimes argued that the spread of global norms in consumption patterns, productivity, trade regimes and investment returns erodes or even destroys the possibility of independent policy action by national governments, since policy actions must inevitably contradict these norms and will therefore be destroyed by

1 TRIPS is the Trade-Related Intellectual Property agreement, and GATS is the General Agreement on Trade in Services, both negotiated inside the WTO.

global market forces. The economic journalists Thomas Friedman and Martin Wolf have made perhaps the most trenchant arguments along these lines. If globalisation implies an inevitable loss of sovereignty then a straightforward policy conclusion follows: that the only possible policies are those that adapt specific countries to the global liberal economic order.

The main argument of this chapter is that innovation policies remain strongly viable in the context of globalisation, but require a 'systems' approach. Liberalisation has certainly changed the economic landscape, as has the growth of WTO trade and investment rules. But this does not mean that innovation policy is exhausted, mainly because innovation is not as affected by the changes of liberalisation as many often assume. By contrast, this chapter argues that if we recognise that innovation has systemic features – that is, that it occurs in the context of institutions, organizations and infrastructures that shape the innovation behaviour of firms – then there is considerable scope for policy action. This scope lies not in market-modifying tax/subsidy policies, or industry creation policies, but rather in actions that shape the components of the innovation system. It will be argued that key components of national innovation systems – including education, knowledge infrastructures, the income distribution and risk management systems – not only remain open to discretionary policy interventions but should be seen as location-specific assets that shape how globalisation affects particular national contexts.

DATA

What kinds of general evidence do we have that bears on trends in globalisation? Potentially, there is a wide range of economic statistics and indicators, which are summarised with definitions in Table 4.1. A key point here is that not all of the data is either as relevant or as high-quality as we might wish, and it is common to find quite restricted uses of data in the literature on globalisation and innovation. These indicators are of varying relevance for the study of globalisation, and it is often the case that the literature relies on indicators (notably patent data) that can be questioned. Some indicators – such as trade data – are highly relevant for studying the growth of interdependent globalisation when they focus on inter-industry trade, but of limited relevance when focussed on arms-length transactions. Others, such as portfolio investment, require careful use. In this case the deregulation of capital markets has not just promoted foreign direct investment (FDI), it has also led to very substantial portfolio shifts, often led by large pension funds or mutual funds. It is now common for stock markets worldwide to have very substantial percentages of shares held by non-nationals, and in particular by large companies. On the one hand this introduces issues concerning the effects of these shifts on corporate governance: it is often claimed that there is a trend towards globalisation of Anglo-Saxon corporate governance institutions as a result. But on the other it raises wide questions about the implications of ownership for national policies.

Table 4.1 Data on globalisation

Indicator	Definition	Data availability	Quality and relevance
Commodity trade	Cross-border transactions in goods	Excellent	Indicator of cross-border links between national entities
Trade in services	Cross-border transactions in services	Good	Limited by recording problems; possibilities of large-scale underreporting
Foreign direct investment	Investment in plant, equipment, material assets	Good	Highly relevant for 'interdependence' globalisation
Foreign portfolio investment	Investment in financial assets	Good	Relevant for institutional interdependence
R&D flows	BERD (Business Expenditure on R&D) investments by firms domiciled elsewhere	Questionable	'Arms-length' indicator of knowledge flows
R&D by affiliates	BERD by foreign-owned affiliates	Good	Highly relevant for interdependence
Patents by location of applicant	Patents by foreigners in domestic patent offices, or by domestic patent holders abroad	Excellent	Measure of invention; patents often unused
Technology balance of payments	Royalty payments and payment for patent licences	Poor	Relevant but limited by data quality
Migration, especially of the highly skilled	Cross-border movements of highly qualified people	Poor	Strongly relevant, with some signs of improvement in data quality

DEFINITIONS AND TRENDS

The general field of globalisation is a definitional mess. On the one hand, a central problem is to distinguish clearly between forms of international economic links that have existed for very long periods – such as trade, migration flows, foreign investment or colonisation – and current forms of such links. Do current forms represent something qualitatively new, and if so, how do we know this and when did it emerge? On the other hand there are problems in distinguishing between different domains – cultural, political, ideological, economic, etc. – to which globalisation definitions might refer. Scholte (2000) distinguished five separate levels or domains of globalisation: as internationalisation (the creation of an international system beyond the relations of separate states); as liberalization (the creation of common policy norms worldwide); as universalisation in the sense of spreading cultural meanings and technologies (such as television or the internet); as westernisation, sometimes seen as coterminous with modernisation; and as deterritorialisation (meaning a diminishing importance of geographical location in cultural or economic life). In dealing with innovation, most definitions emphasise interdependence. For example, in exploring innovation and policy, the work of Archibugi and Iammarino follows the well-known definition of Giddens (1990): 'the intensification of worldwide social relations which link distant localities in such a way that local happenings are shaped by events occurring miles away and vice versa' (Archibugi & Iammarino 1999).

A step beyond general definitions is to classify types of innovative behaviours in terms of their locational implications, and then to trace their relative importance via relevant economic data. Archibugi and Michie (1997) did this by distinguishing three broad types of activity: international exploitation of nationally produced innovations, the global generation of innovations, and global techno-scientific collaboration (Table 4.2).

This kind of classification corresponds closely to the distinctions between 'asset exploiting' and 'asset augmenting' firms' strategies, proposed by Dunning and Narula (1995). Asset-exploiting strategies imply domestic production of a new product, followed by geographical extension of the market through trade or foreign production as the product life-cycle proceeds. This is essentially the product-cycle model of the multinational firm, first proposed by Vernon (1965). Forms of the 'asset exploiting' model have been until recently the central interpretation of the internationalisation of R&D. Patel and Pavitt argued that the technological capabilities of firms were more geographically 'sticky' than marketing, investment or production facilities, for three main reasons. Firstly, the knowledge-creating activities of the firm are its central asset. Secondly, such knowledge has a high tacit dimension, making transfer difficult. Finally, firms are the product of their environments, and are shaped by the innovation systems and policy environments in which they develop:

Table 4.2 Forms of globalisation of technology and innovation

Categories	Actors	Forms
International exploitation of nationally produced innovations	Profit-seeking firms and individuals	Exports of innovative goods Cession of patents and licences Foreign production of innovative goods
Global generation of innovations	Multinational firms	R&D and innovative activities in both home and host countries Acquisition of R&D labs
Global techno-scientific cooperation	Universities Public research organizations Multinational firms	Joint scientific projects International flows of students Joint ventures Exchange of technical information and equipment

Source: Archibugi & Michie 1995.

> ... the technological competitiveness of firms inevitably depends on national systems of innovation, and national systems of innovation inevitably depend on government policy. The level of business-funded R&D is influenced by national policies (e.g. competition, macroeconomics), and also by the behaviour of national institutions (e.g. agencies funding basic research, banks and stock markets, systems of corporate governance). (Pavitt & Patel 1999)

Moreover the firms in which such capabilities reside appear to develop competences and capabilities within specific national contexts, even when the firms are multinational in terms of production and operations (Archibugi & Pianta 1992; Patel & Pavitt, 1994a; 1994b). These considerations suggested that neither globalisation nor economic integration were replacing the national entity as an economic space; rather they were supplementing it, and industrial R&D was relatively immobile.

Asset-augmenting strategies refer to location shifts that seek to complement existing knowledge assets with assets that are located elsewhere and that are not readily available in the home base. OECD (2006) speaks of the resulting R&D arrangements as

> internationally interdependent labs ... whose role is in the long-term basic research of the group, and who will have close collaboration with other similar labs. In such investments, firms aim either to improve their existing assets, or to acquire (and internalise) or create completely new technological assets by locating R&D facilities abroad. Knowledge relations between the foreign laboratory and the central home laboratory become far more interdependent.

As a result of these changes, Cantwell has argued that while home bases remain very important, nevertheless

> [t]echnology leaders have altered the nature of international technology creation by pioneering the international integration of MNC [Multinational Corporation] facilities into regional or global networks. Globalisation in this sense involves the establishment of new international structures for technology creation. In the past, foreign technological activity exploited domestic strengths abroad ... By contrast, today, for companies of the leading centres, foreign technological activity now increasingly aims to tap into local fields of expertise, and to provide a further source of new technology that can be utilised internationally in the other operations of the MNC. In this respect, innovation in the leading MNCs is now more genuinely international or, in the terminology used here, it has become 'globalized'. (Cantwell 1997)

Is there reasonable evidence to support such claims concerning new forms of globalisation? In fact there now seem to be some fairly clear statistical evidence suggesting that this new form of globalisation is indeed a real trend at the present time. The first data-point is simply the massive growth of FDI and affiliate sales around the world in recent decades. Table 4.3 shows some of the dimensions of this. Perhaps more interesting evidence relates to the distinction between R&D located abroad by firms who do not establish subsidiaries and affiliates, and those who do. In the first case, the R&D performed by such firms is recorded as a transnational flow; in the second case, the R&D is recorded as part of BERD (Business Expenditure on R&D) for the performing country, but is identified as R&D carried out by an affiliate of a foreign firm. The first type of R&D is depicted in Figure 4.1, in terms of flows between the USA, the EU and Japan.

The key point to note about these flows is that they are small in relation to the total amounts of R&D being performed and the net flows are even smaller. We could conclude from this that 'arms-length' R&D conducted internationally is not a particularly significant element in the world economy. This, however, is not the case if we turn to R&D performed by foreign affiliates. In the first place, as Figure 4.2 shows, we see a substantial growth in the total volume of R&D performed by such affiliates. R&D has more than doubled in a ten-year period, with compound growth rates of around 12% per year, far higher than the growth of GDP.

The data suggests a substantial increase in foreign affiliate R&D: such R&D more than doubled in the nine years covered here. Approximately 17.5% of BERD in the US, the EU and Japan is now being performed by affiliates of foreign firms. For individual countries the proportions can be very significant. In the larger EU economies these shares are around 25% or more. Figure 4.3 shows the dimensions of this, over time, for a number of OECD economies.

Taken together, this data supports the Cantwell argument noted above, with sustained growth of foreign affiliate R&D and rising shares of such R&D in domestic BERD. Affiliate R&D is not only growing, it is accounting for a very large part of R&D as a whole across OECD countries. For smaller economies, such as Ireland, Belgium, Sweden and Australia, the shares are in excess of 40%, and this is also the case for recent EU-accession countries such as the Czech Republic and Hungary. The

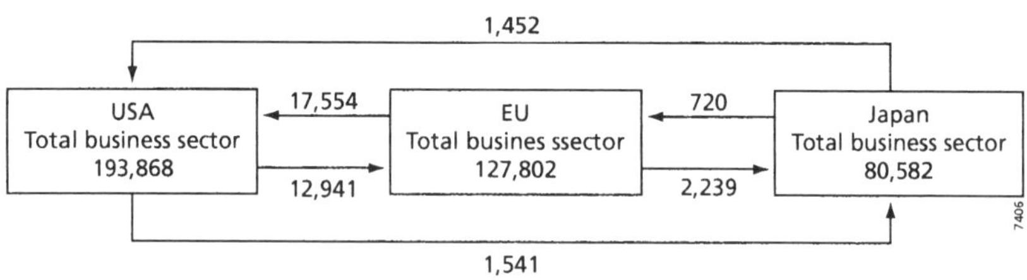

Source: OECD, AFA database and OECD estimates.

Figure 4.1 R&D flows between the EU-15, US and Japan, 2002 ($US million PPP)

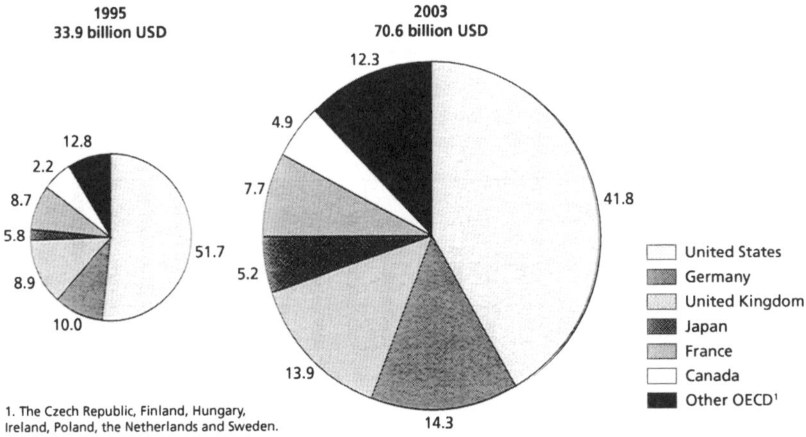

Source: OECD, AFA database and OECD estimates.

Figure 4.2 Trends in R&D expenditure under foreign control 1995-2003

share is relatively low for the USA, although as Figure 4.2 shows the absolute amount is high. This suggests that Cantwell's hypothesis, that the asset-augmenting form of globalisation is a new and dominant form, can be supported by the data.

THE CREATION OF THE GLOBAL ECONOMY

What are the sources of recent globalisation? Since the mid-1970s, the main thrust of economic policy in OECD economies has been towards economic liberalisation

Table 4.3 FDI and international production 1982-2001 ($US billion at current prices and % values)

	1982	2001
FDI Inflows	59	735
FDI Outflows	28	621
FDI Inward Stock	734	6846
FDI Outward Stock	552	6582
Sales by foreign affiliates	2541	18517
Total assets of foreign affiliates	1959	24952
Exports of foreign affiliates	670	2600
Employment of foreign affiliates (000s)	17987	53581
Inward FDI stocks to GDP ratio	6.79%	21.46%
Foreign affiliates exports to total exports	32.2%	34.99

Source: Narula & Zanfei, 2004, based on UNCTAD FDI/TNC database.

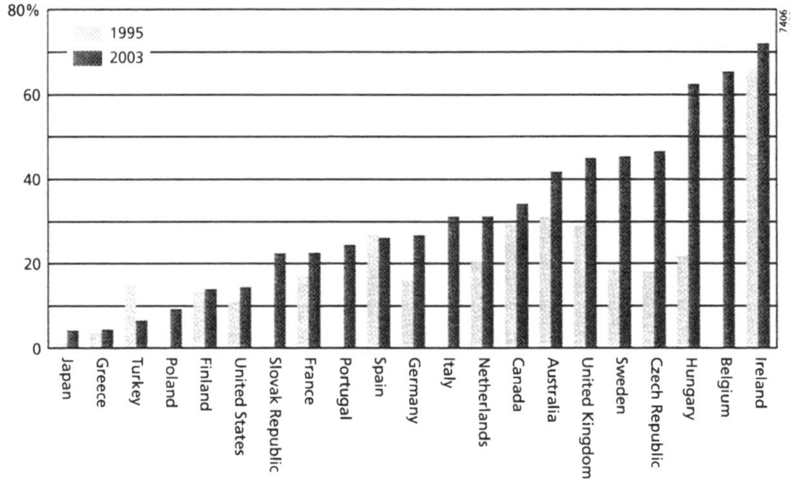

Source: OECD.

Figure 4.3 Shares of BERD performed by foreign affiliates, 1995 and 2003

and a generally more market-oriented approach to policy issues. In some countries, increased marketisation of public sector activities has been identified with 'modernisation', although it is often unclear why modernisation (an ambiguous term in itself) requires this. Be that as it may, in recent years the basic economic policy focus both at national and European levels has been on macro- and micro-economic reform: trade liberalisation, deregulation of financial markets, privatisation and labour market reform. This has been accompanied by attempts to reduce public expenditure as a proportion of GDP, and by shifts in the provision of public services towards more market-oriented approaches. The latter often involve public-private partnerships in which governments contract with private agents to supply 'public' services. This broad policy shift has had two primary aims. The first is cost-reducing efficiency improvements that enhance productivity. The second is the creation of an economic environment that promotes sustained economic growth, where the view is that a liberalised environment is superior in terms of channelling the right volume and composition of investment towards growth opportunities. The combination of these changes has significantly reduced the costs and increased the private benefits of cross-border investment, and thus has driven globalisation.

In its more explicit formulations, such as the so-called 'Washington Consensus' behind the operations of the World Bank and the IMF, the reform shift is often seen as an Anglo-American formulation deriving from the Reagan-Thatcher administrations. However neo-liberal ideas have been very influential across Europe also, leading to diverse reform processes across every European economy. These processes cannot simply be reduced to working out of an Anglo-American model, in particular because European reform processes continue to rest on different conceptions of the role of the state, and different views about the responsibilities of government. Liberalisation measures in France, the Netherlands and the Nordic region, for example, have been very far reaching. They also have a rather long political history – Swedish telecoms deregulation began in the 1960s, for example. These European liberalisation measures often rest on very different political conceptions from those of the Washington Consensus. Europe also has a specific dimension of reform for which the Washington Consensus has shown itself to be a poor guide, namely the reform and reconstruction of the former socialist countries, and the reform issues faced by countries newly acceding to the EU. These processes also have led to significant globalisation effects, in terms of direct and portfolio investment flows and movements of people.

What factors drove the liberalising agendas and their globalising effects? From the technological side, it is sometimes argued that such technological developments as the integration of computing and telecommunications fostered financial integration. It is important to note, however, that many of these changes were driven by discretionary policy decisions that were not the result of inexorable economic forces, or of technological imperatives. In general, they were worked out via domestic policy decisions or inter-government agreements that could readily have taken different forms. In terms of economic imperatives, of course governments usually had potential economic benefits in mind, often related to overcoming the productivity slowdown. But these objectives did not compel the particular reform solutions that were chosen. So rather than seeing this change as a technologically-driven process, it is more appropriate to see the causality the other way around – that integration decisions drove the development of communications systems, integrated trading platforms, etc.

The liberalising policy agenda from the mid-1970s has had many dimensions, the most important of which are:

- Withdrawal of governments from industrial support policies
- Anti-union policies that seek to inhibit union action and to reduce union membership
- Fiscal policy constraints, and reliance on monetary policies for macroeconomic management
- Deregulation across industries, and particularly the removal of barriers to entry
- Privatisation of state-owned industries, especially utilities
- Changes in the governance of the 'knowledge infrastructure' of research institutes and universities
- Removal of exchange controls or other constraints on capital movement
- Active promotion of inward direct foreign investment across countries
- Attempts to promote international mobility of highly skilled people

Taken together, these constitute a major 'regime shift' within modern capitalism. Of these changes, two are probably particularly important for innovation policy, namely privatisation and changes in the governance of the knowledge infrastructure.

INDUSTRY DEREGULATION AND PRIVATISATION

Within the reform agenda, industrial policy has aimed at privatisation and deregulation, with governments focusing on framework issues and in the main withdrawing from direct interventions. It is possible that significant effects of reform on capabilities have come via privatisation of publicly-owned enterprises, or wider processes of industry deregulation. Although privatisation has not always reduced the exploitation of dominant positions (arising from, for example, the control of fixed line networks in telecommunications) it has been associated with greater flexibility, often sharp increases in labour productivity, and willingness to innovate.

However, behind this there is a wider problem that may deserve consideration in the context of research policy. This concerns the nature of knowledge infrastructures, and their wider social and economic effects. Under public ownership, utilities in such fields as energy and telecoms had major technology development capabilities, often maintaining significant research labs and long-term innovation programmes aimed at innovations that could be extremely radical. The most notable case of this in Europe has been modern mobile telecommunications, which emerged in large part from a fifty-year development effort conducted by Nordic state telecoms agencies and their research labs. These agencies developed the basic architectures, network technologies, handset systems and standards that led to GSM (Global System for Mobile communications) and subsequent telecoms systems. This effort has a major impact in placing Europe as a whole in a strong position in this field. Privatisation has affected such capabilities across a range of privatised large firms, with generally falling R&D intensities and a greater focus on innovation related to current lines of business.

The old system of state-owned firms and their research labs could be seen as a kind of knowledge infrastructure, which managed such system-wide innovations as telecom digitalisation. There is therefore a current issue concerning whether the infrastructure concept should be widened, perhaps towards long-term technology platforms and their associated organisations, but certainly considering the nature, coherence and governance of infrastructures as a component of an innovation policy effort.

CHANGES IN THE GOVERNANCE OF KNOWLEDGE INFRASTRUCTURES

Universities and research institutes worldwide present an especially complicated mix of institutional and organisational forms, so it is dangerous to generalize about them. However globally there have been attempts to reform university and institute governance, usually in the direction of making them in some sense more responsive to economic or industrial needs. In at least one European country the entire functioning of the university system has been officially redefined in terms of support for the economic system. Specific measures have included a strong emphasis on industrial collaboration and industrial funding for research, along with a strong emphasis on commercialisation of university research, patenting by universities and the creation of spin-off companies. What is unclear is how these measures and trends affect the long-term research capabilities and profiles of universities, especially in terms of basic research in fields that might potentially have commercial outcomes. The relations between governance changes, knowledge creation and university science capabilities is clearly important for several reasons, foremost among which is its historic role in the development of technology platforms.

POLICY ACTIONS AND GLOBALISATION: THE IMPLICATIONS OF A SYSTEMS APPROACH

The liberalising agenda that has driven globalisation has definitely reshaped and constrained the environment for policy actions. In some policy arenas, governments have simply withdrawn from activity, and in others have been constrained by transnational agreements (such as EU regulation or WTO arrangements). These trends have also been affected by policy ideologies that focus on non-intervention as an allegedly optimal policy approach. This section explores how systems approaches to innovation can be used as a policy framework in the context of globalisation.

Systems approaches to innovation are founded on one of the most persistent themes in modern innovation studies, namely the idea that innovation by firms cannot be understood purely in terms of independent decision-making at the level of the firm. Rather, innovation involves complex interactions between a firm and its environment, with the environment being seen on two different levels. On one level there are interactions between firms – between a firm and its network of customers and suppliers, particularly where this involves sustained interaction between users and producers of technology. Here the argument is that inter-firm linkages are far more than arms-length market relationships – rather, they often involve sustained quasi-cooperative relationships which shape learning and technology creation. The second level is wider, involving broader factors shaping the behaviour of firms: the social and perhaps cultural context, the institutional and organizational framework, infrastructures and the processes which create and distribute scientific knowledge, and so on. These environmental conditions are often seen as specific to regional or national contexts, but they are also dynamic: their forms of operation change with political conditions, changing technological opportunities, economic integration processes and so on. The basic argument of systems theories is that system conditions have a decisive impact on the extent to which firms can make innovation decisions, and on the modes of innovation which are undertaken.

Systems theories involve a very strong hypothesis, which is that diversity in macroeconomic performance can be traced to underlying system differences across the above elements. What factors suggest that a focus on national systems and national policy levels might be relevant? Studies of innovation systems tend to focus on the fact that there are significant differences across national economies in a range of areas relating to innovation and learning. With considerable variations of emphasis, studies tend to focus on the following elements.

Industrial Structures and Technological Specialisations

Even where economies are converging in terms of macroeconomic indicators, it should be noted that this is occurring on the basis of differences in industrial activities and technological capabilities, which can be marked and persistent between the national economies involved.

The technological differences relate to persistent technological specialisation. Macroeconomic convergence appears to involve, at least for many countries, the construction and maintenance of quite specialised technological capabilities, and this is reflected in patterns of R&D expenditure, patenting, scientific publication and so on.

Institutional Structures

Economic behaviour rests on institutional foundations, in the sense of legally or customarily established 'rules of the game', which evolve because of the advantages they offer in reducing uncertainty. Different modes of institutional set-up lead to differences in economic behaviour and outcomes. Some of the most important institutional differences relevant for innovation concern major and persistent variation in systems of governance: both narrowly in the sense of formal regulatory systems of corporate governance, and more broadly in the sense of rules of the game for corporate behaviour. Corporate governance is concerned with the institutions that influence the ways in which business corporations allocate resources and returns. Corporate governance is important for business enterprises because it both enables and proscribes strategic decisions concerning the types of investments they should make and to whom the returns on these investments should be distributed. Corporate governance is also important for national economies, because of the central role of corporations in the allocation of resources and returns on national and international levels. Around the world many different national systems of corporate governance prevail, but both regional economic integration and globalisation are creating pressures toward convergence to an 'ideal' corporate governance system. So the relation between corporate governance and economic performance is therefore one of the key ways in which institutional structures affect system performance. But other key institutional areas include the operation of labour-management relations, cultures of entrepreneurship, and legal frameworks of corporate law.

Administrative and Regulatory Systems

It is well established that public policy systems affect innovation outcomes, but there are important cross-country differences in organisational structures and policy cultures. The differences include the specific arrangements and powers of ministries and agencies responsible for innovation. But there are strong national differences in such areas as accounting standards, and regulatory frameworks affecting health and safety, or environmental change. It is often argued that product regulation inhibits innovation (or, conversely, that some regulation promotes innovation). In fact, regulatory arrangements often have the effect of modifying existing innovation paths, and this leads to differences across national systems.

Education and R&D Systems

Human resources are a key element of innovation systems (a point which will be discussed in more detail below). The OECD (2001) has shown that there has been a substantial secular rise in the qualified labour force, that formal skills are increasingly important and that there is stronger mobility of highly skilled labour. This has directed attention to schooling and university systems as components of the innovation system. The important point here is that there remain major inter-country differences that set national innovation systems apart from one another. It is important to remember that organisations such as universities can differ sharply in terms of financing, scientific research organisation, governance, teaching priorities and capabilities, and so on. So recognizing the increased importance of education in innovation involves complex decisions about the nature and organisation of education systems that remain highly idiosyncratic at national levels.

Physical and Knowledge Infrastructures

There seem to be two broad roles for infrastructures in shaping technological change in innovation systems. On the one hand, such technologies often involve significant accompanying infrastructures. For example, automobile technologies, consumer electric technologies, information and communications technologies and aeronautics all rely on extremely substantial infrastructure investments – highways, electricity distribution networks, cable networks etc. The provision and economics of these infrastructures appear to have powerful effects both in initiating and driving the rise to dominance of new radical technologies. On the other hand, turning to knowledge infrastructures, the major technological innovations that have shaped the modern world mostly originated or were developed in public sector infrastructural organisations. Radar, telecommunications, microelectronics, nuclear power, biotechnology, advanced aircraft, space-based communications, new materials, the Internet – in these core technologies of the modern industrial economy many or most of the important developmental decisions were made, in one way or another, in government or public infrastructural agencies. Of course the decisions came from a variety of organisations – the military, research councils, civil ministries, universities – and the key choices and decisions were not necessarily made in any rational or consistent way. However, none of the qualifications that we might make about the role of the public sector should obscure the extraordinary importance of public-sector decision making at key stages in the evolution of these technologies. It might be argued that radical breakthroughs which have occurred in industrial R&D labs are by contrast noticeable by their rarity, although they may be of great economic importance. Given the prevalence of such infrastructural inputs to modern technology, it seems unlikely that the infrastructural role is merely accidental, and it is therefore worth asking whether there is anything essential (or indeed systematic) about infrastructures and large-scale innovations. If we look at the histories of specific technologies, especially those generally regarded as having a major economic or technological impact, it is surprising how often the fundamentals of the technology are developed in government labs, publicly-owned companies, universities, military R&D programmes, etc. It is often hard to understand how radical technologies in particular emerge or become established unless we take into account the roles of supporting infrastructures.

THE NATIONAL INNOVATION SYSTEM AS A POLICY CONCEPT

In a time of globalisation, it may appear strange to argue that any kind of national approach to innovation or any other domain of policy is relevant. It is sometimes argued that globalization is rendering the state (at whatever level) obsolete, and that the integration of product and capital markets removes the possibility of effective policy interventions by government. However, it is important to note that many of the structural elements of the system described above are either predominantly or wholly national in character: they are built up largely by the use of national resources in national contexts. They are subject to national decision-making, even where there may be increasing transnational elements. For example, how much a country spends on R&D, what kind of education system it constructs and operates, how much it spends on education, how it builds and finances its knowledge infrastructures, how it supports its areas of specialisation, are matters for that country alone. Certainly attempts to create such assets face budget constraints, but if governments wish to develop policies and programmes in the above areas, then there is nothing in either the formal or informal global arena to inhibit them. Moreover, in the area of institutions there is an increasing zone of transnational bargaining and regulation, for example with respect to agreements like TRIPS, GATS, or TRIMS (the WTO Agreement on Trade Related Investment Measures). These considerations suggest that, at least with respect to the core elements of the innovation system, there is no reason to believe that government is any less important in setting the context and framework for economic behaviour than it ever was (see Hirst & Thompson 1996, for a discussion of this). The real problem is how to think about what the desirable performance of the system looks like, and why and how governments should act towards it.

POLICY RATIONALES AND FOUNDATIONS

What are the implications of the ideas described above for the rationale and scope of public policy? The approach developed above suggests that innovation rests on structures of institutions and organizations, and that performance depends on how the structures function operationally. This suggests an approach to policy that rests on assessments of evaluations how specific organizations or structures contribute to system functionality. This offers a perspective that at some points links with the usual 'market failure' approaches to policy, but also goes beyond it.

Systems approaches, at whatever level, tend to see innovation performance as a result of how sets of institutions and organizations work. Both analytical and policy issues then turn on the nature of the components of the system, the nature of the links between them, and how well those linkages work. The links may take various forms. They may be economic, or they may involve the transmission of knowledge, or they may involve the joint use of infrastructures, and so on: the precise connections cannot be specified in advance, and often need detailed empirical investigation to uncover. But the point is that innovation performance can be seen in large part as a form of co-ordination problem – components of the system must work in a coherent way (that is, all moving in more or less the same direction, with more or less compatible objectives) towards the development and use of the new technology that is the object of the innovation process.

Where institutions, infrastructures or inter-firm connections are well established within a particular technological framework, the coordination needed for innovation may be unproblematic (although even stable technologies can run into novel problems). But where a new technology involves a major disruption, coordination can be very difficult. Some innovations are radical with respect to existing procedures, engineering capabilities or technical knowledge bases – they involve major discontinuities and 'shocks' to the existing technological systems. That is why technological historians such as Thomas Hughes write about such innovations largely in terms of 'systems, built by system builders'. The coordination problems that arise are relevant at all levels of innovation. Even if innovation is seen in terms of incremental improvements to existing technology then current organisations and regulation systems can readily run into technological or economic problems that require new forms of coordination to solve. But if we see the task of innovation in a more radical way, as shifting the fundamental technological systems on which the current industrial economy is based, then the coordination problems become really critical. A systems approach would suggest that the identification of co-ordination failures, the design of policy instruments to overcome them, and the development of relevant actors, are likely to be an important rationale for public policy intervention, and important also in deciding its scope and objectives.

The 'system failures' sketched above are all essentially problems of co-ordination, and so the claim made here is that market economies are characterised at the technological level by major co-ordination failures. Such failures are often recognised in economics, particularly in Keynesian macroeconomic theory, where sub-full employment equilibria rest on the existence of coordination failures. Such equilibria are situations in which workers would like to work and to consume (at existing wage and price levels), and firms would like to employ and produce, but no mechanism exists to overcome the inability of labour and product markets to coordinate via price signals. At this point, Keynesian arguments for public intervention come into play.

The argument here is that coordination problems with respect to innovation systems relate primarily to institutional action that precedes the operation of markets, or organizational action that creates a new arena for economic behaviour. There seem to be two types of problems. One is that we might be able to conceptualise a market for some benign outcome, but the real issue is to see how that market could be institutionalised. So issues related to corporate governance, accounting procedures, or risk management devices and so on are issues to do with the shaping of markets or substitution for them, rather than interventions in market operations. A second problem is that it may not be possible, in principle, to create a market at all. That is, the relevant problems are simply not amenable to a market solution, and could only be resolved by some other coordination mechanism. The main alternative is some form of administrative coordination that may be more or less democratic, or more or less hierarchical. Overcoming problems related to knowledge creation, education and training, or lock-in and specialisation, require institution building not market rectification.

REFERENCES

Archibugi, D. and Michie, J. (eds) (1997), *Technology, Globalisation and Economic Performance*, Cambridge, UK: Cambridge University Press.

Archibugi, D. and Pianta, M. (1992), *The Technological Specializations of the Advanced Countries*, Dordrecht, the Netherlands, Amsterdam, The Netherlands, Boston, USA and London, UK: Kluwer.

Cantwell, J. (1997), 'The globalisation of technology: what remains of the product-cycle model?', in D. Archibugi and J. Michie (eds) (1997), *Technology, Globalisation and Economic Performance*, Cambridge, UK: Cambridge University Press, pp. 215-240.

Dunning, J. and Narula, R. (1995), 'The R&D activities of foreign firms in the United States', *International Studies of Management and Organization*, 25 (1-2), 39-73.

Giddens, A. (1990), *The Consequences of Modernity*, Cambridge, UK: Polity.

Hirst, P. and Thompson, G. (1996), *Globalization in Question*, Cambridge, UK: Polity.

Narula, R. and Zanfei, A. (2004), 'Globalization of innovation: the role of multinational enterprises', in J. Fagerberg, et al. (eds), *The Oxford Handbook of Innovation*, Oxford, UK and New York, USA: Oxford University Press, pp. 318-345.

OECD (2006), *OECD Science, Technology and Industry Outlook*, Paris: OECD Publishing.

OECD (2001), *International Mobility of the Highly Skilled*, Paris: OECD Publishing.

Pavitt, K. and Patel, P. (1999), 'Global corporations and national systems of innovation: who dominates whom?', in D. Archibugi, et al. (eds), *Innovation Policy in a Global Economy*, Cambridge, UK: Cambridge University Press.

Patel, P. and Pavitt, K. (1994a), 'National innovation systems: why they are important and how they might be measured and compared', *Economics of Innovation and New Technology*, 3, 77-95.

Patel, P. and Pavitt, K. (1994b), 'Uneven (and divergent) technological development amongst advanced countries: evidence and a framework of explanation', *Industrial and Corporate Change*, 3 (3), 759-788.

Scholte, J. (2000), *Globalization: A Critical Introduction*, London: Palgrave.

Vernon, R. (1965), 'International investment and international trade in the product cycle', *Quarterly Journal of Economics*, 80 (1066), 190-207.

II. Evolution of the innovation systems perspective

5. Rationales for Public Policy Intervention in the Innovation Process: Systems of Innovation Approach

Cristina Chaminade and Charles Edquist

INTRODUCTION

Why government should intervene to support R&D and innovation is a question that can be traced back several decades (Arrow 1962; Nelson 1959), yet is still current. Despite the extensive literature on innovation, the rationales for public intervention in innovation are still subject to intense debate, especially related to new theoretical perspectives such as the system of innovation (SI) approach.

Innovation policy can be defined as 'the public actions that influence innovation processes, i.e. the development and diffusion of (product and process) innovations'.[1] The objectives of innovation policy are often economic ones, such as economic growth, productivity growth or increased employment and competitiveness. However, they may also be of a non-economic kind, such as cultural, social, environmental, or military. As in any policy, the objectives of innovation policies are determined in a political process and not by researchers.

Innovation policy design is a question of the division of labour between, on the one hand, the actions of private firms and, on the other, the actions of public organisations – with regard to factors influencing innovation processes. For example, large-scale and radical technological shifts rarely take place without public intervention, while incremental innovation is normally carried out by firms without any explicit support from the government. To discuss the division of labour between private and public actors in innovation is the same as discussing the rationales, reasons or criteria for public policy intervention. That is, when, how and why should government intervene in the economy supporting certain innovative activities.

Innovation policy is partly influenced by the dialogue between policy and theory. The discussion on the rationales is inherently linked to the theoretical approach that one chooses to explain innovation and technological change. New theoretical insights provide the grounds for new actions whilst old actions are abolished. In this chapter

1 By product innovation we refer to both goods and services; by process innovation we understand both productive and organizational innovations (Edquist 2005).

the central question is how this link between theory, policy and practice changes by the emergence of the innovation system (SI) approach.

Researchers in economics of innovation have, for some time, distinguished between rationales for intervention under the neoclassical and the evolutionary theory (Bach & Matt 2005; Lipsey & Carlaw 1998; Smith 2000). The rationale for public intervention under the systems of innovation perspective has recently received attention among scholars and practitioners (Kohler-Koch 2003; OECD 2001; Smits & Kuhlmann 2004; Woolthuis et al. 2005). However, there has not been yet an attempt to profoundly discuss the implications of the adoption of the SI approach for the design and implementation of innovation policies. We will try to pursue such a discussion both from a theoretical perspective and a practical one, including some examples of innovation system based policies.

Since the emergence of the system of innovation concept in the 1990s in academic arenas (Freeman 1987; Lundvall 1992; Nelson 1993; Edquist 1997) the concept seemed rapidly to attract the interest of policy makers, especially international policy think-tanks such as the OECD (Mytelka & Smith 2002).[2] Its many initiatives on systems of innovation and policy (OECD 1995a; 1995b; 1996; 1999; 2001; 2002) had a strong impact on the way that national governments started to design and implement innovation policies. Today, countries like Finland, Sweden, the Netherlands and Japan have explicitly adopted the system of innovation approach in their innovation policies.

Despite the widespread use of the SI approach in policy-making circles, it remains a fuzzy concept – very difficult to use in practice (Chaminade & Edquist 2006). Therefore much discussion is needed on the implications of the adoption of the SI approach for public policy (what to do, when and how to do it). This chapter attempts to fill this gap by discussing the rationales of innovation policies when the SI approach is adopted.

The chapter is structured as follows. In the first section, we compare the basic assumptions of the neoclassical and evolutionary-systemic theories and the implications of the adoption of one or another for the rationales for public intervention (why to intervene). We then introduce, in the second section, some additional issues of relevance for intervention and some principles that emerge when the system of innovation approach is adopted for innovation policy. Finally, in the last section, we conclude with some open questions and issues for further research.

2 As many studies have argued, the OECD played a significant role in the dissemination of the concept to national governments (Mylteka & Smith 2002; Godin 2004; Sharif 2006).

MAINSTREAM ECONOMICS VS. THE SI APPROACH: CONCEPTUAL FRAMEWORK AND RATIONALES FOR PUBLIC INTERVENTION[3]

The Concept of Knowledge and Innovation in Mainstream Economics

Innovation policies are public actions that influence innovation processes. Innovation policy is the result not only of the interplay between private and public actors but also of the interaction between the innovation scholar community and the policy makers. The way that innovation policy has been designed in a given moment in time partly reflects how innovation is conceptualized at that moment in time, that is, what was the theoretical paradigm used to understand what was happening in practice. There are two largely influential theoretical approaches to innovation: the neoclassical theory (and later the new growth theories) and the evolutionary/systems of innovation approach.

For the neoclassical scholars, the innovation process was initially narrowed down to research (and invention). For these theorists, the process of innovation was often seen as a fixed sequence of phases, where research efforts will turn almost automatically into new products. How to transform the results of the research activity into products or processes that can be used in the economy was considered to be a black box (Rosenberg 1982; 1994). Innovation is thus about the generation of knowledge and knowledge is the same as information, i.e., it is codified, generic, and it is accessible and easily adaptable to the firm's specific conditions.

These tacit assumptions about the properties of knowledge are reflected in the discussion about the process of invention. For Nelson (1959) and Arrow (1962) the knowledge emanating from research has some specific properties: uncertainty, inappropriability and indivisibility (Lipsey & Carlaw 1998).

- Uncertainty refers to the impossibility to fully know the outcomes of the research process and the risk associated to it.[4]
- Inappropriability means that firms cannot fully appropriate the benefits which derive from their inventions. There will always be externalities emanating from the research process. This means that the incentive for research activity by firms is smaller than it would be if it was possible for firms to appropriate all the benefits.

3 In order to simplify this chapter, two main frameworks will be contrasted with each other: on the one hand, neoclassical theory (more specifically the late work in mainstream economics that dealt with R&D and growth), including the most recent additions of the new growth theory and, on the other hand, the evolutionary theory and the related systems of innovation approach. We use the terms neoclassical theory and mainstream economics interchangeably.

4 The problems of uncertainty in the design of innovation policy are discussed in more detail towards the end of this chapter.

- Indivisibility implies that there is a minimum investment in knowledge before any new knowledge can be created.

According to mainstream economics, the three characteristics of scientific knowledge (uncertainty, inappropriability and indivisibility) will lead to an underinvestment in R&D by private actors than what would be desirable from an economic and societal point of view. Policy makers have to intervene because of a market failure: economies will systematically underinvest in R&D not reaching the optimal allocation of resources for invention.[5] This constitutes the main rationale for public intervention in research activities.

The optimal allocation of resources is linked to the central concept of equilibrium in mainstream economics. Markets will always tend to achieve equilibrium under the conditions of perfect information, perfect competition and profit maximization. Economic agents can maximize their profits because they have perfect information about the different options available to them. When this fails to be the case, governments should intervene to mitigate non-desired externalities and asymmetries in information, correct inefficient market structures or eliminate the barriers to entry.

The neoclassical analysis provided governments with strong arguments to invest heavily in fields such as energy, large-scale science and technology projects, defence research, etc. where the public rate of return was expected to be high, the barriers to entry were significant and the externalities were also assumed to be sizeable.

Rationales for Public Intervention in the Neoclassical Theory: Market Failures

The policy implications that emerge from the market failure theory are actually not very helpful for policymakers from a practical and specific point of view. They are too blunt to provide much guidance. They do not indicate how large the subsidies or other interventions should be (as it is not possible to determine the optimum level of investment in R&D or innovation activities) or within which specific area one should intervene. Standard economic theory is not of much help when it comes to formulating and implementing specific R&D policies. It only provides general policy guidance; e.g. that basic research should sometimes be subsidised (Edquist 1994; Edquist et al. 2004). As mainstream theorists tend to ignore the economic structure or institutional frameworks in which the innovation activity takes place, their policies apply across the whole economy (Lipsey & Carlaw 1998; OECD 1998). The market failure approach is simply too abstract to be able to guide the design of specific innovation policies.

---

5 Indeed research conducted for the OECD countries (Norgren & Hauknes 1999) has shown that the social rate of return of investments in R&D and Human Capital largely exceeds the private rate of return, therefore providing strong arguments for public intervention in the supply of R&D and the provision of human capital.

Furthermore, as Bach and Matt (2005) acknowledge, when policy makers intervene to mitigate or solve some market failures, they might end up generating more failures. For example, by introducing IPRs to solve the problems of appropriability and incentives, policy makers introduce barriers to the perfect flow of information, thus generating a second market failure.[6]

The neoclassical theory has been expanded significantly over the last two decades, solving some of the problems that the initial theory had in terms of its narrow conceptualisation of innovation processes.[7] It can be said that the neoclassical theory has expanded in at least two directions relevant for the discussion on innovation policy (Bach & Matt 2005): the new economics of science and technology and the new growth theory. The new economics of science and technology sponsored by Dasgupta and Stoneman among others (Dasgupta 1987; Stoneman & Dasgupta 1987) has focused on the public good properties of technology, concluding that the justification for the state intervention is stronger for science than for technology.

On the other hand, the new growth theory (Romer 1986; 1990; Grossman & Helpman 1991; Aghion & Howitt 1992) takes seriously a number of issues contemplated by the evolutionary theory which were neglected by the neoclassicals. Among others are: the admission of a certain degree of randomness in R&D activities (i.e. there is no perfect information on the outcomes of the R&D process), the focus on the importance for growth of technology flows between agents and the acknowledgement that innovation policy is important for growth (Verspagen 2005).

Despite the significant progress made by the new growth theory from the initial neoclassical premises, it continues to have great limitations (Bach & Matt 2005). The new growth theory assumes low uncertainty, while in reality innovation involves a great deal of uncertainty and risk. It is also based on a very simplistic (and linear) relationship between R&D and growth (Verspagen 2005). And, more fundamentally, it continues to assume that the system can achieve equilibrium. In this sense, alternative theories or theoretical approaches might provide more convincing explanations of innovation that set better grounds for innovation policy.

The Concept of Knowledge and Innovation in the Systems of Innovation Approach

The SI approach has its roots in a compound of theoretical approaches such as the evolutionary theory (Nelson & Winter 1982). But it was also influenced by institutional approaches (North 1990) and sociology (Granovetter 1985). It emerged as a reaction to the perceived inadequacy of the neoclassical theory to explain innovation processes (Lundvall 1992) and, as such, most of its building blocks

6 This can be mitigated by awarding the patents for a limited period of time.

7 The neoclassical theory has been criticised for considering innovation as a linear process which starts with the investment in R&D activities and the strong emphasis on equilibrium and perfect information.

emerged in opposition to the neoclassical theory (i.e. highlighting limitations of neoclassical theory and providing an alternative).

The SI approach emphasises the fact that firms do not innovate in isolation but with continuous interactions with other organisations in the system (at regional, sectoral, national and supranational levels) (Edquist 1997; 2005; Lundvall 1992). The (SI) approach shifts the focus away from actions at the level of individual and isolated units within the economy (firms, consumers) towards that of the collective actions underpinning innovation. It addresses the overall system that creates and distributes knowledge, rather than its individual components, and innovations are seen as outcomes of evolutionary processes within these systems.

The notion of knowledge is also extended from the neoclassical 'information' to embrace also other forms of knowledge, such as tacit knowledge. Knowledge can be both general and specific and is always costly to create and diffuse. Knowledge can be specific to the firm or to the industry (Smith 2000). While in the neoclassical approach information asymmetries are considered to be a market failure, under the evolutionary theory and the SI approach asymmetric information is essential to provide novelty and variety. In this sense, the SI approach links clearly with the evolutionary theory putting the emphasis on the mechanisms of diversity creation and selection (e.g. competition) as the engines of innovation. It also stresses the path-dependency of innovation processes. The SI approach takes the evolutionary theory as one of the points of departure, to focus on the interactive mechanisms that shape the emergence and diffusion of innovations.

Table 5.1 summarises some of the issues related to innovation-related mainstream economics, systems of innovation approaches and the policy implications of these approaches (Table 5.1). It offers a simplistic yet easy to understand summary of the main differences between neoclassical theory on the one hand and the evolutionary theory on the other. For the sake of simplification and clarity, we leave aside in this summary the contributions made by the new growth theory and the new economics of science and technology that, as argued before, have kept most of the neoclassical premises intact (Nelson 2004; Sharif 2006). As Verspagen summarises:

> The evolutionary tradition and the neoclassical tradition have converged somewhat in the phenomena deemed central within each analytical approach. But they disagree on the essential nature of the growth process. The neoclassical theory conceptualises growth as a deterministic process in which causality is clear-cut and policies can be built on an understanding of time invariant determinants of growth patterns. In the evolutionary view, on the other hand, contingencies and specific historical circumstances play a larger role, and casual mechanisms that prevail in one period may be subject to endogenous change in the text. (Verspagen 2005)

Furthermore, we include the systems of innovation approach under the same heading of the evolutionary theory. This does not mean that we neglect that SI had a great

Table 5.1 Two different theoretical frameworks underlying science and innovation policy

	Mainstream	**Evolutionary theory and Systems of Innovation approach**
Underlying assumptions	Equilibrium Perfect information	Non-equilibrium. Asymmetric information
Focus	Allocation of resources for invention Individuals	Interactions in innovation processes Networks and framework conditions
Main policy	Science policy (research)	Innovation policy
Main rationale	Market failure	Systemic problems
Government intervenes to (examples)	Provide public goods Mitigate externalities Reduce barriers to entry Eliminate inefficient market structures	Solve problems in the system or to facilitate the creation of new systems Induce changes in the supporting structure for innovation: support the creation and development of institutions and organizations & support networking Facilitate transition and avoid lock-in
Main strengths of policies designed under each paradigm	Clarity and simplicity Analysis based on long time series of science-based indicators	Context specific Involvement of all policies related to innovation Holistic conception of the innovation process
Main weaknesses of innovation policies designed under each paradigm	Linear model of innovation Framework conditions are not explicitly considered in the model (e.g. institutional framework)	Difficult to implement in practice Lack of indicators for the analysis of the SI and evaluation of SI policies

influence from other theoretical approaches such as institutional economics but we particularly focus on those characteristics of the SI approach that draw mainly from evolutionary theory and thus, offer alternative explanations to the neoclassical theory.

Rationales for Public Intervention in the SI Approach: Systemic Problems

Understanding innovation as a complex interactive process has important implications for the design and implementation of any kind of policy to support innovation. It affects the focus of the policy, the instruments and the rationale for public policy, among other issues (Chaminade & Edquist 2006). The system of innovation approach does not imply that these systems are or can be consciously designed or planned. On the contrary, just as innovation processes are evolutionary, SIs evolve over time in a largely unplanned manner. Even if we knew all the determinants of innovations processes in detail (which we certainly do not, and will never do), we would not be able to control them and design or 'build' SIs on the basis of this knowledge. Centralized control over SIs is impossible and innovation policy can only influence the spontaneous development of SIs to a limited extent.

A main focus of the SI approach is therefore the complex interactions that take place among the different organisations ('actors') and institutions ('rules of the game') in the systems. The SI approach indicates that policy makers should intervene in those areas where the system is not operating well, that is, when there are systemic problems. Hence, one condition for public policy intervention is that such systemic problems exist, i.e. problems that are not automatically solved by private actors.[8] Conducting empirical analyses that explicitly compare systems of innovation can identify systemic problems.

Innovation policy – or other kinds of public intervention – should be a complement to the market, not replace or duplicate it. If there is no 'additionality', the public actions are a substitute for the actions of firms, and other private organisations. The two are overlapping or competing. It is of great importance that there actually is 'additionality' associated with the public intervention[9]. If not the public resources invested will not influence innovation processes, but lead to increased profits for the firms or to increased spending on other things than those targeted by the policy.

In the SI approach the notion of optimality is considered to be irrelevant. 'Market failure' in mainstream economic theory implies a comparison between conditions in the real world and an ideal or optimal economic system. However, innovation processes are path dependent over time and it is not clear which path will be taken. They have evolutionary characteristics. The system never achieves equilibrium. We cannot specify an ideal or optimal system of innovation. Hence, comparisons between an existing system and an ideal or optimal system are not possible. Thereby the notion of 'failure' loses its meaning and applicability. To eliminate completely associations to the notion of optimality, we prefer to talk about systemic problems instead of systemic failures (Chaminade & Edquist 2006; Edquist & Chaminade 2006) Some of these systemic problems mentioned in the literature include the following (Carlsson & Jacobsson 1997; Norgren & Hauknes 1999; Smith 2000; Woolthuis, Lankhuizen et al. 2005):

- *Infrastructure provision and investment problems:* including the physical infrastructure (transport, etc.), the scientific infrastructure (high-quality universities and research labs, technical institutes, etc.) and the network infrastructure (IT, telecoms). Policies aiming at solving infrastructure provision

8 A second condition that has to be fulfilled for public intervention to be pursued is that the public actors have to have, or be able to acquire, the ability to solve or mitigate the problems (Edquist 2001; Chaminade & Edquist 2006).

9 Bach and Matt (2005) distinguish between four types of additionally: outputs, inputs, behavioural and cognitive and indicate that the question to be asked is whether the policy action is changing the outputs, the inputs, the behaviour of firms or the cognitive capacity of the different agents.

problems could focus, as VINNOVA shows, on building competence centres[10] as well as creating new ones and investing in business incubators and a seed capital programme for new companies. It can also involve the creation of a large infrastructure to support innovation in a group of small and medium-sized enterprises (SMEs) such as investing in a modern greenery facility for flower production.

- *Transition problems*: the difficulties that might arise when firms and other actors encounter technological problems or face changes in the prevailing technological paradigms that exceed their current capabilities. Firms might not be capable of foreseeing the emergence of new paradigms, radically new pervasive technologies or significant changes in the markets that require new technological solutions. As we will argue later, the transition from one prevailing paradigm to the next involves a high degree of uncertainty, which might prevent private actors from entering the new technological field or market. Technology foresight exercises might help policy makers to anticipate and prevent potential lock-in and transition problems.
- *Lock-in problems:* derived from the socio-technological inertia, which might hamper the emergence and dissemination of more efficient technologies.[11] Firms and other organizations might be locked into existing technologies (and technology systems). The strength of technology systems might hamper the development of new technologies alien to the prevailing technological system or technology regime. Lock-in problems might lead to transition problems to the extent that the excessive focus on existing technologies might prevent the firms to foresee the emergence of new technological opportunities.
- *Hard and soft institutional problems*: linked to formal rules (regulations, laws) as well as more informal and tacit ones (social and political culture for instance). The system of innovation approach pays special attention to the role of institutions in the systems. Institutions are 'sets of common habits, norms, routines, established practices, rules or laws that regulate the relations and interactions between individuals, groups and organizations' (Edquist & Johnson 1997). The institutional framework plays a very significant role in the production of innovations as well as in the adoption and dissemination of innovations. The government can play a significant role in the development of the formal rules whilst in most cases this role is marginal when the most tacit elements are to be influenced (culture, firm routines, social networks, etc.). Policies targeting, for example, the IPR system in a country might contribute to solve institutional problems such as the inadequacy of the IPR system to encourage innovation.

---

10 Competence centres are the result of the interaction between universities and companies in the field of problem-oriented research of high scientific quality (VINNOVA 2002).

11 One clear example of lock-in is fossil energy. The productive system is so dependent on fossil energy that it is preventing the expansion of new forms of energy (such as solar, aeolic, etc.).

Business support services for SMEs might also help to overcome some soft institutional barriers to innovation.

- *Network problems*: which include those derived from too weak linkages or too strong linkages (blindness to what happens outside the network) in the system of innovation. Although it is easy to understand that the system might suffer from network problems that may require some kind of government response, in practice it is very difficult to assess the adequate degree of strength of the linkages in the system. Both strong and weak linkages are reported to have advantages and disadvantages, in terms of openness and intensity of exchange (Nooteboom 2004). In this sense, policy makers might induce cooperation between the agents. For example, VINNOVA (2001) trains innovation system developers, that is, facilitators that can 'mobilise the level of commitment and resources needed to create efficient groups and processes which will produce concrete results'.
- *Capability and learning problems*: these systemic problems refer to the insufficient competences of firms (human, organisational, technological and so forth) which might limit their capacity to learn, adopt or produce new technologies over time. In other words, the system might have the right infrastructure and institutional framework, but the organisations in the system might have difficulties in accessing or creating new knowledge or in transforming knowledge into innovations. Policy makers might support the acquisition of qualified human resources or the adoption of specific managerial techniques by targeted groups of firms (for examples SMEs).
- *Unbalanced exploration-exploitation mechanisms*: the system might be capable of generating diversity but not having the mechanisms to be able to make the adequate selections or it may have very refined selection procedures but no capability to generate diversity. For example, policy makers might support the emergence of spin-off companies.
- *Complementarity problems*: the competences of the system might not complement each other or they might not be connected so the positive effects that might emerge from the combination of complementary capabilities are not fully exploited.

Policy makers should thus consider intervening when there is a systemic problem. However, this is not so easy in practice as we have argued before (Chaminade & Edquist 2006; Edquist & Chaminade 2006) since the SI approach still has important limitations (Edquist 2005). We still know very little about how the system operates in practice (Chaminade & Edquist 2006). Furthermore, the SI approach has been criticised for being static, for the lack of attention to policy consequences and for weak links between the micro and system level. From a policy perspective, the discussion on SI-based policies has not dealt profoundly with the problems of uncertainty, selectivity and path dependency. The balance of this chapter focuses on the latter issues, while subsequent chapters in this book by Bergek, Jacobsson, Hekkert and Smith; Smits, Kuhlmann and Teubal; Carlsson, Elg and Jacobsson deal with the issues of the dynamic character of innovation, the policy consequences and the links between the micro and the macro level.

UNSOLVED QUESTIONS: ADDITIONAL ISSUES FOR CONSIDERATION UNDER THE SYSTEM OF INNOVATION PERSPECTIVE

Policy Mistakes (and Policy Learning)

The role of the policy maker under the neoclassical theory was to help the market reach equilibrium (again). In the words of Norgren and Hauknes (1999) we could talk about the optimising policy maker. However, the role of the policy maker under the SI perspective is one of adaptation. Policy makers need to adapt their policies to the identified systemic problems in systems of innovation – and these may change over time. We move from the optimising policy maker to the adaptive policy maker. This means the acceptance of mistakes in policy making but also points to the importance of evaluation of policies and policy learning.[12]

Under the SI approach, the formulation of policies is based on existing theory (limited), indicators (limited) and subjective judgements (common sense). It is difficult for the policy makers to know *ex ante* how the system will react to the policy. In a sense, as another building block of the system of innovation, policy is affected by the same systemic problems that were mentioned before. Particularly they are affected by cognitive and learning failures, that is, they might not have the ability to intervene or simply do not know how to do it (Bach & Matt 2005). Policy makers need to experiment and allow some room for mistakes. For this reason, evaluation of policies is very important.

In other words, it is not possible to know for sure – *ex ante* – if public intervention can solve the systemic problem or not. The decision to intervene or not must thus be based upon whether it is likely or not that intervention mitigates the problem. Hence, the decision must be taken in a situation of uncertainty. Then one can afterwards – *ex post* – determine through evaluations whether the problem was solved or mitigated. If this was not the case, we are talking about a policy mistake. Policy mistakes can never be completely avoided because of the uncertainty mentioned. We must accept mistakes in public activity – as we do in private activities. Moreover, in order to be able to evaluate the success or failure of a specific policy, it is necessary to formulate clearly the objectives of the policy *ex ante*.

There may be various reasons why public intervention cannot solve or mitigate a problem. One is that it is not at all possible to solve the problem from a political

12 While the role of learning in a system of innovation has been extensively discussed by, among others, Jensen et al. (2004), Lundvall (1992; 1996), Lundvall and Borras (1999), Archibugi and Lundvall (2000), learning processes in SI-based policy are still very much under-researched.

level, that is, that the problem is out of reach for policy makers. It is an objective that might not be attained by any of the instruments available to the policy makers. Then all types of intervention would be in vain and result in a policy mistake.[13] The other reason is that the state might first need to develop its ability to solve the problem. A detailed analysis of the problems and their causes may be a necessary means of acquiring this ability.[14] The creation of new organisations and institutions to carry out the intervention might also be necessary. A particular body of knowledge may not be represented in the national portfolio and require the establishment of a new research organisation or, a new policy instrument. Finally, the theoretical framework might be insufficient to understand the problems in the system.

It is important to note that a 'systemic problem' that motivates public intervention might concern the future. A 'problem' might be something that has not yet emerged. A 'problem-solving' policy of this kind might alternatively be called an 'opportunity creating' or anticipatory policy.[15] One of the problems to be solved might be that uncertainty prevents new technologies from emerging. One example of such a problem is the case where public funding of basic R&D might be necessary because capitalist actors do not have the incentive to fund it (e.g., because of inappropriability). Another example could be that training people in a certain field could create new opportunities that would not be realised without policy.

The discussion on policy intervention raises two important issues. First, policy intervention is specially needed when uncertainty and risk are very high and private actors do not find incentives to invest in those high risk products or new activities. Second, policy intervention needs to be selective, focusing on specific products, activities or technologies that better fulfil the (economic, social, environmental, etc.) objectives of the government. The issues of uncertainty and selectivity will be discussed next.

Uncertainty

In the previous section we discussed the problems of the uncertainty related to the policy outcomes of an SI-based policy. In this section we discuss another type of uncertainty: the uncertainty linked to innovation processes, and the implications for the design of innovation policies. In other words, we argue that a high degree of uncertainty in the innovation process may be an important cause behind a systemic problem. This would then constitute a reason for public intervention.

Firms and markets perform less efficiently in a context where uncertainty and risk are large. Sometimes they do not perform at all with regard to them. As

13 Hence, the problem is not solvable by private actors or by public intervention.

14 For example, it might be necessary to carry out a detailed comparative empirical analysis.

15 There might be reasons to treat the solving of existing problems and the creation of future opportunities as two different kinds of situations calling for (different kinds of) public intervention.

mentioned earlier, one of the 'problems' to be solved might be that uncertainty prevents new technologies from emerging. For example, public funding of basic R&D might be necessary because private actors do not have the incentive to fund it or they might underinvest (Arrow 1962), or they might even not invest in areas of great social return but low individual return (e.g. some pharmaceuticals). Firms may also assume that educating and training shall be carried out by public organisations, and therefore they might not invest in human capital formation. In the very early stages of the development of new fields of innovation, there is uncertainty whether markets exists or not, and sometimes public organisations have been instrumental in market creation (directly, e.g. through public technology procurement, or indirectly, e.g. through regulation (creating or changing institutions)). Incubating activities have been carried out in public or semi-public science parks to facilitate commercialisation of knowledge in recent decades.

We have just mentioned four examples of activities (R&D, education and training, market creation, incubating)[16] where there are reasons for policy makers to intervene because uncertainty and risk are large and, therefore, private organisations weak or unwilling to act. As a matter of fact, innovation as such is plagued by uncertainty. Such uncertainty is often largest with regard to innovation in new fields of production. Although the discussion of uncertainty tends to be dominated by examples of new technology fields (where the uncertainty problem is more acute), uncertainty also affects mature sectors or small and medium firms. SMEs might lack resources (human capital, financial capital) and the information about innovation opportunities thus increasing their uncertainty about the innovation process. Policy makers might want to stimulate innovation in the system by addressing these systemic problems, providing incentives to innovation through capacity building and information provision.

Historically we have also seen that a minor public intervention in an early stage of the innovation process may have a very large impact.[17] Risks and uncertainty are largest with regard to the emergence of new products. This indicates that 'systemic problems' are more common in such a context (but not exclusively as indicated before). It also seems to be an empirical fact that large-scale and radical technological shifts rarely take place without public intervention (as opposed to incremental innovations in established sectors). This has been indicated in Carlsson and Jacobsson (1997), where they go through the cases of electronics, semiconductors and genetic engineering in the USA and Sweden. David Mowery has also clearly shown that publicly funded R&D in combination with public technology procurement has played a crucial role for the development of new high technology sectoral systems of innovation in the USA (and thereby in the world). Some examples are the early

16 We systematically discuss ten such activities in a policy perspective in Chaminade and Edquist (2006) and in Edquist and Chaminade (2006).

17 The public support of the development of the NMT 450 mobile telecom standard in the Nordic countries is an example. The support amounted to a few hundred man-years.

phases of the development of numerically controlled machine tools, commercial aircraft, semiconductors, computer hardware, computer software, and the Internet in the USA. Hence, the innovation policy support to new products and sectors has been very strong in the USA. The objectives of this public innovation polices have often been of a military character (Mowery 2005).

Smits and Kuhlmann (2004) suggest the use of what they call systemic instruments to cope what we call systemic problems. Systemic instruments do not focus on elements of the system but try to take the whole system, or at least part of the system into account. Two of these instruments are in particular associated with the uncertainty of any innovation process: supplying the information that the different actors need to define their innovation strategies (Strategic Intelligence, see Chapter 16 by Smits, van Merkerk et al. in this book) and providing the actors with the instruments, facilities and environments for experimenting and learning. This includes the provision of markets for new products (public technology procurement), the provision of incentives for research in certain priority areas (R&D incentives) or the provision of information, business services or capacity building to SMEs.

In sum, innovation involves a high degree of uncertainty and risk. The higher the uncertainty, the lower incentives private actors have to invest in innovations – and public intervention will be motivated.

Selectivity

We have argued that innovation policies should be focused upon solving or mitigating certain 'systemic problems'. This means that the policies cannot be neutral but are necessarily selective. When designing innovation policy, policy makers have to select not only the objectives of the policy (why to intervene) and the instruments (how to intervene), but also which problems to address by the intervention. Ideally, policies should be designed on the basis of a thorough analysis of the system, i.e. of how the system is operating and what the systemic problems to be addressed are. Such analyses should be based on systematic comparisons between existing systems of innovation (since no optimal system of innovation can be identified) (Edquist & Chaminade 2006).

However, in practice the final policy is often the result of not only the analyses of the system, but often also of ideology, of the imitation of policy 'models' from other systems or of the influence from pressure groups (lobbyism) (Edquist & Chaminade 2006). Lobbyists are special interests groups. They seldom find general subsidies or general support worth pressing for. Instead they often push for sector- and firm-specific public support, i.e. they enhance selective policies. Since lobbyists normally represent established interests and industries, they normally argue for policies supporting these established industries (for example, ship-yards, automotive and ICT) – and they are sometimes successful in achieving such public support.

One example indicating that innovation policy is generally selective is public investments in R&D. Analysis may reach the conclusion that *x* billion Euros shall be reallocated to research of relevance for the biomedical industry (from somewhere else). This is automatically a selective policy, since it favours the biomedical sectors of production and, more specifically, the products and firms active in this industry – at the expense of others. The analysis of the system might also show that pharmaceutical companies are underinvesting in R&D related to some drugs that might have an extraordinary social impact (e.g. a drug to cure malaria) but that are rejected by private companies in favour of more profitable drugs (e.g. Viagra). The government might decide to allocate funds to research in those socially needed drugs.

In both the cases of innovation policy formed by lobbyism (which is common) and strictly based on the comparative analyses of systems of innovation, the resulting policy is selective rather than neutral. However, in the two cases we might expect that policies are selective in different directions. For example, the identification of new sectors where uncertainty is large will not come as a result of lobbyism (as there is not yet a critical mass of lobby groups) but from the analysis with the purpose of identifying systemic problems.[18]

As Norgren and Hauknes (1999) argue, one of the basic choices that policy makers need to make is between strengthening existing systems and facilitating the creation of new systems. In general, we argue that government support is most needed when uncertainty and risk are high and there is a risk that the private sector will not act – thus neglecting opportunities for change and renewal. Such support to new activities and products can encourage the emergence of brand new sectors as well as support to the transformation of more traditional sectors by supporting their adoption of new products and processes. The focus should be on supporting new products (goods as well as services) and new processes (technological as well as organisational), and they can be in new or existing sectors. Furthermore, it should be noted that support might be needed for the emergence of a new product or process but also for the dissemination of it, for example across sectors. This is especially relevant for generic technologies whose wider dissemination might increase the number of applications through complementary innovations.

The issue of selectivity needs to be further discussed in relation to different kinds of systems of innovation: sectoral, regional and national. Policies targeting sectoral systems of innovation tend to aim at promoting specific sectors that are considered crucial for growth (or some other policy objective). Whether policy makers should select new and emerging sectors or support new activities and products in existing sectors is still the subject of hot debate among academics and policy makers. In reality, what most governments have done is to choose a combination of emerging new sectors (bio-tech, ICT, nanotechnology) with other sectors deeply rooted in the country's economic structure (transport or materials).

18 There are reasons to limit the degree of selectivity in innovation policy – and maybe accept it only for sectors and products in very early stages of their development. However, this must be analysed in more depth in future work. Then a distinction could be made between selectivity with regard to sectors, products, activities and firms respectively.

Policies aiming at regional innovation systems are basically concerned with the growth of a specific region and its integration in international markets by means of mobilising 'all relevant players involved in the process of becoming internationally competitive within specific areas of growth' (VINNOVA 2001). Government intervenes to 'create' or develop the regional system of innovation, by facilitating interaction between the different actors and the development of a common growth strategy.

Finally, policies aiming at national innovation systems often aim at generating national competences for learning and growth (competence building) and are concerned, for example, with the development of a skilled workforce, a strong research capacity, etc.

Inertia or Path Dependency in Policy Making

It is important to note that a systemic innovation policy brings together a variety of policies that have traditionally been separated (education policy, industrial policy, etc.). In this sense, innovation policy can be seen as a (policy) system itself, integrating traditionally individual and independent policies into a new systemic policy with new rationales, new (systemic) instruments and new governance bodies. Adopting the SI approach implies the adoption of new rationales that might collide with former rationales. In other words, policy makers might adopt the system of innovation approach in their discourse while still using 'market failure' arguments for allocating resources for innovation.

One way to overcome this path-dependency is to create new government structures or organizations responsible for the design, implementation and evaluation of innovation policies that explicitly adopt a system of innovation approach for policy-making. One example of this is the case of VINNOVA in Sweden (see Chapter 7 by Carlsson et al. in this book).

CONCLUSIONS

In this chapter we have tried to discuss rationales for public intervention based on neoclassical theory and on the systems of innovation approach. We have argued that governments should intervene when a systemic problem that is not spontaneously solved by private actors exists, i.e. when private actors do not achieve the objectives. Furthermore, the public agencies must have the ability to solve or mitigate the problem.

We have argued that innovation policy is normally, and should be, selective. The crucial question is in which direction the policy is selective and whether the choice of direction is based on lobbyism or on a rigorous analysis as a basis for designing the policy. We have argued that the selection should also be made on the basis of a rigorous analysis of the system of innovation – which has to be empirical and comparative between existing systems. We have also argued in favour of prioritising those areas where there is a greater degree of uncertainty and risk or where the collective returns might be very high (for example in environmental or social terms). This is an important issue for further analysis. Such an analysis of directions of selectivity could preferably make a distinction between selectivity with regard to sectors, products, activities and firms respectively.

In all policy making one has to accept that mistakes are being made – just like in private activities. Policy makers are embedded in the system of innovation and, as such, they are subject to systemic problems, including the lack of capacity to identify a problem or the lack of information or knowledge on how to solve it. In this sense, policy learning becomes a fundamental element in innovation policy and the continued interaction between academia and practice, in this sense between researchers and policy makers, is not only recommendable but necessary.

More research is needed in collaboration with policy makers. It seems clear that the discussion on the rationales of policy intervention is rather theoretical. It is important to acknowledge that policy making is not always a rational process (rationales are often sought after the decision has been taken). The rationale emerges as a result of an *ex post* analysis and not as an *a priori* exercise (Elg 2006)

Furthermore, discussion until now has been at a rather general level. More research is needed on the impact of the adoption of the system of innovation approach in relation to specific policies or instruments, such as IPRs, the support of university-industry relationships, etc. That is, we need to zoom in and discuss how these institutions and relationships of the system should be dealt with when policy makers adopt a SI approach. More insight into the dynamics of the system are a necessary precondition for this (see Chapter 6 by Bergek et al.). In this context also the development of instruments that take (parts of) the system into account, instead of instruments that focus on elements of or bilateral relations in the system (as is is the case now), is necessary (see Chapter 17 by Smits et al.).

REFERENCES

Aghion, P. and Howitt, P. (1992), 'A model of growth through creative destruction', *Econometrica*, 60 (2), 323-351.

Archibugi, D. and Lundvall, B.-Å. (eds) (2000), *The Globalising Learning Economy: Major Socio-Economic Trends and European Innovation Policy*, Oxford, UK and New York, USA: Oxford University Press.

Arrow, K. (1962), 'Economic welfare and the allocation of resources for invention', in R. Nelson (ed.), *The Rate and Direction of Inventive Activity*, Princeton: Princeton University Press, pp. 609-629.

Bach, L. and Matt, M. (2005), 'From economic foundations to S&T policy tools: a comparative analysis of the dominant paradigms', in M. Matt and P. Llerena, (eds), *Innovation Policy in a Knowledge-Based Economy: Theories and Practices*,

Berlin, Germany, Heidelberg, Germany and New York, USA: Springer-Verlag, pp. 17-45.

Carlsson, B. and Jacobsson, S. (1997), 'Diversity creation and technological systems: a technology policy perspective', in C. Edquist (ed.), *Systems of Innovation: Technologies, Institutions and Organizations*, London, UK and New York, USA: Pinter, pp. 318-341.

Chaminade, C. and Edquist, C. (2006), 'From theory to practice. The use of the systems of innovation approach in innovation policy', in J. Hage and M. De Meeus (eds), *Innovation, Learning and Institutions*, Oxford, UK and New York, USA: Oxford University Press, pp. 141-160.

Dasgupta, P. (1987), 'The economic theory of technology policy: an introduction', in P. Stoneman and P. Dasgupta (eds), *Economic Policy and Technological Performance*, Cambridge, UK: Cambridge University Press, pp. 7-23.

Edquist, C. (2005), 'Systems of innovation: perspectives and challenges', in J. Fagerberg (ed.), *The Oxford Handbook of Innovation*, Oxford, UK and New York, USA: Oxford University Press, pp. 181-208.

Edquist, C. (2001), 'Innovation policy: a systemic approach', in B.-Å. Lundvall and D. Archibugi (eds), *Major Socio-Economic Trends and European Innovation Policy*, Oxford, UK and New York, USA: Oxford University Press, pp. 222-237.

Edquist, C. (1997), *Systems of Innovation: Technologies, Institutions and Organizations*, London, UK and New York, USA: Pinter.

Edquist, C. (1994), 'Technology policy: the interaction between governments and markets', in G. Aichholzer and G. Shienstock (eds), *Technology Policy – Towards an Integration of Social and Ecological Concerns*, Berlin, Germany and New York, USA: Walter de Gruyter, pp. 67-93.

Edquist, C. and Chaminade, C. (2006), 'Industrial policy from a systems of innovation perspective', *European Investment Bank (EIB) Papers*, 11 (1), 108-133.

Edquist, C. and Johnson, B. (1997), 'Institutions and organizations in systems of innovation', in C. Edquist (ed.), *Systems of Innovation: Technologies, Institutions and Organisations*, London, UK and New York, USA: Pinter, pp. 41-63.

Edquist, C., Malerba, F., Metcalfe, J., Montobbio, F. and Steinmueller, E. (2004), 'Sectoral systems: implications for European innovation policy', in F. Malerba (ed.), *Sectoral Systems of Innovation: Concepts, Issues and Analysis of Six Major Sectors in Europe*, Cambridge, UK: Cambridge University Press, pp. 427-464.

Freeman, C. (1987), *Technology Policy and Economic Performance: Lessons from Japan*, London, UK and New York. USA: Pinter.

Elg, L. (2006), Personal e-mail communication, 11 August 2006.

Godin, B. (2004), 'The New Economy: what the concept owes to the OECD', *Research Policy*, 33 (5), 679-690.

Granovetter, M. (1985), 'Economic action and social structure', *American Journal of Sociology*, 91 (3), 481-510, reprinted in M. Granovetter and R. Swedberg (eds), *The Sociology of Economic Life*, Oxford: Westview Press.

Grossman, G. and Helpman, E. (1991), *Innovation and Growth in the Global Economy*, Cambridge, MA, USA and London, UK: MIT Press.

Jensen, M., Johnson, B., Lorenz, E. and Lundvall, B.-Å. (2004), 'Forms of knowledge and modes of innovation', working paper, Institut for Erhvervsstudier, Aalborg Universitet, Aalborg.

Kohler-Koch, B. (ed.) (2003), *Linking EU and National Governance*, Oxford, UK and New York, USA: Oxford University Press.

Lipsey, R. and Carlaw, K. (1998), 'A structural assessment of technology policies: taking Schumpeter seriously on policy', Industry Canada Research Publications Program, 123, Ottawa: Industry Canada.

Lundvall, B.-Å. (1996), 'The social dimension of the learning economy', Druid Electronic working papers.

Lundvall, B.-Å. (ed.) (1992), *National Systems of Innovation: Towards a Theory of Innovation and Interactive Learning*, London, UK and New York, USA: Pinter.

Lundvall, B.-Å. and Borrás, S. (1999), *The Globalising Learning Economy: Implications for Innovation Policy*, Luxembourg: Office for Official Publications of the European Communities.

Mowery, D. (2005), 'National security and national innovation systems', mimeo, paper presented at the PRIME/PREST workshop on re-evaluating the role of defence and security R&D in the innovation system, University of Manchester, Manchester.

Mytelka, L. and Smith, K. (2002), 'Policy learning and innovation theory: an interactive and co-evolving process', *Research Policy*, 31 (8-9), 1467-1479.

Nelson, R. (ed.) (1993), *National Innovation Systems: A Comparative Analysis*, Oxford, UK and New York, USA: Oxford University Press.

Nelson, R. (2004), 'The market economy and the scientific commons', *Research Policy*, 33 (3) 455-471.

Nelson, R. (1959), 'The simple economics of basic scientific research', in N. Rosenberg (ed.), *The Economics of Technological Change*, London, UK and New York: Penguin, pp. 478-486.

Nelson, R. and Winter, S. (1982), *An Evolutionary Theory of Economic Change*, Cambridge, MA, USA: Harvard University Press.

North, D. (1990), *Institutions, Institutional Change and Economic Performance*, Cambridge, UK: Cambridge University Press.

Nooteboom, B. (2004), *Interfirm Collaboration, Learning and Networks: an Integrative Approach*, London, UK and New York, USA: Routledge.

Norgren, L. and Hauknes, J. (1999), 'Economic rationales of government involvement in innovation and the supply of innovation related services', RISE project report, Wp3.http://www-centrim.bus.bton.ac.uk/open/we/do/proj/rise/risewp/wp3synth.pdf

OECD (2002), *Dynamising National Innovation Systems*, Paris: OECD Publishing.

OECD (2001), *Innovative Clusters: Drivers of National Innovation Systems*, Paris: OECD Publishing.

OECD (1999), *Boosting Innovation: The Cluster Approach*, Paris: OECD Publishing.

OECD (1998), *Technology, Productivity and Job Creation*, Paris: OECD Publishing.

OECD (1996), 'National innovation systems: report on pilot case studies', mimeo, Room Document, OECD.

OECD (1995a), 'National innovation systems: where are we and where do we need to go?', mimeo, Room Document, OECD.

OECD (1995b), 'National innovation systems', mimeo, Work plan for pilot case studies, Room Document, OECD.

Romer, P. (1990), 'Endogenous technical change', *Journal of Political Economy*, **98** (5), 71-102.

Romer, P. (1986), 'Increasing returns and long run growth', *Journal of Political Economy*, **94** (5), 1001-1037.

Rosenberg, N. (1994), *Exploring the Black Box: Technology, Economics and History*, Cambridge, UK: Cambridge University Press.

Rosenberg, N. (1982), *Inside the Black Box*, Cambridge, UK: Cambridge University Press.

Sharif, N. (2006), 'Emergence and development of the national innovation systems concept', *Research Policy*, 35 (5), 745.

Smits, R. and Kuhlmann, S. (2004), 'The rise of systemic instruments in innovation policy', *International Journal Foresight and Innovation Policy*, 1 (1-2), 4-32.

Smith, K. (2000), 'Innovation as a systemic phenomenon: rethinking the role of policy', *Enterprise and Innovation Management Studies*, 1 (1), 73-102.

Stoneman, P. and Dasgupta, P. (1987), *Economic Policy and Technological Performance*, Cambridge, UK: Cambridge University Press.

Verspagen, B. (2005), 'Innovation and economic growth', in J. Fagerberg, D. Mowery and R. Nelson (eds), *The Oxford Handbook of Innovation*, Oxford, UK and New York, USA: Oxford University Press, pp. 487-513.

VINNOVA (2002), 'Effective innovation systems and problem-oriented research for sustainable growth', VINNOVA Strategic Plan 2003-2007. VINNOVA. Available on www.vinnova.se, March 2006, Stockholm.

VINNOVA (2001), 'Regional growth through dynamic innovation systems', VINNOVA, www.vinnova.se, March 2006.

Woolthuis, R.K., Lankhuizen, M., et al. (2005), 'A system failure framework for innovation policy design', *Technovation*, **25** (6), 609-619.

6. Functionality of Innovation Systems as a Rationale for and Guide to Innovation Policy

Anna Bergek, Staffan Jacobsson, Marko Hekkert and Keith Smith

INTRODUCTION

This chapter discusses an approach to the rationale and operations of innovation policy from a 'systems' perspective, focusing on functions performed by an innovation system. The systems approach stems from a key insight of innovation studies, which is that innovation by firms cannot be understood purely in terms of independent decision-making at the level of the firm. Firms' strategies are central to innovation, but strategic options are shaped and constrained by environmental factors such as collaborative patterns, regulatory systems and customary practices which persist in systemic ways and which influence how innovation may occur. These environmental conditions are often specific to technological, regional or national contexts, but they are also dynamic: their forms of operation change with political conditions, changing technological opportunities, economic integration processes and so on.

The basic argument of systems theories is that system conditions have a decisive impact on the extent to which firms make innovation decisions, on the modes of innovation which are undertaken and on the success or failure of these. Much of the literature focuses on differences in the structure of systems and trans-national variations including across institutions and organizations. However, from an innovation point of view, what also matters is functionality: that is, what do the components of an innovation system actually do, and what do they achieve? We argue that innovation processes involve various functions that need to be fulfilled; they must be initiated (usually by some perception of an opportunity), they must be financed, and they must deploy capabilities that in turn have been created and maintained. Any evaluation of system performance should therefore rest in part on an account of what a system actually does in terms of such functions.

A principal aim of this chapter is to formulate a taxonomy of system functions, as a step towards an understanding of how systems affect innovation. But we go further, suggesting that the functions of a system can form a framework for understanding the foundations and shaping the content of policies in support of innovation. The objective of this chapter, therefore, is to develop a framework that adds a functional level to the structural one, which may provide a tool for policy makers seeking a rationale for – as well as a guide to – policy interventions. Illustrative empirical examples will be drawn from analyses of technological innovation systems (TIS) – systems centred on specific technologies, products or industries, such as microwave technology, solar cells or salmon farming (see Carlsson 1997).[1] The approach does not, of course, have anything to say about the desirability of a particular TIS but assumes that such a decision has already been taken by policy makers.

The chapter is structured as follows. In the first section, we discuss rationales for public policy, and how they are affected by the evolutionary dynamics of innovation systems, following from the chapter by Chaminade and Edquist earlier in this volume. The second section outlines a set of functions in innovation systems that act as an intermediate between the structure of the system and its ultimate performance. We exemplify with the case of 'biomass gasification' in the Netherlands. In section three, we proceed to analyse how a functional analysis can be used as a very practical tool to find not only a (partial) rationale for policy intervention, in terms of system weaknesses, but also a guide for finding the key policy issues to tackle. We illustrate here with the case of IT (information technology) in home care in Sweden. This section also further develops a rationale for intervention by assessing the nature of system weaknesses in relation to the opportunities for system-building activities by entrepreneurs. The fourth section contains our conclusions.

POLICY RATIONALES: BEYOND MARKET FAILURES

For a long time, the rationale for and scope of policy intervention in industry has been highly controversial. Yet it is clear that in industrially advanced countries policy intervention is widespread and part of larger transformation processes.[2] Contemporary policy intervention, however, rarely consists of 'orchestration' of overall development by governments, nor 'industrial policy' as conventionally understood (consisting of subsidies to specific industries to promote growth or offset decline). Rather, modern policy incorporates a wide range of industrially-relevant disconnected measures including those found within science and technology policies, tax policies (including CO_2 and other pollutant taxes, but also tax credits to R&D), venture capital provision, standardization measures, formation of early markets via e.g. procurement policies,

1 This is so for two reasons. First, policy makers often have a particular product group/knowledge field as their responsibility. Second, at a higher system level, the framework probably loses some of its power to guide policy makers.

2 See, for instance, Computer Science and Telecommunications Board (1999) for an exciting analysis of the US case and Jacobsson and Bergek (2004) for the case of renewable energy technology in Germany.

and measures to broaden the search space of firms.[3] In contrast to subsidies for dying industries, such measures are, on the whole, deemed to be legitimate. In a developing country context, it is also 'increasingly recognized that developing societies need to embed private initiative in a framework of public action that encourages restructuring, diversification, and technological dynamisms beyond what market forces on their own would generate' (Rodrik 2004). In such a framework, it has not only been argued that the appropriate public action needs to be selective, i.e. differ across industrial fields (Katz, 1983; Carlsson & Jacobsson, 1996; Rodrik 2004), but also that it needs to change its content over time (Galli & Teubal 1997; Katz 2004).

Against the background of this complex array of measures, there is a need for a coherent conceptual approach that provides policy makers with a rationale for intervention as well as a guide to the central issues that a selective and time-specific policy should aim at resolving. Scholars on innovation and technology have almost completely rejected the market failure approach as a basis of policy action. It is argued repeatedly in the literature (e.g. Metcalfe 1992; 2004; Malerba 1996; Carlsson & Jacobsson 1996; 1997; Smith 2000b) and by Chaminade and Edquist in this book, that the approach is flawed and insufficient. A systems approach to innovation is seen as more useful. Systems approaches have four basic conceptual underpinnings:

- Technologies should be seen as systemic at the technical level – that they consist of complex systems, and that the creation of complementary technologies, such as in electricity production and distribution, should be seen as a process of system building. The nature of the artifacts may thus matter, which is not recognized as a problem in the market failure approach.
- Economic behaviour rests on institutional foundations, in the sense of legally or customarily established 'rules of the game', which limit and constrain choices but at the same time reduce uncertainty, which is important for investments. Different modes of institutional set-up lead to differences in economic behaviour and outcomes. The market failure approach neglects the importance of institutions.
- Technological (and other) knowledge is generated by interactive learning, and technological knowledge in general takes the form of 'distributed' knowledge bases among different types of economic agents who must interact in some way if technological knowledge is to be applied. The market failure approach, however, assumes strong independence between firms (and indeed all economic agents), and hence clear boundaries between firms. In contrast, the study of innovation constantly presents phenomena of interdependence and blurred boundaries between actors.
- Competitive advantage results from variety and specialization, and that this has path-dependence-inducing effects characterized by increasing returns. That is,

3 A good overview of the range and complexity of relevant measures and instruments can be found on TrendChart, the European Commission's monitoring project on innovation policy: www.trendchart.cordis.lu

successful specializations are self-replicating, with system-building as an outcome – particularly around specific industrial structures. Increasing returns are, thus, an inherent feature of system formation and not a 'failure' – indeed, without increasing returns, systems will fail to develop.

These conceptual underpinnings have serious effects on the rationale for policy intervention.

A central proposition in the systems literature is that just as the nature of actors/markets may block or obstruct the formation of a TIS, so can institutions and networks (e.g. Malerba 1996; Carlsson & Jacobsson 1997; Edquist 1999; Jacobsson & Johnson 2000; Unruh 2000; Rotmans et al. 2001). Additionally, some scholars (e.g. Hughes 1983; Sandén & Jonasson 2006) point to the exclusion effects of specific artifacts (particularly in the energy field). The concept of 'system failure' (or better, system weaknesses) emerged to capture this broader rationale for policy intervention (Smith 2000b).[4] Carlsson & Jacobsson (1997), for instance, argued that: firms, institutions and networks become 'locked in' to the 'old' technologies. Thus, the cumulativeness and path dependency of innovation lead to risks of lock-in into technological, institutional and networking cul-de-sacs.

Here, all the four underpinnings are present and, combined, these may lead to a 'failure' at the level of the entire system, i.e. a system fails to develop or does so in a stunted way only.[5] Edquist (1999) suggested, similarly, that 'system failures' can be ascribed to characteristics of organizations, institutions or interactions (networks), i.e. in the nature of the components of a system. Metcalfe (2004) argued that: 'government' takes responsibility for the ecology of organizations and institutions that facilitate business experimentation but recognizes that without the necessary interconnections the ecology is not a system – because systems are defined by components interacting within boundaries, it follows that a system failure policy seeks to address missing components or missing connections.

It follows that policy makers may find a rationale for intervention in a specific innovation system in the form of system weaknesses in structure.[6] Yet, the mere existence of system weaknesses is not necessarily a rationale for policy intervention since private actors may have the scope for repairing problems independently of government. As Smits and Kuhlmann (2004) put it: '[O]ne should not overestimate the instrumental power of public policy vis-à-vis other actors in complex policy-making arenas.' Individual firms, groups of entrepreneurs, industry association and other interest organization may very well identify and act upon system weaknesses in their own self-interest. For example, referring to the work of van de Ven (2005), Aldrich and Fiol (1994) emphasize the possibility for actors to jointly influence the institutional framework:

4 See Woolthuis et al. (2005) for a recent overview of the system failure literature.

5 This is also emphasized in Smith (1996).

6 Chaminade and Edquist in this book use the term 'systemic problems'.

> [A] small handful of actors can generate networks that, in the aggregate, result in institutional-legitimating events. If founders can overcome the barriers to effective collective actions, they can rise above the level of their individual ventures and run together 'in packs'. (Aldrich & Fiol 1994; cf. also van de Ven 2005)

However, such system-building activities may be in conflict with the interests of individual actors. For example, although variety creation may be necessary on a system level, individual firms will usually have to focus on developing one (or few) technological alternative(s) at a time and cannot be expected to support the development of competing substitutes. In addition, first-mover system builders may become victims of the 'burnout of pioneers' phenomenon (Olleros 1986), where the benefits of their system-level investments in an early phase are reaped by follower firms while they themselves perish. Indeed, 'there is an ongoing tension for each industry participant to organize its own proprietary function and distribution channels as opposed to contributing to the creation of the industry's resources and institutional arrangements' (van de Ven 1993). There are, therefore, strong reasons to expect that public policy makers may need to make interventions in order to secure developments on a system level.

Focusing on system weaknesses means that we have left the old and sterile debate over the ability of the state to 'pick winners' in the form of individual firms (Carlsson & Jacobsson 1996). As Stewart and Ghani (1991) rightly pointed out many years ago, the systems view on the innovation process makes us instead focus on the conditions and processes whereby winners are created. However, even when the rationale for public policy intervention is clear, the question remains how to identify system weaknesses that should be tackled by public policy. Some have tried to address this policy problem by specifying a set of policy issues pertaining to possible 'generic' system weaknesses that may be particularly important to tackle in the process of emergence of a new system (Carlsson & Jacobsson 1997; Jacobsson & Johnson 2000; Jacobsson & Bergek 2004), and require the application of systemic instruments, as distinct from those which primarily focus on individual organizations or bilateral relations (Smits & Kuhlmann 2004, see also their chapter in this book as well as Chapter 17 by Smits, Kuhlmann and Teubal). The safeguarding of variety is an example of a key generic policy issue in the face of uncertainty.[7] The formation of 'prime movers' is vital in an early phase, the formation of new networks may be required to enable an alignment of actors' expectations and coordination of their

7 See also Malerba (1996) for a useful discussion on variety.

investment,[8] and articulation of demand[9] is required to form markets and induce firms to enter.[10]

Potentially 'generic' policy issues are, however, of little guidance for policy makers dealing with a specific TIS. As Rodrik (2004) argues, policy has to focus on specific activities (e.g. a new technology, a particular kind of training, a new good or service) rather than on a sector *per se* and should be thought of as 'a process designed to elicit areas where policy actions are most likely to make a difference'.[11]

As noted above, these areas differ between TIS and change over time and, therefore, require the application of non-uniform and often a wide range of policies. Indeed, Katz (2004) concludes from a study of the successful evolution of salmon farming in Chile that 'it is the diversity of roles the State has played affecting industry's behaviour that strikes as the major lesson'. The relevant issue then is how policy makers can identify those activities/areas that are of critical importance to the dynamics of a specific TIS.

In the received literature on 'policy and innovation systems', we find little guidance on how to do so – it stops at outlining general policy challenges. In the few cases where more concrete advice is given, policy makers and researchers are recommended to compare different innovation systems with each other: 'In this rationale to government involvement the recommended methods of identifying failures are the use of comparative analysis, benchmarking and best practice' (Norgren & Hauknes 2000). However, very little guidance is given on which dimensions to compare. In the following section, we suggest that a functional analysis of a TIS constitutes a framework that allows us to identify system weaknesses and to develop (more systemic) policies to tackle these weaknesses in an emerging TIS.

SYSTEM FUNCTIONS AS KEY PROCESSES IN THE DYNAMICS OF INNOVATION SYSTEMS

In order to be able to find a rationale for policy intervention and to identify the central policy issues in a specific innovation system, we need to develop an

---

8 Carlsson and Jacobsson (1993) discussed in terms of the role of networks in 'blending visions' or technological expectations.

9 This point is emphasized in Frenken et al. (2004). See also Malerba (1996) and Edquist (1999).

10 Others pointed to 'general' policy issues coming out of the application of an innovation system approach. Smith (1996) pointed to four (infrastructural provisions, 'transition failures', lock-in failures and institutional failures). Edquist (1999) mentions the last two and adds problems associated with interactive learning (networks) and demand side issues. Some of these policy issues are found in Chapter 5 by Chaminade and Edquist in this book.

11 This does not imply that more generic systemic policy challenges are useless or that they can not be found at the level of national innovation systems.

understanding of the key processes in system evolution. A focus on key processes, rather than states, implies, of course, that our approach is about the dynamics of innovation systems (see chapters in this volume by Kuhlmann et al. and Smits, Kuhlmann and Teubal for a critique of the literature on innovation systems being static). The key processes can be identified at two levels. The first is structural and refers to the three components in a TIS: actors, networks and institutions. Dynamics can be analysed in terms of how actors enter into a TIS, networks are formed and institutions are changed (or not). The second level refers to 'what is achieved in the system' in terms of processes that have a more direct and immediate impact on the 'goal' of the system, which could be stated as to generate, diffuse and utilize new technology (Carlsson & Stankiewicz 1991). These processes are here labelled 'functions'. We propose that if an innovation system is to successfully evolve, it has to perform well in seven dimensions. We derive these functions from contributions from the literature on innovation systems and evolutionary economics but also from political science, sociology, organization theory and related fields. Earlier versions of the framework have been applied to a number of TIS.[12] These functions are:

1. Knowledge development and diffusion
2. Influence on the direction of search and the identification of opportunities
3. Entrepreneurial experimentation and management of risk and uncertainty
4. Market formation
5. Resource mobilization
6. Legitimation
7. Development of positive externalities

In what follows, we outline the content of these seven functions.[13] A brief illustrative case of biomass gasification in the Netherlands ends the section.

Knowledge Development and Diffusion

This is the function that is normally placed at the heart of an innovation system in that it is concerned with the knowledge base of the TIS (globally) and its evolution. The function captures the breadth and depth of the (scientific and technical) knowledge base of the TIS and how that knowledge is diffused and combined in the system. It is, thus, closely related to the concept of 'learning', which is at the core of the innovation system approach.

Learning and knowledge creation happen across innovation systems at different levels. Firstly, there is the level of the firm, where highly specific forms of expertise and knowledge are created, relevant to the products and process around which the

12 See Bergek and Jacobsson (2003), Jacobsson and Bergek (2004), Jacobsson et al. (2004), Bergek et al. (2005) and Hekkert et al. (in press).

13 The description of the seven functions draws heavily but not exclusively on Bergek et al. (2005).

firm seeks to compete. Secondly, there is knowledge shared within an industry or sector: broader forms of knowledge (often codified in the form of disciplines such as chemical engineering). Thirdly, there are general scientific or social forms of knowledge that make up an even broader background to production and innovation. Enterprises very rarely innovate on the basis solely of their internal knowledge bases. Key reasons for this include technological complexity and technological dynamics. Technological complexity means that firms can rarely command all relevant knowledge internally. Technological dynamics means that knowledge is often evolving in directions that firms cannot master as a whole. A prevalent solution to this is technological cooperation or collaboration.

A key site of such collaboration and support is the sub-system of organizations that make up the knowledge infrastructure: universities, government labs, standards-setting organizations, research institutes, etc. A major issue in system functioning is how these organizations maintain and diffuse elements of scientific knowledge bases or specific industry-relevant knowledge bases. Moreover, as argued elsewhere in this book (see chapter by Smits, van Merkerk, et al. in this book), such organizations form part of a Strategic Intelligence infrastructure that is concerned with perceptions of technological opportunities, possible applications as well as consequences of realising these and what strategies are required to do so.

Influence on the Direction of Search and the Identification of Opportunities

If a TIS is to develop, a range of firms and other organizations have to perceive and exploit new opportunities. In much economic theory, the problem of recognizing an innovation opportunity is in effect assumed away – the assumption is that the market mechanism leads to the taking up of all profitable opportunities. This idea rests on assumptions of costless and instantaneous access to new information and of identical interpretation of the very same information by all firms.

However, opportunities rarely present themselves in a clear and transparent way and this applies in particular to those associated with technologies that disrupt existing technological knowledge. As Metcalfe (2004) argues, firms (individuals within these) read and interpret the same information in very different ways and 'here we find one of the principle sources of variation in the innovation process, innovations are conceived in individual minds and these minds differ'.[14]

---

14 In the classic book on Korean industrialization, Jones and Sakong (1980) explained how government attempted to increase variety and experimentation through influencing the direction of search. The means was to manipulate the perceived opportunity set of business – 'field augmentation' – so that they would *enter into new areas for business*: 'Field augmentation ... operates through expanding information about existing opportunities. The controller/the firm/considers his perceived opportunity set that includes only a finite number of feasible alternatives, due to limited information. The controller/the government/can expand the decision-maker's perceived opportunity set by filling this information gap.'

Identification of opportunities and related demand articulation is therefore a problematic characteristic of a TIS and there must be sufficient incentives and/ or pressures for firms to search for opportunities and undertake investments in an emerging TIS.

The second function is the combined strength of factors influencing this search and investment behaviour. This function also covers the mechanisms influencing the direction of search within the TIS, in terms of different competing technologies, applications, markets, business models etc. Examples of influencing factors are beliefs in growth potentials (e.g. Raven, 2005), changes in the 'landscape' (Geels 2002), e.g. the climate debate, regulations, articulation of demand by leading customers (e.g. von Hippel 1988; Carlsson & Jacobsson 1993) and technical bottlenecks (e.g. Rosenberg 1976; Hughes 1983). It ought to be recognized that this is not simply a matter of firm behaviour. We noted above that 'infrastructural' organizations have often played an important role in shaping major new technologies: these too have a role in identifying opportunities. For example, the development of mobile telephony in the Nordic countries was mainly the result of an opportunity envisaged by state telecoms laboratories that carried out an extended development effort, which was only transferred to firms at a late stage.

Entrepreneurial Experimentation and Management of Risk and Uncertainty

The origin of a TIS can be traced back to a whole range of factors and circumstances, such as an abundance of skilled labour (Breschi & Malerba 2001; de Fontenay & Carmell 2001), unique university research expertise, competence in related industries (Porter 1998), advantageous geographic location (Feldman & Schreuder 1996) and abundance of natural resources (Katz 2004). These 'triggering factors' operate, however, only if there are entrepreneurs that conduct experiments, delving into uncertain markets and technologies and challenging institutions.

Entrepreneurial uncertainties are a fundamental feature of technological and industrial development and are not limited to early phases in the evolution of a TIS but is a characteristic of later phases as well (Rosenberg 1996). After all, innovation is by definition a novelty, and we do not know what the nature and impact of novelty might be. Innovation, thus, involves both serious risk and serious uncertainty, both in technological and in economic terms. Risk here refers to the fact that potential outcomes of an innovation may be known but show very large variability in results. Uncertainty means that we do not even know what the potential outcomes might be – indeed, it has very rarely been possible to predict the path of innovation, even in general terms. The identification and management of uncertainty and risk is therefore central to any TIS and systems differ sharply in how they do that. The policy issues here relate to the longstanding and well-recognized role of government of reducing, shifting or diversifying risk.

From a social perspective, a fundamental way to handle uncertainty and risks is to ensure that many entrepreneurial experiments take place. These refer, of course, to the activities of the entrepreneurial firm, but include also actions by others (e.g. policy makers) that experiment with new ways to promote a technology. These experiments imply a continuous probing into new technologies and applications, where many will fail, some will succeed. A TIS without vibrant experimentation will stagnate and, indeed, without the initial experiments, it will not be formed.

As emphasized in Strategic Niche Management (e.g. Raven 2005) a multi-dimensional social learning and search process will unfold through the course of these experiments. As mentioned above, Strategic Intelligence (see Chapter 16 by Smits et al. in this book) plays an important role here. Strategic Intelligence provides the tailor-made information about the potential of a new technology, possible applications and the consequences of turning options into concrete products and services that actors need in order to develop their strategies and action plans.

Where experiments take place, knowledge formation is of a more applied nature than that captured under the first functional heading. Entrepreneurial activity always involves a range of activities related to business that have little to do with technological capabilities – this includes business planning, investment programming, strategic marketing and human resource policies. In turning inventions into economic outcomes the crucial 'middle term' is business development, which involves a range of complementary assets (skills and capabilities) that are crucial for, but not necessarily directly related to, innovation. These include the ability to finance capital investment programs, the ability to create efficient production systems on an adequate scale, the ability to recruit and coordinate appropriately skilled labour forces, the ability to construct and use marketing channels, and the ability to create and deploy logistics systems.

From a company perspective, experimentation always leads to risk and uncertainty problems. Whereas these are also the source of business opportunities, profit-seeking companies may be very hesitant to invest in longer-term and high risk projects, in particular those requiring large-scale commitments. Historically, such projects have been managed by such means as government subsidies (e.g. for biotech development), military procurement systems (e.g. for semiconductor development and computing, or for the GPS system), income-contingent loans (e.g. the launch-aid system for Airbus),[15] through research institute funding (for the world-wide web), or through the independent labs of state-owned companies (for mobile telephony). Risk and uncertainty management is central to innovation, but is rarely openly acknowledged or addressed. An economic system that cannot bear or diversify risk, or meet the challenges of uncertainty, is likely to fail to innovate.

15 The creation of Airbus, for example, required major long-term financing in the face of serious technological and economic risk, financing that was well beyond the scope of even the largest banks or consortia of banks. In effect governments assumed this risk, financing the company via income-contingent loans that diversified risk across governments, reduced risk to the shareholders, and through this made a major technological development possible.

Market Formation

For emerging TIS, markets may not exist, or be greatly underdeveloped. Marketplaces may be absent, potential customers may not have articulated their demand, or have the competence to do so, price/performance of the new technology may be poor, standards may not exist or uncertainties may prevail in many dimensions. As regards the last, Rosenberg (1976) puts it like this: most innovations are relatively crude and inefficient at the date when they are first recognized as constituting a new innovation. They are, of necessity, badly adapted to many of the ultimate uses to which they will eventually be put; therefore, they may offer only very small advantages, or perhaps none at all, over previously existing techniques.

Consequently, it is important that smaller 'nursing markets' (Erikson & Maitland 1989) evolve so that a learning space is opened up in which the TIS can find a place to be formed (Kemp et al. 1998).[16] Within such an environment, actors can learn about the new technology and expectations and beliefs be formed. Nursing markets may eventually give way to 'bridging markets' (Andersson & Jacobsson 2000) that allow for volumes to increase and for an enlargement of the TIS in terms of number of actors. Finally, in successful TIS, mass markets may evolve, often decades after the formation of the first market.

Resource Mobilization

Innovating is much more than creating knowledge or learning. As noted above, it always involves a range of assets and abilities that are not necessarily related directly to innovation. These include the ability to finance investments, to create efficient production systems and to recruit appropriately trained staff. As a TIS evolves, a range of different resources therefore need to be mobilized including human and financial capital. In some areas, resource mobilization also entails building a physical infrastructure, in addition to a knowledge infrastructure. Clearly, because of the diversify of resources required for a major new technology (or major upgrade of an existing technology) there are potential coordination failures in resource mobilization.

Legitimation

Legitimacy is a matter of social acceptance and compliance with relevant institutions; the new technology and its proponents need to be considered appropriate and desirable in order for resources to be mobilized, for demand to form and for actors in the new TIS to acquire political strength. Legitimacy also influences expectations among managers and, by implication, the function 'influence on the direction of search'.

16 Strategic Intelligence is, of course, also relevant here.

As is widely acknowledged in organization theory, legitimacy is a prerequisite for the formation of new industries (Rao 2004) and, we would add, for the formation of new TIS. Legitimacy is not given but is formed through conscious actions by organizations and individuals in a process of legitimation, which eventually may help the new TIS to overcome its 'liability of newness' (Zimmerman & Zeitz 2002). This process may take considerable time and is often complicated by a requirement of the new TIS to relate to and often compete with a number of established TIS (van de Ven & Garud 1989). Regardless of whether the new TIS may become part of the incumbent regime or threaten to overthrow it, vested interests may oppose this force of 'creative destruction' and defend the existing TIS' and the institutional frameworks associated with them. In that case, 'political networks' or advocacy coalitions (e.g. Smith 2000a) may function as catalysts by engaging in wider political debates in order to secure institutional alignment. The scale and success of such coalitions are directly dependent on the available resources and prevailing expectations of the future of the new technology.

Development of Positive Externalities

As markets go beyond the first niches, there is an enlarged space in which the emerging system can evolve and the functions be strengthened via a generation of positive external economies. Structural change in the form of entry of firms is central to this process. First, new entrants may resolve at least some of the initial uncertainties with respect to technologies and markets (Lieberman and Montgomery 1988), thereby strengthening the functions 'influence of the direction of search' and 'market formation'. Second, new entrants may, by their very entry, legitimate the new TIS (Carroll 1997). New entrants may also strengthen the 'political' power of advocacy coalitions that, in turn, enhances the opportunities for a successful legitimation process. An improved legitimacy may positively influence four functions: 'resource mobilization', 'influence of the direction of search', 'market formation' and 'entrepreneurial experimentation'.

Further externalities may arise due to the co-location of firms. Marshall (1920) discussed economies that were external to firms but internal to a location:[17]

- Emergence of pooled labour markets, which strengthens the function 'knowledge development and diffusion' in that subsequent entrants can recruit staff from early entrants (and vice versa as time goes by).

17 To these, we may add that the greater the number and variety of actors in the system, the greater are the chances for new combinations to arise, often in a way which is unpredictable (Carlsson 2003). An enlargement of the actor base in the TIS therefore enhances both the opportunities for each participating firm within in the system to contribute to 'knowledge development and diffusion' and for the firms to participate in 'entrepreneurial experimentation'.

- Emergence of specialized intermediate goods and service providers; as a division of labour unfolds, costs are reduced and further 'knowledge development and diffusion' is stimulated by specialization and accumulated experience.
- Information flows and knowledge 'spillover', contributing to the function 'knowledge development and diffusion'.

Hence, new entrants may contribute to a process whereby the functions are strengthened, benefiting other members of the TIS through the generation of positive externalities. This function is therefore not independent, but rather one which indicates the dynamics of the system.

Drawing on Negro et al. (2008), the next subsection applies the functions approach to the development of the TIS for biomass gasification in the Netherlands as a brief illustrative case in point.

The Case of Biomass Gasification in the Netherlands

Biomass gasification technology converts biomass into electricity (and heat) or into feedstock for the production of chemicals and liquid bio fuels. Despite the promises of high-energy conversion efficiency and the wide variety of applications, biomass gasification has not yet been successfully developed and implemented in the Netherlands.[18]

When biomass gasification came onto the political agenda for large-scale production of electricity in the early 1990s, the development of the knowledge base could have build upon experience with gasification of coal in the Netherlands and with building small-scale biomass gasification plants for developing countries. However, the first initiatives in 'knowledge development' were not so much focused on technology development as on energy systems studies to identify the potential contribution of biomass gasification to the energy sector and the available biomass resources. These desktop studies presented such a positive picture for biomass gasification that it greatly raised expectations. This influenced the 'direction of search' towards biomass gasification. Further desktop studies confirmed the initial expectations, which increased expectations and positively 'influenced the direction of search' even more. In a very short time period, biomass gasification was considered to be on a par with other renewable energy technologies. The high expectations were reflected in reports in media that large-scale demonstration projects could be realized quickly and that no unsolvable technological problems were expected.

The expectations led to an interest by some parties to actually build a biomass gasification plant ('entrepreneurial experimentation'), which triggered the initiation of R&D projects ('knowledge development' as well as some 'resource mobilization'). The EU contributed financially to several projects, energy companies invested in feasibility studies and experiments and the Dutch government included biomass

18 Also in other countries this technology is still in an emerging stage.

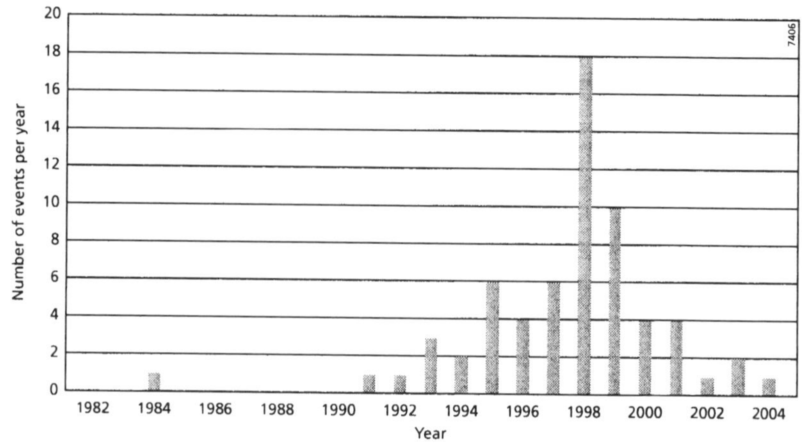

Figure 6.1 Activity pattern of the function 'knowledge development and diffusion'

gasification in their R&D program to produce energy from biomass and waste. Also, biomass gasification projects were able to obtain financial resources from more general (renewable) energy R&D subsidies.

Two major 'entrepreneurial experiments' with biomass gasification dominated. The first was an energy company that added a biomass gasification unit to a coal-fired power plant. The unique feature was that it used waste wood as a biomass resource. This combination (waste wood in a biomass gasification plant) resulted in many technological problems. The other project was a coalition of local authorities, energy companies and R&D institutes. The two experiments attracted much attention in

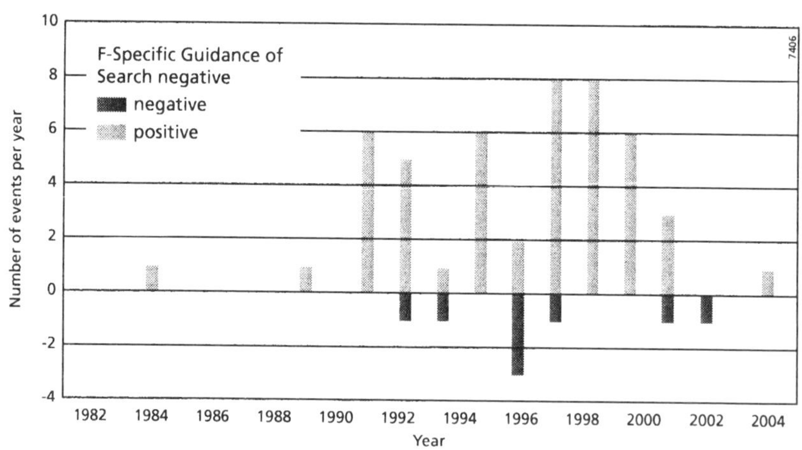

Figure 6.2 Activity pattern of the function 'influence the direction of search'

the energy community and the first phase of the projects boosted many other system functions: in particular they 'influenced the direction of search' and 'knowledge creation' since they initiated several feasibility studies and enhanced the 'process of legitimation' of biomass gasification.

The process of legitimation led to a drastic change in perception of this technology; from a relatively unknown technology to a building block in the transformation of the energy sector. Since very influential actors became part of the 'advocacy coalition', successful lobbying activities took place, leading to the start of a specific publicly financed research program on biomass gasification.

The function 'market formation' was, however, never well fulfilled. The high technological risks were not compensated by specific market stimulation measures and no 'nursing market' was created. The effect of this poorly developed function became immediately visible after the liberalization of the Dutch energy market in 1998. The key words for electricity companies after the liberalization were 'low costs'. Without specific market formation policies in place, the technological uncertainties resulted in an abortion of all projects. Without these, positive externalities could not emerge and the TIS around biomass gasification never became really entrenched in the larger energy innovation system. Indeed, the emerging TIS stalled and declined rapidly, as indicated in Figures 6.1 and 6.2 which show the rise and fall in the functions 'knowledge formation' and 'influence the direction of search'. Note that in Figure 6.2 negative events are recorded. This implies that guidance away from biomass gasification took place.

FUNCTIONAL ANALYSIS AS A GUIDE AND RATIONALE FOR POLICY

Having described and illustrated the seven functions, we will now proceed to analyse how a functional analysis can be used as a tool to find the key policy issues to tackle and a further rationale for doing so.

System Weaknesses in Functional Terms: A Guide for Policy

As mentioned above, most of the literature discussing innovation system weaknesses (or failures) has tended to focus on perceived weaknesses in the structural composition of a system. For example, all the four types of system failures identified by Woolthuis et al. (2005) in their recent synthesis and re-categorization of previous system failure literature are related to structural components: infrastructural failures (related to actors and artefacts), institutional failures (related to institutions), interaction failures (related to networks) and capabilities' failures (related to actors).

However, it is difficult, if not impossible, to evaluate the 'goodness' or 'badness' of a particular structural element or combination of elements without making references to its effects on the innovation process.[19] For example, how do we know whether the existence of a strong network constitutes a weakness rather than a strength of a particular system, e.g. a source of synergy rather than of lock-in or 'group-think' (cf. Woolthuis et al. 2005), without identifying its influence on the innovation process and its key sub-processes? Therefore, an attempt to find system weaknesses at the structural level only is fraught with difficulties. Hence, an assessment of rationale and options for policy in a specific TIS should rest, in part, on the dynamics of what the system achieves rather than on the nature and evolution of its structural components.

This is, indeed, the main benefit of the functional analysis: it allows us to separate structure from content by adding the second level – functions – to the structural one. With a functional analysis, the prime focus is on the dynamics of 'what is actually achieved' in the TIS rather than on the dynamics in terms of the structural components of the TIS. This does not mean that a policy analysis should neglect structural factors. These contribute to how the functions are performed and any intervention has, of course, to focus on structural elements. It is, however, through a functional analysis that specific system weaknesses in a given TIS can be sought out.

The functional pattern, i.e. what is achieved in terms of the seven key processes, can be analysed empirically. This was first demonstrated in Johnson and Jacobsson (2001).[20] In later work, we used this framework to trace the dynamics of functional patterns. In Bergek and Jacobsson (2003) and Jacobsson et al. (2004), for example, analyses of functional patterns were used to understand the evolution of two TIS in the energy field (wind power and solar power). Empirical analyses of this nature not only improve our understanding of dynamics, but can provide policy makers with a specification of which the weak functions are. It follows that we can also formulate policy problems in functional terms, i.e. 'what do we think should be achieved that is not', which implies that system weaknesses can be expressed in functional terms.[21] A functional analysis can therefore not only provide a rationale for intervention but also constitute the focusing mechanism for policy makers (and others who desire to influence the dynamics) in their efforts to specify the key policy issues. In here we differ from more structural approaches focusing on system failures or systemic problems (e.g. Chaminade & Edquist in this book; Woolthuis et al. 2005).

How can we then explain why certain functions are weak in a particular TIS? Each TIS has a degree of uniqueness in terms of its structural components. Yet, for

---

19 Mowery's (2005) analysis of the differences in paths followed by Korea and Taiwan in the semiconductor industries clearly demonstrates that industrial development does not involve following one structural path only but is achieved in different ways in different contexts.

20 In that analysis, we restricted ourselves to five functions.

21 Assessing the goodness of a specific functional pattern – the functionality of a TIS – is a challenge that requires more work to meet. For initial discussions, see Bergek and Jacobsson (2003) and Bergek et al. (2005).

emerging TIS, many of the components are shared with other TIS, incumbent as well as other emerging ones (Johnson & Jacobsson 2001). For instance, institutional change in the form of tradable emission permits influences many TIS using or supplying energy.[22] Within the boundary of the TIS there are, thus, both technology-specific components and those that are part of also other TIS.

An innovation system evolves in part as a result of its internal dynamics. As noted by Myrdal (1957), in a rapidly developing system a chain reaction of positive feedback loops may materialize, setting in motion a process of cumulative causation. In such a process, the boundary alters and impacts on the dynamics in functional terms (what is being achieved). However, the internal dynamics is only part of the picture. Factors that are exogenous to a TIS may have an impact on its structural evolution and functional dynamics. Myrdal (1957) showed a keen understanding of the interplay between internal and external sources of dynamics and even suggested that 'the main scientific task is ... to analyse the causal inter-relations within the system itself as it moves under the influence of outside pushes and pulls and the momentum of its own internal processes'. Hence, the determinants of system development are both endogenous and exogenous (Sandén & Jonasson 2006).[23]

An endogenous driving force, or inducement mechanism, may be demand from a leading-edge customer and an exogenous one may be the climate debate or accidents like that in Chernobyl.[24] From the perspective of an emerging TIS, it is particularly vital to identify blocking mechanisms, i.e. factors that provide obstacles to the development of functions and, therefore, tilt the selection environment in favour of incumbent technologies.[25] An endogenous blocking mechanism may, for instance, be poorly developed learning and 'political' networks that limit knowledge diffusion and legitimation. An exogenous blocking mechanism may come in the form of highly organized incumbents that defend their markets and investments and make sure that institutions continue to be aligned to the dominant technologies.[26]

---

22 Indeed, the whole notion of functions was developed to handle an integration of technology-specific and more general influencing factors, see Johnson and Jacobsson (2001). See also Geels' (2002) discussion of a multi-level framework.

23 This means that within our framework, we incorporate determining factors found at the levels of 'regimes' and landscape' in Strategic Niche Management.

24 Raven (2005) has a useful discussion of 'destabilization' of regimes which opens up windows of opportunities for new 'niches'. This was, for instance, the case in Germany where the current regime (power sector) lost much legitimacy after the Chernobyl accident (Bergek & Jacobsson 2003). The stabilization and destabilization of the regime (and the dominant TIS making up that regime), should perhaps be included as a key process.

25 See Johnson and Jacobsson (2000), Unruh (2002) and Jacobsson and Bergek (2004).

26 Setting the boundary of a system that is emerging is very difficult. It could be argued that incumbents should be included in the system as their behaviour maybe influenced by the growth of the new TIS. In functional analysis, the issue of setting boundaries (in terms of the structural elements of the TIS) is, however, of lesser importance. The focus is on the functions

Further blocking mechanisms may be traced to the emergence of other TIS that compete for space both in the market and in the political arena. A case in point is that of fossil gas in Sweden where a 'battle of institutions' is taking place with the proponents of biomass CHP.[27]

The functional pattern and the dynamics of a TIS can, thus, be empirically linked to and explained by a set of driving forces and blocking mechanisms. System weaknesses in the emerging TIS, as expressed in functional terms, can then be traced to structural features of the TIS or to exogenous factors. This means that the key policy issues are readily definable as and when such an analysis has been completed.

We will illustrate this with the example of the emerging Swedish TIS for 'IT in home care' which is defined by the application of a generic technology (IT) to a particular application: care of elderly and ill people in their homes instead of in a hospital. For a number of reasons (demographic, public sector funding restrictions, technological opportunities etc.), this is a TIS which is thought of as having a large growth potential. However, it is still in an early phase, as judged by, e.g., the following structural features:[28]

- There are many competing experiments that are linked to specific IT platforms (no standards and high technical uncertainty).
- The number of firms supplying IT solutions is small.
- Markets are small, with high uncertainties, e.g. with respect to applications and choice of software
- The 'advocacy coalition' for the TIS is weak
- The demand is poorly articulated by customers with poorly developed competences

In this early phase, the functional pattern can be summarized as follows:

- Knowledge development and diffusion: pilot projects in only some of the 290 counties and 21 county councils
- Market formation: local pilot projects constitute a small and fragmented nursing markets
- Influence on the direction of search: some government R&D funding, opportunities, although unclear, to find new markets, awards

---

and how these are influenced by structural elements found at different system levels (Johnson & Jacobsson 2001).

27 We do not exclude that the relationship between emerging TIS can be complementary. For instance, small-scale hydropower proponents and the advocacy coalition for wind power teamed up to influence the German legislation so that it would favour renewables (Jacobsson & Lauber 2006). Sandén and Jonasson (2006) discuss how competition and collaboration evolved among emerging TIS in alternative fuels in Sweden.

28 This case was developed together with Cecilia Sjöberg at VINNOVA.

- Entrepreneurial experimentation and management of risk and uncertainty: a few IT firms only have developed solutions – risks are high
- Resource mobilization: EU and government R&D funding, some co-funding by firms, poor adjustment by the higher educational sector
- Legitimation: partly underdeveloped legitimacy, especially among care providers
- Development of positive externalities: early stage of cluster formation in three cities

This pattern is shaped by both inducement and blocking mechanisms (see Figure 6.3). There are two significant inducement mechanisms: belief in growth potential and government R&D policy. The former is driven by a range of factors, as was mentioned above[29] and has a bearing on the function 'influence on the direction of search' among both care providers (e.g. county councils) and suppliers (IT firms), as well as on the functions 'market formation' (nursing markets) and 'entrepreneurial experimentation'. The latter inducement mechanism both signals attractiveness and provides some resources for research and experiments. Hence, it strengthens the functions 'influence on the direction of search' and 'legitimation' as well as 'resource mobilization' and 'knowledge development and diffusion'.[30]

In spite of these two inducement mechanisms, there are several system weaknesses in functional terms. In particular, we would single out 'market formation', 'entrepreneurial experimentation', 'influence on the direction of search' and 'legitimation'. The blocking mechanisms here are strong and manifold. 'Market formation' is blocked by an absence of standards (which leads to a fragmented market), two factors that reflect poor awareness and competence among potential customers (leads to poor articulation of demand) and an associated lack of knowledge among suppliers of IT solutions of customer needs.

'Entrepreneurial experimentation', 'influence on the direction of search' and 'legitimation' are blocked by two factors each. These three have a common blocking mechanism in the form of a 'lack of competence and a poor articulation of demand'. This is strengthened by an additional but different factor in each case (uncertainties of customer needs, lack of markets for standard software solutions and a weak advocacy coalition).

Functions are not independent, but tend to reinforce each other (dotted lines in Figure 6.3). A poor 'market formation', thus, affects negatively both 'entrepreneurial experimentation' and 'influence on the direction of search', whereas little 'entrepreneurial experimentation' negatively influences 'resource mobilization' and 'knowledge development and diffusion'. This means that the impact of blocking mechanisms is magnified by such interdependencies. Clearly, it could be argued that

29 These include demographic changes with a larger share of elderly people in the population, public sector funding restrictions and emerging technological opportunities.

30 Indirectly, it also strengthens 'entrepreneurial experimentation' as a consequence of its positive influence on the direction of search.

policy must focus on reducing the strength of the blocking mechanisms that have such a pervasive effect. In the fourth column in Figure 6.3, we list six specific policy issues connected to removing or reducing the strength of the blocking mechanisms. The first three of these focus on the potential customers (care providers) and are aimed at removing the most pervasive blocking mechanisms. These policy issues can be specified as:

- How to raise user competence so that demand is articulated and uncertainties reduced for potential suppliers
- How to support users in order to (1) increase their knowledge of the benefits of IT in home care and of ways to distribute the costs and benefits over organizational boundaries and (2) diffuse knowledge of the outcome of early experiments in order to reduce uncertainties further
- How to support experimentation with new applications in order to reduce the level of uncertainty of needs[31]

A fourth policy issue refers to:

- How to develop standards in order to move from a fragmented market of 290 local councils and 21 county councils

In addition to these four issues, we can also deduce two additional ones which are:

- How to alter research and education at universities in order to allow for 'resource mobilization' in terms of staff with relevant background
- How to support a weak advocacy coalition so that it can improve the process of 'legitimation'

Hence, by analysing the functional pattern of the TIS ('what is actually going on'), we can point out the key system weaknesses in functional terms. By specifying the main endogenous and exogenous determinants of these weaknesses, in particular the main blocking mechanisms, we arrive at the central policy and/or strategy issues.

The Nature of Blocking Mechanisms: An Additional Rationale and Widened Scope for Policy

As noted above, the mere existence of system weaknesses is not necessarily a rationale for policy intervention; individuals and groups of entrepreneurs may act on system weaknesses without any support from policy (and often do, as shown in the case of solar cells in Germany, see Jacobsson and Lauber 2006). However, the very nature of

31 It would be interesting to analyse how a well developed Strategic Intelligence infrastructure could support the identification of solutions to these three issues.

blocking mechanisms may set limits to entrepreneurial influence on system dynamics and, thus, justify public policy intervention.

First, entrepreneurial experimentation is not automatic. Hausmann et al. (2005) argue that markets are good at signalling the profitability of already existing activities but not of uncovering profitability of those that might exist – firms may get 'locked in'. Hence, the blocking mechanism may be of a nature where there are no or very few entrepreneurial experiments going on. This severely restricts the opportunities for these firms to influence system functionality.[32]

Second, blocking mechanisms may be many and occur with wide variety. In the case of 'IT in home care', blocking mechanisms ranged from poorly articulated demand to institutional misalignment, such as the lack of standards. Individual actors, as well as networks of actors, can only be expected to allocate a smaller part of their investment budgets to improve system functionality and, therefore, to take on a limited number of roles and influence a limited number of mechanisms simultaneously. This implies that policy makers may need to lessen the burden of firms and other actors. This is particularly so in early phases in system evolution where firms are often small and lack resources.

Third, overcoming some blocking mechanisms may simply be out of reach of entrepreneurial actors. A few examples can be given:

1. Entrepreneurs – on their own or in 'packs' – do not have the opportunity to reduce uncertainties and shift risks. The importance of this is currently seen in the success of the feed-in law used in Germany (and Spain) in promoting the development of TIS centred on wind and solar power. A key feature of that institutional change is that it reduces risks for investor.
2. Even though firms may act through advocacy coalitions to stimulate institutional alignment in a case where market formation is blocked by current legislation (or lack thereof), legislative changes in the end still require the participation of policy makers.
3. Other blocking mechanisms than those requiring legislative changes may also be beyond reach. In the case of 'IT in home care', an alteration of research and education at universities is needed in order to allow for 'resource mobilization' in terms of staff with relevant background. This is clearly not a problem easily solved by entrepreneurial actors alone. Nor would we expect that the issue of lack of software standards could be solved by entrepreneurial IT suppliers alone,

32 This means that there are ample opportunities for policies that aim at 'influencing the direction of search' of firms. For example, Katz (2004) argues that: 'In the case of salmon, the *perception* that large natural rents are potentially present ... required the public sector to take a proactive stance in favour of inducing the erection of salmon farming production capacity ... it certainly exercised a crucial catalytic role ... showing that "*it could be done*"' (our emphasis). This involved, inter alia, starting the first commercial salmon farming operation in Chile.

although it is conceivable that care providers could agree on standards without policy intervention.

4. There may be a need for coordination of parallel initiatives by different actors and actor groups. Such coordination may be very difficult for firms to achieve – especially when many actors or types of actors are involved – and policy makers may, therefore, need to step in. For instance, a shift to fuel-cell-powered automobiles requires a simultaneous investment in development and production of fuel cells, fuel-cell-driven cars, production of energy carriers for fuels cells, 'petrol stations' for fuel cells, etc. Coordination then requires that a range of firms supplying complementary products/services are influenced in their respective search and investment processes (Frenken et al. 2004).

By analysing the nature of blocking mechanisms we may specify the rationale for public policy interventions and shed light on possible interventions. The ability of firms and other market actors to influence the functionality of a TIS may be particularly limited when (1) they do not perceive the opportunities; (2) functions are blocked by factors that for some reason are out of reach for firms, (3) there is a wide range of blocking mechanisms, or (4) there are many different initiatives that need to be coordinated. Policy makers, thus, need to understand the nature of the blocking mechanisms that shape the functional dynamics and identify those blocking mechanisms that individual actors or any foreseeable collaboration between actors cannot be expected to handle.[33]

These four issues do not only specify the rationale for public policy intervention in TIS formation but may also point at a broadened scope for policy intervention in comparison with a 'conventional' innovation system perspective. We have already argued that an analysis of functions in innovation systems can develop the notion of 'system weaknesses' from its conventional focus on structural factors only, to an identification of weak functions and the mechanisms causing weaknesses. It is reasonable to ask though if a 'conventional' innovation system analysis would not have identified the same functions (although termed differently), blocking mechanisms and associated policy issues, particular since our list of functions draws heavily on previous innovation system literature. Reviewing the 'Policy and

33 This is, however, not an easy task. As was evident from the case of 'IT in home care', the 'picture' of a functional patterns and the mechanisms influencing it may become quite complex. It is more common than not for functions to be influenced by several blocking mechanisms and for blocking mechanisms to influence several functions simultaneously. The prevalence of feedback loops between functions increase complexity further (Jacobsson & Bergek 2004). The co-existence of endogenous and exogenous (in the sense described above) blocking mechanisms may make the picture even more confusing. As a consequence of this complexity, even what may appear to be a 'simple' issue (such as what blocks the formation of markets) may be obscure (Jacobsson & Bergek 2004) and the task of identifying, tracing and assessing the strength of various blocking mechanisms requires highly competent staff.

innovation system' literature of the 1990s (largely) generates some interesting observations.

First, although many of 'our' functions were indeed mentioned in this literature, there was a clear focus on three functions only: 'knowledge development and diffusion' (especially learning),[34] 'resource mobilization' (especially knowledge and competence building in individuals)[35] and 'influence on the direction of search' (especially general incentives for innovation[36] and the phenomenon of lock-in,[37] but also the building of visions and awareness[38]).[39] The functions 'market formation' and 'entrepreneurial experimentation' were mentioned by a few authors only, primarily in terms of demand articulation and variety creation respectively.[40]

Second, two of 'our' functions were either mentioned only in passing or completely left out. That 'positive externalities' were only mentioned by a few (and primarily in terms of learning) was perhaps most surprising, since we consider this to be one of the central features of a well-functioning innovation system. The main exception was Malerba (1996), who discussed 'dynamic complementarities' at some length. With the exception of Carlsson and Jacobsson (1997), 'legitimation' was not mentioned, which is also somewhat puzzling since its importance has been noted in several strands of literature, including organizational science and sociological analysis of technology development.

The main observation, however, was that none of the policy literature we reviewed mentioned all functions at the same time. Instead, policy researchers seemed to focus on a few functions or general policy problems to be solved.[41] They, moreover, tended to mix functions with structural elements. Here lies the main difference between the functional analysis and 'conventional' innovation system analyses: explicitly stating and including all functions allows for the systematic identification of policy problems, as illustrated by the case of 'IT in home care', at the same time as it avoids the pitfall of focusing on the structural characteristics of the system. As a consequence of the latter, policy makers are informed on the basis

34 See, e.g., Malerba (1996) and Smits and Kuhlmann (2004).

35 See, e.g., Hauknes (1999), Malerba (1996) and Metcalfe (2004).

36 See, e.g., Malerba (1996), Smith (1996) and Teubal (2001).

37 Malerba (1996), Smith (1996), Carlsson and Jacobsson (1997).

38 Malerba (1996), Smits and Kuhlmann (2004), Teubal (2001).

39 Related to this function were also discussions of the trade off between variety creation and selection (see Malerba, 1996). The first two functions were often mentioned together, for example in terms of 'too low level of R&D' or as a 'learning failure' (Malerba, 1996).

40 See, e.g., Malerba (1996) on trial-and-error and variety creation and Malerba (1996), Smits and Kuhlmann (2004) and Teubal (2001) on market formation issues. Carlsson and Jacobsson (1997) put more emphasis on entrepreneurial experimentation.

41 Malerba (1996) and Carlsson and Jacobsson (1997) are the main exception, covering most of our functions.

of what is actually achieved by the system, and in explaining the pattern exogenous blocking mechanisms are included in addition to endogenous ones.

The wide range of possible policy challenges coming out of a functional analysis implies that traditional policy instruments such as R&D support, education programmes or taxes are not enough. Depending on the functional patterns and associated blocking mechanisms in particular cases, public policies may need to

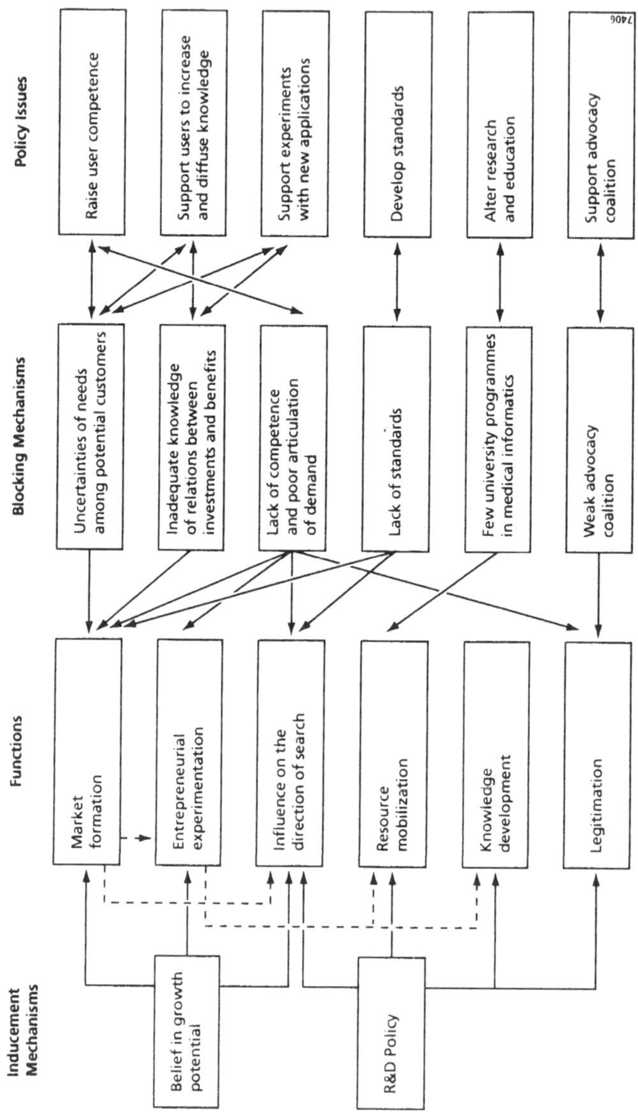

Figure 6.3 Inducement and blocking mechanisms, functions, and policy issues in the case of 'IT in home care'

work through all types of structural components simultaneously, including different types of actors, different types of networks within and between actor groups (e.g. standard setting groups, user-supplier learning networks and advocacy coalitions) and different types of institutions (not only regulatory, but normative and cognitive as well), in order to strengthen the system's functionality. Due to this widened focus, there may be a need to develop novel policy instruments, in particular aiming at e.g. legitimation (e.g. support of emerging advocacy coalitions). Indeed, the functional analysis further strengthens the innovation system perspective's tendency to open up for more rich and difficult innovation policies (see also the chapter by Smits, Kuhlmann and Teubal in this book).

CONCLUSIONS

The purpose of this chapter was to develop a framework that provides a tool for policy makers seeking a rationale for as well as a guide to policy interventions in a TIS. It rests on a set of systemic interdependencies in terms of technologies (artefacts and knowledge), institutions, networks and industrial structures that co-evolve in a process characterized by increasing returns. Whereas the literature on 'policy and innovation systems' has made progress in terms of focussing on system weaknesses, it has largely been limited to defining these in terms of the nature of the structural components in a system and to identifying an associated set of 'generic' policy challenges. What also matters, though, for policy is functionality: that is, what is achieved by the system and, moreover, policy makers in charge of a specific TIS need a tool that identifies the unique policy issues pertaining to the dynamics of that system.

We suggested that policy makers may focus on a set of key processes in the evolution of a system. In addition to those at the structural level (entry of actors, formation of networks and institutional alignment), we argued that there are seven additional ones at the functional level. This second level refers to processes that have a more direct and immediate impact on the 'goal' of the system and can be seen as indicators of 'what is achieved by the system'. The functional pattern, i.e. what is achieved, can be analysed empirically and system weaknesses can, therefore, be specified in functional terms. These weaknesses (and strengths) can then be empirically linked to and explained by a set of driving forces and blocking mechanisms for each TIS. This means that the key policy issues are readily definable as and when such an analysis has been completed.

Some of these blocking mechanisms may be, and often are, addressed by individual firms or groups of firms. Yet, the very nature of blocking mechanisms may set limits to entrepreneurial influences on system dynamics and justify public policy intervention. The nature of these therefore needs to be carefully assessed by policy makers. Policy makers also need to think much more broadly than in terms of issues pertaining to R&D support, education and tax changes. With a functional analysis, a wide range of policy challenges is identified, some of which may be in need of new types of instruments (see chapter in this book by Smits, Kuhlmann and Teubal). The approach also opens up a systematic process for policy learning. Although it is clear that there are no 'one size fits all' policy implications, there is presumably no infinite variety in the evolution of different TIS. It is therefore possible for policy-making bodies to repeat the analysis, with a given framework, and develop a typology of the evolution of functional patterns and associated policy challenges. A typology of that nature would be expected to be a useful guide for policy makers in their search to find the key policy issues in specific TIS.

ACKNOWLEDGEMENTS

We are grateful to VINNOVA and KSI for funding most of the work for this paper. Constructive comments were given by workshop participants in Karlsruhe, June 2006. Special thanks to Ruud Smits for a thorough review of an earlier draft.

REFERENCES

Aldrich, H. and Fiol, C. (1994), 'Fools rush in? The institutional context of industry creation', *Academy of Management Review*, **19** (4), 645-670.

Andersson, B. and Jacobsson, S. (2000), 'Monitoring and assessing technology choice: the case of solar cells', *Energy Policy*, **28** (14), 1037-1049.

Bergek, A. and Jacobsson, S. (2003), 'The emergence of a growth industry: a comparative analysis of the German, Dutch and Swedish wind turbine industries', in S. Metcalfe and U. Cantner (eds), *Change, Transformation and Development*, Heidelberg and Berlin: Physica-Verlag, pp. 197-228.

Bergek, A., Jacobsson, S., Carlsson, B., et al. (2005), 'Analysing the dynamics and functionality of sectoral innovation systems: a manual for policy makers', report delivered to VINNOVA.

Breschi, S. and Malerba, F. (2001), 'The geography of innovation and economic clustering: some introductory notes', *Industrial and Corporate Change*, **10** (4), 817-833.

Carlsson, B. (2003), 'The new economy: what is new and what is not?', in J. Christensen and P. Maskell (eds), *The Industrial Dynamics of the New Digital Economy*, Cheltenham, UK and Northampton, MA, USA: Edward Elgar.

Carlsson, B. (ed.) (1997), *Technological Systems and Industrial Dynamics*, Dordrecht, The Netherlands, Amsterdam, The Netherlands, Boston, USA and London, UK: Kluwer.

Carlsson, B. and Jacobsson, S. (1997), 'In search of a useful technology policy: general lessons and key issues for policy makers', in B. Carlsson (ed.), *Technological*

Systems and Industrial Dynamics, Dordrecht, The Netherlands, Amsterdam, The Netherlands, Boston, USA and London, UK: Kluwer, pp. 299-315.

Carlsson, B. and Jacobsson, S. (1996), 'Technological systems and industrial dynamics: implications for firms and governments', in E. Helmstädter and M. Perlman (eds), *Behavioral Norms, Technological Progress, and Economic Dynamics*, Ann Arbor: The University of Michigan Press.

Carlsson, B. and Jacobsson, S. (1993), 'Technological systems and economic performance: the diffusion of factory automation in Sweden', in D. Foray and C. Freeman (eds), *Technology and the Wealth of Nations*, London, UK and New York, USA: Pinter, pp. 77-94.

Carlsson, B. and Stankiewicz, R. (1991), 'On the nature, function, and composition of technological systems', *Journal of Evolutionary Economics*, **1**, 93-118.

Carroll, G. (1997), 'Long-term evolutionary changes in organizational populations: theory, models and empirical findings in industrial demography', *Industrial and Corporate Change*, **6** (1), 119-143.

Computer Science and Telecommunications Board (1999), 'Development of the internet and the World Wide Web', Washington, DC: National Academy Press, pp. 169-183.

Edquist, C. (1999), 'Innovation policy: a systemic approach', TEMA-T working paper, Linköping University, Linköping.

Erikson, W. and Maitland, I. (1989), 'Healthy industries and public policy', in M. Dutton (ed.), *Industry Vitalization*, New York: Pergamon Press.

Feldman, M. and Schreuder, Y. (1996), 'Initial advantage: the origins of the geographic concentration of the pharmaceutical industry in the Mid-Atlantic region', *Industrial and Corporate Change*, **5** (3), 839-862.

Fontenay, C. de and Carmell, E. (2001), 'Israel's Silicon Wadi: the forces behind cluster formation', SIEPR Discussion Paper No. 00-40, Stanford Institute for Economic Policy Research, Stanford.

Frenken, K., Hekkert, M. and Godfrij, P. (2004), 'R&D portfolios in environmentally friendly automotive propulsion: variety, competition and policy implications', *Technological Forecasting and Social Change*, **71** (5), 485-507.

Galli, R. and Teubal, M. (1997), 'Paradigmatic shifts in national innovation systems', in C. Edquist (ed.), *Systems of Innovation*, London, UK and New York, USA: Pinter, pp. 342-370.

Geels, F. (2002), 'Technological transitions as evolutionary reconfiguration processes: a multi-level perspective and a case study', *Research Policy*, **31** (8-9), 1257-1274.

Hauknes, J. (1999), 'Innovation systems and capabilities of firms', literature review for the RISE project, Brighton: University of Brighton.

Hausmann, R., Rodriguez-Clare, A. and Rodrik, D. (2005), 'Towards a strategy for economic growth in Uruguay', Inter-American Development Bank, Economic and Social Study Series, February.

Hekkert, M., Suurs, R., Negro, S., Kuhlmann, S. and Smits, R. (in press), 'Functions of innovation systems: a new approach for analysing technological change', *Technological Forecasting and Social Change*, forthcoming.

Hippel, E. von (1988), *The Sources of Innovation*, Oxford, UK and New York, USA: Oxford University Press.

Hughes, T. (1983), *Networks of Power: Electrification in Western Society, 1880-1930*, Baltimore: The John Hopkins University Press.

Jacobsson, S. and Bergek, A. (2004), 'Transforming the energy sector: the evolution of technological systems in renewable energy technology', *Industrial and Corporate Change*, **13** (5), 815-849.

Jacobsson, S. and Johnson, A. (2000), 'The diffusion of renewable energy technology: an analytical framework and key issues for research', *Energy Policy*, **28** (2000), 625-640.

Jacobsson, S. and Lauber, V. (2006), 'The politics and policy of energy system transformation: explaining the German diffusion of renewable energy technology', *Energy Policy*, **34** (3), 256-276.

Jacobsson, S., Sandén, B. and Bångens, L. (2004), 'Transforming the energy system: the evolution of the German technological system for solar cells', *Technology Analysis and Strategic Management*, **16** (1), 3-30.

Johnson, A. and Jacobsson, S. (2001), 'Inducement and blocking mechanisms in the development of a new industry: the case of renewable energy technology in Sweden', in R. Coombs, K. Green, A. Richards and V. Walsh (eds), *Technology and the Market: Demand, Users and Innovation*, Cheltenham, UK and Northampton, MA, USA: Edward Elgar, pp. 89-111.

Jones, L. and Sakong, I. (1980), *Government, Business, and Entrepreneurship in Economic Development: the Korean Case Studies in the Modernization of the Republic of Korea: 1945-1975*, Cambridge, MA, USA: Harvard University Press.

Katz, J. (2004), 'Economic, institutional and technological forces inducing the successful inception of salmon farming in Chile', mimeo, University of Chile and Conicet, Argentina.

Katz, J. (1983), 'Technological change in the Latin American metalworking industries: results from a programme of case studies', CEPAL Review April, No.19, Santiago, Chile, pp. 85-143.

Kemp. R., Schot, J. and Hoogma, R. (1998), 'Regime shifts to sustainability through processes of niche formation: the approach of strategic niche management', *Technology Analysis and Strategic Management*, 10 (2), 175-195.

Lieberman, M. and Montgomery, C. (1988), 'First-mover advantages', *Strategic Management Journal*, **9** (1), 41-58.

Malerba, F. (1996), 'Public policy and industrial dynamics: an evolutionary perspective', manuscript submitted to the Commission, December.

Marshall, A. (1920), *Principles of Economics*, Vol. 8, London, UK and New York, USA: MacMillan.

Metcalfe, S. (2004), 'Policy for innovation', mimeo, ESRC Centre for Research on Innovation and Competition, Manchester: University of Manchester

Metcalfe, S. (1992), 'The economic foundation of technology policy: equilibrium and evolutionary perspectives', mimeo, Manchester: University of Manchester.

Mowery, D. (2005), 'The role of knowledge-based "public goods" in economic "catchup": lessons from history', paper presented at the Industrial Development Report 2005 Workshop, UNIDO, May 11th-12th, Vienna.

Myrdal, G. (1957), *Economic Theory and Underdeveloped Regions*, London: Ducksworth Publishers.

Negro, S.O., Suurs, R.A.A. et al. (2008), 'The bumpy road of biomass gasification in the Netherlands: explaining the rise and fall of an emerging innovation systems', *Technological Forecasting and Social Change*, 75 (1), 57-77.

Norgren, L. and Hauknes, J. (2000), 'Economic rationales of government involvement in innovation and the supply of innovation-related services', report from the RISE work programme, University of Brighton, Brighton.

Olleros, F. (1986), 'Emergent industries and the burnout of pioneers', *Journal of Product Innovation Management*, 1, 5-8.

Porter, M. (1998), 'Clusters and competition: new agendas for companies, governments and institutions', in M. Porter (ed.), *On Competition*, Boston: Harvard Business School Press, pp. 197-287.

Rao, H. (2004), 'Institutional activism in the early American automobile industry', *Journal of Business Venturing*, **19** (3), 359-384.

Raven, B. (2005), *Strategic Niche Management for Biomass*, Eindhoven: Eindhoven University Press.

Rodrik, D. (2004), 'Industrial policy for the twenty-first century', mimeo, Cambridge, MA, USA: John F. Kennedy School of Government, Harvard University, Cambridge.

Rotmans, J., Kemp, R. and Van Asselt, M. (2001), 'More evolution than revolution: transition management in public policy', *Foresight*, 3 (1), 15-31.

Rosenberg, N. (1996), 'Uncertainty and technological change', in R. Landau, T. Taylor and G. Wright (eds), *The Mosaic of Economic Growth*, Stanford: Stanford University Press, pp. 334-355.

Rosenberg, N. (1976), *Perspectives on Technology*, Cambridge, UK: Cambridge University Press.

Sandén, B. and Jonasson, K. (2006), 'Variety creation and co-evolution of contenders: the case of alternative transport fuels in Sweden 1974-2004', paper submitted to a scientific journal.

Smith, K. (2000a), 'Policy networks and advocacy coalitions explaining policy change and stability in UK industrial pollution policy?', *Environment and Planning C: Government and Policy*, **18** (1), 95-118.

Smith, K. (2000b), 'Innovation as a systemic phenomenon: rethinking the role of policy', *Enterprise and Innovation Management Studies*, 1 (1), 73-102.

Smith, K. (1996), 'Systems approaches to innovation: some policy issues', report from the ISE project, submitted to the Commission, December 1996.

Smits, R. and Kuhlmann, S. (2004), 'The rise of systemic instruments in innovation policy', *International Journal of Foresight and Innovation Policy*, **1** (1-2), 4-32.

Stewart, F. and Ghani, E. (1991), 'How significant are externalities for development?', *World Development*, **19** (6), 569-94.

Teubal, M. (2001), 'Enterprise restructuring and embeddedness – a policy and systems perspective', *Industrial and Corporate Change*, **1** (9), 87-11.

Unruh, G. (2002), 'Escaping carbon lock-in', *Energy Policy*, **30** (4), 317-325.

Unruh, G. (2000), 'Understanding carbon lock-in', *Energy Policy*, **28** (12), 817-830.

Ven, A. van de (2005), 'Running in packs to develop knowledge-intensive technologies', *MIS Quarterly*, **29** (2), 365-378.

Ven, A. van de (1993), 'The development of an infrastructure for entrepreneurship', *Journal of Business Venturing*, **8**, 211-230.

Ven, A. van de and Garud, R. (1989), 'A framework for understanding the emergence of new industries', in R. Rosenbloom and R. Burgelman (eds), *Research on Technological Innovation, Management and Policy*, Vol. 4, Greenwich: JAI Press, pp. 195-225.

Woolthuis, R.K., Lankhuizen, M. and Gilsing, V. (2005), 'A system failure framework for innovation policy design', *Technovation*, **25** (6), 609-619.

Zimmerman, M. and Zeitz, G. (2002), 'Beyond survival: achieving new venture growth by building legitimacy', *Academy of Management Review*. **27** (3), 414-431.

7. Reflections on the Co-evolution of Innovation Theory, Policy and Practice: The Emergence of the Swedish Agency for Innovation Systems

Bo Carlsson, Lennart Elg and Staffan Jacobsson

INTRODUCTION

In 1987 the Swedish National Board for Technical Development (STU) invited a group of Swedish scholars in economics, engineering, management and sociology to do a joint study of 'Sweden's Technological System and Future Competitiveness' (the 'STS Project'). The research group continued its work for 15 years (on a part-time basis),[1] produced numerous publications summarized in three volumes (Carlsson 1995; 1997; 2002)[2] and developed a framework that is 'close' to the needs (and in part practice) of VINNOVA (The Swedish Agency for Innovation Systems). During this period, the Swedish technology policy changed, as reflected in the reorganization of the technology policy agency from STU to NUTEK (The Swedish National Board for Industrial and Technical Development) and eventually VINNOVA. At the same time, the thinking and focus of the research group shifted in light of its own experience, the changing policy environment and the research findings in related areas globally.

This chapter traces how our academic work and policy practice co-evolved and to reflect on what has been learned. Even though the STS project was initiated and funded by VINNOVA's predecessors, the links between the project and the formation of VINNOVA are not as clear and direct as might be supposed. The chapter is structured as follows. The next section outlines the economic, intellectual and political context of the project. After this, we then 'tell the story' of the STS project until 2002. Much of the story refers to how the group improved its

1 The annual budget for the group as a whole was on the order of 1 million SEK (around $130,000).

2 In addition, several PhD theses were written in conjunction with and with partial funding via the STS project (Tryggestad 1995; Ehrnberg 1996; Johansson 2001; Holmén 2001; Bergek 2002; Fridh 2002).

understanding of the structural components of an innovation system, how such systems are formed, and how the development processes vary between specific systems. It is also a story about improved methodological understanding and of some interactive learning between researchers and practitioners. We then analyze the formation of VINNOVA and consider the impact of the academic research in a wider political context. This is followed by a section which continues the story of interactive learning post 2002. Whereas our work until then had been focused on developing concepts and methods, it subsequently developed in a direction that was closer to the practical needs of the policy makers. Finally, we reflect on the whole process and what can be learned from it.

THE CONTEXT OF THE STS PROJECT

As long as Sweden has existed as a nation state, policymakers have shown an active interest in improving manufacturing and trade, and Swedish trade unions early adopted a positive attitude to structural renewal (Thorslund et al. 2005). However, the term 'innovation policy' was first used in official Swedish documents as late as a government bill in 1999. Moreover, with the exception of an occasional homage to Joseph Schumpeter, the earliest explicit use of research on innovation in policymaking appears to have been a parliamentary review of the organization of STU, which included a review of current research on innovation published in 1977. Soon after that, the Royal Swedish Academy of Engineering Sciences launched a study of Sweden's technical and industrial competence chaired by Erik Dahmén (see below in this chapter). A third early example was a joint project between STU and the Center for Policy Alternatives at MIT, which introduced the concept of 'new technology-based firms' into the Swedish discourse (Reitberger & Utterback 1982).

The initiative for the STS project clearly came from the policy side. In 1987 Göran Friborg, division chief at STU, contacted Bo Carlsson, professor of Industrial Economics at Case Western Reserve University in Cleveland, Ohio (previously associate director of the Industrial Institute for Economic and Social Research, IUI, in Stockholm), to explore the possibility of organizing and directing a group of Swedish researchers to study 'Sweden's Technological System and Future Competitiveness'. The basic charge was to gain a better understanding of the role of technology in economic growth, to raise (on a permanent basis) the level of competence, to analyze such questions and thereby to build a better foundation for technology policy. A comprehensive, interdisciplinary study was envisioned. Three leading Swedish research institutes were invited to participate: the IUI, the Department of Industrial Management and Economics at Chalmers University of Technology in Gothenburg, and the Research Policy Institute at the University of Lund.

The background to the initiative was the following. In the late 1970s the Royal Swedish Academy of Engineering Sciences, in collaboration with STU, had conducted a study of Sweden's technical and industrial competence. This was one of the largest research projects ever undertaken in Sweden and had a broad representation of academics, practitioners (especially engineers) and policy makers. One of the working groups within the project was focused on the reasons why Sweden suddenly found itself in an economic crisis after more than 100 years of rapid growth. The group was made up of industrialists, engineers, economists and policy makers (including Göran Friborg) and was chaired by professor Erik Dahmén. Bo Carlsson was the editor and main author of the report (Carlsson et al. 1979)

The project as a whole was regarded as a success. But while the findings were widely acclaimed, they had limited impact on economic policy. To illustrate, the five-year forecast for the Swedish economy that was issued by the government in 1980 was similar to its predecessors in that the neoclassical macroeconomic model upon which it was based reflected little insight about the central role of technological innovation, entrepreneurship and industrial competence in economic growth. There was no room left for policy (e.g., with respect to technology and innovation) other than traditional macroeconomic policies. Technological change was still treated as an exogenous factor rather than being viewed as taking place *within* the economic system. From a theoretical perspective this meant that in order to indigenize technological change, a more dynamic, systematic approach was needed. From a technology policy perspective, there was a need to better articulate the role of technology in order to build a stronger foundation and support for policy. The question became: what are the characteristics of an environment conducive to technological innovation? This is where the notion of technological systems entered. According to interviews with those involved, several strands of thought led to STU's initiative to set up the STS project.[3] One was the strategic challenge to small, advanced countries formulated as the 'SMALL COUNTRY SQUEEZE' by Kristensen and Levinsen (1983) and later Walsh (1987). The perceived challenge was that while the lower tech manufacturing base increasingly relocated to newly industrializing countries, the growing complexity of new core technologies forced small countries to focus on limited niches, with less opportunity to hedge strategic bets.

A second motive was STU's policy need to find a platform for discussing the economic effects of investments with the Ministry of Finance. Linking up research on innovation with economic research was an attempt to accomplish this.

Finally, there was an operational need at STU to understand the impact of its considerable investments in research on manufacturing technology, in the context of its linkages to other parts of Swedish industry. According to Friborg, it was these linkages that led to the concept of a 'technological system', while the work of Lundvall and colleagues was not yet appreciated.[4]

3 Interviews with Göran Friborg, Staffan Håkansson and Gunnel Dreborg, at that time senior analysts at STU's planning department.

4 This was brought on the policy agenda mainly through OECD's 'Technology-Economy-Productivity' project in the early 1990s.

The STS project started at the same time as work was getting under way at the OECD (Organization for Economic Cooperation and Development) in Paris on what came to be known as National Innovation Systems (NIS). Christopher Freeman (professor at the Science Policy Research Unit, SPRU, at the University of Sussex in England) had presented a paper at the OECD's expert group on Science, Technology and Competitiveness in 1982 in which he mentioned the concept of 'National Innovation System'. Drawing on Friedrich List's concept of 'national systems of production' (List 1841), Freeman discussed List's advice to Germany in the mid-19th century regarding how to catch up with the UK. The paper was considered by the expert group to be too provocative in that it identified a role in technological innovation by factors outside the neoclassical theoretical framework (Sharif 2006) and was not published until 2004 (Freeman 2004). But via the network of scholars connected to both the OECD and SPRU, the concept of 'national innovation system' began to catch on. Lundvall (1985) used the term 'innovation system' (but without the adjective 'national'). The first publication to actually use the term 'national innovation system' was Freeman's book *Technology and Economic Performance: Lessons from Japan* (1987). This was followed by a volume edited by Dosi et al. (1988) that contained four chapters on national innovation systems.[5]

Although it has not previously been articulated in the literature, Sharif (2006) shows that the concept of national innovation systems arose in the context of economic stagnation ('Eurosclerosis') and concern about competitiveness in Europe combined with the rapid rise of Japan. However, there are divergent views as to whether the concept came from economic theorists or from policy makers (Sharif 2006). What is clear is that the STS project started at the same time as the term 'national innovation system' was first published and that the economic contexts for the two approaches were similar.

From the point of view of the technology policy agency (STU) in Sweden, there was another strand of research that was relevant and important. In the early 1980s, STU had come in contact with the 'network' school of industrial marketing, formulated by Håkan Håkansson et al. (1982; 1987; 1989) at Uppsala University. This bottom-up approach to system thinking started from the observation that relationships between suppliers and users (of industrial products) were of a long-term nature, and that marketing should focus on building such relationships rather than closing individual deals, as emphasized in the 'standard' US marketing literature.[6] They also observed that firms were involved in multiple such relationships that constituted relatively stable 'networks' of firms – and other actors, such as research institutes.

---

5 It is interesting to note that there was an article entitled 'The Technology Innovation System in Japan' published in 1984 (MacDowall 1984) that does not appear to have been known or cited by other authors on innovation systems.

6 Perhaps not coincidentally, Bengt-Åke Lundvall's work (1985; 1988) on national innovation systems started out with a focus on user-producer interaction.

This view of industrial markets fitted much better than the textbook approach with the intuitive experience at STU, gained from extensive practice in influencing the formation of networks, and led to a long-term collaboration between STU (and its successor organization NUTEK) and the Uppsala group. This involved both inducing the researchers to look at new, emerging networks (e.g., by funding Anders Lundgren's PhD thesis (1991) on image analysis – which combined the network approach with a Schumpeterian perspective) as well as asking researchers from this group to evaluate STU programs in terms of how they contributed to the formation of new industrial networks.

Another foundation was Erik Dahmén's work on development blocs. His monumental study of Swedish industrial entrepreneurial activity (Dahmén, 1950) had become influential among technology policy makers and industrial economists in Sweden. His theory emphasized the disequilibrium nature of economic development, resulting in continually changing relationships among economic agents (see Dahmén 1989) and the importance of entrepreneurial activity. This dynamic micro-based theory of economic development was a starting point for the STS project.

While STU's collaboration with the network school helped develop the internal thinking within the agency, STU perceived a need to improve its position in the policy network. It had hitherto not been effective in communicating with the (national) political level which allocated resources to STU/NUTEK and decided on which terms of reference would be used to evaluate performance. One reason was that it was not obvious how their micro-focus could be aggregated to assess the impact on macro-economic performance.

Some local circumstances increased the need to find new ways of communicating with the policy level. First, mainstream neoclassical economists had a stronger hold on policy makers in Sweden (and particularly on the Ministry of Finance) than in many other countries. These policy makers were prepared to discuss the need for innovation policy only in the narrow framework of market failure, and were prepared to measure policy effects only in terms of changes in the *volume* of industrial R&D, not in terms of changes in direction or organization of R&D efforts or in any other way.

Secondly (and not unrelated), the 'linear model' of innovation – where increases in basic research somehow automatically lead to industrial development – was, and to some degree still is, the primary mental framework of Swedish policy makers. This has been fuelled by both self-interest from the academic lobby and mistrust of government policy by industry. One reason it has survived so long is the structure of Swedish industry, where most of R&D has been concentrated in a small number of large firms. These firms were able to internalize some of the innovation system's translation functions between discipline-based academic knowledge and multidisciplinary industrial competence needs.

THE STS PROJECT 1987-2002 – DEVELOPMENT OF BASIC FRAMEWORK AND METHODS

As indicated already, the concept of technological innovation system came from policy makers; it was the task of the research group to define more precisely what such a system was and how it could be used in innovation policy and practice. It was clear from the beginning that a comprehensive, interdisciplinary and systematic approach was required, going far beyond the relatively simple notion of adding a 'technology factor' to the aggregate production function as was being done concurrently by 'new growth' theorists such as Romer (1986; 1990) and Lucas (1988), essentially by indigenizing technological change in the analysis of the economic growth process. The main policy implication of that line of research was to increase knowledge creation via support of R&D.

An important feature of the STS project was that a representative of the technology policy agency, Lennart Elg, took an active part in our discussions throughout the course of the project. Not only did this facilitate communication between the researchers and the policy makers; in retrospect it was essential, because the initial charge to the group was very loosely formulated. The more precise problem formulation was worked out collaboratively as the work progressed. This required a high degree of mutual trust. It also made it somewhat difficult to distinguish which ideas came from the theory side and which from the policy and practice side.

One of the first questions addressed by the project group was whether or not it is appropriate to think of a single (national) innovation system, or whether it is more useful to think of a whole set of systems, each based on a different technology, generic or otherwise. We concluded that even though there are clearly national features and institutions shaping the rate and direction of innovation, the relevant actors and institutional arrangements in each area of technology are sufficiently different to justify examining each technology area separately. Thus, the English translation of the project title became 'Sweden's Technological Systems' (plural) and 'Future Development Potential'.[7] On this fundamental point we were influenced by the practice of STU which for long had been organized to support specific knowledge fields (e.g. semiconductor technology). This meant that in order to be relevant in the Swedish technology policy context, we needed to provide a basis for considering questions beyond the simple notion of supporting the national R&D effort. Eventually, we had to be able to deliver frameworks and tools that were valuable at the level of a single technological innovation system.

---

7 Actually, as already indicated, the original title referred to *konkurrenskraft* (competitiveness). This was changed to Dahmén's concept *konkurrensförmåga* (roughly translated as 'development potential') in order to avoid invoking static macroeconomic concepts and instruments such as exchange rates, interest rates, and relative wage rates preferring instead to refer to more dynamic concepts such as innovation and entrepreneurship that influence long-term growth.

We started by picking a relatively mature area of great importance to Swedish industry, factory automation, in which we knew *a priori* that Sweden had a strong position internationally and in which we could therefore expect to find a fully developed innovation system. By studying such a system, we thought we would be in a better position to identify important characteristics (including the role of public policy and other institutional features) than if we had selected a less fully developed system. Subsequently we broadened our analysis to three additional systems, namely the Swedish technological systems for electronics and computers, powder technology and pharmaceuticals, and biotechnology. Later on, in order to better understand the role of institutions in different settings, we extended our work to international studies of the biomedical cluster. With different funding, we also analyzed renewable energy technology, microwave technology and mobile data communication. Hence, in the course of the project we focused the systems studied on different units of analysis: knowledge fields, products, industries and sectors.

Some difficult theoretical issues confronted us from the start. In particular, it quickly became evident that we needed a clear definition of what we meant by 'technological system' and a theoretical formulation of its nature, function and composition. This led to the following definition[8]:

> A technological system may be defined as a network of agents interacting in a specific economic/industrial area under a particular institutional infrastructure or set of infrastructures and involved in the generation, diffusion and utilization of technology. Technological systems are defined in terms of knowledge/competence flows rather than flows of ordinary goods and services. They consist of dynamic knowledge and competence networks. In the presence of an entrepreneur and sufficient critical mass, such networks can be transformed into development blocs, i.e., synergistic clusters of firms and technologies within an industry or group of industries. (Carlsson & Stankiewicz 1991)

With this definition it became clear that we were expanding the national innovation systems framework beyond focusing primarily on institutions at the national level into a broader micro-macro perspective: it is necessary to understand not only the institutional framework but also the micro underpinnings of macroeconomic growth. Thus, individual actors – be they entrepreneurs or established firms – and their competence and interactions (networks) need to be incorporated in the

8 The definition was based on our reading of the literature in the 1980s in industrial dynamics – the economics of innovation, new technology, and structural change in industries and firms – and in the emerging field of evolutionary economics.

analysis.[9] There also needs to be an analysis of the economic impact of the systems, not just a description of their components.

Although there was much academic literature on networks that influenced us, we can also trace the effects from practice to theory. As indicated above, STU had a long tradition of working with network formation, often taking a 'broker' role. For instance, this was plainly seen in an evaluation of the Swedish textile and clothing policy that one of us participated in. This tradition has, of course, continued through the reorganization and renaming of STU. A more recent case is the numerous 'centers of excellence' that VINNOVA funds and that have contributed to shaping university-industry networks.

These were the structural components of a system that emerged in our early discussions. They were incorporated in our first collected volume, entitled *Technological Systems and Economic Performance: The Case of Factory Automation* (Carlsson, 1995). In addition to presenting the conceptual framework for the study of technological systems there were chapters on its various components: infrastructure, user-producer interaction and analyses of sub-sectors (suppliers) of factory automation. We also attempted to formalize the analysis (through a micro-based macro model) and empirically estimated the economic impact of economic competence and factory automation. The book concluded with a discussion of the policy implications and an international comparison.

The policy chapter was aimed at moving beyond the traditional discussion of market failure in a static context and into the realm of policymaking in a dynamic and systemic framework instead. Since this was a new arena for policy discussion without precedent (at least known to us) in the literature, and since we had only one case to draw upon, we took a cautious approach. Our recommendations were to focus on molding the system as a whole, recognizing that all the constituent parts can be subjected to the influence of policy, that policy intervention is not always necessary, that increased connectivity (networking) among the constituent parts may be beneficial, that policy intervention might be directed at weak parts of the system, and that it might be necessary to influence the visions of management of various entities. Thus, our recommendations involved the *direction* rather than the *content* of policy – away from a focus on R&D support (either generally or targeted for individual firms and projects) and towards an evolutionary systems perspective: 'recognizing the difficulties of identifying future technological opportunities, the emphasis shifts from the pursuit of individual technologies or approaches to either pursuit of alternative strategies simultaneously or emphasis on early identification of important developments and on increasing the absorptive capacity of the economy' (Carlsson & Jacobsson 1995). This represented a necessary conceptual step in a new policy direction, but it was still a long way from being practically useful.

9 The inclusion of individual actors, especially in the form of entrepreneurs, turned out to be unique to our approach as compared to national, regional and sector innovation system approaches.

At the end of this first phase of the project we had described and analyzed a fully developed innovation system in a particular domain of technology.[10] But it was still only a single case, and it was not clear that the findings could be generalized at all. For that reason we turned in the second phase (summarized in Carlsson 1997) to a study of three additional systems (electronics and computers, pharmaceuticals and powder technology), at different stages of development, all at an earlier stage than factory automation. The powder technology case was a system still in its infancy, centered on two knowledge fields (involving the production of extremely hard metallic and ceramic compounds under high pressure). The pharmaceuticals case was originally intended to be a study of biotechnology, but we soon found that there was not sufficient activity in Sweden in the early 1990s to justify such a narrowly focused study. Instead, we focused on the development of beta blockers (a product-centered system) and the transition from chemistry to biology/biochemistry as a basis for the pharmaceutical industry (changes in the knowledge base of a system centered on an industry). We also tried to gain a better understanding of general purpose technologies and their impact on economic growth. In addition, we simulated the economic effects of knowledge spillovers.

We found many similarities among the systems – especially, the importance of interaction among actors and organizations in the system, the characteristics of knowledge and spillover mechanisms, the absorptive capacity of actors within the system and the mechanisms that create variety. We found that path dependence plays a major role in every system. We also found important differences, especially with respect to where strength lies in each system – partly as a result of different historical background. As a result, it became obvious that public policy must be adjusted to each specific innovation system, that policy should focus on the weakest links in each system, and that absorptive capacity is an important constraint.

The policy discussion that resulted from our analysis distinguished between strengthening or preserving existing systems and creating new systems. In a dynamic perspective, the latter is far more important but also more difficult. We found three roles of public policy to be particularly important: to improve the receiver competence or absorptive capacity within the system – and here the prime movers in the form of existing firms or entrepreneurs play a role, along with the academic sector, in creating the necessary knowledge and skills; to promote connectivity among the various actors in the system; and to create variety. Again, we were quite influenced by practice. We have already referred to the 'broker' role of STU (promote connectivity). Additionally, by sharing the experience of a 'policy entrepreneur' in the area of microelectronics, Sven-Ingmar Ragnarsson, we came to understand the

10 The notion of technology was applied quite broadly here. Others may have labeled our approach at this time a sectoral innovation system.

need for a proactive science and educational policy that forms the competence base (including receiver competence) required to break path dependence.[11]

Thus, the policy discussion in the second phase of the project became both more dynamic and somewhat more concrete than in the prior phase. But it was still at a general and conceptual level. With hindsight, much of the discussion aimed at identifying generic system weaknesses, defined in terms of the nature of the structural components of an emerging technological system. Of course, from the perspective of a policy maker responsible for intervention in a specific system, a general discussion of this nature can only act as a source of inspiration, at best. Translation into specific policies and from there into practice was still a long way and many difficult steps away.

In the third phase of the project (see Carlsson, 2002) we focused on technological change in the rapidly evolving bio industries (loosely defined). In order to address issues involving path dependence and institutional differences in settings other than in Sweden, we wanted to make an in-depth international comparison. We chose Ohio. The choice of geographic focus reflected both practical considerations (part of the research team was located in Ohio, part in Sweden; given the nature and volume of the necessary data collection, physical presence in each location was required) and certain similarities in demographic size, industry composition[12] and economic history.[13]

The broad nature of the system under study forced us to confront a number of difficult conceptual and methodological problems:

- How should an emergent technological system be delimited?
- What should be the main sublevels and subunits of such an analysis? Should they be products, technologies, firms, or industries?
- How should we reconcile the longitudinal and synchronous aspects of the analysis – the fact that different (parts of) systems are at different stages of development?
- What yardsticks should be used to measure the performance of systems that are immature and rapidly evolving?
- How should a comparative analysis in such a heterogeneous field be done?

11 Government funding of academic research in Sweden includes both block grants to universities (around 45% of total funding in 2005) and competitive funding at both the program and project level, involving both peer-reviewed 'basic' research, and strategic or mission-oriented research. It is in the latter category that STU/Nutek/VINNOVA plays a role.

12 Both Ohio and Sweden have a large share of output and employment in manufacturing, especially transportation equipment and industrial machinery and equipment.

13 Both Ohio and Sweden performed well in international comparison in terms of economic growth until the mid-1970s but have since suffered a relative decline.

Questions such as these posed both analytical and methodological challenges. Analytically, we needed an even more dynamic approach, focusing not only on the technology generation and the organizational/institutional arrangements conducive to diffusion and utilization of technology (the supply side) but also on conversion of technical possibilities into business opportunities that are successfully exploited via entrepreneurial activity in the market (the demand side). This led to a clearer articulation of the three dimensions of technological innovation systems: the cognitive dimension (design space), the organizational and institutional dimension (actor networks) and the economic dimension (competence bloc).

Whereas in other approaches to innovation system analysis the object of study is given (a nation, a region, or a sector), the delimitation of a technological innovation system is itself a matter of careful deliberation and raises interesting methodological issues. In each dimension, where are the boundaries, and what techniques can be used to determine them? In terms of evaluating system performance we had to develop our own tools, as the existing literature provided little guidance.

Before we proceed to discuss the innovation policy environment and the interaction between innovation theory and policy we summarize below the theoretical, methodological and policy contributions of the project through the publication of our 2002 volume.

Theoretical Contributions

The main theoretical contributions of our study can be summarized as follows:

1. Technological systems emerge and evolve; the composition of actors and their roles vary over time as the magnitude and direction of the underlying driving forces shift, these changes do not necessarily follow any particular pattern or trajectory, linear or otherwise. Therefore, we need to move beyond static analyses.
2. There are several dimensions of innovation systems: cognitive (i.e., technology), organizational and economic.
3. In order to understand the dynamics and evaluate the *economic* performance of an innovation system, it is necessary to view the innovation process from two angles simultaneously: the cognitive or technology side, and the product or market side, respectively. A technological innovation system is defined primarily from the input (technology) side and focuses on the knowledge base of the system. Innovations are generated, diffused and utilized within the system. The market-demand side is represented by the notion of development and/or competence blocs, which are defined primarily from the product or user side. A development bloc is a synergistic cluster of firms and technologies that together constitute an industry or set of industries.

The market selection of technologies and products is intertwined and dynamic; it is not a linear process. Not all technical possibilities are converted into business opportunities, and not all business opportunities are successfully exploited in the market. Sometimes the impetus for change comes from the product side, sometimes from the technology side. Their confrontation in the market is what generates industrial dynamics in the form of economic transformation and growth. Thus, combining the two approaches allows us to examine why a particular system is or is not successful in converting technical possibilities into business opportunities that create economic growth. To our knowledge, ours is the only innovation system approach to integrate entrepreneurial activity into the analysis.

Methodological Contributions

In the area of methodology, our findings are the following:

1. A multidisciplinary approach to the analysis of innovation systems is often necessary. For solid empirical work in multifaceted fields, one needs people who understand each of the dimensions of innovation systems and who can also communicate with each other sufficiently to have a common understanding.
2. The content of each innovation system needs to be operationalized in each dimension (cognitive, organizational, and economic).
3. Identification of the actors and delimitation of the system are important research issues.
4. It is important to evaluate system performance and identify appropriate measures.

Policy Contribution

In the policy arena, the main insight had to do with the unique, dynamic, and multidimensional nature of each system. No policy can possibly fit all cases or any specific case over time. It is necessary to identify and address the weaknesses in each system, from each period to the next. But not all weaknesses need to be addressed through public policy. Sometimes systems emerge spontaneously and actors within them take appropriate action to avoid or overcome hurdles. In other cases, policy intervention such as building networks and infrastructure (perhaps in the form of absorptive capacity and legitimacy) or providing incentives and resources is necessary.

Our research showed that policy making requires high competence, a large measure of adaptability and continuous learning. We realized that as academic researchers we could only provide a conceptual framework – more precise and fully developed at the end of our project than at the beginning, to be sure – but still leaving the translation into specific policies and conversion into practice to policymakers who have the necessary political influence and to practitioners with sufficient specific competence. We identified, however, a set of 'generic' policy challenges (Carlsson & Jacobsson 1997); connected to these is a wide range of systemic policy instruments that may be deployed. These include science policy measures, organizing platforms or meeting places and forming early markets via e.g. procurement policies (see Chapter 16 by Smits et al. on the toolkit of policy makers pursuing systemic policies).

THE FORMATION OF VINNOVA

During the 1990s, many Swedish flagship industries became foreign-controlled through mergers or acquisitions: Asea became part of ABB, Volvo Cars – the automobile part of Volvo – was bought by Ford, Saab Automobile was acquired by General Motors, Astra became part of Astra Zeneca and Pharmacia merged with Upjohn. The task of policy could no longer be seen as supplying a small number of national flagship firms with 'good enough' R&D competence covering all their needs. Instead, Sweden had to compete globally for each new R&D (and production) investment, which involved the need to be world class in at least some areas. Can that be achieved without both stronger focusing of R&D efforts and the capability to refocus rapidly as priorities change?

Even if Sweden succeeded in retaining many successful large firms, their future growth will not necessarily result in new jobs, since many are operating in relatively mature sectors where relentless rationalization is a key means to profitability, and new production is located close to emerging markets.

For this reason, a second policy challenge is to build the foundations for new, high value-added businesses. This requires a much broader, systemic innovation policy approach. Perhaps in response to these challenges, there has been a gradual shift of focus at the Ministry of Industry from 'growth policy' towards 'innovation policy' and towards a more systemic view, starting around 1995.

Around 1996-97 attempts were made to formulate a 'Coordinated Growth Policy' (not yet innovation policy), inspired by OECD's 'Technology-Economy-Productivity' project, which would also have involved the ministries for labour and for finance. In the end, however, the attempt was abandoned.

At the same time, regional policy was being reshaped. The basic idea was that regional policies should be adapted to local needs (and local perception of needs). Regional actors were encouraged to formulate regional development agreements which set out how available resources should be prioritized. The first round of such agreements was reached in 1997-98.

In 1998, the ministries of industry, of labour market and of communications were merged into one large 'ministry of growth'. At the time it was described as a growth policy measure, but it is also consistent with a more long-term process to break down barriers between different policy areas. At about the same time as the new regional policy came into place, there was an increased interest in the role of the entrepreneur (not just 'small firms'), inspired by research on entrepreneurship.[14]

14 In particular the work done by Per Davidsson at the Jönköping International Business School.

More attention was also being focused on innovation issues – but these were still largely understood in a narrow sense as dealing with inventors.

In Sweden, research policy has oscillated between calls for relevance and defence of academic independence, but, as was noted above, the 'linear model of innovation' has retained a strong hold on policymakers. This was again demonstrated in 1998 when a high-profile committee on research funding, appointed by the Ministry for Education and Research, recommended that all government research funding should be based on academic excellence alone, and awarded on the sole basis of academic peer review. This provoked strong reactions from industry and the Ministry for Industry, who saw funding for industrially relevant research threatened. To engage in this battle, a credible alternative policy rationale was needed.

The innovation systems paradigm had already been adopted by several other countries, and filled two important requirements:

1. It gave a rationale for other kinds of policy intervention than research funding based on academic excellence.
2. OECD's extensive work on National Innovation Systems legitimized this perspective as a valid policy rationale.

A new committee set up by the Ministry for Industry recommended the creation of a funding agency for industrially relevant research, based on the innovation systems concept. This became VINNOVA, the Swedish Agency for Innovation Systems.

The process which led to VINNOVA was, thus, initiated by a political crisis which had little to do with advances in the theory of innovation systems: decision-makers looked for a way to legitimize a reorganization, and the innovation systems concept 'was in the air'. The fact that the innovation systems concept had been legitimized by OECD was crucial to its acceptance: 'For every academic theory there is always a counter-theory, but this was state-of-the-art according to OECD'. It also helped that the concept fitted well with the personal experience of some of the actors involved in the process.

The creation of VINNOVA can perhaps best be described in terms of Cohen et al.'s (1972) 'Garbage can theory', formulated to give a more realistic description of organizational decision making. In this theory, problems and solutions are to a large extent formulated independently of each other, and collected in a 'garbage can' of possibilities. When a problem demands the attention of a decision-maker, he looks into the garbage can for a solution which reasonably fits his interpretation of the problem. Note that this theory has a lot of elements in common with evolutionary theory: history matters, limited rationality, local search, satisfying behaviour, etc. Hence, the impact of theory on policy came through this type of indirect and non-linear process.

What does it then mean to base innovation policy on a systems perspective? At the simplest level, it could mean doing 'business as usual', only cloaked in a new rhetoric. A higher level of ambition might involve retaining the policy goal of producing industrially relevant research results, but making this process more effective by adopting a systemic view. However, we believe that in order to achieve the true potential of the systems perspective, policy objectives should be formulated in terms of more effective innovation systems.

As suggested by the STS project, a policy aimed at leveraging system performance needs to identify weak parts of each innovation system, in order to strengthen these, and enhance connections between the constituents of each system. Policy results should be measured in terms of impact on system performance. Since each innovation system is unique, and subject to genuine uncertainty, there is no hope of designing an 'optimal' policy. Instead policy learning becomes important. This learning can come from various sources: past experience (sometimes formalized as evaluation), policy analysis, foresight and intelligent benchmarking.

At VINNOVA, all programs involve cooperation between industry, academia and other actors, and must be supported by an analysis of the specific innovation system involved, its strengths and weaknesses. Funding is allocated along two dimensions: a Division for Competence Areas designs programs for specific technology fields and sectors.[15] A second Division for Actors designs programs aimed at supporting the connections between various groups of actors in the system. To support strategy development VINNOVA was set up with a strong policy analysis and evaluation capacity. This has been enhanced though long-term funding of four competence centers for 'Innovation system research on R&D and growth'.

With this background, it is necessary to discuss the relationship between theory and policy on two different levels: (1) theory as a basis for policy rationale and (2) theory as a basis for policy practice. As we have argued above, the innovation systems concept used to motivate the creation of VINNOVA was very much of the NIS flavour, as interpreted by the OECD. On the other hand, when the new organization tried to interpret its mandate, the experiences from the STS project were very important, not least in emphasizing that the innovation system looks different in different fields of technology or industries. STS showed that not only differences in national institutions were important (a result of the work on NIS) but that technological systems had different characteristics, and as a result that policy design had to be based on an understanding of the specific conditions of each unique system. As described above, part of this understanding was based on a mutual development between theory and policy practice, which long predates the creation of VINNOVA.

It has taken more than 20 years from 'industrial networks' to an Agency for Innovation Systems and the government's innovation strategy. For a number of years, it is fair to say that policy practice started to develop ahead of a policy rationale, which was still founded (at best) on a market failure argument. Academic research

---

15 This also accommodates legacy programs from predecessor organizations, which were mainly structured along this dimension.

learned a great deal from practice but also added to it.[16] However, with the creation of VINNOVA, it appears as if theory has run ahead of practice: VINNOVA has a mission framed in innovation systems terms but is still struggling with what that means in practice. We will now turn to the next stage in interactive learning, one in which policy makers can learn more from academics in applying a system perspective.

THE CO-EVOLUTION OF POLICY, PRACTICE, THEORY AND METHODS 1987-2002 – THE DEVELOPMENT OF FUNCTIONAL ANALYSIS

At the establishment of VINNOVA in 2001, some of its activities were organized, as mentioned above, into a number of competence areas, each representing a system, although these were delineated in different ways (some broad sectors, e.g. telecommunications and other applications of a generic technology (e.g. IT in home care)). This led immediately to the question of how to operationalize the notion of innovation system for public policy purposes.

Meanwhile, some members of the STS team (Johnson 1998; Jacobsson & Johnson 2000; Johnson & Jacobsson 2001, subsequently developed in Jacobsson & Bergek 2004, Jacobson et al. 2004 and Bergek et al. 2005) had applied system analysis to the energy field. The starting point of that work was our 1997 policy discussion on blocking mechanisms in system formation (Carlsson & Jacobsson 1997).

Initiated by a PhD student (Anna Bergek, formerly Johnson), we developed a 'functional analysis of innovation systems'. With this framework, we go beyond analyzing technological systems at the structural level only and added a second level: functions in innovation systems.[17] These functions capture what is being achieved by the system (and what is not achieved). System weaknesses can then be described not only in structural but also in functional terms. Seven such functions were eventually identified (Bergek et al. 2005; see also Bergek et al. in this book): (1) knowledge development and diffusion; (2) guiding the direction of search; (3) entrepreneurial experimentation; (4) market formation; (5) legitimation; (6) resource mobilization (financial and human capital in particular); and (7) development of positive externalities.[18]

16 An example is the adoption of the 'network' concept of Håkansson et al. Indeed, Hans Weinberger gave his history of STU the title 'The network entrepreneur' (Weinberger, H. Nätverksentreprenören (The network entrepreneur) Stockholm Papers in the History and Philosophy of Technology 1997).

17 See for a more elaborate discussion of the functions approach the chapter of Bergek et al. in this book.

18 The functional analysis was developed further in collaboration with researchers in Utrecht.

With funding from STEM (Swedish Energy Agency), we tentatively applied an early version of this framework to the evolution of wind and solar power in Germany (and the Netherlands) but only as a way to analyze the historical evolution of these systems (Jacobsson & Johnson 2003; Jacobson et al. 2004). In 2003, there was chance meeting with a senior official of VINNOVA, S.G. Edlund. Edlund, with a background in control engineering, asked one of us if the notion of 'function' could be useful in innovation systems analysis and he was shown our early attempts to apply such a framework to renewable energy technology. This led to a small experiment whereby one of us applied our framework to a case already developed at VINNOVA, wood manufacturing. The response was enthusiastic and the case study was presented to most analysts at VINNOVA – legitimizing the approach. This positive experience led to a further project involving the application of this framework to three more case studies. Also these were presented to most analysts at VINNOVA.

In these experimental cases, we went from arranging the empirical material in functional terms to specifying the policy problem in such terms. The objectives of policy were formulated in terms of 'process goals' that were specified as functions that need to be fulfilled, rather than in structural terms or in final outcomes such as the diffusion of a particular technology in a particular application. Current functional patterns can be explained by linking functions to the structural components of a system (actors, institutions, networks), identifying, in particular, mechanisms that 'block' the development of powerful functions. A functional analysis can then act as a 'filter' for policy makers in finding the precise points where an intervention may help the system to move in the desired direction. Of course, for a policy maker who has the responsibility to design policies for a specific innovation system, a framework like this may prove to be quite valuable.

Although VINNOVA only wanted us to solve some methodological problems in connection with the experimental applications, we decided jointly that the best way to do this was to write a policy 'manual' that would synthesize and make available our findings and that could also serve as a teaching tool. In this manual we included not only 'functional analysis' but much of what we had learned in the course of the whole STS project, including methodological issues (Bergek et al. 2005). The manual was then seen by a member of the Swedish Environmental Protection Agency who requested another experimental application – organic milk. Again, the framework seems to meet the needs of policy makers.

SOME REFLECTIONS ON THE PAST AND FUTURE

Although the STS project was set up with a loosely formulated charge, it was expected to contribute to an improved understanding of technology in economic growth and, therefore, build a better foundation for technology policy as well as to raise competence. The former does not mean that researchers should provide specific policy suggestions, although many of us always try to do that. Policy making is a 'political' process where the political context needs to be understood, and where knowledge is needed of which policy suggestions have been put forward previously, and been rejected, and which therefore are not sensible to put forward. Policy makers are keener to see new perspectives, or 'glasses' through which they may interpret the world. In that connection, it is a strength that researchers from different disciplines come to similar conclusions – in our case the similarities between the Uppsala school of industrial networks and our approach which both end up at a 'meso' level of some kind. Our work was, therefore, only one of several that used a system perspective applied to a meso level.

Yet, it was not any of these that most influenced the formation of VINNOVA (at least the choice of its name) but it was the National Systems of Innovation concept which, although less suitable to an organization like VINNOVA, due to its main focus on policies for specific technological systems, was legitimized by the OECD and used in a political struggle over the nature of science and technology policy. In that respect, the OECD was much more influential than the STS project. Nor was the STS project successful in finding means for VINNOVA and its predecessors to communicate to the politicians at the national level, but it is fair to say that the politicians have taken notice of the combined efforts of research using different system approaches. A case in point is the current vogue to design cluster policies.

We have probably been successful, although the future will tell, in developing an analytical framework and methods for analyzing systems centred on a particular knowledge field/technology or product. We got a head start here as we quickly left the NIS approach. There are also some attempts to apply the 'functions in innovation systems' approach at VINNOVA although it is much too early to say whether or not it will be widely implemented (there is also an international interest in the approach, see e.g. Hekkert et al. 2006).

We were, however, only partly successful in developing competence. In addition to senior researchers, only two PhDs graduated within the STS project, although several others worked with the same framework using additional funding from other sources. The emphasis on senior researchers means that there is now a clear shortage of younger researchers who can undertake analyses using our approach. It is an old lesson that needs to be re-learned: research and doctoral training should go hand in hand.

Over the whole period, there was an interplay between 'needs driven' or directed research and 'curiosity driven' academic work. Both VINNOVA and STEM fund directed research but both do it with much space for free thinking by the academics. Indeed, the early formulations of the 'functional approach' were entirely the result of 'curiosity driven' research although, as underlined above, it rested on earlier work done in the STS project. These efforts proved to be highly complementary. Yet, it would have been impossible to assess the utility of the STS project beforehand. In particular, it was impossible to even imagine what would come out of it 15 years after its inception. As with much other knowledge development, the time frame is long and uncertainties profound. Funding of directed research therefore need not only to allow much space for 'curiosity driven' work but also to be patient.

Yet, the process of developing and applying new knowledge could have been more effective. A number of lessons can be drawn for future work. First, although the research was followed by one or two members of STU and Nutek, more, and more direct, interaction with policy makers would have been useful. A better mutual understanding of the policy issues between the research team on the one hand and the decision makers within STU/Nutek/VINNOVA as well as the policy makers at the ministry level on the other might have led to better communication and therefore both better theory (greater relevance of the research) and better policy formulation. Second, the conversion of policy into practice could also have been more effective. As it was, there was almost no communication between the researchers and those responsible ('practitioners') at STU and Nutek for implementation of policies in specific domains. A more intense interaction could have been a channel from academic research to the policy making community and instrumental in developing the receiver competence among policy makers. Perhaps collaborative research efforts would have been part of the solution in the form of doctoral level courses by academic researchers for policy makers. Third, although we developed a manual for policy makers, more synthesis work is required in order to make science useful – efforts are required to 'translate' the ever growing academic literature to useful methods and tools. It is in the interest of both policy makers and academics to organize this 'translation' and it is the responsibility of both parties to act upon this opportunity to use science as a tool to shape a useful policy.

REFERENCES

Bergek, A. (2002), 'Shaping and exploiting technological opportunities: the case of renewable energy technology in Sweden', Ph.D dissertation, Department of Industrial Dynamics, Chalmers University of Technology, Gothenburg.

Bergek, A., Jacobsson, S., Carlsson, B., Lindmark, S. and Rickne, A. (2005), 'Analyzing the dynamics and functionality of sectoral innovation systems: a manual for policy makers', report delivered to VINNOVA.

Carlsson, B. (ed.) (2002), *Technological Systems in the Bio Industries: An International Study*, Dordrecht, The Netherlands, Amsterdam, The Netherlands, Boston, USA and London, UK: Kluwer.

Carlsson, B. (ed.) (1997), *Technological Systems and Industrial Dynamics*, Dordrecht, The Netherlands, Amsterdam, The Netherlands, Boston, USA and London, UK: Kluwer.

Carlsson, B. (ed.) (1995) *Technological Systems and Economic Performance: The Case of Factory Automation*, Dordrecht, The Netherlands, Amsterdam, The Netherlands, Boston, USA and London, UK: Kluwer.

Carlsson, B., Dahmén, E., Grufman, A., Josefsson, M. and Örtengren, J. (1979), 'Teknik och industristruktur: 70-talets ekonomiska kris i historisk belysning' (Technology and Industry Structure – The Economic Crisis of the 70s in Historical Perspective), IUI, Stockholm.

Carlsson, B. and Jacobsson, S. (1997), 'In search of a useful technology policy: general lessons and key issues for policy makers', in B. Carlsson (ed.), *Technological Systems and Industrial Dynamics*, Dordrecht, The Netherlands, Amsterdam, The Netherlands, Boston, USA and London, UK: Kluwer, pp. 299-315.

Carlsson, B. and Jacobsson, S. (1995), 'Factory automation and government policy', in B. Carlsson (ed.), *Technological Systems and Economic Performance: The Case of Factory Automation*, Dordrecht, The Netherlands, Amsterdam, The Netherlands, Boston, USA and London, UK: Kluwer, pp. 417-440.

Carlsson, B. and Stankiewicz, R. (1991), 'On the nature, function, and composition of technological systems', *Journal of Evolutionary Economics,* 1 (2), 93-118.

Cohen, M.D., March, J.G. and Olsen, J.P. (1972), 'A garbage can model of organizational choice', *Administrative Science Quarterly*, 17 (1), 1-25.

Dahmén, E. (1989), 'Development blocs in industrial economics', in B. Carlsson (ed.), *Industrial Dynamics: Technological, Organizational, and Structural Changes in Industries and Firms,* Dordrecht, The Netherlands, Amsterdam, The Netherlands, Boston, USA and London, UK: Kluwer, pp. 109-121.

Dahmén, E. (1950), 'Svensk industriell företagarverksamhet', Stockholm: IUI. Also published in 1970 by the American Economic Association Translation Series under the title 'Entrepreneurial Activity and the Development of Swedish Industry, 1919-1939', Vol. 1 and 2, Stockholm.

Dosi, G., Freeman, C., Nelson, R., Silverberg G. and Soete, L. (eds) (1988), *Technical Change and Economic Theory*, London, UK and New York, USA: Pinter.

Ehrnberg, E. (1996), 'Technological discontinuities and industrial dynamics', Ph.D. dissertation, Department of Industrial Management and Economics, Chalmers University of Technology, Gothenburg.

Freeman, C. (2004), 'Technological infrastructure and international competitiveness', *Industrial and Corporate Change,* 13 (3), 541-569.

Freeman, C. (1987), *Technology Policy and Economic Performance: Lessons from Japan,* London, UK and New York, USA: Pinter.

Fridh, A.-C. (2002), 'Dynamics and growth: the health care industry', Ph.D. dissertation, Department of Industrial Economics and Management, Royal Institute of Technology, Stockholm.

Håkansson, H. (ed.) (1989), *Corporate Technological Behavior: Cooperation and Networks*, London, UK and New York, USA: Routledge.

Håkansson, H. (ed.) (1987), *Industrial Technological Development: A Network Approach*, London: Croom Helm.

Håkansson, H. (ed.) (1982), *International Marketing and Purchasing of Industrial Goods: An Interaction Approach*, Chichester: John Wiley.

Hekkert, M., Suurs, R., Negro, S., Kuhlmann, S. and Smits, R. (2007), 'Functions of innovation systems: a new approach for analyzing technological change', *Technological Forecasting and Social Change,* Vol. 74, 4, pp 413-432

Holmén, M. (2001), Emergence of Regional Actor Systems. Generic Technologies and the Search for Useful or Saleable Application of Industrial Dynamics, Chalmers University of Technology, Göteborg, Sweden, ISBN 91-7291-038-0, ISSN 0346-718x.

Jacobsson, S. and Bergek, A. (2004), 'Transforming the energy sector: the evolution of technological systems in renewable energy technology', *Industrial and Corporate Change,* **13** (5), 815-849.

Jacobsson, S. and Johnson, A. (2003), 'The development of a growth industry: the wind turbine industry in Germany, Holland and Sweden', in S. Metcalfe and Cantner, U. (eds), *Change, Transformation and Development,* Heidelberg and Berlin: Physica-Verlag, pp. 197-228.

Jacobsson, S. and Johnson, A. (2000), 'The diffusion of renewable energy technology: an analytical framework and issues for research', *Energy Policy,* **28** (9), 625-640.

Jacobsson, S., Sandén, B. and Bångens, L. (2004), 'Transforming the energy system: the evolution of the German technological system for solar cells', *Technology Analysis and Strategic Management,* **16** (1), 3-30.

Johansson, D. (2001), 'The dynamics of firm and industry growth: the Swedish computing and communications industry', Ph.D. dissertation, Department of Industrial Economics and Management, Royal Institute of Technology, Stockholm.

Johnson, A. (1998), 'Functions in innovation system approaches', mimeo, Department of Industrial Dynamics, Chalmers University of Technology, Gothenburg.

Johnson, A. and Jacobsson, S. (2001), 'Inducement and blocking mechanisms in the development of a new industry: the case of renewable energy technology in Sweden', in R. Coombs, K. Green, A. Richards and W. Walsh (eds), *Technology and the Market, Demand, Users and Innovation,* Cheltenham, UK and Northampton, MA, USA: Edward Elgar, pp. 89-111.

Kristensen, P. and Levinsen, J. (1983), *The Small Country Squeeze,* Roskilde: Førlaget før Samfundsøkonomi.

List, F. (1841), *Das Nationale System Der Politischen Ökonomie,* Basel: Kyklos (translated and published under List, F. (1841), *The National System of Political Economy,* London: Longmans, Green and Co).

Lucas, R., Jr. (1988), 'On the mechanics of economic development', *Journal of Monetary Economics,* **22** (1), 3-41.

Lundgren, A. (1991), *Technological Innovation and Industrial Evolution: The Emergence of Industrial Networks,* Stockholm: EFI.

Lundvall, B.-Å. (1988), 'Innovation as an interactive process: from user-producer interaction to national innovation systems', in G. Dosi, et al. (eds), *Technical*

Change and Economic Theory, London, UK and New York, USA: Pinter, pp. 349-369.

Lundvall, B.-Å. (1985), *Product Innovation and User-Producer Interaction*, Aalborg: Aalborg University Press.

MacDowall, J. (1984), 'The technology innovation system in Japan', *Journal of Product Innovation and Management*, **1** (3), 165-172.

Reitberger, G. and Utterback, J. (1982), 'Technology and industrial innovation in Sweden: a study of new technology-based firms', Center for Policy Alternatives, MIT.

Romer, P. (1990), 'Endogenous technological change', *Journal of Political Economy*, **98** (5, part 2), 71-102.

Romer, P. (1986), 'Increasing returns and long-run growth', *Journal of Political Economy*, **94** (5), 1002-1037.

Sharif, N. (2006), 'Emergence and development of the national innovation systems concept', *Research Policy*, **35** (4), 745-766.

Thorslund, G.J., et al. (2005), 'The end of an era? Governance of Swedish innovation policy', in OECD (ed.), *Governance of Innovation Systems, Vol. 2, Case Studies in Innovation Policy*, Paris: OECD Publishing, pp. 245-282.

Tryggestad, K. (1995), 'Teknologistrategier og postmoderne Kapitalisme: introduksjon av computerbasert produksjonsteknikk', Ph.D. dissertation, Lund University, Lund.

Walsh, V. (1987), 'Technology, competitiveness and the special problems of small countries: a survey of the literature', IKE workshop on Technological Change and the Competitiveness of Small Countries, Denmark.

III. Ongoing and new issues of innovation policy

8. Innovation and Small and Midsize Enterprises: Innovation Dynamics and Policy Strategies

Philip Shapira

INTRODUCTION

A central component of any effort to encourage economy-wide innovation has to be attention to small and medium-sized enterprises (SMEs). Such firms are the modal type, comprising about 99 percent of all enterprises in developed economies (OECD, 2002, Table A1). As indeed many panels and studies have declared in recent years, for an economy to be successful in a world of increased international competition, innovation – the creation and use of new products, process, services, or organizational forms – is critical, and small firms and entrepreneurs (along with large firms and other institutions) have themselves to be innovative to contribute towards this goal (see, for example, European Commission 2005; JSBRI 2007; Kaufmann Foundation 2007).

The difficulty is that many small and midsize enterprises lack the capabilities, resources and incentives for innovation, are concerned about the risks associated with change, and – even if disposed towards innovation – are skeptical about the role of government in its efforts to foster innovation. Moreover, the scale and diversity of the small business world makes policy design to encourage innovation in SMEs anything but a simple process, despite – or perhaps because of – the fact that government is a major factor in establishing frameworks, institutions and incentives which affect innovation in SMEs. In other words, SMEs and policymakers are variously influenced, aided and at times stymied by one another in the directions that each takes, in the context of the multiple further factors that bear upon decisions, strategies and performance in the worlds of business and public policy.

This chapter explores the relationships between innovation dynamics in SMEs and innovation policy strategies for SMEs. There are three main aims: first, there is a review of evidence and thinking about innovation and SMEs. Second, the chapter probes the relationships between innovation system dynamics and the development of policy strategies for promoting innovation in SMEs. This part of the chapter focuses attention to the evolution of models used to understand and influence innovation. Third, the chapter concludes with a synthesis and discussion of insights that might be useful for future rounds of policymaking for SME innovation.

All small and medium-sized enterprises (SMEs) are small; but, not all SMEs are innovative. Yet, this apparently simple statement belies the diversity and complexity of the small business sector and its connection to innovation. Hence, the chapter opens with an 'orientation' – to discuss how SMEs are defined and to examine how innovation differentially affects, and in some cases is driven by, contrasting types of SMEs.

THE WORLD OF SMES: ORIENTATION

What exactly is an SME? What is an innovative SME?

Observably, an SME is an enterprise which is not large. Intuitively, an SME as an object for policy assistance is also an enterprise which lacks market power. In other words, what might be regarded as a small firm in sectors such as telecommunications or chemicals might be one that is larger than a small firm in less concentrated sectors like beautician services or restaurants. This combination of small size and market power relative to the reference sector is a useful working concept of an SME, but it is hard to measure without further definition. Hence, official statistical definitions usually focus on employment size, sales, or capital assets in defining SMEs. These are measurable, but not necessarily consistent across countries, as a comparison of US, European and Japanese definitions of SMEs demonstrates. In the US, a small business is defined (SBA 2005a) as a business concern 'that is independently owned and operated and which is not dominant in its field of operation'. Financial and size definitions vary by industry – in many industries, the small business employment threshold is 500 or fewer employees, but can diverge (e.g. 100 or fewer employees in wholesale trade or 1,500 or fewer employees in telecommunications). In Japan, SMEs are generally defined as enterprises with 300 or fewer regular employees (JSBRI 2004), but again with variations (e.g. 100 or fewer employees in wholesale and services; 50 or fewer employees in services). Small enterprises in Japan are defined as those with 20 or fewer employees (5 or fewer in commercial and service sectors). For Europe, there are differences in definitions by country, however the European Union's Observatory of European SMEs (2004) uses the classification (with employment size range): medium-sized enterprises (50-249 employees); small enterprises (10-49 employees); and very small (micro) enterprises (9 or fewer employees).

Notwithstanding these definitional differences, it is apparent that in the major economic blocs of the US, Japan and Europe, SMEs constitute – as noted in the introduction – about 99 percent of all enterprises. However, there are significant differences in the employment and economic role of SMEs. For example, in the US, less than 50 percent of enterprise employment is in SMEs, compared with about 70 percent (on average) in Europe and Japan (Table 8.1). Within Europe, there are

Table 8.1 Enterprise size classes and employment, US, Japan and EU-19

	SME	Large	% SME	% Large	N*
Enterprises (USA 2000)	21,164	59	99.7	0.3	21,223
Persons (USA 2000)	63,593	66,042	49.1	50.9	129,635
Enterprises (Japan 2001)	4,690	13	99.7	0.3	4,703
Persons (Japan 2001)	29,963	12,692	70.2	29.8	38,277
Enterprises (EU-19 2003)	19,270	40	99.8	0.2	19,310
Persons (EU-19 2003)	97,420	42,300	69.7	30.3	139,710

SME = Small and Medium Enterprise; Large = Large Enterprise. * x 1,000.

Sources: Observatory of European SMEs (2004), Table 3.5, updated by author with Japanese data from JSBIR (2004), Supplementary Statistical Tables 1(2) and 3(3). There are some differences in SME size classes (e.g. EU = 250 employees or fewer; Japan = 300 employees or fewer).

also considerable differences in SME roles by country: by employment share, the UK, Germany and the Netherlands are examples of countries where large firms are relatively more dominant, whereas SMEs are relatively more dominant in France, Spain or Italy (Observatory of European SMEs 2004, Table 3.2). These differences signal important variations in the structural composition of sectors by enterprise size. For example, in Germany there is a well-developed group of highly-innovative and often globally-prominent medium-size manufacturing firms (Simon 1996). The manufacturing sector in Japan contains a high proportion of very small (4 employee or less) companies, making sub-components in extensive, vertically-integrated large company supply chains (Shapira 2008). This tiered pattern of very small manufacturing suppliers is relatively less common in the US. In 2004, 11 percent of US manufacturing employees worked in enterprises with fewer than 20 employees; this was almost half the 21 percent of Japanese manufacturing workers who were employed that same year in enterprises with under 20 employees (OECD 2008).

Although many SMEs in advanced economies are highly productive and sophisticated, on average across all advanced countries, small firms tend to lag larger firms by standard measures of per capita output. For example, across all OECD member countries, manufacturing labor productivity is lower in SMEs than in the largest enterprises. In the UK, the normalized productivity of workers in manufacturing enterprises in the smallest size class of 1-9 employees is 61 percent of that of workers in manufacturing enterprises with more than 250 employees. The comparable figures for the US and Japan are 42 percent and 27 percent respectively, with the overall OECD median at 41 percent (OECD 2008; data for most recent year through 2005). On average, small firms also lag large ones in terms of wages, working conditions, aggregated R&D expenditures and the pace of new technology adoption (see, for example, Parker 2001; Kennedy & Hyland 2003). Factors that potentially influence these differences in outcomes across size classes include capital intensity, scale economies and managerial expertise. Yet, at the same time, several studies in the US and Europe suggest that SMEs are responsible for a disproportionate share of net new job creation (these studies include: Birch 1979; Storey & Johnson 1987;

see also discussion in Observatory of European SMEs 2004). In addition to these US and European studies, at least one Japanese study has emphasized the central role of small firms in Japan's post-World War II economic development (Friedman 1988). However, at least for quantitative US and European studies of SME job creation, there has been much debate about the methodologies, reliability and implications of such studies (see, for example, Harrison & Bluestone 1994; Davis et al. 1993), including the extent to which SME employment growth is due to indigenous entrepreneurialism or to sectoral shifts and corporate outsourcing. Some studies note that while small firms may contribute disproportionately to new job creation, over time they also have higher job destruction rates relative to their employment share (Broersma & Gautier 1997; Hijzen et al. 2007).

It is in this uncertain context about the basic dynamics of the SME sector that discussion and policy formulation about innovative SMEs takes place. Clearly not all SMEs are innovative, nor does innovation automatically imply enterprise stability or even growth. However, in considering what an innovative SME is, the stylized view is often that of a high-technology start-up. In the US, this is idealized by the San Francisco Bay Area 'garage start-ups' of the likes of Hewlett and Packard (HP) or Jobs and Wozniak (Apple Computers).[1] Here, entrepreneurial innovators establish a small technology-oriented enterprise which subsequently grows into a major corporation. There have been many examples in recent years of entrepreneurial and innovative start-ups, although it is apparent that these represent only a small proportion of SMEs. Yet, this stylized view of an innovative SME is far too narrow (especially from an innovation systems perspective). There is much more to innovation than the development of new high technology products (undeniably important as this is). More formally, innovation can involve the introduction of new or improved products, processes, organizational and business methods, and marketing approaches; innovations can be radical or incremental in nature; and innovations can be associated with both goods and services (see, for example, discussion in OECD 2005a). Recent work has reminded us that while innovations codified in patents and new product introductions are more readily identifiable, there is a wide spectrum of other business innovation forms that are important albeit hard to measure (NESTA 2007).

This broad definition of innovation is helpful in distinguishing inventions (new discoveries) from innovations (new applications), and entrepreneurship from innovation. Entrepreneurship, which is the process of founding and running a business, may certainly involve risk but it does not necessarily result in innovation. In many instances, entrepreneurs follow tried paths and do not experiment or innovate (Staber 2005). Additionally, this broad definition allows us to see the variety of ways in which innovation can occur in SMEs. The mix of (1) innovation types (e.g. product, process, organization, etc.) and (2) innovation forms (e.g.

---

1 This is nicely epitomized today by the Silicon Valley start-up venture capital fund Garage Technology Ventures (http://www.garage.com/).

radical v. incremental or embodied in goods or services) is mediated by (3) sector characteristics and enterprise strategic orientation and (4) national and regional framework conditions.

In short, the world of small business is diverse, and the ways in which innovation will occur are similarly differentiated. While many SMEs are not especially innovative, others are – and often in ways that are not easily monitored or measured. For example, in a high-technology start-up SME, there will be a current and future focus on innovation, most likely through undertaking internal R&D and developing a novel product for which intellectual property protection (a pre-requisite for venture financing) can be obtained. On the other hand, an existing and mature manufacturer may or may not have a current focus on innovation. A mature manufacturing firm could be pursuing a niche market, by introducing innovative products; it could be attempting to reduce costs to customers through process innovation or organizational improvements; or it could be offering value-added services (such as design, maintenance, or training) to customers to distinguish itself from competitors. Alternatively, a knowledge-intensive business service company, such as a managerial consulting firm, may seek to formulate innovative ideas and service concepts and make these available to potential customers. In this process, formal R&D may be much less important than practices of learning and improving by doing or acquiring. Particularly in advanced services and creative industries but also across other types of enterprises, Miles and Green (2008) observe that innovations today may occur in a series of diffuse and interrelated ways, including through developments in management, business models, value-chain location and positioning, communications, and marketing and customer relations.

EVOLUTION OF THOUGHT ABOUT INNOVATION IN SMES: A SELECTIVE SURVEY

How has the world of scholarly research mirrored these developments in the forms and pathways through which innovation occurs at the enterprise level? One would, of course, expect an inter-relationship between the practice of innovation in enterprises, including SMEs, and ideas and concepts about those practices. Indeed, there has been a significant growth in recent years in scholarly activity related to small and medium-sized enterprises. Within this, attention paid to innovation has also increased. An analysis of publications in the Social Sciences Citation Index (SSCI) suggests that the fraction of publications with an emphasis on innovation has increased – from about 9 percent of articles related to SMEs in the period 1990-1993 to a mean of 23 percent in the period 2005 to mid-2008 (see Figure 8.1).[2] The number

2 Author's Vantage Point analysis of search conducted of the Web of Science, Social Science Citation Index (SSCI), on August 2, 2008. Search terms for 'All SME Publications' = [(SME), NOT ('simulated moments estimat*' OR 'subject matter expert*' OR 'subject-matter expert*'

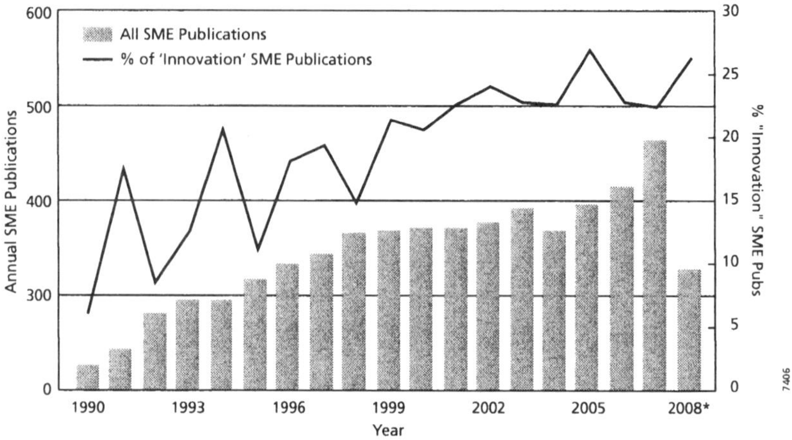

Source: Social Science Citation Index (SSCI), Author's Vantage Point analysis of SME publication records. N = 4,253 (limited to articles, reviews, and notes and excluding book reviews, editorials, and other document types) and SME publications with a focus on innovation (N=811). For definition of search terms, see Note 2 in this chapter.

Figure 8.1 Growth of SSCI Publications on SMEs with a focus on 'innovation', 1990-2008 (Data for 2008 for part year (7 months))

of articles published annually on innovation topics related to SMEs has increased, from under 15 in the early 1990s and 50 in 2000 to 85 in 2007. The leading journal categories for SME innovation publications are Management, Business, Economics, Operations Research & Management Science, Planning and Development, and Industrial Engineering (journals can be listed in more than one category). Articles in the Economics journal category showed the fastest growth – SME innovation publications were up 191 percent for the period 2000-2007 compared with 1993-1999, although articles in the Management journal category are still published

OR 'secure messaging exchange' OR 'SME Server')] AND [('small compan*') OR ('small firm*') OR ('small business') OR ('small enterprise*') OR ('small manufactur*') OR ('small establishment*') OR ('small entrepren*') OR ('small employer*')] AND [('medium size* compan*') OR ('medium size* firm*') OR ('medium size* business') OR ('medium size* enterprise*') OR ('medium size* manufactur*') OR ('medium size* establishment*') OR ('medium size entrepren*') OR ('small employer*')] AND [('mid size* compan*') OR ('mid size* firm*') OR ('mid size* business') OR ('mid size* enterprise*') OR ('mid size* manufactur*') OR ('mid size* establishment*') OR ('mid size entrepren*') OR ('mid size employer*')] AND [('small and medium size*' OR 'small to medium size*' OR 'small and mid size*' OR 'small to mid size*') AND (compan* OR firm* OR business* OR enterprise* OR manufactur* OR establishment* OR entrepren* OR employer*)]. N = 4,253 (limited to articles, reviews, and notes and excluding book reviews, editorials and other document types). 'Innovation SME Publications' = 'All SME Publications' with 'innovat*' in key term. N = 811.

most frequently (45 percent of all SME innovation articles published in 2007 are in journals listed under the Management category).

What does this burgeoning literature say about innovation in SMEs? And, how well does it reflect the diversity of the worlds and innovation strategies of SMEs discussed above?

A starting point for much of the modern discussion about SMEs and innovation is Schumpeter's (1934) argument that the broader innovation process is driven by entrepreneurial innovation. In this view, innovation is advanced through waves of 'creative destruction' stimulated by the activities of progressive small business entrepreneurs. The quandary is that subsequently Schumpeter (1942) appeared to change his position, arguing that innovation had become dominated by big corporations with large R&D budgets and that the role for small entrepreneurs was much reduced. It has been suggested that Schumpeter's position is more consistent than often recognized (Langlois 2003). Nonetheless, Schumpeter's dichotomy has stimulated considerable research to assess the innovative role of SMEs.

Thus, in recent decades, a series of studies has found that SMEs are responsible for a disproportionately high proportion of innovations relative to their employment size or compared with larger firms. These include what might now be regarded as 'classic' 1980s firm size and innovation studies, such as Gellman Research Associates (1982) and Pavitt et al. (1987), and more recent work (see review in Acs & Audretsch 2005). Explanations for strong small firm performance in innovation include their flexibility, lack of hierarchy, willingness to exploit new niches and motivation. This body of work has sought to address deficiencies in conventional micro-economic theory (which is felt to have little to say on sources of innovation), management theory (including influential arguments about the technological pre-eminence of big firms due to advantages of large research budgets, scale economies, and managerial and system integration capabilities, as in Chandler 1990), and specific critiques that suggest the innovative and economic role of SMEs has been overestimated (e.g. Harrison & Bluestone 1994). Yet, there remains debate. For example, Tether (1998) makes the point that while 'counts' of innovation may be disproportionately high for SMEs, if assessments take into account the 'value' of innovations then large firms prevail as stronger performers.

Notwithstanding the importance of this small v. large firm innovation debate, scholarly thinking about SME innovation has emerged in more complex (and perhaps more interesting) ways. This is indicated in an analysis of the knowledge sources (more than 28,000 citations) referenced in recent social science literature (811 SSCI articles, published 1990 to mid-2008) focusing both on SMEs and innovation. Certainly, Schumpeter's 1934 and 1942 classics remain among the most frequently cited (see Table 8.2), in third and fourth positions among citations in the publication sample analyzed, although with only a moderate proportion of recent (2004-2008) cites. Another frequently referenced classic is Penrose's (1959) *Theory of the Growth of the Firm*, which highlighted the importance of sources of growth and innovation within firms as they draw upon managerial and entrepreneurial capabilities, learning,

Table 8.2 Most frequently cited references in SSCI Publications with a focus on SMEs and innovation, published 1990-2007

Reference (Short Title)	Total citations	Cited 2004-2008* (share of total)
All Publications		
PORTER ME, 1990, COMPETITIVE ADVANTAG	85	0.33
SCHUMPETER JA, 1934, THEORY EC DEV	81	0.38
COHEN WM, 1990, ADMIN SCI QUART, V35,	78	0.69
SCHUMPETER JA, 1942, CAPITALISM SOCIALISM	76	0.40
PIORE MJ, 1984, 2 IND DIVIDE POSSIBI	72	0.21
ACS ZJ, 1990, INNOVATION SMALL FIR	68	0.30
ACS ZJ, 1988, AM ECON REV, V78	58	0.34
NELSON RR, 1982, EVOLUTIONARY THEORY	58	0.36
STOREY DJ, 1994, UNDERSTANDING SMALL	54	0.43
ROGERS EM, 1983, DIFFUSION INNOVATION	56	0.54
SAXENIAN A, 1994, REGIONAL ADVANTAGE C	51	0.51
BARNEY J, 1991, J MANAGE, V17	45	0.69
VONHIPPEL E, 1988, SOURCES INNOVATION	42	0.29
WILLIAMSON OE, 1975, MARKETS HIERARCHIES	41	0.24
PENROSE ET, 1959, THEORY GROWTH FIRM	41	0.55
PORTER ME, 1980, COMPETITIVE STRATEGY	40	0.40
TEECE DJ, 1986, RES POLICY, V15	38	0.41
GRANOVETTER M, 1985, AM J SOCIOL, V91	37	0.68
MALECKI EJ, 1991, TECHNOLOGY EC DEV	35	0.27
PAVITT K, 1984, RES POLICY, V13	35	0.31
PORTER ME, 1985, COMPETITIVE ADVANTAG	35	0.54
Published since 2000		
EISENHARDT KM, 2000, STRATEGIC MANAGE J, V21	17	0.76
FREEL M, 2000, ENTREP REGION DEV, V12	13	0.77
HANNA V, 2002, R&D MANAGE, V32	11	1.00
SHANE S, 2000, ACAD MANAGE REV, V25	11	0.73
MARTIN R, 2003, J ECON GEOGR, V3	10	0.90
ROMIJN H, 2002, RES POLICY, V31	10	1.00
ZAHRA SA, 2002, ACAD MANAGE REV, V27	10	0.80

*Data for 2008 for part year (7 months).

Source: Social Science Citation Index (SSCI), Author's Vantage Point analysis of 28,008 cited references in 811 SSCI 1990-2008* publications on SMEs and innovation (limited to articles, reviews, and notes and excluding book reviews, editorials, and other document types). Lists references published since 1990 with 35 or more citations and since 2000 with 10 or more citations. For definition of search terms, see Note 2 in this chapter. Later editions of volumes included under date of first edition (also, for Schumpeter's works, German and English volumes combined using first English publication date). In the analysis above, self-citations and own publications are excluded from column counts and percentages. First author only listed for multiple-authored publications. It is recognized that SSCI does not cover all publications.

and the dynamic discovery and exploitation of underused resources. However, the first, third and fifth most cited references are: Porter's (1990) *Competitive Advantage* (whose emphasis on factor conditions, demand, complementarities and strategy has accelerated recent interest in clusters and enterprise development); Cohen and Levinthal's (1990) article on absorptive capacity (noting the importance of learning and external information linkages in innovation); and Piore and Sable's (1984) *Second Industrial Divide* (highlighting flexible-specialization, small firm linkages and economies of scope). Although published in 1990, Cohen and Levinthal have garnered a high share of citations since 2004, influencing recent work on innovation which stresses the value not just of producing new knowledge but also of assimilating and applying it to SMEs. Other well-cited work includes Nelson and Winter's (1982) *Evolutionary Theory of Economic Change*; Rogers' (1983) *Diffusion of Innovations*; and Acs and Audretsch's (1990) *Innovation and Small Firms*, with its emphasis on the dynamics of the innovation process.

Among more recent articles (published since 2000), articles with high levels of citation from the SME and innovation literature continue to refer to dynamic managerial capabilities (Eisenhardt & Martin 2000) and absorptive capacities (Zahra & George 2002) in enterprise innovation. Another highly-cited article – Shane and Venkataraman (2000) – highlights not only organizational but also cognitive and individual elements in understanding entrepreneurship. All three of these contributions are far more highly cited outside of the domain of innovation and SMEs than within it, indicating they are broader contributions to management theory which have been brought into the innovation and SME world. Larger shares of the articles that highly cite other recent articles are published within the innovation and SME literature. These include Freel (2000), Hanna and Walsh (2002) and Romijn and Albaladejo (2002), all of which discuss the importance of networks and collaborative linkages in small firm innovation. This is consistent with the increased attention paid in recent years to issues of enterprise collaboration involving SMEs, often in a regional context. However, in their well-cited article, Martin and Sunley (2003) offer a caution as to the issues that can arise if such insights are naively turned into cluster policies.

Overall, these frequently cited sources suggest that there is a thrust in current literature on SMEs and innovation that takes an innovation systems view and which emphasizes the role of organizational concepts such as learning, capability, competence, flexibility and collaboration. What seems to have emerged as important is organizational and managerial competence within the SME (irrespective of industry or sector) and the positioning of the SME within larger learning, knowledge and transactional systems – be they supply-chain relationships with large customers, inter-firm networks and districts, regional innovation clusters, local-global interactions, or university spin-offs and commercialization pathways. The next section turns to probe the question of how such research insights are interpreted in innovation policymaking.

THE WORLD OF POLICY AND THE INTERFACE WITH SMES

There has long been scholarly interest in enterprise development, small firms and innovation, stretching back to the nineteenth century. However, interest in the small and medium-size enterprise, as a distinct *policy target*, is relatively recent. For much of the twentieth century, particularly in the 1950s and 1960s, small firms were often neglected (albeit benignly) as a policy category, frequently viewed as backwards or as peripheral players in the processes of technological development. In the immediate post-World War II decades, European and Japanese policymakers were more enamored with fostering large 'national champions' able to compete with large US multinationals (Servan-Schreiber 1967; Johnson 1982). In turn, US policymakers, notwithstanding early efforts in establishing the Small Business Administration in 1953 and Small Business Investment Companies in 1958, were not greatly driven to focus on SMEs, arguably because of the very success of large US companies.

This all began to change in the 1970s – an era wracked by macro-economic instability, oil-price shocks, industry restructuring, the acceleration of corporate off-shoring and technological change (including the introduction of the microprocessor). Small firms were seen both as the 'victims' of these changes and, particularly for new technology firms, increasingly as 'saviors'. Each perspective demanded increased policy attention. The UK's 'Bolton Report' (Bolton 1971) on small firms was one of the first official studies to specifically focus on the role of small firms and their problems (one of which was government itself, in terms of regulatory burden, but financial and other issues were also highlighted). In the same decade, Birch's (1979) work in the US on the job generation process highlighted the contribution of new small enterprise growth to economic and business development, leading to a significant boost of policy interest in SMEs.[3] Other studies, such as those by Piore and Sabel (1984) on the emergence of small firm industrial districts (especially in the Third Italy) and flexible specialization or by Freiberger and Swaine (1984) on the start-up phenomenon in Silicon Valley's computer industry, variously reinforced further attention to the nexus between small firms, innovation and economic performance.

Hence, in addition to 'traditional' arguments about jobs and economic development, there are today new arguments about the importance of innovative SMEs in fostering radical innovations, including the role of SME entrepreneurial innovation in fostering more effective use and appropriation of knowledge and competitiveness (Observatory of European SMEs 2004; SBA 2005b). In established (i.e. non-high-tech start-up sectors), innovation is also seen as important in maintaining and strengthening manufacturing and supply-chain capabilities (in the face of global competition and to avoid 'a race to the bottom' in terms of

---

3 Early uses of the acronym 'SME' are by *The Financial Times* in 1982 and the EEC in the 1983 European Year of Small & Medium-sized Enterprises (*Oxford English Dictionary*, OED Online, 2nd ed., Oxford University Press, <http://dictionary.oed.com>, accessed June 22, 2007).

costs and wages). Yet, attention to SME innovation can also be attributed to other factors. These include societal opportunity, as noted in Saxenian's (1999) work on the role of skilled immigrants in creating new Silicon Valley start-ups, which has stimulated greater awareness of the advantages of attracting and retaining highly-skilled immigrants to national and regional innovation systems. Politically, attention to SMEs has increased perhaps because, in a world of transnational corporations and global finance, they represent tangible objects which policymakers at national and regional levels are more readily able to influence. This is evident in Japan, for example, where policymakers can no longer sway Japan's now global large companies through the administrative guidance that was one of the pillars of post-World War II industrial policy (Johnson 1982) and now focus considerable attention to SMEs (OECD 2005b; Shapira 2008). Elsewhere, including in older industrial areas of the US, Western Europe and the former Soviet bloc, where large firms have downsized, collapsed or moved away, endogenous small business development has emerged as important economic development strategy (Bock 1994; Hassink 2002).

Today, as policy interest in SMEs has grown, it is apparent that SME innovation policy has multiple aspects, including framework policies which address national or state regulatory, tax, bankruptcy, labor market and related regimes within which SMEs have to operate, and direct policies with goals and resources targeted specifically towards SMEs (such as fostering innovative start-ups, promoting innovation in existing SMEs, or increasing innovative procurement from SMEs). Additionally, there are many other policies which have indirect or side-effects on SME innovation, for example policies aimed at targets such as research and development or university reform. Indeed, to the extent that countries have explicit or implicit 'SME innovation' policies, these are often found at the interstices of other policy categories, including those oriented to general small business development, science and technology, industry and competitiveness, industry adjustment, the promotion of growing sectors, economic and regional development, information technology, government purchasing, and education and training.

These multiple policies have led to the promulgation of numerous public and quasi-public programs, schemes and incentives that offer support to innovation in SMEs. Cataloging all the small business innovation support measures that are available is beyond the scope of this chapter.[4] However, in terms of broad classification, looking across the array of small business innovation support initiatives, perhaps the largest cluster of activities comprise those which aim to further the growth of dynamic new start-up ventures, for example through technology business incubators, university commercialization initiatives, or small

4 See, for example, Trend Chart, which is cataloguing innovation policy measures in Europe. Over 230 innovation policy measures targeted at SMEs across Europe are listed (http://trendchart.cordis.lu/index.cfm, innovation policy knowledge base, accessed June 21, 2006). This effort probably only captures a fraction of all SME innovation policy measures in Europe, but it does confirm a wide variety in the types of measures.

business innovation research programs. But there are other important activity arenas. One activity arena that has gained strength in recent years involves efforts to strengthen innovation in existing small businesses, through access to finance, business support services and through extension services for small and medium-sized manufacturers. Another cross-set of activities involves initiatives at the regional level to foster innovative clusters or 'hot spots' within which SMEs are encouraged to interact with other public and private partners to foster knowledge exchange, business linkages and spillovers.

Yet, although policy attention to SMEs has increased, and multiple programs have been implemented, coherence in both policy and program development is an elusive goal. Effectively, SME innovation policy and programming is a porous arena comprised of multiple spheres and layers of activity with many sponsors and stakeholders pursuing diverse targets and interests. This is clearly seen in the US, where there are numerous federal programs, state and local agencies, and non-profit organizations offering a bewildering array of programs and services aimed at fostering small business innovation. There is a complex institutional and political history underlying the development of these policies and programs in the US (see, for example, Bean 1996). In Japan, small business innovation policy is similarly complex, particularly with recent efforts to decentralize policymaking and program development (see Aoyama 1999). In the European case, diverse national and local activities are further compounded by multiple European-wide initiatives. Yet, if this was not complicated enough, not only is the world of SME innovation policy populated by multiple and diverse actors and interests, but there are a variety of quite different models on offer as to how SME innovation policy should proceed, as the next section discusses.

MODELS OF INNOVATION: TRANSLATIONS BETWEEN SMEs AND POLICY

In the SME and innovation literature, as well as in related literature on business strategy and entrepreneurship, a variety of models are put forward about the innovation process. In considering the relationships between the dynamics of SME innovation and policy development, it is suggested that such models are crucial mediating or 'translation' mechanisms. There are great numbers of enterprises and a great diversity in enterprise characteristics, resources, motivations, sectoral attributes, and other factors. The innovation process itself is also often complex and, intrinsically, innovation for external worlds begets innovation in internal corporate strategies (and vice versa).

In this context, models are representations and abstracts which simplify the complexities of enterprise innovation processes. At the same time, models are also central components in discourse about innovation and are often the ways through which policymakers interpret the enterprise innovation process in order to manipulate policy levers and tools to influence subsequent enterprise innovation rounds. Policymakers tend to mediate and act through models of how they understand the world (Young 1977; Marshall 1988). Such models can have implicit as well as explicit aspects. Moreover, as Keynes (1936), among others, has observed, the models that policymakers operate on are often based on old rather than new ideas, and the power of those tightly-held ideas cannot be underestimated even relative to the perceived influence of advocates who try to get policymakers to adopt different ideas.

Of course, there are many models of the enterprise or firm-level innovation process. Indeed, it has been suggested that we are now at the fifth stage (since the 1950s) in the evolution of such models (see discussion in Hobday 2005). What is remarkable, at least from the view of this chapter, is that many of these models – particularly older ones – are frequently derived from observations of, or applications to, large firms. Whether they are appropriate – or can be made appropriate – for SMEs is an issue. At the same time, there are some newer models – particularly from a regional or networked perspective – that do relate more close to SMEs.

The inevitable starting point for any discussion of innovation models is the linear version.[5] This model implies that innovation occurs towards the end of a process that begins with research, development, production and sales in a sequential manner. Public and private sector policies based around this model tend to fund research and hope that innovation will emerge at the other end. Although viewed typically as a '1950s-1960s' model of research and technology 'push' (Rothwell, cited in Hobday, 2005), this is a model – much to the dismay of contemporary innovation theorists, but as Keynes would have predicted – that remains popular and prevalent among many of today's policy leaders. The linear model has been criticized for ignoring feedback loops, market signals and iterative learning, and for over-narrowly equating research and development with innovation. To the extent that it still has value, this model is oriented toward the innovation activities of technology-based SMEs in scientific industries and does not take into account innovation in mature industries, particularly 'soft' process and organizational innovation.

There are a variety of models in the management literature that also depict sequential development processes of innovation. One of the most well-known is Cooper's Stage-Gate Process® (Cooper 2001). Stage-Gate takes product ideas through a progression of steps that include discovery, scoping, building a business case, development, testing and validation, launch, and post-launch review. Several other methodologies are available to systematize innovation including Phased Development, Product and Cycle-time Excellence, and Total Design (Jenkins et al. 1997; Kumar & Krob 2007). These models emphasize the role of management in the innovation process (rather than seeing innovation simply as a function of science). Such models are influential in designing public programs that provide management

5 The discussion in this section draws upon a larger review of innovation models in Shapira et al. (2006).

expertise to assist enterprise innovation. Reduced versions of product innovation models have been developed for SMEs, for example, ATOM-SME which is based on Stage-Gate concepts.[6] While they have many strengths, the weaknesses of such models include being based on expectations of systematization and resources which many SMEs do not possess, being focused on product development to the exclusion of other innovation forms, and underemphasizing the iterative and collaborative nature of innovation among SMEs and between SMEs and larger customers.

Between first generation linear innovation models (and their management offshoots) and the present are several other rounds of innovation model development. These include technology-pull models (1960s-1970s), combined models with feedback loops (1970s-1980s) and integrated models (1980s-1990s) with concurrent development, supplier-customer linkages and other forms of integration. (See Hobday, 2005, for further details.) In a way, particularly from the development of early integrated models (in the 1980s), this evolution has paved the way for the current state of thought on innovation models. Two contemporary examples in the series are open innovation models and networked innovation models.

The open innovation model proposes that enterprises look to a variety of external sources (e.g. private firms in complementary or competitive markets, trade associations, and university and government laboratories) for new ideas that can improve, extend and accelerate internal innovation. In addition, firms should be willing to yield innovations they cannot fully exploit – e.g. orphan patents – to external firms and markets (Chesbrough 2003). A core underpinning of the open innovation approach is that strategies to keep good ideas and innovations inside a single enterprise (particularly small firms but especially for large corporations) will limit returns. No given company has a monopoly on good ideas. Similarly, no company can develop and exploit an innovation entirely by itself. Open innovation proponents highlight the importance of developing and organizing lead users of existing or related products into a user-centered innovation system (von Hippel 2005). Innovations can also be developed by observing or interacting with customers as they use a product or service to identify problems and improvement opportunities (Leonard & Rayport 1997; Kelley 2000). On the side of expanding markets for internally developed innovations, the focus is on crafting of a business model that provides sufficient value for the innovator through licensing and spin-offs. The open innovation model also argues that different types of firms hold different positions in the value chain. Some will create new ideas, others will put them together, and still others will use them in novel ways.

6 ATOM-SME (Accelerate to Market for Small- and Medium-sized Enterprises) is a new product development methodology, based on Stage-Gate, customized for smaller companies. In the US, it is available through state and local affiliates of the US Manufacturing Extension Partnership which provide management and technical assistance services to small and medium-sized manufacturers.

For SMEs, the open innovation model promises to take into account limits to a small enterprise's ability to generate and capitalize on its own innovations and highlights the need for SMEs to develop connections with other actors in the innovation system. To the extent that open innovation has become a practiced model among policymakers, this is a significant development. It signals to SMEs that they do not, by themselves, have to develop all the internal capabilities for innovation. Similarly, the model signals to policymakers that it is helpful to innovation in evolutionary or ecological terms. As opposed to 'straitjacket' models that view SMEs in isolation and expect each enterprise to be innovative in either a sequential or integrated way, the open innovation model encourages policymakers to consider how diverse types of SMEs can contribute to innovation and how policy can best provide a conducive framework environment for innovation. However, a weakness of most of the current models of open innovation is that there is a dependence on intellectual property management, which orients the approach to larger corporations or small specialized technology-based firms. Although many of the principles of the model are applicable, the model may need to be modified to ensure value creation for the average SME. In addition, Chesbrough and Teece (2002) note that open innovation based on virtual networks and alliance formation is not always appropriate for every effort and can cause late entry and loss of market position. Finally, it is important for firms to recognize that knowledge from customers or suppliers is available to a number of firms and hence may not create much of a competitive advantage.

A complementary approach to innovation modeling is found in networking, cluster and agglomeration theories, which highlight the internal and external economies of scale and scope that can occur when firms locate in regional concentration with other firms. Again, this is not a new a new line of thought. Nineteenth century economist Alfred Marshall, in noting the role of organization, knowledge and learning in economic development, emphasized the importance of geographically-concentrated clusters of firms ('industrial districts') in facilitating specialization and technological development. Geographical concentration allows a supply of skilled labor to develop, provides opportunities for specialized subcontractors and services, and fosters skill development and the exchange of knowledge (Marshall 1890). However, in recent years, researchers have re-emphasized the significance of enterprise clustering. Examples include the flexibly-specialized textile industries of northern Italy (Piore & Sabel 1984), broker-facilitated networks in Denmark (Indergaard 1996), and the flexible and non-hierarchical networks of the information and computing technology industry in California's Silicon Valley (Saxenian 1994). Michael Porter's 'Diamond of Competitive Advantage' identified clusters based on four drivers of competitiveness: context for firm rivalry, input factor conditions, demand conditions (such as level of sophistication), and related and supporting industries (Porter 1990). In economics, there has also been a rediscovery of the returns to spatial agglomeration (Krugman 1991). In theory and modeling, it has been recognized that innovation and technological development depend on knowledge, that knowledge spillovers across agents and firms lead to increasing returns, and that there are geographic boundaries to information flows and knowledge spillovers especially of tacit knowledge. Hence, innovative industries (including small firms) which depend on knowledge spillovers can gain advantages by clustering geographically (for an example of this approach, see Audretsch & Feldman 1996). Others (non-economists) have emphasized the significant of other geographically-clustered factors in fostering development and enterprise, for example the role of social capital and 'civicness' (Putnam 1993) or the three 'Ts' of talent, technology and tolerance (Florida 2002).

The expansion of interest among scholars in clustering and its relationships to enterprise development, innovation and creative industry growth has been mirrored in the policy world. At local, regional, national and international levels, initiatives to promote innovation clusters have multiplied. For example, in Japan, a regional industry cluster program is promoted by the Ministry of Economy, Trade and Industry, which in 19 localities supports cluster activities (including business-academic-government association, knowledge exchange, workshops and projects) to improve innovation in existing SMEs and stimulate new business formation (OECD 2005b). In the US, a more decentralized approach is pursued, with numerous innovation cluster initiatives established over the last few years by state, local and non-profit entities (NGA 2007). Similarly, in Europe, multiple cluster initiatives have been fostered at regional, national and supra-national (European Union) levels. Most European clusters are dominated by SMEs in the sector that defines the cluster, including science-based, high-technology, traditional industry and creative sector clusters (European Commission 2002). The Competitiveness Institute – a non-profit international practitioners network for competitiveness, clusters and innovation – identifies cluster initiatives in 40 countries (and this is by no means a complete listing).[7] Most of these clusters have an explicit SME focus, seeking to accelerate innovation in SMEs by fostering inter-firm linkages, knowledge exchange, capacity building and collaborative projects among groups of SMEs and between SMEs and larger firms, universities, technology transfer institutes and other organizations. Intellectually, policy interest in clusters has been underpinned by the work not only of Porter (1990, 1998) and other researchers (such as Cooke & Morgan 1993; Rosenfeld 1997; or Storper 1993) but also by national and international organizations that have actively and widely disseminated cluster information and best practice insights (examples include: Council on Competitiveness 2005; OECD 2007; European Commission 2007).

This brief account simplifies a large body of literature on regional innovation, enterprise, and clusters (for a more detailed synthesis, see Moulaert & Sekia 2003). The regional literature has disparate roots from the innovation management models discussed earlier. Yet, today, these literatures have converged in innovation models which emphasize the role of the region and geographical proximity in fostering

7 Search of the Cluster Initiative Database, The Competitiveness Institute, August 8, 2008 (http://www.competitiveness.org).

innovation and creativity, especially for SMEs. As discussed above, these models are popular with policymakers, naturally at regional and local levels, but also at national and international levels. While the most common SME innovation model that policy makers use today is arguably one based (or freely-translated) on a Silicon Valley world populated by high technology venture start-ups where the individual entrepreneur dominates the frame, it is significant that more collaborative models have risen highly in importance. These include innovation models based on clusters (where the strong role for governance and public-private partnership appeals to the policy toolbox) and perhaps also those aiming to replicate creative environments for innovation (stimulated by Florida 2002, and others). From the view of SMEs, the choice of innovation model does make a difference. A high-tech start-up model favors leading-edge technologies and talented entrepreneurs able to attract venture capital, underpinned by investment in R&D and technology transfer. This mode favors high-technology start-ups; a cluster model allows for both science-driven and mature SMEs to have roles, linked with larger firms and public institutions; while a creative-class model gives an edge to SMEs in creative industries as well as in 'conventional' high technology.

Such important details notwithstanding, there are some larger points to be made about the rise of these regionally-based innovation models and the actual practice and performance of innovation in SMEs. As in open innovation, regional innovation models encourage policymakers to adopt a systematic view of their region, considering framework issues (such as the quality of human capital or social environment in the region), complementary assets (including the availability of finance for innovation) and relationships (for example, university-industry linkages). This is an evolution in policy thinking, encouraged both by real changes in innovation opportunities and needs and by developments in ideas about innovation.

Yet, while these regionally-based models present good examples of interactive knowledge transfer between the scholarly and policy communities, there are risks in this translation process. One risk is that policymakers idealize the model and/or the region from which that model emanates; but conditions and circumstances vary, and what works in one region often does not work as planned in another. In short, the translation of nuanced scholarly concepts and complex business processes into the policy arena will fall short if it is done mechanically or faddishly. It has already been suggested that many policy efforts to promote clusters are being developed chaotically, more as fashion than as substantive intervention (Martin & Sunley 2003). Moreover, even if policies are customized, economic-structural and institutional path dependencies may make it difficult to ensure that such end goals as creating clusters of highly-innovative enterprises are successfully achieved (see Fuchs & Shapira 2005).

A second kind of risk is where translation is not updated or accompanied by learning to adapt to fast-moving changes in business practice, technology, or globalization. In the 1970s and 1980s, the success story of Italian industrial districts – with their tightly-knit clusters of subcontracting SMEs with high-levels of intra-district collaboration – was widely told (Brusco 1982; Pyke et al. 1990). This story remains influential in policy circles, even though these industrial districts have undergone major restructuring over the last decade as SME links have weakened or moved beyond the district in the face of tough competition from highly-capitalized multi-national corporations, off-shoring and globalization (Whitford 2001). A danger here is that policymakers may overemphasize the region as the sole site for SME innovation. Knowledge developed within regions may be self-reinforcing and limit learning from outside (Gertler 2005), the exploitation of innovation may require strong external linkages (Gallaud & Torre 2005), and innovation and learning are not always localized, meaning that non-local collaborations may be necessary for SMEs (Malmberg & Power 2005). More fundamentally, just as clusters have risen highly on the policy agenda, leading-edge debate has turned towards non-proximate enterprise linkages aided by developments in communications technology, integration software and economic globalization (Friedman 2005). Attention is beginning to turn to how SMEs can benefit from internationalization and develop transnational networks that can support innovation (see, for example, Ratten et al 2007; Tiwari & Buse 2007). For organizations that have invested in cluster promotion founded on local or regional proximity, it remains to be seen how readily adaptation can be made to a significant rise in the importance of international SME innovation linkages based on technological connectedness.

SOME IMPLICATIONS

This chapter has surveyed the development of thought and current debates about the relationship between innovation in small and midsize enterprises and explores the relationships between innovation system dynamics and innovation policy strategies that include SMEs. The field of literature and the arena of policy activity is very large (even considering only developed countries), and inevitably the discussion has been selective. Within this, it has been posited that a key mediating mechanism used by policymakers (and managers) to understand innovation dynamics is the model. The chapter has focused attention to the evolution of models used to understand and influence innovation in SMEs. A core analogue of the chapter is the idea of 'worlds' – where the diverse and dynamic worlds of small business interact with the complex worlds of policymaking. Models of innovation, more or less applied to SMEs, are a translation device, used to codify the state of knowledge and theory about innovation and which are drawn upon by policymakers as well as managers in developing strategies. SME innovation policies are surely influenced by stakeholder interests, inertia, broader political and economic changes, and other factors. But, they are also powerfully influenced by ideas and models about the innovation process itself. We have seen that, in successive generations, innovation models moved from linear to feedback to integrated approaches, with a particular stress now on innovation ecologies (through open innovation and cluster models).

It has also been observed that while policy interest and activity related to SME innovation has expanded in recent years, this has not necessarily led to policy coherence. With the growth of SME innovation policies and programs, inefficiencies have become apparent. Not everything tried has worked, and at times what has been tried has conflicted with or duplicated aims elsewhere in the innovation policy system. New efforts in to 'join-up' and integrate policies have emerged, particularly in Europe, where the European Commission seeks to develop a single and unified 'policy framework' for SMEs (European Commission 2005). This is laudable: it can clearly help SME 'users' (and promote program accountability) if policy goals and support mechanisms are clearly charted. Yet, there is likely to be a limit to the usefulness and relevance of top-down directives, particularly given the variety of national and regional conditions across Europe.

In contrast to their European Union counterparts, US federal policymakers have been satisfied to eschew national consistency and to devolve leadership on small business innovation policies to states, as part of what has been termed 'laboratories of democracy' (Osborne 1988). In this perspective, states and localities can experiment with different approaches and learn from experience as to which policies work best. Sabel (1996) has suggested that the decentralized network of centers to advance technological development among US small manufacturers represents an exemplary case of such experimental federalism, where centers are able to learn what is effective (and what is not) in a manner that parallels collaborative learning mechanisms among firms. The tension, which has deepened since that study was published, is that decentralized policy learning systems essentially also require resources, communication support, brokerage and long-term horizons, and the US federal government has only patchily provided this in its support to small manufacturing enterprise policies in recent years.

If there is a fundamental insight from these broad contrasts, it is this: given the diversity of small business worlds and the dynamics of innovation, perhaps what is more important is that SME innovation policy measures are developed in the context of a learning system, through which there is a co-developed capacity to experiment, assess, improve and advance policies. National or supra-national governments cannot direct this from the center; they can facilitate and catalyze, they can also seek to foster favorable institutional, regulatory and macro-economic frameworks for enterprise development and innovation. Yet there is also a caution: as business, technological and societal conditions change, the models through which policy is mediated and guided can and do become obsolete. We have seen this in the successive generations of innovation models advanced over the past five decades. In part, there may be an element of academic 'niche' searching here. But more deeply, these evolutions in models reflect real changes in underlying conditions. Operational models tend to lag actual circumstances, particularly when they become institutionalized in policies and programs that are then hard to change. Hence, an ongoing challenge for the co-evolution of policy – in the world of SME innovation just as in other arenas of policy – is to tackle and replace the old ideas as well as to advance new ones.

ACKNOWLEDGEMENTS

An earlier version of this chapter was presented at the Second Workshop on the Co-Evolution of Innovation Policy, Fraunhofer Institute for Systems and Innovation Research, Karlsruhe, Germany, June 22-23, 2006. The author appreciates comments received from workshop participants and from an external reviewer.

REFERENCES

Acs, Z. and Audretsch, D. (2005), 'Entrepreneurship and innovation', paper prepared for Summer Institute of the Entrepreneurship, Growth and Public Policy Division of the Max Planck Institute of Economics, Jena, Germany.

Acs, Z. and Audretsch, D. (1990), *Innovation and Small Firms*, Cambridge, MA, USA and London, UK: MIT Press.

Aoyama, Y. (1999), 'Policy interventions for industrial network formation: contrasting historical underpinnings of small business policy in Japan and the United States', *Small Business Economics*, **12** (3), 191-202.

Audretsch, D. and Feldman, M. (1996), 'R&D spillovers and the geography of innovation and production', *American Economic Review*, **86** (3), 630-640.

Bean, J. (1996), *Beyond the Broker State: Federal Policies Toward Small Business, 1936-1961*, Chapel Hill: University of North Carolina Press.

Birch, D. (1979), *The Job Generation Process: Program on Neighborhood and Regional Change*, Cambridge, MA, USA and London, UK: MIT Press.

Bock, J. (1994), 'Innovation as creative destruction: the role of small businesses in the Commonwealth of Independent States (CIS)', *International Journal of Technology Management*, **9** (8), 856-863.

Bolton, J. (1971), *Report of the Committee of Enquiry on Small Firms*, London: HMSO.

Broersma, L. and Gautier, P. (1997), 'Job creation and job destruction by small firms: an empirical investigation for the Dutch manufacturing sector', *Small Business Economics*, **9** (3), 211-224.

Brusco, S. (1982), 'The Emilian model: productive decentralisation and social integration', *Cambridge Journal of Economics*, **6** (2), 167-184.

Chandler, A. (1990), *Scale and Scope: The Dynamics of Industrial Capitalism*, Cambridge, MA, USA: Harvard University Press.

Chesbrough, H. (2003), *Open Innovation*, Boston: Harvard Business School Press.

Chesbrough, H. and Teece, D. (2002), 'Organizing for innovation: when is virtual virtuous?', *Harvard Business Review*, **80** (8), 127-135.

Cohen, W. and Levinthal, D. (1990), 'Absorptive capacity: a new perspective on learning and innovation', *Administrative Science Quarterly*, **35** (1), 128-152.

Cooke, P. and Morgan, K. (1993), 'The network paradigm: new departures in corporate and regional development', *Environment and Planning D: Society & Space*, 11 (5), 543-564.

Cooper, R. (2001), *Winning at New Products*, Reading. MA, USA; UK: Perseus.

Council on Competitiveness (2005), 'Clusters of innovation initiative: regional foundations of U.S. competitiveness', Washington, DC.

Davis, S., Haltiwanger, J. and Schuh, S. (1993), 'Small business and job creation: dissecting the myth and reassessing the facts', working paper No. 4492, National Bureau of Economic Research, Cambridge, MA, USA.

Eisenhardt, K. and Martin, J. (2000), 'Dynamic capabilities: what are they?', *Strategic Management Journal*, 21 (10-11), 1105-1121.

European Commission (2007), 'Innovation clusters in Europe: a statistical analysis and overview of current policy support', Europe Innova/PRO INNO Europe paper No. 5, Office for Official Publications of the European Communities, Luxembourg.

European Commission (2005), 'Implementing the Community Lisbon Programme: Modern SME Policy for Growth and Employment', Com (2005) 551, Commission of the European Communities, Brussels.

European Commission (2002), 'Regional clusters in Europe', Observatory of European SMEs, Office for Official Publications of the European Communities, Luxembourg.

Florida, R. (2002), *The Rise of The Creative Class, And How It's Transforming Work, Leisure, Community and Everyday Life*, New York: Basic Books.

Freel, M. (2000), 'External linkages and product innovation in small manufacturing firms', *Entrepreneurship and Regional Development*, 12 (3), 245-266.

Freiberger, P. and M. Swaine (1984), *Fire in the Valley: The Making of the Personal Computer*, Berkeley: Osborne and New York: McGraw-Hill.

Friedman, T. (2005), *The World is Flat: A Brief History of the Twenty-first Century*, New York, USA: Farrar, Straus and Giroux.

Friedman, D. (1988), *The Misunderstood Miracle: Industrial Development and Political Change in Japan*, Ithaca: Cornell University Press.

Fuchs, G. and Shapira, P. (eds) (2005), '*Rethinking Regional Innovation and Change: Path Dependency or Regional Breakthrough?*, Berlin, Germany, Heidelberg, Germany and New York, USA: Springer-Verlag.

Gallaud, D. and A. Torre (2005), 'Geographical proximity and the diffusion of knowledge', in G. Fuchs and P. Shapira (eds), *Rethinking Regional Innovation and Change: Path Dependency or Regional Breakthrough?*, Berlin, Germany, Heidelberg, Germany and New York, USA: Springer-Verlag, 127-146.

Gellman Research Associates (1982), 'The relationship between industrial concentration, firm size, and technological innovation', report prepared for the Small Business Administration, Jenkintown, PA.

Gertler, M. (2005), 'Tacit knowledge, path dependencies, and local trajectories of growth', in G. Fuchs and P. Shapira (eds), *Rethinking Regional Innovation and Change: Path Dependency or Regional Breakthrough?*, Berlin, Germany, Heidelberg, Germany and New York, USA: Springer-Verlag, pp. 23-41.

Hanna V. and Walsh, K. (2002), 'Small firm networks: a successful approach to innovation?', *R&D Management*, **32** (3), 201-207.

Harrison, B. and Bluestone, B. (1994), *Lean and Mean*, New York: Basic Books.

Hassink, R. (2002), 'Regional innovation support systems: recent trends in Germany and east Asia', *European Planning Studies*, **10** (2), 153-164.

Hijzen, A., Upward, R. and Wright, P. (2007), 'Job creation, job destruction and the role of small firms: firm-level evidence for the UK', GEP Research Paper 07/01, Leverhulme Centre for Research on Globalisation and Economic Policy, University of Nottingham, Nottingham, UK.

Hippel, E. von (2005), *Democratizing Innovation*, Cambridge, MA, USA and London, UK: MIT Press.

Hobday, M. (2005), 'Firm-level innovation models: perspectives on research in developed and developing countries', *Technology Analysis and Strategic Management*, **17** (2), 121-146.

Indergaard, M. (1996), 'Making networks, remaking the city', *Economic Development Quarterly*, **10** (2), 172-187.

Jenkins, S., Forbes S, Durrani T. and Banerjee, S. (1997), 'Managing the product development process: an assessment', *International Journal of Technology Management*, **13** (4), 359-378.

Johnson, C. (1982), *MITI and the Japanese Miracle: The Growth of Industrial Policy, 1925-1975*, Stanford: Stanford University Press.

JSBRI (2007), 'White Paper on small and medium Enterprises in Japan', Japan Small Business Research Institute, Ministry of Economy, Trade and Industry, Tokyo.

JSBRI (2004), 'White paper on small and medium enterprises in Japan', Japan Small Business Research Institute, Ministry of Economy, Trade and Industry, Tokyo.

Kaufmann Foundation (2007), *On the Road to an Entrepreneurial Economy: A Research and Policy Guide*, Version 2.0. Kansas City: Ewing Marion Kauffman Foundation.

Kelley, T. (2000), *The Art of Innovation*, New York: Currency/Doubleday.

Kennedy, J. and Hyland, P. (2003), 'A comparison of manufacturing technology adoption in SMEs and large companies', Paper presented at 16th Annual Conference of Small Enterprises Association of Australia and New Zealand, Ballarat, pp. 1-10.

Keynes, J. (1936), *The General Theory of Employment, Interest and Money*, New York, USA and San Diego, USA: Harcourt Brace Jovanovich.

Krugman, P. (1991), 'Increasing returns and economic geography', *Journal of Political Economy*, **99** (3), 483-499.

Kumar, S. and Krob, W. (2007), 'Phase reviews versus fast product development: a business case', *Journal of Engineering Design*, **18** (3), 279-291.

Langlois, R. (2003), 'Schumpeter and the obsolescence of the entrepreneur', *Advances in Austrian Economics*, **6**, 287-302.

Leonard, D. and Rayport, J. (1997), 'Spark innovation through empathetic design', *Harvard Business Review*, 75 (6), 102-113.

Malmberg, A. and Power, D. (2005), 'Role of global demand in local innovation', in G. Fuchs and P. Shapira (eds), *Rethinking Regional Innovation and Change: Path Dependency or Regional Breakthrough?*, Berlin, Germany, Heidelberg, Germany and New York, USA: Springer-Verlag, pp. 273-290.

Marshall, A. (1890), *Principles of Economics*, London, UK and New York, USA: Macmillan.

Marshall, C. (1988), 'Bridging the chasm between policymakers and educators', *Theory into Practice*, 27 (2), 98-105.

Martin, R. and Sunley, P. (2003), 'Deconstructing clusters: chaotic concept or policy panacea?', *Economic Geography*, 3 (1), 5-35.

Miles, I. and Green, L. (2008), *Hidden Innovation in the Creative Industries*, National Endowment for Science, Technology and the Arts, London.

Moulaert, F. and Sekia, F. (2003), 'Territorial innovation models: a critical survey', *Regional Studies*, 37 (3), 289-302.

Nelson, R. and Winter, S. (1982), *An Evolutionary Theory of Economic Change*, Cambridge, MA, USA: Harvard University Press.

NESTA (2007), *Hidden Innovation. How Innovation Happens in Six 'Low Innovation' Sectors*, Research Report, National Endowment for Science, Technology and the Arts, London.

NGA (2007), 'Cluster-based strategies for growing state economies', National Governors Association, Washington, DC.

Observatory of European SMEs (2004), 'SMEs in Europe 2003, Report 2003/7', Enterprise Directorate-General of the European Commission, Office for Official Publications of the European Communities, Luxembourg.

OECD (2008), *OECD Economic Factbook 2008: Economic, Environmental, and Social Statistics*, Paris: OECD Publishing.

OECD (2007), 'Competitive regional clusters: national policy approaches', OECD Reviews of Regional Innovation, Organisation for Economic Cooperation and Development, Paris.

OECD (2005a), *Oslo Manual: Guidelines for Collecting and Interpreting Innovation Data*, Vol. 3, Organisation for Economic Cooperation and Development and the Statistical Office of the European Communities, Paris: OECD Publishing.

OECD (2005b), 'OECD territorial review: Japan', Organisation for Economic Cooperation and Development, Paris.

OECD (2002), *Small and Medium Enterprise Outlook, 2002*, Paris: OECD Publishing.

Osborne, D. (1988), *Laboratories of Democracy*, Boston: Harvard Business School Press.

Parker, R. (2001), 'The myth of the entrepreneurial economy: employment and innovation in small firms', *Work, Employment and Society*, 15 (2), 373-384.

Pavitt, K., Robson, M. and Townsend, J. (1987), 'The size distribution of innovating firms in the UK: 1945-1983', *Journal of Industrial Economics*, **35** (3), 297-316.

Penrose, E. (1959), *The Theory of the Growth of the Firm*, Oxford: Blackwell.

Piore, M. and Sabel, C. (1984), *The Second Industrial Divide: New Possibilities for Prosperity*, New York: Basic Books.

Porter, M. (1998), 'Clusters and the new economics of competition', *Harvard Business Review*, November-December, 77-90.

Porter, M. (1990), *The Competitive Advantage of Nations*, New York: The Free Press.

Putnam, R. (1993), *Making Democracy Work: Civic Traditions in Modern Italy*, Princeton: Princeton University Press.

Pyke, F., Becattini, G. and Sengenberger, W. (eds) (1990), *Industrial Districts and Inter-Firm Co-operation in Italy*, Geneva: International Institute for Labour Studies.

Ratten, V., Dana, L.-P., Han, M. and Welpe, I. (2007), 'Internationalisation of SMEs: European comparative studies', *International Journal of Entrepreneurship and Small Business*, **4** (3), 361-379.

Rogers, E. (1983), *Diffusion of Innovations*, New York: The Free Press.

Romijn, H. and Albaladejo, M. (2002), 'Determinants of innovation capability in small electronics and software firms in southeast England', *Research Policy*, **31** (7), 1053-1067.

Rosenfeld, S. (1997), 'Bringing business clusters into the mainstream of economic development', *European Planning Studies*, **5** (1), 3-23.

Sabel, C. (1996), 'A measure of federalism: assessing manufacturing technology centers', *Research Policy*, **25**, 281-307.

Saxenian, A. (1999), *Silicon Valley's Skilled Immigrants: Generating Jobs and Wealth for California*, San Francisco: Public Policy Institute of California.

Saxenian, A. (1994), *Regional Advantage: Culture and Competition in Silicon Valley and Route 128*, Cambridge, MA, USA: Harvard University Press.

SBA (2005a), 'Small business size standards', U.S. Small Business Administration, http://www.sba.gov/size/index.html (accessed December 28, 2005).

SBA (2005b), 'The small business economy: a report to the President', U.S. Small Business Administration, Washington, DC.

Schumpeter, J. (1942). *Capitalism, Socialism, and Democracy*, New York: Harper and Brothers.

Schumpeter, J. (1934), *The Theory of Economic Development*, Cambridge, MA, USA: Harvard University Press.

Servan-Schreiber, J.-J. (1967), *Le Défi Américain*, published in English as *The American Challenge*, London: H. Hamilton (1968).

Shane, S. and Venkataraman, S. (2000), 'The promise of entrepreneurship as a field of research', *Academy of Management Review*, **25** (1), 217-226.

Shapira, P. (2008), 'Putting innovation in place: evolution of Japanese strategies for industrial services, regional clusters, and manufacturing SMEs', *Prometheus*, **26** (1), 69-87.

Shapira, P., Youtie, J., Lamos, E., Bhaskarabhatla, A., Mohapatra, S. and Cheney, D. (2006), 'Product and service innovation', report to the Manufacturing Extension Partnership, National Institute of Standards and Technology, Georgia Tech Program in Science, Technology, and Innovation Policy and SRI International, Atlanta, GA and Arlington, VA.

Simon, H. (1996), *Hidden Champions: Lessons from 500 of the World's Best Unknown Companies*, New York: McGraw-Hill.

Staber, U. (2005), 'Entrepreneurship as a source of path dependency', in G. Fuchs and P. Shapira (eds), *Rethinking Regional Innovation and Change: Path Dependency or Regional Breakthrough?*, Berlin, Germany, Heidelberg, Germany and New York, USA: Springer-Verlag, pp. 107-126.

Storper, M. (1993), 'Regional "worlds" of production: learning and innovation in the technology districts of France, Italy and the USA', *Regional Studies*, 27 (5), 433-455.

Storey, D. and Johnson, S. (1987), *Job Generation and Labour Market Change*, London, UK and New York, USA: MacMillan.

Tether, B. (1998), 'Small and large firms: sources of unequal innovations?', *Research Policy*, 27 (7), 725-745.

Tiwari, R. and Buse, S. (2007), 'Barriers to innovation in SMEs: can the internationalization of R&D mitigate their effects?', research project Global Innovation, working paper No. 50, Germany: Hamburg University of Technology, Hamburg.

Whitford, J. (2001), 'The decline of a model? Challenge and response in the Italian industrial districts,' *Economy and Society*, 30 (1), 38-65.

Young, K. (1977), 'Values in the policy process', *Policy and Politics*, 5 (1), 1-22.

Zahra, S. and George, G. (2002), 'Absorptive capacity: a review, reconceptualization, and extension', *Academy of Management Review*, 27 (2), 185-203.

9. Co-evolution of Intellectual Property Protections and Innovation: The Case of 'Continuation' Patents in the United States

Stuart J.H. Graham

INTRODUCTION

This chapter considers the co-evolution of intellectual property (IP) protections and innovation policy within the context of a particular IP institution: the 'continuing' patent application available in the United States. While not generally available elsewhere in the world, the continuation procedure allows a US patent applicant to postpone the granting of a patent, affording inventors several strategic opportunities, among which are delay and secrecy. This chapter also demonstrates the perverse effects of so-called 'submarine patents' – continuation patents that surface in a marketplace in which the patented technology has been widely embraced by adopters unaware that a valid patent was pending and hidden from view. In addition to examining several specific cases of submarine patents in which the patentee was able to extract extraordinary economic rents, the chapter investigates the technological and historical drivers of the development of the continuation rules. The chapter examines how and why innovators in the semiconductor and pharmaceutical technologies have employed the continuation patent application procedure, demonstrating that in each of these important sectors the continuation procedure has been widely used, offering the opportunity for strategic gain.

At least since North (1990), we have understood that institutions (or 'the rules of the game') co-evolve with other economic processes. Innovation is one such economic process that has co-evolved with the institutions that support and constrain it. In this chapter, the term 'innovation' is used in its Schumpeterian sense. Innovation is the process by which inventions – new ideas – are successfully introduced by entrepreneurs to the marketplace (Schumpeter 1942). Innovation thus involves adapting inventions to the market in the form of new products, coordinating production, and modifying or developing new organizations, among other activities.

In the modern knowledge economy, innovation is seen as a primary driver of national competitive advantage. Accordingly, the institutional environment in which technologies advance has drawn increased attention (Mowery & Nelson 1999; Nelson & Nelson 2002; Graham et al. 2003). Because economic growth, particularly in the industrialized world, relies increasingly upon the development and exploitation of 'intangible' knowledge assets, the development and operation of intellectual property protections has likewise become a basis for active study (Hall & Ziedonis 2001; Lamoreaux & Sokoloff 2002; Graham & Mowery 2004).

Policy plays an active role in mediating the relationship between institutions and innovations – sometimes with unintended consequences. Any IP regime is the outcome of policy formulation and implementation. IP policy has been a mix of public and private concerns: encompassing both the 'public good' nature of invention (allowing inventors a limited monopoly in order to provide adequate economic incentives) and the political economy of industry capture. Thus, the rules that constitute the modern IP regimes of industrialized nations have metamorphosed into institutions that are much more complex than the simple 'monopoly for disclosure' *quid pro quo* contemplated by Nordhaus (1969).

In the IP policy sphere, one of the most active areas over the last several decades has been the international harmonization of rules. Policymakers in the industrialized world have argued for, and implemented, agreements aimed at bringing the markedly different IP regimes of various nations into accord. In the run-up to the General Agreement on Tariffs and Trade (GATT) negotiations during the 1995 Uruguay Round, and subsequently through the World Trade Organization (WTO) and Trade Related Aspects of IP Rights (TRIPS), harmonization was a major driver of IP policy among the TRIAD nations (Europe, Japan and the US). In the United States, two major pieces of legislation were motivated by a desire to bring US laws into harmony with those of Japan and Europe; the 1995 change from a 17-year-from-issue patent term to a 20-year-from-filing patent term, and the 1999 requirement that (most) patent applications be published after 18 months.

While Arrow (1962) suggested that invention, and by extension innovation, is unlikely without the rewards provided by the intellectual property monopoly, we cannot expect that the manner in which IP protections have developed, or the ways in which they operate, necessarily produces efficient outcomes. As institutions develop, these 'rules of the game' spawn, and in turn are modified by, organizations that arise to take advantage of the opportunities that the institutions present (North 1990). The evolution of a nation's 'institutions of innovation' (including IP protections) may be thus closely related to the evolution of organizations and technological paths alike. This co-evolution may, as a result of endogeneity in the process, lead to outcomes that are difficult to comprehend without an understanding of the technological, political and historical context in which the institutions developed. In an effort to highlight the complexities involved in understanding how the 'institutions of innovation' evolve, this chapter uses a case study of the development and use of one such protection – the 'continuing application' available in the United States, patent system.

It is often surprising to observers of the US patent system that an applicant for a US patent can prevent the officer responsible for deciding on the patentability of the invention (the examiner) from finally denying the patent application. It comes as an equally great surprise that the applicant can also prevent the officer from finally *granting* the patent. The institution that allows this odd result is the availability in the US of the 'continuing' patent application. The continuation is a procedure available under the US patent laws that gives a patent applicant the right to re-file a patent application at will, at any point in the application process. Thus, the patent applicant may choose to file the application on the day that the patent has been granted – or rejected – and thus start the process anew. This right is an institution peculiar to the United States.

The continuation procedure effectively affords a patent applicant the right to extend the period of pre-issue pendency prior to the patent examiner's final patentability decision can take effect. This procedure thus gives an applicant effective control over the timing of the patent's grant. The first in what may become a long string of applications wins for the applicant a 'priority date', this date being subsequently applied to all later continuation applications. Thus the applicant captures with the first filing a presumption of having been 'in the Patent Office door' early. Furthermore, during the period of extended continuation delay, the applicant has the opportunity to file amendments to the original application, preserving for an applicant the opportunity to engage in strategic gamesmanship.

Thus, the 'continuing patent application' may be representative of a broader phenomenon in the current policy discussions over national systems of innovation: the broken patent system. Jaffe and Lerner (2004) provide a perspective about this issue in the US, suggesting that the strengthening of patent rights along with the weakening of the Patent Office have created a crisis that threatens the operation of the innovation system. Other institutional rules, like the continuing application, may be empowering the few at the expense of overall social welfare.

The development of the 'continuing' application process in the United States offers a good exemplar of the co-evolution of intellectual property protections, innovation policy and interest group dynamics. The unique rule on 'continuing' applications in US law shows elements of path dependence – the 'stickiness' and resistance to change that often characterizes existing institutions when other rules, norms, and interests build around the institution over time – and elements of political economy, as interest groups that have come to profit from the institution exert influence to maintain it. This co-evolution of intellectual property protections, incentives for innovation and economic interest has involved public (agencies, legislatures and courts), private (inventors and industry) and quasi-public (lawyers) actors alike.

The 'continuing' patent application offers an interesting case study in the co-evolution of IP protection and innovation policy because in operation it produces some rather surprising economic distortions. The 'continuation' is the institutional rule that allows a patent applicant to take advantage of a 'submarine' patent application. The archetypical submarine patent, characterized by an unusually long period of pre-issue secrecy and delay, is granted into a marketplace in which the patented technology has been widely adopted, thus making these prior uses now 'infringing' uses of the applicant's recently issued patent. The patent is thus like the naval submarine; it surfaces from a secret place to menace the lanes of commerce.

Circumstances surrounding a number of patents invented by Jerome Lemelson's covering 'bar coding' and 'machine vision' have become exemplars of the submarine patent phenomenon. Originally filed in the 1950s, several of Lemelson's patent applications were kept secret inside the US Patent and Trademark Office (USPTO or Patent Office) while Lemelson and his patent attorney altered the description of the original patents to cover later-created products and adoptions of the technologies, which they were observing in the marketplace. When Lemelson finally allowed his patents to issue, some had been hidden almost 40 years. For instance, his US patent 5.283.641, 'apparatus and methods for automated analysis' was granted on 1 February 1994, but its original filing date was 24 December 1954. The method used by Lemelson to 'submarine' these patents was the 'continuation' application: in the case of US patent 5.283.641, Lemelson had used no fewer than eleven continuation applications during the 39 years that the application was in process within the Patent Office.

The potential economic consequences of the most notorious type of gamesmanship, submarine patenting, are illustrated by the actions taken by Lemelson and his attorney concerning 'bar coding' patents: they practiced economic hold-up on large firms in major industries that had adopted these technologies. Armed with these late-emerging patent rights, and with the threat of court action and possible injunction, Lemelson and his attorney demanded licensing fees first from Japanese and then from US automakers, firms that by the 1990s were using the technology extensively throughout the value chain. Lemelson and his attorney are estimated to have collected $1 billion in fees and settlements from firms in many industries (Albright 2004). By practicing just such economic hold-up, threatening costly infringement litigation, large damage awards and injunctions used to shutter operations, the 'continuation' can offer significant strategic advantage.

Because the 'continuation' procedure is unavailable in other major industrialized nations (Europe and Japan), it comes as a surprise to many observers that Lemelson was allowed to use the US patent laws to practice such a hold-up strategy. This strategy, permitted by the courts and by Congress, appears to run afoul of the basic *quid pro quo* that is often used to justify the patent regime: the public will offer to inventors limited monopolies over their inventions, creating incentives to encourage investment in research, while inventors will in turn give to the public a timely disclosure of these ideas so that follow-on innovation will not be unduly retarded, and the invention itself will ultimately be made available for public use after the limited monopoly expires. To an economist, it is no surprise that Lemelson engaged in a submarine strategy – he appears to have been rationally maximizing the expected returns from the patent right. When general adoption of a technology is comparatively slow, and firms may be expected to deploy substantial complementary assets alongside the adopted technology (thus 'locking-in' to the technical regime), the type of delay and hold-up strategy practiced by Lemelson can be quite beneficial, as evidenced by the enormous rents he captured from industry. More confounding to observers, however, is the recognition that the courts and Congress created the 'continuation' procedure in the first place, and allowed it to operate for well over a century, unimpeded by meaningful equitable limits on its use by applicants.

Despite the fact that the continuation procedure (as described) is available only in the United States, innovators and researchers around the world should be familiar with its ramifications. First, the United States is a large and important market for many goods, particularly in technology products, and innovations from outside the US are often embodied in products sold in the US, thus opening non-US innovators to infringement liability under the US patent laws. Second, the US is also an important intellectual property market, with a substantial share of US patents now issuing to non-US firms. This demand for US patents opens foreign entities to the operations of the US patent laws, and thus the continuation procedure is both a strategic opportunity and a strategic risk to any firm that either sells, or patents, in the US. Third, similar mechanisms are available (with the 'division') in both Europe and Japan.

The US continuation could historically be used to hide the original application and all subsequent amendments for extended periods of time, and it was the method used by Lemelson and others to keep applications secreted. In 1995, however, Congress changed the Patent Act in ways that appear to have removed some of the patentees' incentives for engaging in continuation practice. By changing the patent term from 17 years from date of grant to 20 years from date of application filing, the Congress intended to force those patentees using the continuation to choose between *ex ante* grant delay and *ex post* property rights (see section 'Changes in the US Patent Law, 1995'). While the new legislation reduced the incentives motivating patentees, it did not completely eliminate incentives to use, or the actual use by, patent applicants of continuation practice.

The use of continuation practice by patentees, both before and after the 1995 regime change, raises a number of questions. First, what roles have the historical development of different technologies and the interests of different groups played in the evolution of this institutional rule? Second, from an empirical standpoint, what effect have the 1995 and 1999 legislations had on the use by patentees of the continuation procedure? Third, are there significant differences in the incidence of continuation practice by industry or technology sector and, if so, have these differences persisted since the incentive changes brought about by the 1995 and 1999 regime changes? Fourth, are there demonstrable affects of the policies and policy changes on the amount or pace of innovation more generally? I employ data on the use by patent applications of 'continuation' practice in the USPTO from 1975 to 2002 to shed light on these questions.

To that end, this chapter analyzes the continuation's use by applicants in several important industrial sectors, and empirically examines the determinants of the incidence of the continuation application's use in US patenting. The chapter is organized as follows. The next section describes the historical development of continuation practice, and the 1995 and 1999 policy changes meant to reduce the incentives for its use. The following section examines the available data on continuation application use in several important sectors, notably pharmaceutical and semiconductor technologies. Part four conducts regression analysis to shed light upon the determinants of patentees' use of the continuation application. The final section offers observations and conclusions.

CONTINUATION APPLICATION PRACTICE UNDER THE US PATENT LAW

In the United States, inventors enter into a complex application process when seeking a patent. Inventors may claim a utility patent[1] on processes, machines, manufactures, or compositions of matter by making application to the United States Patent and Trademark Office (USPTO or Patent Office). Procedurally, the application must be filed within one year of the invention's public use or publication[2] and must contain an adequate description with one or more claims.[3] Before the patent issues, the Patent Office ensures that the invention covers patentable subject matter,[4] is useful,[5] novel[6] and not obvious.[7] The patent examiner, a Patent Office employee with specialized technical knowledge, is the initial arbiter of these requirements.[8]

1 While the vast majority of US patents – and the focus in this chapter – are the so-called Utility patents authorized by 35 USC §101, patents are also available on Plants (35 USC §161) and Designs (35 USC §171).

2 35 USC §102(b).

3 35 USC §112, 113, 114. The 'written description' requirement is intended to allow any person skilled in the art to either make or use the invention.

4 35 USC §101. See *Diamond v. Chakrabarty*, 447 US 303 (1980)(determining that man-made living micro-organisms are patentable subject-matter).

5 35 USC §101. See *Brenner v. Manson*, 383 US 519 (1966)(upholding examiner's determination that the output of a chemical process was not useful if merely similar to a useful compound).

6 35 USC §101, 102. See *Jamesbury v. Litton Industrial*, 756 F.2d 1556 (CAFC 1985)(finding that an invention was 'novel' when no prior art was precisely equivalent).

7 35 USC §103. See *Graham v. John Deere Co.*, 383 US 1 (1966)(finding an invention invalid on grounds that the improvement would have been obvious to a person of ordinary skill in the art).

8 While courts are given the opportunity to decide on the legitimacy of these determination by the examiners, so few patents are litigated in practice that these initial findings by examiners

Prosecution of the patent has been characterized as a 'give-and-take-affair,' with negotiation and re-negotiation between the patentee and the patent examiner that ordinarily continues for an average of 2-3 years (Merges et al. 1997).[9]

There is a wide variance in the elapsed time between application date and issue, however, and Patent Office processes can be 'gamed' to use time to advantage. Data show that applications pend, in reality, for as little as several months to as much as several decades and, while technological complexity and examiner expertise are certainly major determinants, applicants exert substantial control over their applications' pendency periods. Such control would be advantageous in light of the absolute secrecy that applications historically enjoyed (prior to 1995) before issue. While the 'give and take' of negotiation with the Patent Office provides an applicant with a limited opportunity to delay the examiner's decision, patent applicants may be using the 'continuation' mechanism to achieve longer pendency for their applications – and thus longer periods of secrecy. The continuation, an institutional mechanism, allows the patentee to keep his invention under a cloak of secrecy for longer than the 2-3 year average for first applications, and provides a means for the applicant to exert considerably more control over pre-issue secrecy than simple negotiation with the examiner.

The 'continuation' patent regime in the United States appears to be a result more of historical path dependence than of sound economic and social policymaking. Continuation applications have been available to patentees in the United States at least since 1863. In *Godfrey v. Eames*, 68 US 317 (1863), the US Supreme Court interpreted the Patent Act of 1836 to allow continuation applications, in that case when the original application was abandoned on the same day that the new continued application was filed. The court appeared to be offering the patentee an opportunity to keep an application alive in a situation that, otherwise, would have led to injustice. It is a feature of the form of common law used in the United States, with its adherence to precedent, that singular cases – even if correctly decided – can produce general rules that are inappropriate for judging the broader class of phenomena to which the rule applies. Thus, the continuation procedure in US patenting became a feature of the law, being upheld decades later by the Supreme Court in *Crown Cork & Seal Co. v. Gutmann Co.*, 304 US 159 (1938) and *General Talking Pictures Corp. v. Western Electric Co.*, 304 US 175 (1938). The continuation remained judge-made law for over a century, but the procedure was finally codified in statute by the US Congress in the Patent Act of 1952. It appears that Congress made relatively few judgments about the propriety or welfare consequences of its actions in passing the 1952 Patent Act: it is clear from legislative histories that Congress simply intended to codify the existing state of judge-made law into federal law.

---

often are the final determinations on these questions of patentability.

9 Lemley (1994) found an average length of 864 days (2.37 years) for 2,081 patents issued on 27 December 1994.

Continuations are currently authorised under the patent statute[10] and allow an applicant under certain conditions[11] to adopt the date of an earlier application still pending within the patent office. The essential condition for the filing of a continuation is that both the first and the new applications must disclose the same invention. An application may remain hidden for extended periods of time if the original application is abandoned by the patentee. Since there are no limits to the number of times this abandonment may occur, chains of continuations may develop, as evidenced by Lemelson's bar-coding patent with its string of eleven continuations and abandonments. A strategic applicant can, in this manner, keep inventions secret for an extended period of time.

Absent other benefits, however, a rational strategic applicant would have no incentive to seek a period of added secrecy through continuation. On its face, the continuation appears to offer no advantage over extended and exclusive trade secret protection, while imposing costs through early payment of fees, expenditures for prosecution and added risks of valuable information 'leaking' from the Patent Office to competitors.[12] But the patent applicant does receive a significant benefit from a chain of continuations – an early 'filing date'. In the US 'first to invent' patent system, the filing date is a government-sanctioned recognition that the applicant had reduced an invention to practice by that date. A competitor claiming an earlier invention dates must show why it was not diligent in reducing the invention to practice or had otherwise delayed the filing.[13] Additional benefit is offered to the applicant through a legally recognized protection of the common law rights to trade secret protection during pendency of the application.[14] Thus, the patent applicant historically retained secrecy of invention throughout the application period.

---

10 35 USC § 120 (1991). Benefit of earlier filing date in the United States.

11 Under 35 USC § 120, a patent application is entitled to adopt the filing date of a 'parent' application when (1) both applications disclose the same invention; (2) both applications are filed by the same inventor; (3) both applications are simultaneously co-pending; (4) the earlier application meets the disclosure requirements of 35 USC § 112; and (5) the later application contains a specific reference to the earlier application. *Sampson v. Ampex Corp.* (1971, DC NY), 333 F. Supp. 59, aff'd. (2nd Cir. NY) 463 F2d 1042.

12 While formally the contents, and even the existence, of the patent application were historically kept secret, any revelation of corporate information outside the confines of the firm raises the risk that the information will 'leak' to competitors.

13 'Public policy favors the early disclosure of inventions. This underlies the requirement for 'reasonable diligence' in reducing an invention to practice.' *Naber v. Cricchi*, 567 F.2d 382 (CCPA, 1977), quoted in *Griffith v. Kanamaru*, 816 F.2d 624 (CAFC, 1987).

14 The applicant is 'promised that merely by soliciting, before the Patent Office, [patent protection for his secret invention] he is not giving up his common-law rights. Though the Patent Office may refuse to issue a patent, it must do nothing to jeopardise even an unsuccessful applicant's common-law rights.' *Irons & Steel v. Dann* 606 F.2d 1215 (D.C. Cir. 1979).

Table 9.1 Comparison of the benefits to patentees of using the continuation application, pre-1995 and post-1995 regime changes.

Time period	Patent protection	Application secrecy	Trade-off when considering applying for patent
Pre-1995	17 years from date of patent grant	No time limits in the law	None (apart from delaying the beginning of patent protection)
Post-1995	20 years from date of application filing	20 years maximum, after which the application sunsets	Each additional day of secrecy chosen results in one lost day of patent protection

The patent regime in place prior to 1995, moreover, enabled the applicant to extend this period of secrecy virtually without limit. Because the patent term was 17 years after issue and there were no requirements that applications be published prior to issue, patentees could protect their priority dates based upon filing for many years before beginning their patents' monopoly protections. Continuation allowed the strategic applicant to manage this process, both in terms of timing and technological change.

Changes in the US Patent Law, 1995

As part of a series of statutory modifications meant to harmonize US law with its international trading partners, Congress altered the patent regime's 17-year term in 1995. Measuring the patent term from issue rather than application date had put US law out of step with most of the world while also creating incentives for 'submarine' patenting (Lemley 1994). The 1995 legislative changes which followed the GATT trade talks[15] altered the US patent laws that had supported the submarine practices. By increasing the patent term from 17 to 20 years, the new legislation offered patentees some additional *de jure* length, but began measuring the patent term from the application date instead of the issue date. The Act's legislative history demonstrates that this change was intended to eliminate some of the incentives to engage in 'submarine' patenting by forcing the patentee to 'trade off' *ex ante* secrecy for *ex post* protection. Users of the continuation, and particularly submarine patentees, would henceforth have a maximum of 20 years to delay the patent's grant, and each extra day of delay would come at the cost of one day of protection (see Table 9.1).

While the 1995 legislation reduced a strategic applicant's incentives to engage in continuation practice, it nevertheless did not entirely eliminate these incentives. The statute was engineered to reduce 'submarine' incentives while also offering a longer

15 The legislation was offered and passed in 1994 as part of the fast-track vote on the Uruguay Round of the General Agreement on Tariffs and Trade (GATT). The law's provisions did not take effect until 1995 so it will be termed the '1995 Act'.

Table 9.2 Comparison of publication requirements in the United States, pre-1999 and post-1999 regime changes

Time period	Patent published by the USPTO	Exceptions
Pre-1999	Never	Not necessary
Post-1999	18-months after application filing	Yes, if applicant declares there is no intent to patent in another country that requires publication[1]

1 35 USC 122(B): '(i) If an applicant makes a request upon filing, certifying that the invention disclosed in the application has not and will not be the subject of an application filed in another country, or under a multilateral international agreement, that requires publication of applications 18 months after filing, the application shall not be published .... (ii) An applicant may rescind a request made under clause (i) at any time.'

patent term, by granting each patentee a maximum 20-year window in which to both prosecute his application and enjoy the benefits of patent protection. So long as patent prosecution lasted less than three years, a patentee could enjoy an added benefit – a longer patent term than had been available under the 17-year term provided by the previous law. For inventors considering a continuation strategy, the legislation required that the applicant weigh the present value of early priority, coupled with extended *ex ante* non-disclosure, against added patent protection at the end of the patent term. Patentees whose value in early priority and a period of relatively lengthy pre-issue secrecy outweighed that of late term patent protection retained an incentive under the new statutory scheme to resort to continuation practice.

Changes in the US Patent Law, 1999

The disincentives faced by patent applicants using the continuation were increased in 1999 with the passage of the American Inventors Protection Act.[16] This amendment to the patent laws generally requires that applications be published after 18 months, thus bringing the US law into harmony with the patent laws of most of the major patenting jurisdictions.[17] Effective lobbying by independent inventors, however, resulted in one important loophole: patent applications made in the United States, but not also in another country requiring 18-month publication, are exempted from the US publication requirements. Therefore, although the 1999 publication requirement may prove a substantial disincentive for the continuation applicant motivated to capture a longer period of non-disclosure, a class of applicants is exempted from the requirements of the law (Table 9.2).

Because the continuation process is expensive and the rational applicant would avoid the investments in monetary fees and time that it entails, a non-strategic

16 Pub. L. No. 106-113, 1000(a)(9), 113 Stat. 1501, 36.

17 Japan and the signatory nations to the European Patent Convention, including Germany, France and Great Britain, abide by an 18-month application publication rule.

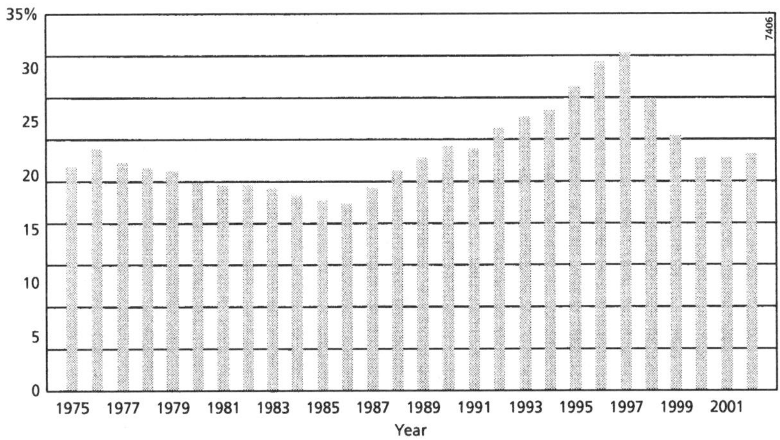

Figure 9.1 Patents issued after a continuation application, as a share of annual issued patents, 1975-2002

continuation is likely the result of information asymmetries between applicants and examiners concerning the technology (that the applicant may understand better) or the application process (that the examiner may understand better). These asymmetries are more likely in the presence of technological newness. When the subject matter of the patent is a new technology, applicants and their agents – as well as the examiners – have had limited time in which to learn the technology, and all parties face increased uncertainty about its patentability. Information asymmetries can also be expected to increase with technological complexity.

Data Analysis: Continuation Patents

To shed light upon patentees' uses of the continuation, as well as its importance as an economic phenomenon, the overall trend in the use by patentees of continuations from 1980 to 1999 is presented in graphic form below. Figure 9.1 plots the fraction of patents issued in each of the listed years that had at least one continuation[18] in application lineage. The data thus represent only the terminal patent, the patent that emerges at the end of the chain of continuations. Because some applications may have been abandoned, this terminal patent may be the only evidence of a continuation chain. The data in Figure 9.1 are based upon issue date and not upon application date because, under continuation practice, the recorded application date for what is essentially the final application in a chain of continuations is a somewhat artificial measure. Issue dates are a more meaningful metric of the use by patentees

18 My definition of the term 'continuation' includes also the 'continuation-in-part' and 'division' processes by which particular claims or groups of claims can be continued as opposed to an entire application and all disclosures therein.

of the procedure because the issue date is one over which the applicant exercises a greater deal of control than in the 'normal' single-application patent case, and give an indication of when, within a certain distribution, the patentee intends the period of secrecy to end.[19]

These numbers suggest that the continuation remains an important economic phenomenon, accounting for at least 17 percent and as much as 31 percent of annual issued patents during the 1975-2002 period. Continuation practice was commonly used by patentees before 1995 and such use remains significant after the Congressional changes to the patent law. The downward trend after 1997 may be an indication that the disincentives intended in the 1995 regime change have had an impact on behaviour, when one corrects for some identifiable lag in the prosecution of patents already in the system. For these overall data, continuation practice has fallen from a high of 31 percent in 1997 to 24 percent in 1999-2002. The overall use of continuation has remained roughly flat after 1999, which suggests that the 1999 regime change has not had an immediate effect on the use of the procedure.

It is important here to note that the growth in the use of the continuation procedure beginning in 1986 reflects continuation applications that were filed several years earlier – certainly prior to the 2-3 year mean delay for non-continuation applications. The increased use of continuations by patentees began well before the data reflects an upturn in their issuance, and may well correspond with the pro-patent legal regime that attended the founding of the Court of Appeals for the Federal Circuit in 1983 (Kortum & Lerner 1997). Nonetheless, between 21-23 percent of the patents issuing 2000-2002 involved continuations and, as will be discussed below, these 'post-reform' shares are much higher in specific patent classes.

CONTINUATIONS: DESCRIPTIVE STATISTICS WITHIN SECTORS

The trend presented in Figure 9.1 demonstrates that the use of continuation was widespread across all patenting through the period 1975-2002, and thus of economic interest. To answer the questions posed in the section above ('Continuation application practice under the US Patent Law'), however, it is necessary to supplement this aggregate trend with data from individual technology classes, thus providing a window into the patterns of use across different technologies and sectors. As a benchmark, I used a reasonably mature technology – combustion engines.

19 There will always be a certain amount of uncertainty about the date of issue, although the patentee can minimize this uncertainty through the negotiation process with the USPTO examiner and the deft use of the continuation process itself.

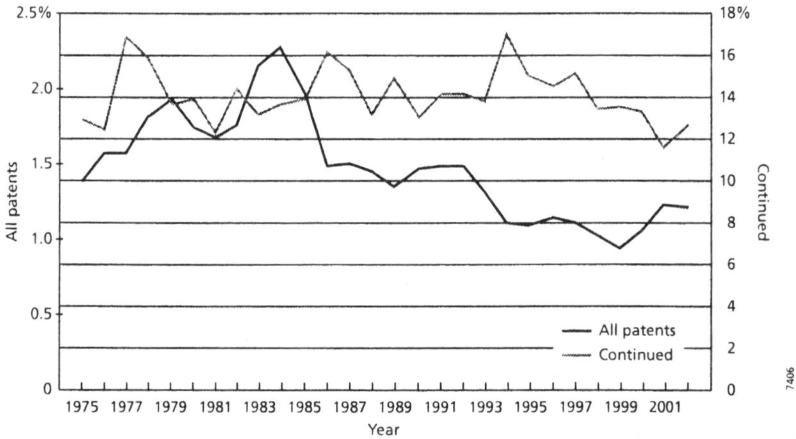

Figure 9.2 Motor-generator technologies, share of all patenting and 'continuations' share, 1975-2002

Combustion Engines

Combustion engine technologies were chosen as a benchmark on two grounds. First, the technology is mature – Daimler and Benz produced the first working gasoline combustion engines in the mid-1880s.[20] Because the technology has been progressing incrementally for over a century and underpins a large, mature industry (automotive), the technology has been more insulated from rapid, discontinuous change and greater uncertainty than have newer technologies.[21] Second, the technology sector into which this technology falls, motor-generator, appears to be one in which secrecy is not an important consideration. In fact, the technology group is characterized in Cohen et al. (2000) as having effectiveness ratings for both secrecy and patenting below the mean of all sectors, for both products and processes.

Using the International Patent Class identifier for 'Combustion Engine' technologies (F02-; COMBUSTION ENGINES and HOT-GAS OR COMBUSTION-PRODUCT ENGINE PLANTS), Figure 9.2 (left axis) plots the share of combustion engine patents within all utility patents issued in the years 1975-2002, demonstrating that over time this technology has become increasingly marginalized in the post-80s surge in patenting, one indicator that it is not a rapidly changing technology area. Figure 9.2 (right axis) shows that the use of continuations by combustion engine patentees over time lacks the dynamics shown in the aggregate of all technologies summarized in Figure 9.1. The share of all F02- patents showing a continuation application lineage after 1975 appears to fluctuate between 11 percent

20 Coal gas models predated these improvements, with Lenoir and Otto early pioneers.

21 Although the patterns of radical (versus incremental) change in technologies can create discontinuities in even the most mature technologies.

and 17 percent, and does not exhibit the sharp downturn after 1997 characterized by the trend of all patents.

Pharmaceuticals

Because of the vital role that patents play, and the regulatory hurdles imposed by the US government, the pharmaceutical industry is demonstrably different from other industries in terms of its innovation characteristics – and therefore meaningful comparisons with other industries are difficult, if not impossible. This observation is also true in terms of continuation practice, because the pharmaceutical industry has historically used the continuation application to aid in solving the thorny problems of patenting within the regulatory regime surrounding new drug applications (NDAs). Thus, the pharmaceutical industry, unlike other sectors, enjoyed an added and potentially dominant motivation for using the continuation strategy: it allowed firms to 'fit' the beginning of a product's patent protection to the timing of a drug's exit from the extended regulatory approval process, thus allowing maximum patent protection during the time when the drug is actually marketable.[22] Thus regulatory demands may have been the primary motivator for the use of the continuation application in this industry.

The drugs industry is characterized by scholars as one in which patents are economically valuable and important in appropriating returns to innovation (Henderson & Cockburn 1996; Cohen et al. 2000). The characteristics of the industry and pharmaceutical products make a long patent term particularly valuable. Longer patent terms are particularly valuable in the pharmaceutical industry, especially when compared to industries with relatively quick invention-to-manufacture intervals and short product life cycles; products emerge from regulatory procedures quite late, and a large portion of industry revenues are generated from drugs that generate massive sales and have long lives (Comanor 1986).[23] The 1962 Kefauver amendments requiring stronger proof of the 'efficacy' of new pharmaceutical products shortened the patent term for pharmaceuticals by extending the period of clinical trials and regulatory approval (Grabowski 1976). Partly in response to lobbying from the pharmaceuticals industry and the desire of pro-consumer lawmakers to facilitate the entry by producers of generic drugs

22 In the United States, the Federal Food and Drug Administration (FDA) requires that all pharmaceuticals be tested for efficacy and safety in a lengthy drug approval process. So, for instance, if a company's patent on a new drug's molecular make-up were to issue prior to the drug completing this government approval process, then patent protection would be 'wasted.' Because of the long life of drug products in the marketplace, and to the extent that patent protection suppresses competition from 'generic' drug producers, it is a benefit to the innovator to extend patent protection as long as possible.

23 Lemley (1994) suggests that pharmaceutical and biotechnology patents may be more valuable at the end of their patent terms than are software patents.

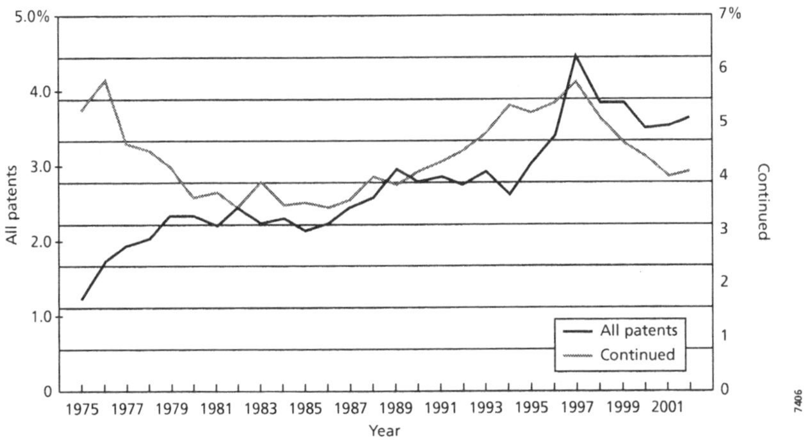

Figure 9.3 Pharmaceutical technologies, share of all patenting and 'continuations' share, 1975-2002

following the expiration of pharmaceutical patents, the 1984 Drug Price Competition and Patent Term Restoration Act extended the patent term for pharmaceuticals by up to five years.[24] While there was some empirical support for this argument leading up to the passage of the act (Schwartzman 1976; Grabowski & Vernon 1983), the effects of continuation practice in extending the effective patent term were underplayed in the literature (Comanor 1986). Under the pre-1995 patent regime pharmaceutical companies were using continuation practice to engage in 'serial prosecution'. Serial prosecution allowed the pharmaceutical company to patent different components of the invention – often beginning first with the molecule – over a long period of time, thus extending the *de facto* patent term to much longer than the 17 years permitted under the then existing law (Burchfiel 1995). Firms may have also used the continuation to coincide with their extended regulatory approval periods, maximizing the patent term by launching the patent close in time to their drugs emerging from the regulatory approval process.

In order to gain some insight into the use of continuations in the pharmaceutical industry, I examined trends in the share of continuations within international patent classification (IPC) A61K (PREPARATIONS FOR MEDICAL, DENTAL, OR TOILET PURPOSES; excluding A617/7 'cosmetics or similar toilet preparations', following Harhoff and Reitzig (2004)). This patent class is shown in Figure 9.3 (left axis), growing from 2.1 percent to 3.9 percent of overall US patenting 1975-2002. IPC A61K has been used to define pharmaceutical and biotechnology patents in several studies (Lanjouw & Schankerman 1997; Harhoff & Reitzig 2004).

The share of pharmaceutical patents that are the terminal patent in a chain of continuation applications are plotted from 1975 to 2002 in Figure 9.3 (right axis).

24 The act also reduced entry barriers to generic drug providers.

A large percentage of pharmaceutical and biotechnology patents are the product of continuation practices ranging from an annual low of 34 percent to a high of 60 percent, well above the 12-17 percent exhibited by the mature technology combustion engines. As with the broader population of patents, some downward trending is exhibited after 1997, with continuations in this sector falling by 12 percentage points through year 1999 from the 1997 high of 51 percent. Continuations, however, rebound in use, actually increasing in use between 2000 to 2002 accounting for 52 percent of issued drug patents by 2002, several years after the disincentives of the 1995 and 1999 Acts had been introduced.

Unlike some other sectors, the comparatively high level of continuation practice exhibited in this industry in the late 1980s continued after the 1995 Act. Because the 1995 Act forced a 20-year term inclusive of prosecution, it eliminated the alternative and potentially dominant motivation for pharmaceutical firms' use outlined above – maximizing the drug patent term. The figures thus lend evidence to support a role for other strategic motives, such as protecting technology secrecy, for example, in addition to maximizing the patent term, as a motivating factor for continuation practice in this sector.

Semiconductor Products

I identified semiconductor technology classes by referring to the patenting from 1983 to 1994 of five major semiconductor firms – Texas Instruments, Intel, Advanced Micro Devices, Micron Technology and National Semiconductor.[25] Analysis of patenting for these five firms allowed me to identify two international patent classes, H01L (SEMICONDUCTOR DEVICES; ELECTRIC SOLID STATE DEVICES NOT OTHERWISE PROVIDED FOR) and H03K (PULSE TECHNIQUE) which constitute a large percentage of overall patenting by these firms (23.4 percent and 10.0 percent, respectively) while at the same time covering product (for example, circuit) process technology. The trend in Figure 9.4 (left axis) shows that these electronic technologies have captured a steadily increasing share of patenting through 1980-99, growing from two percent to five percent of all patenting during this period. This growth suggests that these technologies are more dynamic and fast-changing – and thus more uncertain – than mature technologies like combustion engines.

Figure 9.4 (right axis) plots the share of patents in these patent classes that were the terminal patent in a continuation chain, 1975-2002. In these semiconductor technologies, the use of the continuation in patenting has been common since 1975, with shares of patents issued with some continuation use fluctuating between 17-37 percent through 1996. Like pharmaceutical patenting, there appears to be a discernable pattern in the growth of use by applicants of the continuation during the

25 I thank Rosemarie Ziedonis for allowing me to use her semiconductor patent data to complete this analysis.

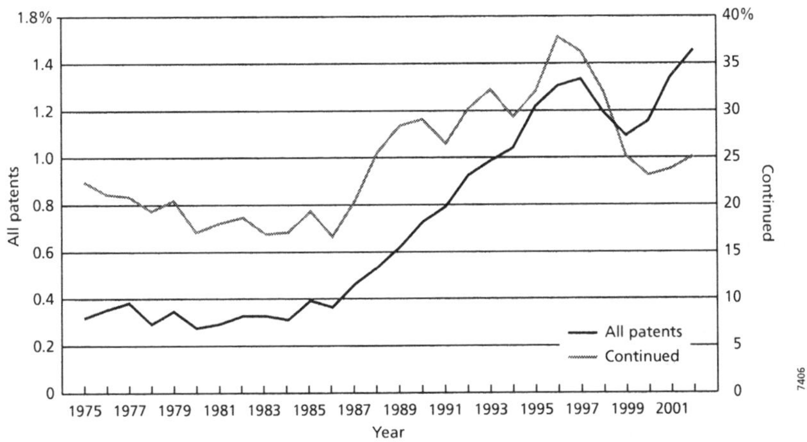

Figure 9.4 Semiconductor technologies, share of all patenting and 'continuations' share, 1975–2002

pre-1995 era: use grew strongly from 17 percent of issued patents in 1987 to 37 percent by 1996.

The trend for continuation use in semiconductor patenting, like pharmaceuticals and the broader sample of patents, does show a marked downturn after 1997. In this sector it is of a similar magnitude, dropping 11 percentage points through year 2000 from a high of 37 percent in 1996. Furthermore, by year 2000, continuations in these technologies account for only slightly more than 23 percent of issued patents, but has rebounded to 25 percent by 2002. This industry embraced continuation practice in the years prior to enactment of the 1995 changes, but while the disincentives wrought by the 1995 law appear to have brought a behavioural change in patenting, effects from the 1999 law are not immediately evident.

My discussions with patent counsel at several large semiconductor firms give me some insight into these continuation patterns. It is common practice in the industry to carve out claims that the examiner deems as objectively patentable and for the patent agents to leave the less-certain claims behind as continuing applications. Some share of these 'second-hand' applications is then later issued as patents. What is unanswered by such behaviour is the quality of these 'second-hand' applications. While some of the patent attorneys I interviewed saw the claims associated with these continued applications as 'junkier' (less deserving of patent protection), it is possible that some of these 'second-hand' applications embody high quality patents, the patentability of which was not, or not capable of being, taught to the patent examiners in a timely manner.

In both pharmaceutical and semiconductor technologies, however, the use of continuations appears to have been negatively impacted by the disincentives envisioned and implemented in the 1995 Act. This pattern in these two technologies is consistent with the dramatic decline in continuation usage that Graham and

Mowery (2004) demonstrate in US software patenting after 1995. Graham and Mowery, however, also show that that in the software industry, characterized like semiconductors by short product life cycles, patentees continue to use the continuation mechanism relatively intensively as of 2000, despite the fact that the marginal value of a 21st year of protection is surely low. This pattern is unlike the one I have shown above for semiconductor technologies, which exhibits continuation practice falling to levels not unlike those found in 'static' mature technologies like combustion engines.

Conversely, the continuation disincentives expected from 'publication' in the 1999 Act appear to have had little impact. The share of all patents issuing after 1999 that include at least one continuation in the application chain have instead increased, year on year. This pattern is also reflected generally in both the semiconductor and drugs technologies, with at best no discernable pattern in the post-1999 trend. Due to grant lags, the data through 2002 may not offer sufficient longitudinal coverage to draw convincing conclusions. But even this limited evidence suggests that the continuation behaviour of patent applicants has not altered substantially after the 1999 Act.

CONCLUSION

If we accept that innovation policy ought to be aimed at producing the greatest welfare to the greatest number of people (however such a thing is calculated), we must accept that innovation policy theory (such as harmonization) and policy implementation (such as changes in and selective enforcement of the laws) have not always worked in concert to accomplish that goal. In fact, any evaluation of policy must comprehend that these two elements of policy may interact to have uncertain effects. We need look no further than the history, politics and economics surrounding the US patent office's 'continuing application' for evidence of 'path dependency and structural sclerosis' so characteristic of innovation systems (Introduction in this book).

For instance, the results of the legislative changes meant to create disincentives to using the 'continuing application' are murky. According to the empirical evidence presented in this chapter, the 1995 Act appears to have lowered the incidence of continuation use, but this trend is not as pronounced in all sectors (in drugs, for instance). Moreover, early evidence does not suggest that the 1999 Act has had a similar effect. And, in fact, there is substantial evidence that a narrow 'winner' in the patenting game – independent inventors – used their considerable lobbying influence to secure an important loophole in the 1999 legislation. This loophole allows domestic inventors to use the continuing application procedure in secret, securing to themselves important advantages in watching others adopt the patented technology in the marketplace, and then, in secret, changing their claims on the patent application to cover those newly adopted uses. At first blush, this outcome appears to offer few benefits to society at large. The 'continuing application,' written into the statutory law and protected by the courts interpreting that law, is a rule that has shown itself, for a particularly vociferous group of inventors, to be resistant to change since it was first introduced 150 years ago.

And what influence have the legislative changes had upon the amount and pace of innovation? While continuation rates fell after 1995, overall patent application rates have continued to rise. But, we do not observe the counterfactual and thus know nothing about whether patent applications would have grown 'more' than they did absent the 1995 or 1999 Acts. Patents are also an imperfect measure of innovation. Because we do not observe whether innovators were shifting away from patenting to other appropriability mechanisms (such as secrecy), or even 'towards' patenting from these alternative mechanisms (if these other methods are substitutes), it is difficult if not impossible to measure the impact that the harmonization policy initiatives have had upon innovation generally.

In his use of the continuation application to engage in 'submarine' patenting, Jerome Lemelson demonstrated how this potent tool could be employed. Because the continuation allowed an applicant to protect an early priority, and coupled this protected priority with secrecy, the pending patent right could be engineered to maximize the applicant's rent transfers from other 'users' of the patented technology. The use that Lemelson made of the continuation application was unlikely to be isolated: this chapter demonstrates that overall usage of the continuation was reasonably high and in several important technology sectors, notably pharmaceuticals and semiconductors, a majority of patents issued in many of the years 1980-1995 were granted after some continuation practice. While I lack conclusive evidence that submarine tactics were used by a large share of other applicants during these years, several patent attorneys and corporate patent counsels active during these years have reported to me that it was common practice for applicants to take advantage of priority and secrecy to ensure maximum value from their patent rights.

Such outcomes provide evidence for the co-evolution over time of the legal institutions that support the innovation system. As demonstrated throughout this book, institutions, politics, policy and economics tend to co-evolve in innovation systems, with substantial endogenous effects characterizing the process. The history of the 'continuing application' in the US patent law offers a case study of this co-evolution, demonstrating both 'path dependency and structural sclerosis' as well as 'new combinations, related chances and options'. The evidence presented in this chapter suggests that shocks such as the General Agreement of Tariffs and Trade offered policymakers options to alter the existing rules, but that these changes were both demanded by powerful forces (large international conglomerate firms and foreign governments) and resisted by influential constituencies (domestic independent inventors). We have seen that the history of these rules in the United States since 1995 has been one of both experimentation and resistance to change. As such, the final outcome is not yet clear, except to say that the process has been characterized by endogenous forces, with the static economics of innovation driving policy adoption, and that adoption in turn setting the environment in which dynamic changes will occur moving forward.

ACKNOWLEDGEMENTS

I would like to thank Philip Shapira for helpful comments on this chapter. I acknowledge research funding by the Ewing Marion Kauffman Foundation, and note that the contents of this chapter do not reflect the views of the Kauffman Foundation. All omissions and mistakes are my own.

REFERENCES

Albright, B. (2004), 'Bar code industry wins case against Lemelson Foundation', *Frontline Solutions*, 5 (4), 40-41.

Arrow, K. (1962), 'Economic welfare and the allocation of resources for invention', in R. Nelson (ed.), *The Rate and Direction of Inventive Activities*, Princeton: Princeton Universtity Press, pp. 609-625.

Burchfiel, K. (1995), 'US GATT legislation changes patent term', *Journal of the Patent and Trademark Office Society*, 77, 222-252.

Cohen, W., Nelson, R. and Walsh, J. (2000), 'Protecting their intellectual assets: appropriability conditions and why US manufacturing firms patent (or not)', NBER working paper, No. 7552.

Comanor, W (1986), 'The political economy of the pharmaceutical industry', *Journal of Economic Literature*, 24 (3), 1178-1217.

Grabowski, H. (1976), *Drug Regulation and Innovation*, Washington, DC: American Enterprise Institute.

Grabowski, H. and Vernon, J. (1983), *The Regulation of Pharmaceuticals: Balancing the Benefits and Risks*, Washington, DC: American Enterprise Institute.

Graham, S. (2004a), 'Patenting in the shadow of secrecy: innovators' uses of US patent office continuation practice, 1975-2002', in *Continuation, Complementarity, and Capturing Value: Three Studies Exploring Firms' Complementary Uses of Appropriability Mechanisms in Technological Innovation, doctoral dissertation*, Berkeley: University of California.

Graham, S. (2004b), 'Hiding in the patent's shadow: firms' uses of secrecy to capture value from new discoveries', in *Continuation, Complementarity, and Capturing Value: Three Studies Exploring Firms' Complementary Uses of Appropriability Mechanisms in Technological Innovation, doctoral dissertation*, Berkeley: University of California.

Graham, S. and Mowery, D. (2004), 'Submarines in software? Continuations in US software patenting in the 1980s and 1990s', *Economics of Innovation and New Technology*, 13 (5), 443-456.

Graham, S., Hall, B., Harhoff, D. and Mowery, D. (2003), 'Post-issue patent "quality control": a comparative study of US patent re-examinations and European patent oppositions', in W. Cohen and S. Merrill (eds), *Patents in the Knowledge-based Economy: Proceedings of the Science, Technology and Economic Policy Board*, Washington, DC: National Academies Press, pp. 74-119.

Hall, B. and Ziedonis, R. (2001), 'The determinants of patenting in the US semiconductor industry, 1980-1994', *Rand Journal of Economics*, **32** (1), 101-128.

Harhoff, D. and Reitzig, M. (2004), 'Determinants of opposition against EPO patent grants: the case of biotechnology and pharmaceuticals', *International Journal of Industrial Organization*, Elsevier, **22** (4), 443-480.

Henderson, R. and Cockburn, I. (1996), 'Scale scope and spillovers: the determinants of research productivity in drug discovery', *RAND Journal of Economics*, **27** (1), 32-59.

Jaffe, A. and Lerner, J. (2004), *Innovation and Its Discontents: How Our Broken Patent System is Endangering Innovation and Progress, and What to Do About It*, Princeton: Princeton University Press.

Kortum, S. and Lerner, J. (1997), 'Stronger patent protection or technological revolution: what is behind the recent surge in patenting?' NBER working paper No. 6204. Cambridge, MA: National Bureau of Economic Research.

Lamoreaux, N. and Sokoloff, K. (2002), 'Intermediaries in the US market for technology, 1870-1920', NBER working paper, No. W9017.

Lanjouw, J. and Schankerman, M. (1997), 'Stylised facts of patent litigation: value, scope and ownership', NBER working paper, No. 6297.

Lemley, M. (1994), 'The implications of GATT on US intellectual property laws: an empirical study of the 20-year patent term', *AIPLA Quarterly Journal*, **22** (3), 369-388.

Merges, R., Menell, P., Lemley, M. and Jorde, T. (1997), *Intellectual Property in the New Technological Age*, New York: Aspen Press.

Mowery, D. and Nelson, R. (eds) (1999), *Sources of Industrial Leadership*, Cambridge, UK: Cambridge University Press.

Nelson, R. and Nelson, K. (2002), 'Technology, institutions, and innovation systems', *Research Policy*, **31**(2), 265-272.

Nordhaus, W. (1969), *Invention, Growth, and Welfare*, Cambridge, MA, USA and London, UK: MIT Press.

North, D. (1990), *Institions, Intitutional Change and Economic Performance*, Cambridge, UK: Cambridge University Press.

Schumpeter, J. (1942), *Capitalism, Socialism, and Democracy*, New York: Harper and Brothers.

Schwartzman, D. (1976), *Innovation in the Pharmaceutical Industry*, Baltimore: John Hopkins University Press.

10. The Use of the Regulatory Framework for Innovation Policy

Knut Blind

INTRODUCTION

Topics of regulation, innovation and competitiveness in global markets have been discussed for several decades. However, little has been done in a comprehensive way to understand the effect of regulation on the ability of industries to innovate and be competitive.[1] The debate has taken place at the level of anecdotal evidence and with poor systematic empirical foundations. Moreover, surveying those studies which are available provides no clear picture as to whether the negative impacts of regulation outweigh the positive effects. For countries that are members of the Organisation for Economic Co-Operation and Development (OECD), Bassanini and Ernst (2002) find a negative correlation between the intensity of product market regulations and the intensity of research and development expenditure. Swann (2005) shows that for a significant share of British companies the content of regulations is both an important source for innovation, but also a severe obstacle for the innovation activities. Following the seminal paper of Porter and van der Linde (1995), several further contributions (e.g. Jaffe et al. 1995; Jaffe & Palmer 1997) produce ambivalent results regarding the influence of regulation on the development of new environmental technologies. Furthermore, most of the approaches assume a static framework, not recognising the long-term dynamic feedback loops between regulation and innovation (an exception is Kemp 1998) or even the interactions of the institutions and actors in the innovation systems.

There are discussions among innovation policy makers to extend their focus on the regulatory framework as a possible policy instrument (Blind et al. 2004). Due to restrictions in public budgets, policymakers often have little leeway to increase public spending on research and development to promote innovation. Therefore, policies to improve the framework conditions relevant for innovation become more important. Consequently, regulatory impact assessments are an important instrument for regulatory policies, not only do they consider the burden for companies to fulfil

1 Blind (2007) presents a recent collection of studies about the interaction between innovation and regulation ranging from conceptual approaches to specific industry case studies.

specific regulations, but they also try to take into account possible impacts on innovation.

Besides limiting the negative impacts of regulation on the innovative activities of industry, policy makers are starting to look more systematically for options to use specifications of the regulatory framework to promote these activities. Static efficiency concepts are not sufficient to achieve this aim, instead dynamic frameworks are required which allow at least for temporary static inefficiencies. However, the extension of the traditional objectives of regulatory policies for innovation-related goals needs to be systematically checked for possible conflicts or synergies. For example, the protection of the environment can often be achieved by innovative environmental technologies which can produce win-win results, whereas the introduction of new drugs invariably raises safety considerations with regulation and innovation finding themselves in competition.

In addition to the internal consistencies or inconsistencies between the traditional and the innovation-related objectives of a regulation or a regulatory body, the various regulatory policies along the innovation cycle have to be coordinated in order to exploit possible synergies and to avoid unnecessary friction or even contradictions. For example, regulatory activities focused on the research and development activities in a specific technological area have to be coordinated and timed with regulations fostering the demand side, either by increasing the purchasing power via subsidies for new products or by promoting their acceptance through safety regulations which take into account new emerging risks.

Successful innovations need a coordinated interplay of all relevant stakeholders of the innovation system concerned. Shaping regulatory policies to foster innovations within the systems of innovation approach requires the involvement of the relevant stakeholders and institutions from the very beginning of the emergence of the regulation until its implementation and revision. However, the role of the numerous regulations relevant for successful innovation has not been integrated in the innovation systems approach.

Finally, the relationship between regulation and innovation has to take into account that the regulatory framework is also influenced or pressurised by the progress in science and technology, and is therefore an endogenous and not exogenous factor in the innovation system. This dynamic dimension represents a further challenge both for regulatory and innovation policy, but also an opportunity to use the regulatory framework for the promotion of innovation.

Considering together all the varied and complex dimensions which are required for the successful use of the regulatory framework to promote innovation makes very obvious that this relationship represents a great challenge for theoretical concepts and analysis, empirical investigations and practical policy implementation.

This chapter aims to address the options various regulatory policies provide for the support of innovation. The chapter starts with an overview of types of regulations and their ambiguous impacts on innovation. The second section presents a new taxonomy of product market regulations. The third section discusses the role of the various product market regulations within the innovation system approach by taking the various functions of innovation systems into account. In the fourth section, the dynamic dimension of innovation processes is highlighted by discussing the role of regulations in innovation phases. The fifth section addresses the co-evolution of regulatory economics, regulation policy and practice. The final sections outline proposals to improve the relationship between innovation and regulatory policy and discuss selected future challenges for regulatory policy, promoting innovation.

TYPES OF REGULATION AND THEIR IMPACT ON INNOVATION

The term 'regulation' generally refers to the implementation of rules by public authorities and governmental bodies to influence market activity and the behaviour of private actors in the economy. Such intervention in the market is justified by the goal to maximise collective welfare, including reaching some distributive goals. In economic literature by the OECD (1997), economic, social and institutional regulations are usually distinguished, but will be explained in the context of their impact on innovation in the next section.[2]

The impact of regulation on innovation is difficult to assess, because most assessments perform an evaluation based on static efficiency criteria, although innovation is intrinsically a dynamic process. Furthermore, innovation is a complex phenomenon and it is difficult to grasp all of its – sometimes contradictory – aspects. The OECD and Eurostat (2005) 'Oslo Manual' defines innovation to include product innovations, covering both goods and services introduced to the market which are either new or significantly improved with respect to fundamental characteristics. However, often process innovations in the form of new or significantly improved technology are crucial for the production or the supply of new goods and services. The impact of regulation on innovation should consider both product and process innovations.

Typically, the assessment of regulation is based on static efficiency criteria instead of dynamic ones, since the latter approach would require long-term observations of the behaviour of the affected economic agents. Economic regulations are judged on the basis of the generated improvement of social welfare estimated by changes of production costs, prices and turnover. The efficiency of social regulation is estimated on the basis of the remaining level of externality, e.g. environmental pollution, or of the volume (and costs) of available resources. Additionally, the efficiency of administrative regulations in the sense of institutional and legal framework conditions is assessed through the level of the so-called 'transaction costs', which often cannot be directly measured except by proxies. Often, assessments of

2 Further information can be found in Blind et al. (2004), which is mainly based on Brousseau (1998) and the literature cited there.

economic regulations are based on the analysis of theoretical models without a direct application to concrete regulatory changes. An exception is the regulatory impact assessment programme of the United States of America (OECD 1999), which has also been adopted in the European Union.

In general, the impact of regulation on innovation can be measured either by the input or the output of the innovation process. Innovation also requires physical and human resources like a classical production process. It is performed in a first stage by formal processes of R&D and by informal and sometimes unintentional processes of learning-by-doing or learning-by-interacting. In order to generate economic effects, the results of the R&D process have to be adapted to the needs of potential users and have to be promoted by marketing activities. As a consequence for the analysis, the inputs into the innovation process cannot be reduced just to R&D expenditures. Especially in the service sector, new services are generated by a close interaction between the consumers and the service providers, whereas R&D activities are of minor importance. Finally, innovation processes are characterised by a high level of uncertainty, which makes it difficult to identify a close link or causality between inputs into the innovation process and its final result.

The measurement of innovation by output has to take into account that it can take place at the level of new products, new services or even new markets, but may include also process innovations like the application of new technologies and organisational innovations caused by changes in the management. Some of these aspects can be more easily observed than others, but often they are interconnected. Furthermore, radical innovations can be observed better than incremental innovations, which are characterised by gradual changes of products, processes or organisations. However, the latter type of innovation is much more frequent than the first, more spectacular type. A final problem for the observation of process or organisational innovations is the fact that they are often kept private and not disclosed by a product announcement or a patent application. Summarising the characteristics of the output of innovation, it is easier to observe completely new products and services than improved processes or organisational changes.

In addition, studies about the impact of regulation on innovation should take into account how regulations affect the various aspects of the complex process of innovation. Since innovation is a process of search, discovery, development, improvement and adoption of new products and processes, it is also a cumulative process that produces an increasing amount of knowledge.

This has consequences for the analysis of the impact of regulation, which cannot be based on a simple linear and mechanistic stimulus-response model, since this does not take into account the complex interdependencies between regulatory policy measures and the manifold aspects of innovation. First, since an innovative process requires the co-ordination of very diverse operations, its efficiency is influenced by the dynamic articulation of these various tasks as much as by the individual performance. It must therefore be pointed out how regulations affect the ability of economic agents to articulate these tasks within and among firms, and finally how it impacts on the speed of the innovation process, as well as on its quality in terms of the fitness of the results to the needs of the users and of the appropriability by producers. Second, since an innovation process leads not only to new products or processes, but also to knowledge, it is essential to determine how regulation affects the production and diffusion of knowledge. Third, innovation implies the coordination of various parties within and among firms, the impact of regulation on this coordination including inter-firm collaboration has to be taken into account. Fourth, since innovation is a dynamic process, the causal relationship between regulation and innovation may change over time. Therefore, the assessment of the impact of regulation has to acknowledge its various impacts on the diverse phases of an innovation process.

The Impact of Economic Regulation on Innovation

Most of the existing literature on this issue is dedicated to the analysis of the impact of antitrust regulation on innovation. Recently, the approaches have become more differentiated and go beyond simple monopoly regulation towards more efficient incentive regulation, also taking into account stimulating effects on innovation.

In general, it has to be underlined that the effect of economic regulation on innovation is very controversial. Therefore, one cannot analyse the relationship on an abstract level, but must investigate concrete variants of economic regulation. Nevertheless, regulatory constraints generate contradictory effects for innovation. It is central to take into account that a very pro-competitive regulation scheme forces companies to innovate in order to reach a competitive advantage. However, this scheme may forbid strategies and organisational arrangements, like close collaborations between firms, which are necessary to introduce new technologies and new products into the market successfully. However, if the regulatory regime tries to protect infant industries and firms in order to develop new markets, then the consequence could be the persistence of a protected industry, which will not become competitive with foreign, unprotected industries.

These two examples make clear that it is necessary to analyse the impact of the various types of economic regulation on the various aspects of innovation and innovative processes.

Antitrust or Pro-competition Regulation

The link between pro-competitive regulation and innovation has been discussed since Schumpeter's seminal contributions. Four issues have to be taken into account. First, regulation and innovation mutually affect each other. Second, there are two very contradictory positions about the impact of regulation on innovation. Third, this antagonism is mainly caused by a divergent understanding of competition that leads to very contrasting implementations of pro-competition or antitrust regulation.

It is generally accepted that not only does regulation influence innovation, but that innovation also has an impact on regulation, since in many industries the so-called de-regulation or liberalisation processes have been induced by technological innovations. The best example is the telecommunications industry, which was formerly a natural monopoly, but triggered by technological change (e.g. digitalisation of transmission) into an industry with 'normal' production cost structures. Therefore, the regulatory framework adequate for a natural monopoly had to be adapted to the new situation caused by technological innovations. We have observed not only in telecommunications, but also in other industries, that the technological development drove the evolution of regulation towards highly pro-competitive regulation.

The other way round, it is confirmed that pro-competitive policies affect the likelihood to innovate. Deregulation of markets leads in most cases to a higher pressure on costs and consequently to related process innovations and finally to decreasing prices. Reduced prices force companies either to make production processes more efficient by searching for new innovative technologies, or companies have to introduce new products into the markets, which may be superior to the existing one. Consequently, de-regulation often allows formerly regulated companies to expand their product and service assortment by the introduction of new products and services. This kind of economic regulation fosters innovations, which are reducing costs or which lead to new products opening new markets and relaxing the pressure on costs.

There are two antagonistic views about the impact of regulation on innovation. On the one hand, pro-competitive frameworks are supposed to be favourable to innovation. Since companies are allowed to choose their strategies and actions without restrictions, innovation activities should be easier to perform. In addition, in a highly competitive environment, firms are strongly forced to innovate, because both product and process innovations allow them to survive in the long run by being ahead of the competitors. In a Schumpeterian perspective, innovating firms enjoy a temporary monopoly that provides them with rents. In sum, competitive environments caused by pro-competitive regulations are positive for innovation.

However, the empirical evidence is rather ambivalent, not only for methodological reasons but because it is based on theoretical approaches, which assume that innovation requires both the close cooperation between firms and significant resources, i.e. profits derived from monopolistic positions. Therefore, on the other hand, pro-competitive policies reduce the ability of firms to form strong alliances, especially in the R&D stage. In addition, innovation is costly and leads to temporarily inefficient resource allocation. Only in the long run can the broad distribution of new products or the broad application of new technology make it possible to recover the investments in the early stages. Consequently, innovation processes should deserve some kind of protection in their infancy stage, which is realised by patent protection. However, protection does not necessarily mean a direct public intervention, but only that innovators should be able to protect their innovation as they want, including the formation of strong company alliances – often the target of antitrust investigations. Besides this liberal approach, evolutionary economists even plead for a direct protection of industries and innovators in their infancy by subsidising R&D, erecting barriers to entry or admitting some anti-competitive practices, including mergers leading to high market concentration. In total, all these trains of thought postulate that pro-competitive regulation damages innovation.

The explanation for these two antagonistic views is based on differences regarding the understanding of competition. Based on neo-classical economics, markets should be competitive in the sense that as many firms as possible should supply a market with their goods. This excludes a supplier structure with only a few or just even one dominant firm – a monopolist. Consequently, regulatory bodies have to prohibit an integration between firms via mergers and acquisitions, which leads to a high level of concentration and just a very small number of suppliers inclined to adopt monopolistic behaviour, like price increases or little innovation activities. Or if markets are dominated by very large firms, regulatory bodies have to break them down into many small firms.

From a dynamic point of view, competition is a selection process that in the long run selects the most efficient techniques, commercial strategies and organisations. During this selection process, firms with dominant, even monopolistic positions can emerge. This is not a problem from the dynamic efficiency criteria, as long as these positions are contestable by new entrants (Baumol et al. 1982). This entry threat leads the dominant firms to be efficient and not to exercise monopoly power. Consequently, dominant positions do not have to be forbidden, but their abuses.

In relation to innovation, pro-competitive policies have to take into account that in their infancy innovation processes are not efficient and 'competitive', but they are necessary to ensure competition in the long run (Teece & Coleman 1998). As a consequence, during the formation phase of a technological or product life cycle, while companies build up new capacities, pro-competition authorities should tolerate most firm practices including cross-subsidisation, market restriction and barriers to entry. In the long run, such practices should be forbidden, since they enable firms to escape from competition, leading to inefficient resource allocation, a slowdown of technical progress, a reduction of innovative activities and a rationing of consumers.

Monopoly and Public Utilities Regulation

Under the regulatory framework in the 1960s and 1970s, monopolies and public utilities had no strong or biased incentives to innovate. Therefore in the 1980s, the United States started to implement regulations in order to motivate them to make productivity gains and realise innovations. However, these new regulatory principles reduced the rents of the regulated firms they formerly captured and often used for large R&D projects and other innovation activities. Therefore, an incentive-financing dilemma emerged for some public utilities.

Network-based services, especially like telecommunications, water and energy supply, were regulated under the old regulation principles, which consisted of either rate of return regulation or pricing at marginal costs. Under the rate-of-return to keep its regulation, the monopoly was forced profitability no higher than the average firm in the industry. Under marginal cost pricing, the monopoly was forced to price its products according to two-part tariffs (Ramsey Pricing).

In relation to innovation, these regulatory schemes were responsible for the low or biased technical progress towards capital-intensive production (Averch & Johnson 1962) and little innovation in some of the regulated industries, like telecommunications and the energy sector. Based on the progress of the economics of information (Stiglitz 1975), appropriate incentive schemes have been developed to overcome the information asymmetries between regulated companies and regulatory bodies. This led to the implementation of new regulatory schemes based on the idea that there is a 'revealation-incentives' dilemma that can be solved by a fine tuning via 'price cap' regulation. Price cap regulations are based on contracts between the regulator and the regulated firm, which require minimum quality and fixed maximum prices. If productivity gains can be appropriated at least partly by the regulated firm, then this scheme causes incentives for innovation, especially in the direction of productivity gains but not necessarily to new products. If the regulatory body wants to capture all productivity gains, the regulated firms have no incentive to innovate. The same is true if the regulatory framework tries to implement a competition which allows multiple suppliers with inefficient cost structures in their production. Consequently, they try to increase their market shares by price competition, which reduces their profits and does not allow them to invest in R&D and innovation.[3]

The Impact of Social Regulation on Innovation

The impact of social regulations on innovations is not as often analysed as the impact of economic regulations. However, it is also less controversial.

Most of the existing literature on social regulations and their impact on innovation is focused on the analysis of the impact of environmental regulation caused by the increasing importance of environmental issues (Kemp 1998). In addition, new environmental regulations have discarded existing machinery and equipment and enabled new entrants to introduce new production techniques in industry. Environmental regulations have caused the emergence of new industries, as in the case of the 'environmental industry' and of new products with fewer or almost no negative impacts on the environment. Consequently, environmental regulations drive the industry explicitly towards technologies protecting the environment or producing fewer environmental damage. For example, Kemp (1998) proposes to use

3 The highly competitive pressure facing deregulated electric utilities have led to concerns about energy shortages in the USA, due to too little investment in new power plants.

regulation as a modulator of technical change, i.e. social regulation may change the direction of technical change into innovations with fewer negative impacts on the environment.

The counter-hypothesis postulates that environmental regulations restrict the firms in their innovative activities and cause additional costs, which have a negative impact on their competitiveness and consequently also on their capability to innovate. It is consensus that the regulation of end-of-pipe technologies has these negative effects, whereas the regulation of integrated environmental protection may be ambivalent for innovation.

In empirical studies, Jaffe et al. (1995) find no support either for the conventional wisdom that environmental regulations have large adverse effects on competitiveness or that they stimulate innovations.

In sectors with strong ethical dimensions and a high importance of externalities as in the matter of health, the activities and strategies of the involved actors are so bound by regulations that the link between regulation and innovation is obvious and close. Safety regulations may on the one hand prohibit innovations, if the public authorities forbid presumably risky products and therefore likely radical innovations. On the other hand, safety regulations increase the acceptance of new products and services among consumers, since they can rely on some minimum product safety. However, the health sector is especially affected by various other means of intervention (Day & Frisvold 1993). Consequently, the perspective has to be broadened from the single regulation to the institutions that surround the regulatory framework.

Concerning public goods, most of the public policies fit into the logic of direct intervention rather than the logic of regulation. Nevertheless, the provision of public goods fosters innovation in general if they represent physical (road system) or intangible (education) infrastructures. However, individual economic agents are deprived in their innovative activities, because they have to transfer funds via taxes to public institutions.

The Impact of Institutional Regulation on Innovation

Besides single economic and social regulations, the institutional framework implemented by administrative regulations is essential for the analysis of regulation and innovation. Two approaches have been developed to link the legal framework to innovation.

The economic analysis of law has also focused on the way the legal environment influences economic efficiency, including innovation. The impact of liability rules on innovation, particularly in the domain of product safety, has been especially analysed. Too-strong liability rules lead innovators not to introduce new products and services in the market – especially radical innovations, because the risks are higher, the expected revenues decrease and the users of the products decrease their self-protection efforts, leading to more accidents. Viscusi and Moore (1993) are able to confirm empirically that very high levels of liability have negative effects on product innovation. However, without product liability, the acceptance of new products among consumers is reduced and that may prohibit their success at the market.

The best-analysed link between administrative or institutional regulation and innovation is the impact of Intellectual Property Rights (IPRs), especially patents and copyrights, on innovation. As Besen and Raskind (1991) pointed out, the fundamental dilemma lies between invention and diffusion. On the one hand, a strong patent protection encourages innovation. On the other hand, a weak one favours a rapid and wide diffusion of inventions, which leads then to innovations and to growth for the whole economy. Appropriate licensing schemes may be a good way to reach the two goals simultaneously. Since the innovation processes differ by industries, optimal IPR rules should also vary from industry to industry from a purely economic point of view, but this is not practicable for the legal system. The latter institutional regulations can provide in general positive incentives for innovative activities, but force suppliers of new products and services to introduce less risky products and services into the markets.

The Impact of Standards on Innovation

The analysis of impacts of standards was started only in the middle of the 1990s. Swann et al. (1996) studied the impact of technical standards on trade flows. Subsequent efforts have sought to assess the impacts of standards on economic growth (see Blind and Jungmittag 2005; 2008). Based on first findings by Blind (2004), it became obvious that the output of research activities requires or generates new standardisation activities, but also that standards promote the diffusion of new technologies and provide the platform for subsequent innovation activities. In network industries, like telecommunication and software, the diffusion of new technologies and services is almost impossible without standards implemented. Existing standards may also present hurdles for new technologies and products, because they compete with existing technologies and products, which are more familiar to the users and in which additional human and physical capital has been invested. In the following sections, we will differentiate them into functional types and their roles in various innovation phases.

A classification of standards can be performed in several ways, not only depending on the respective scientific discipline, but also within the same discipline. David (1987) proposed a categorisation based on the economic effect of a standard. This is a useful approach in order to analyse both the economic driving forces for standardisation and the economic impact dimensions. It has been widely accepted and used (Nicolas & Repussard 1988; Swann 2000), though some researchers have extended the number of categories (Tassey 2000). In our classification we use the dimensions of compatibility, quality, variety-reducing and information standards. However, even if standards are developed just to serve one purpose they often fulfil multiple functions. Therefore, it is not likely that each standard will fall exactly

Table 10.1 General effects of standards

Type of standard	Positive effects	Negative effects
Compatibility/ interface	• Network externalities • Avoiding lock-ins • Increased variety of systems products	• Monopoly
Minimum quality/ safety	• Correction for adverse selection • Reduced transaction costs • Correction for negative externalities	• Regulatory capture 'raising rival's costs'
Variety reduction	• Economies of scale • Building focus and critical mass	• Reduced choice • Market concentration
Information	• Facilitates trade • Reduced transaction costs	• Regulatory capture

Source: Blind et al. (2004) modified after Swann (2000).

and exclusively into a single category, but very often several economic aspects are touched by a single standard. However, a distinction is important for the theoretical discussion, because standards have different economic effects and impacts on innovation.

Table 10.1 summarises these four different purposes of standardisation and highlights their positive and negative effects, which also have implications for innovation. The usefulness of this distinction lies in what it contributes to our understanding of the variety of standard types and their sometimes ambivalent effects. Compatibility and interface standards are crucial for the development of network industries like the telecommunication and computer sectors, since they allow the communication between different users or the interplay between hard- and software. Minimum quality and safety standards prohibit that inferior and even risky goods and services are offered by suppliers, which increase the confidence of consumers in new innovative products and services. The concept of 'regulatory capture' describes the problem that some producers may lobby so skilfully that they persuade the standardisation development organisations to define regulations in the interest of the producers rather than in the interest of the customer (as originally intended). Since standards define very often a certain specification, they automatically exclude other options and reduce the variety, if the standard succeeds, but represent critical masses attracting new investors and companies contributing to the development of new markets. Standards of information and product description are usually treated as a distinct category from the above, but for many purposes it is sufficient to treat these as a hybrid of the above three categories. A standardised product description often has the function of a quality standard, eases compatibility with other products and is an expression of product variety.

Besides taking into account the various functions of standards, we introduce a further dimension into our general framework, which makes explicit that standards play a crucial role in the research and innovation process with its various feedback loops (see Figure 10.1). More specifically, we distinguish the various roles of different

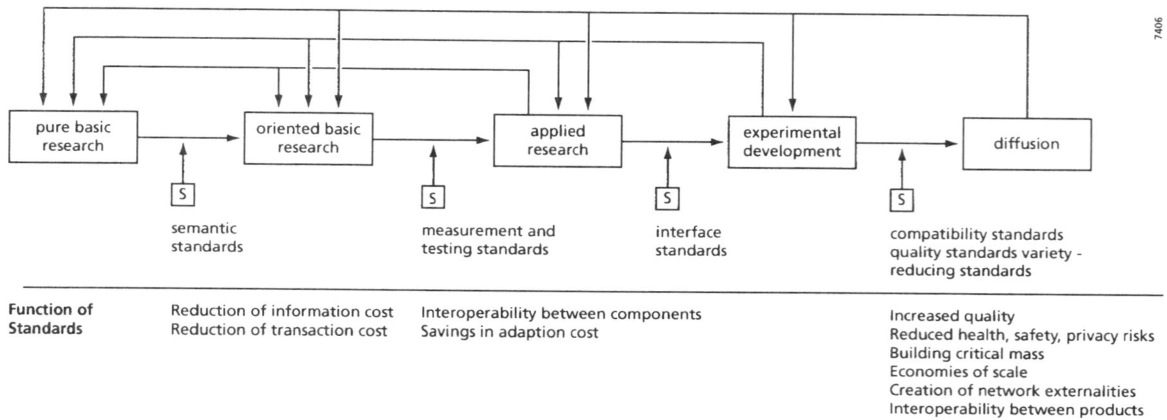

Source: Blind and Gauch (2008).

Figure 10.1 Various roles of different types of standards in the innovation process

types of standards within and between the different phases, especially of the research process while still bearing in mind the whole innovation process. The approach provides a further differentiation of the already presented functions by David (1987), who discuss the general economic functions of standards especially in the market or between actors in the market, or the analysis of David and Greenstein (1990), who focus on the economic impacts of compatibility standards, which is applied to the telecommunications market by David and Steinmueller (1994). This concept was developed by Blind and Gauch (2008) in the context of the analysis of standardisation in the emerging field of nanotechnology.

Terminology standards are already required in basic research investigating new technologies, like nanotechnology, in order to allow or facilitate efficient communication, but they play a crucial role in the transfer of knowledge from basic to oriented-basic and applied researchers. The transfer of knowledge from basic to applied research requires in addition measurement and testing standards, which allow progress towards first product-related developments. The gap between applied research and experimental development of new products and processes is facilitated by interface standards, which allow the interoperability of components integrated into product or process technology. Finally, the transition of pilot products into mass markets requires, besides compatibility standards to ensure the interoperability between products or whole systems, above all quality standards which guarantee that the products comply with some minimum safety regulations. The role of compatibility standards as a mediator between research contexts is additionally moderated by the scientific and technological field of the researchers. Fields that incorporate an underlying logic of networking of products – information and communication technology being the prime example, or other fields that facilitate large technical systems (Bijker et al. 1993) – moderate the relative importance of compatibility and interface standards at early stages in the research process.

Requirements such as quality and safety standards are either set by a regulator (Blind et al. 2004), who refers to a set of specific standards, as in the New Approach, or are demanded by lead users and early adopters who are interested in the new products, but not willing to accept rather large risks by using these new, unknown products. Finally, the diffusion of new products is fostered by variety-reducing standards which allow the exploitation of economies of scale and by compatibility standards generating positive network externalities among users.

In general, the self-regulation of the actors in the innovation system promotes existing trajectories of technologies, but is not an instrument to push radical innovations, i.e. completely new technologies. The involvement of all relevant stakeholder groups leads to a consensus of interests taking all preferences of the involved parties into account. This procedure has implications for the released specifications and consequently for the impacts on innovation. If we assume that besides the technology providers, i.e. the supply side, the users and consumers are adequately represented, but also representatives of environmental groups and other stakeholders in society, then we can expect standards which promote incremental innovations balancing the interests of the involved stakeholders. This means that they also promote user-driven or customer-oriented solutions and take environmental concerns into account as well, which is different from simple technology-driven specifications. However, despite the openness of standardisation processes, weak stakeholders like consumers, but also small and medium-sized enterprises, have problems to participate adequately in these processes, which is likely to lead to specifications biased to the preferences of large companies having different implications on innovation.

Summary

The discussion of the different types of regulation, the difficulty in identifying and defining innovation and the ambivalent impact of regulation on innovation has shown that simple approaches are not adequate to analyse this issue and no clear-cut results of such an analysis can be expected.

Table 10.2 summarises the different types of regulation which have been discussed in the previous sections and their possible impacts on innovation, in the sense of the development of new products and services. If we resume the implications on the direction of innovation activities, economic regulations at first put pressure to realise process innovations in order to succeed in price competition, but also generate incentives to introduce successfully new products and services into the market in order to escape a fierce competitive pressure. Social regulations have strong impacts on the direction of innovation activities towards the protection of the environment and the health and safety of citizens in general, but more specifically of consumers and the labour force. Institutional regulations just provide stronger incentives for innovation activities and in case of product liability towards safer products and services. Self regulation via standardisation promotes the development of technologies and products compatible or complementary with the specifications set in the released standards. The various dimensions of both the regulatory framework and the innovation activities being raised in the theoretical discussion on the relation between regulation and innovation underlines the complexity of the interrelation, which calls for rather differentiated and detailed approaches to analyse this relationship.

THE REGULATORY SYSTEM SHAPING MARKETS FOR INNOVATION

For a more detailed and comprehensive analysis of regulations responsible for shaping new markets in the narrower sense, the following new taxonomy of product market regulations was developed by screening the directory of the relevant European

Table 10.2 Types of regulations and their impacts on innovation

Type of regulation	Positive impact on innovation	Negative impact on innovation
Economic regulation		
Antitrust or pro-competition regulation	eases and enforces innovation	prohibits (R&D) alliances
Protection of infant industries (R&D subsidies, barriers to entry)	allows costly and risky innovations	continued protection does not enforce innovative activities
Public utilities: rate of return regulations; pricing at marginal costs	rents available for R&D and innovation	little and biased incentives to innovate
Public utilities: price cap	incentives to reach productivity gains	-
Public utilities: competition	-	price pressure and low profits do not allow to invest in innovation
Protection of selected industries (e.g. aerospace)	funds available for large R&D projects and innovation	no competitive pressure to innovate
Social regulation		
Environmental regulations	create incentives for new processes creating less environmental damage and for the development of new products	restrict innovative activities and hamper the competitiveness and therefore their innovative capacity regarding end-of-pipe technologies
Labour market regulations	create incentives for new processes creating less risks for the labour force	restrict implementation of innovative processes and hamper the competitiveness and therefore their innovative capacity
Safety regulations	increase acceptance of new products among consumers	additional restrictions for innovators
Institutional regulation		
Product liability	producer liability increases the acceptance of new products among early adopters	too high liability reduces the incentive for producers of innovative goods
Intellectual property rights	additional incentives to innovate	additional protection for monopolies hinders the diffusion of new technologies and products
Self regulation		
Compatibility or interface standards	promote positive network effects in network industries	promote lock-in in existing technologies
Minimum quality or safety standards	attract lead users	restrict spectrum of product quality
Variety reducing standards	foster economies of scale and critical masses in new technologies	promote monopolies and reduce variety
Information standards	reduce transaction costs	restrict product spectrum

Source: modification of Blind et al. (2004).

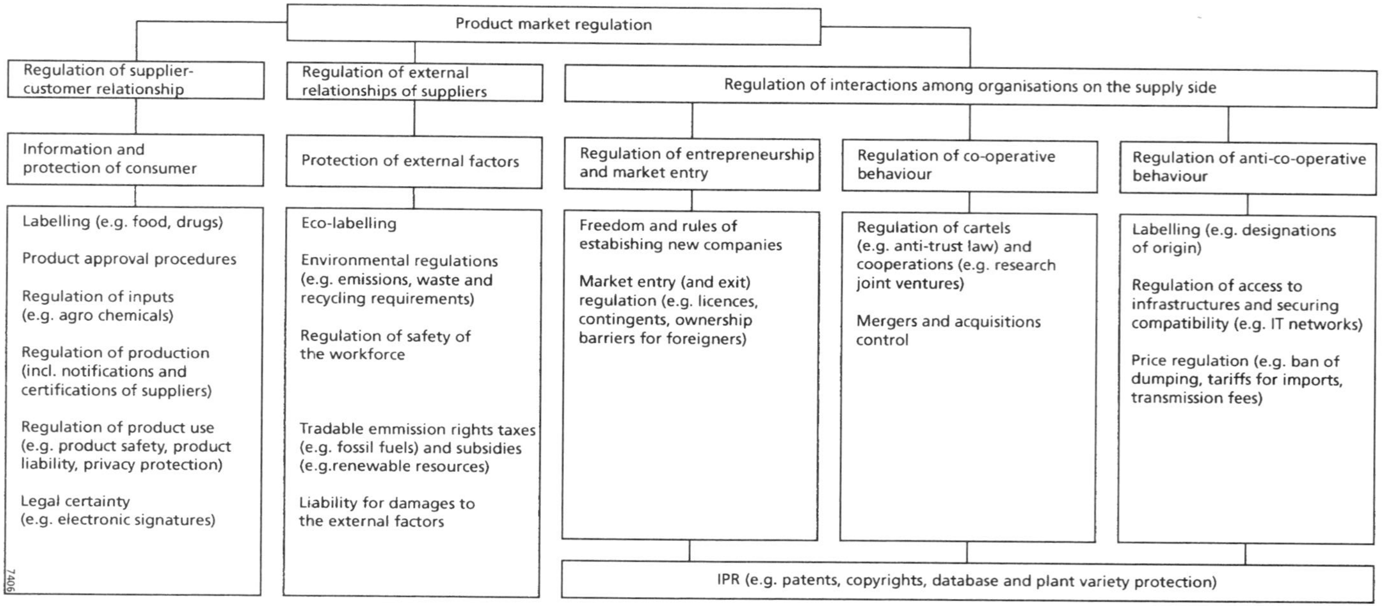

Source: own taxonomy following Blind et al. (2004).

Figure 10.2 Taxonomy of Product Market Regulations

Community legislation and clustering the regulations and directives.[4] The taxonomy is based on the assumption that regulations concern actors or relations between actors, which fits very well into the innovation systems approach being based on various actors and institutions, and defines the following types of relations: supplier-customer relationships; external relationships of suppliers; and interactions among organisations on the supply side.

The main objective of the regulation of the supplier-customer relationship is to ensure freedom of choice of the consumers by adequate information as well as protection of consumers' health and well-being, while the protection of the suppliers does not play a key role, although fraud by consumers is a problem e.g. in the insurance industry. The relevant areas and instruments cover the labelling of products, approval procedures, rules for the input of production, regulation of the production of goods, e.g. meeting quality standards, regulation of the usage, especially in case of accidents through product safety and liability rules, protection of privacy, e.g. of individual user behaviour, and legal certainty, e.g. in contract issues.

The second type of relationship is the one between the supplier of goods and services and external factors, which also includes the work force, besides environmental resources. All these regulations have again the objective to reduce negative externalities, i.e. to protect the production factors, like the environment and the workers, by: eco-labelling; promoting the usage of environmentally friendly products; environmental regulations, e.g. ceilings for pollution; regulation of occupational safety; taxes, e.g. on fossil fuels; and subsidies for the usage of renewable resources, tradable emission rights and liability for damages to the external factors.

Besides the regulation of the relations to customers and the environment, the interactions among organisations of the supply side also need to be regulated in order to prevent negative impacts for the consumers and society as a whole. One can distinguish three types of activities which are relevant in this aspect. First, the establishment of companies and their entry to the market need some rules which secure the freedom and support of establishing new companies, and regulate market entry and exit, e.g. licenses.

The interaction between companies can be characterised as cooperative or anticooperative. In order to prevent negative consequences for customers, e.g. high prices and reduced variety, cooperative behaviour has to be regulated by regulation of cartels (e.g. antitrust law) and cooperations (e.g. research joint ventures) and by mergers and acquisitions control.

However, anti-cooperative behaviour also needs to be regulated in order to secure a sufficient level of competition and to prohibit the distortion of competition by labelling requirements; e.g. origin of destination; regulation of access to infrastructures; e.g. IT networks; and securing compatibility; e.g. interoperability between networks or software, price regulation; e.g. ban on dumping, tariffs for imported goods and transmission fees.

---

4 For comparison see the OECD taxonomy published in Nicoletti et al. (2000).

Across all the three categories, property rights are essential, not only for the creation of new companies, but also for the cooperative and anti-cooperative activities of companies. Assuming that property rights in the general sense are understood and accepted, intellectual property rights (IPR), like patents and copyrights, are especially important not only for the development, but also for the market introduction for new products and services.

This taxonomy (Figure 10.2) is independent of sectoral aspects and covers the different instruments available for regulation of product markets, ranging from information requirements to strict command and control regulations. Not all of these regulations have an impact on the introduction of new products and services. If one considers cartel policies, they focus more on the activities of established companies competing in restricted markets, such as cement, vitamins and amino acids, with a small number of large companies. Others, like labelling policy, are more relevant for the introduction of new products and services, since they can increase the acceptance of new products among consumers. In the following section we present an overview of the most important European regulations relevant for shaping new markets.

REGULATIONS IN THE INNOVATION SYSTEM APPROACH

The regulatory framework is an important component in the innovation system approach. Especially in economies characterised by ongoing innovation and fundamental uncertainty, the institutional setting has a major impact upon how economic agents behave and also upon the conduct and performance of the system as a whole (Carlsson et al. 2002). Among the formal institutional arrangements necessary and supportive for innovation, we observe a strong focus on well defined and implemented property rights of different kinds, including intellectual property rights, contract laws, corporate law, but also arbitration institutions and collective bargaining and other labour market institutions. However, we have seen a much broader variety of institutional or regulatory arrangements, which have a major impact on innovation activities in innovation systems (see Figure 10.3).

Based on the taxonomy of product market regulations, we discuss the major regulations, which have to be considered in shaping a successful innovation system, considering the various functions of innovation systems (Hekkert et al. 2007).[5] The relationship between supply and demand has to be regulated in a way that besides mainstream economic efficiency criteria legal certainty and trust between suppliers and consumers are fostered as well as more efficient producer-user interactions. Here, the promotion of knowledge diffusion by setting information requirements is an important side aspect of this kind of regulation. Consequently, very biased regulations only focusing on the interests of the consumer or solely trying to

5 Walz (2007) analyses specifically the differences in wind energy regulation between Germany and the United States according to the functions of the innovation systems.

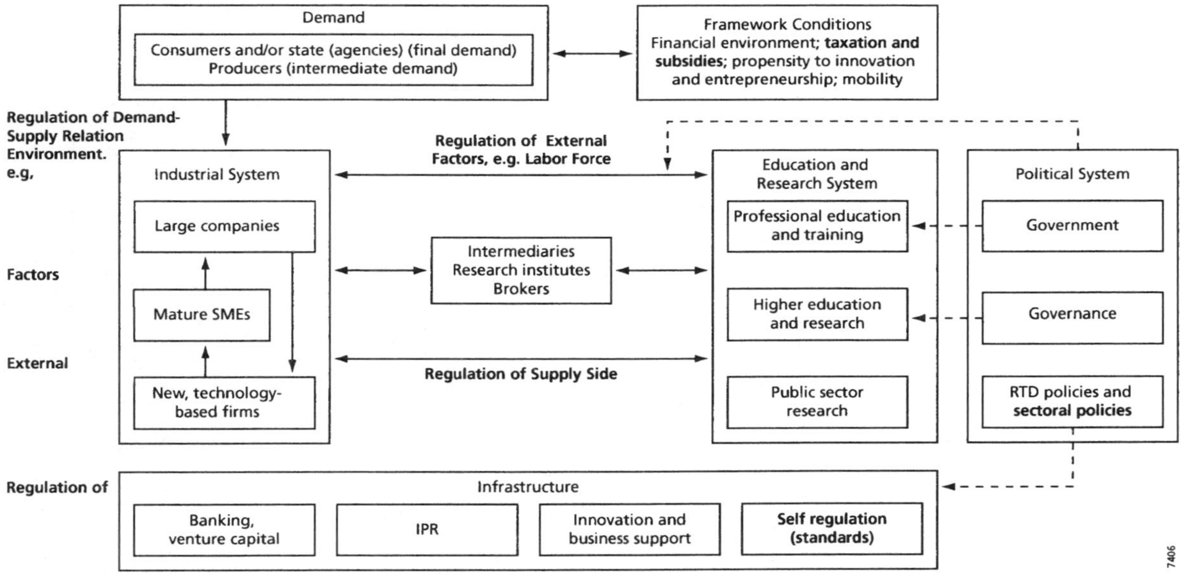

Figure 10.3 Regulation in the innovation system: a more comprehensive view (modification of Kuhlmann and Arnold)

maximise the short-term profits of suppliers will not lead to long-lasting innovations in the relevant market. Therefore, careful arbitraging processes are required to identify and level out the possible conflicting interests in the short run in order to find sustainable solutions for the long run. Besides the supply-demand relation the private and public demand for innovative products can be encouraged either supporting private demand or by focusing public demand (see overview in Chapter 12 in this book by Edler).

The regulation of the external factors covers two aspects. First, the work force has to be protected from risks for health and safety. Here again the economic-oriented interests of companies have to be balanced with the more comprehensive needs of the work force, which not only encompasses wage consideration, but also health and safety aspects. Then trustful relationships can be established, which are a productive basis for innovation activities including sometimes also risks. Second, the environment is the major second external factor of the production process, which needs regulation or better protection. Since the environment has no explicit representative in the negotiations, environmental groups, political parties or even the government represents its interests. The other actors involved are the companies and sometimes also the consumers. In order to develop regulatory frameworks fostering innovation, a balance has to be found between the economic interests of the companies, which already started to change from rather short-term interests to sustainable paths, and the rather long-term generation overarching requirement for the protection of the environment and natural resources. Here, one can argue that the Porter hypotheses (Porter & van der Linde 1995) of promoting the competitiveness of the national companies producing environmental technologies is complementary to the goal to protect the environment. The regulatory policies have to take such complementarities into account in setting the environmental standards, which may overcompensate both short-term burdens of companies and the burdens some actors, like consumers, have to bear by the generation of further employment or reduction of taxes. Furthermore, environmental regulations focusing on labelling requirements do not significantly restrict the supply side, but supports the exchange of information within the innovation system.

The last major category of product market regulations focuses on the interactions of the supply side. The first subcategory deals with entrepreneurship and market entry, also identified as a central success factor of innovation system by Johnson (2001). According to the functions of the innovation system, the framework has to be open and has to provide incentives for entrepreneurial activity and the market entry of new companies as one major category of innovative work. In addition to securing the openness of markets for entry and providing incentives to take entrepreneurial risks, entry and entrepreneurial regulation has also to provide a certain level of security. One option mentioned by Hekkert et al. (2007) is an insolvency regulation, which does not threaten to destroy the whole wealth of an individual accumulated during his/her life time like in the United States in contrast to the former rules applied in Germany.

The second and third subcategories focus on the relationships between companies at the supply side. The interaction between companies can be characterised as cooperative or anti-cooperative. In order to prevent negative consequences for customers, e.g. high prices and reduced variety, cooperative behaviour has to be regulated by regulation of cartels (e.g. antitrust law) and cooperations (e.g. research joint ventures) and by the control of mergers and acquisitions. In the context of the functions of the innovation systems the mainstream rationale of preventing market dominating company collaborations or even cartels has to be balanced against the role of strong incentives to engage in innovative work, the necessity to build critical masses for doing research and development, the opportunity to exchange information and knowledge in early stages of emerging markets and the reduction of uncertainties for companies involved in risky activities. Consequently, the rather narrow focus of traditional antitrust regulation has to be widened in order to allow the development and exploitations of collaborations between companies for the sake of innovation. However, if company collaboration leads to rigidities and lock-ins in rather static markets, regulatory bodies have to break up these static and closed structures in order to allow and promote market entry.

However, anti-cooperative behaviour needs also to be regulated in order to secure a sufficient level of competition and to prohibit the distortion of competition by labelling requirements, e.g. origin of destination, the regulation of access to infrastructures, e.g. IT networks, and securing compatibility, e.g. interoperability between networks or software, and finally even price regulation, e.g. ban on dumping, tariffs for imported goods, transmission fees. Regarding the functions of innovation systems, these regulations are not only securing the incentives to innovate, but are important for the sustainable development and expansion of markets by also providing a secure legal framework for the competitors.

Across all the three categories of supply-side regulation, property rights are essential, not only for the creation of new companies, but also for the cooperative and anti-cooperative activities of companies. Assuming that property rights in the general sense are understood and accepted, intellectual property rights (IPRs), like patents and copyrights, are not only incentives for research and development, but also for instruments for the market introduction and diffusion of new products and services. Furthermore, successful innovation systems depend on an efficient division of labour or differentiation in the value chain. IPRs allow the definition, the exchange and diffusion of technological know-how and realise not only the static efficiency of the diffusion of knowledge as public goods in the mainstream economic sense, but also promote the functioning of innovation systems by providing incentives for innovative activities, facilitating the exchange of knowledge and reducing legal uncertainty and therefore generating also trust between actors in the innovation system

In summary, regulations in the context of the innovation system approach have several additional functions besides the economic function of achieving static or

Table 10.3 Types of regulations and their functions in the innovation system

Type of regulation	Functions in the innovation system
Regulation of supply-demand relation	Exchange of information between supply and demand; promotion of user-producer interaction; provision of secure legal framework for pioneering companies and lead users
Protection of the work force	Provision of secure legal framework for employers and employees
Protection of the environment	Exchange of information; provision of incentives for innovation in environmental technologies; creation of markets for environmental technologies; provision of secure legal framework for the supply side; option to promote radical innovative ideas for regulatory bodies
Promotion of entrepreneurship and market entry	Provision of incentives for innovative entrepreneurs; creation of markets by entrepreneurs and new entrants; provision of secure legal framework for entrepreneurs
Regulation of cooperative behaviour at the supply side	Exchange of information between innovative companies; provision of incentives for collaborative innovation; creation of markets by allowing critical masses; provision of secure legal framework for company collaborations
Regulation of anti-cooperative behaviour at the supply side	Exchange of information between supply and demand side; provision of incentives for innovation; creation of markets by new entrants; provision of secure legal framework for competitors
Intellectual property rights	Exchange of information about innovative technologies; creation of markets by defining new IPR; reduction of legal uncertainty for rights owner

dynamic efficiency. Table 10.3 presents a rough summary of the various functions of the main categories of regulations.

We have seen that the multiplicity of functions of the numerous regulations also generates various tensions between the different types of regulations, e.g. between securing the openness of markets or breaking-up rigid structures and providing stable framework conditions. This conflict points to the different roles of regulations in the various phases of the innovation processes, which will be discussed in the following section.

REGULATIONS IN INNOVATION PHASES

After analysing the role of regulations in the sense of correcting static economic inefficiencies, but also trying to achieve dynamic economic efficiency, we have discussed the role of product market regulations in the context of the innovation system approach. Finally, it is obvious that the types of regulations play different roles within the innovation phases. However, a systematic analysis of the timing of regulation in the context of a whole innovation cycle is missing. Especially, the co-evolution of innovation and the regulatory framework has so far not been addressed. In the following overview we attribute the various types of regulations in a very rough phase model by distinguishing research, development, market introduction, market expansion and finally market maturity.

By starting with research, we have to mention the crucial role of intellectual property rights for the development of a new research and later technology field.

The main output of basic research, mainly scientific publications, is protected by copyright law, which is rather robust against the development of new technologies with the exception of the digitalisation of contents and its wide distribution through the internet. The patent regime experienced more serious changes by adaptations to new technologies (e.g. biotechnology and software inventions in the United States), but also by institutional changes[6] forcing basic research organisations more into patenting activities (Mowery et al. 2001). The main role of IPRs is their incentive, disclosure and coordination function, which are crucial for the success of an innovation system. Besides the basic regulation of IPR, regulations of research joint ventures are crucial for the research activities of companies. In the European Union, the block exemption for R&D cooperation allows companies to perform research together as long as it concerns no specific products or markets and is therefore no pre-stage of a cartel. In general, regulatory bodies also in the United States are fostering this kind of cooperation especially in order to avoid the duplication of efforts.

In addition to the role of IPR in the research phase, patents especially are crucial as incentives for development activities targeted to specific technologies or even products. Patents are in this stage not only an incentive instrument, but also a barter chip for the different actors, which allows further cooperation, but also differentiation in the value chain (Hall & Ziedonis 2001). Further roles of patents (Blind et al. 2006) are its function of signalling competencies in research and development, but also its misuse of blocking competitors in their activities leading to further transaction costs or even patent thickets (Shapiro 2001). The regulation of cooperations in development activities moves closer to market activities, which have to be more carefully proved whether they have negative implications for the concentration of markets. Finally, one has to mention all other regulations influencing possible market shares and profits. They have an indirect incentive effect on research and development activities, if the companies anticipate high market shares or profits to be derived from the results of their research and development activities. One example is the orphan drug regulation, which devotes exclusive market rights for very rare diseases for one single company in order to make large and risky research and development activities attractive (see the case study in Blind et al. 2004).

In addition to these relevant regulations for the research phase, a series of further regulations are relevant for the development of products and processes. First, approval procedures have an impact on the incentives to innovate not only

6 In this overview, we do not consider general institutional settings, which can also have major impacts on innovation, e.g. by forcing public research institutions to raise funds from industry.

be generating the risk of innovative products not being approved by the responsible regulatory agencies, but also by the approval times. This problem is most relevant for the pharmaceutical and also for the food sectors, because of the high R&D costs and the very long-lasting R&D processes (compare the case studies in Blind et al. 2004). In general, the regulations of inputs to be used for the production of goods and of the production processes themselves have to be considered in the respective development processes. This includes specific environmental regulations and the regulation of occupational safety, and also general environmental and further liability rules. Furthermore, taxes on materials and energy and also the prices of tradable emission rights are taken into account in the development and construction of goods, and also machinery relying heavily on these inputs. Finally, the regulation of access to infrastructures, e.g. IT networks, and securing obligatory compatibility (e.g. interoperability between networks or software), have to be considered in the development of complementary products and services.

All the previously listed regulations are also relevant for the market introduction phase.[7] In addition, we have to consider the following regulatory framework conditions besides general legal certainty especially in contract issues. Among the set of IPRs, trademarks are crucial for the introduction of new products and services (Schmoch 2003). They do not protect technological know-how, but the reputation and uniqueness of brands. This information function for consumers is also realised by the obligatory regulation of labelling requirements for products, e.g. origin of products, but also the voluntary rules in the context of eco-labels. More serious for the health and safety of consumers are regulations of usage, especially in case of accidents through product safety and liability rules, which may even prohibit the market introduction due to too high risks for the suppliers (Viscusi & Moore 1993). In the last decades the protection of privacy has gained importance in the regulatory debate due to the generation of huge amounts of information about individual user behaviour due to the dominating use of information and communication systems. Closely related with the introduction of new products and services is often the market entry of new companies or even the foundation of completely new companies. Consequently, regulatory bodies try to keep the barriers for the establishment of new companies and entrepreneurship at a very low level (see, for an overview of the efforts to establish a company in various countries, Worldbank 2003).[8] Furthermore during market expansion and even more important in maturing markets, market entry has to be guaranteed and incumbents have to be prohibited from building-up entry barriers. The support of co-operations between companies in research and development has to be reconsidered in the stages of market introduction and expansion, since the implications of antitrust behaviour may in this phase outweigh possible cost saving effects in the context of innovation activities. This

7 The demand side is addressed by Edler (this volume).

8 The insolvency law is particularly crucial for entrepreneurship, although it only becomes relevant at market exit.

requires the control of mergers and acquisitions and in this early market phase, and also during market expansion and most likely in mature markets. Finally, price regulations are already relevant in the market introduction, since on the one hand price dumping strategies may distort the market entry of competitors, but on the other hand are the requirement to pay high transmission fees in network industries, like telecommunication or electricity networks. During market expansion and especially in mature markets, price becomes the most important parameter for competitors, which increases both the incentives to establish price cartels among the incumbents and to deter market entry of newcomers by price dumping, which has to be monitored and sanctioned based on specific regulations.

This brief overview of the various types of regulations in the innovation cycles has demonstrated the following main observations. First, specific types of regulations are relevant for specific innovation phases. For example price regulation only becomes relevant in market expansion and maturity, whereas it has no direct relevance at all in research and development. Based on this example, one can illustrate the second observation of the changing role of specific regulations during the innovation cycles. The indirect impact of price regulation on research and development activities calls for rather high prices, whereas after the market introduction prices based on monopolistic price-setting strategies hinder the broad diffusion of innovative products or the development of complementary markets in case of access pricing in the context of telecommunication or electricity networks. Similarly, the patent right has at first an incentive function, then after its application its disclosure and therefore diffusion function becomes more prominent, before it serves also as a coordination and knowledge exchange mechanism between actors in knowledge markets. Furthermore, we see a kind of accumulation phenomenon in the sense that the regulations relevant for the research and development phase are still relevant for the market phases, which are in addition confronted with further market-specific regulations, like entry or price regulations. Finally, we have treated the regulations separately and not in combination. From case studies like Walz (2007), focusing on the regulation of wind energy, we know that the effective and efficient combination of regulations affecting the research and development phase with rules applicable in the emerging and expanding market phase is required in order to achieve a longlasting success regarding the creation of new markets. However, the analysis of combinations of regulations has to be done in a specific context depending on the specific character of the relevant technology, on the actors and structures of the respective innovation system and on the appropriate phases in the innovation cycle.

THE CO-EVOLUTION OF REGULATORY ECONOMICS, REGULATION POLICY AND PRACTICE

This chapter has shown that the intentional use of the regulatory framework to promote innovation is a rather recent phenomenon within economic policy and, more specifically, within innovation policy, since the original objective of regulation is to avoid or at least to reduce economic inefficiencies. The economic concept of efficiency was for a long time referring to the so-called static inefficiency caused by the exploitation of market power or insufficiently defined property rights. Innovation does not fit into a static framework and is rather linked to a dynamic framework. Only in the last decades have economists moved their focus from static to dynamic efficiency assessments in their considerations about regulatory intervention. Then, innovation gained some, but still little, importance in theoretical economic analysis about the necessity of regulatory intervention and the comparative efficiency of different regulatory instruments.

From the perspective of innovation policy, it is only in the last few years that regulation has moved into the focus of policy makers. However, several other policy areas, like health, safety and environmental policy and competition policy, have implemented regulations which have pushed innovative activities, although the policy objectives are completely different. Examples are the liberalisation of many network and service industries, like telecommunications, and the establishment of more rigorous environmental standards. The impacts of the latter implemented in California on the innovation activities of the producer of environmental technologies in the United States was the basis for the seminal contribution of Porter and van der Linde (1995), who postulated the innovation-promoting effect of environmental regulations. Whereas the hypothesis is rather trivial from a theoretical perspective, several empirical analyses tried to test the hypothesis for various environmental regulations, but found ambivalent results in total. This ambivalence has consequences for the acceptance and use of regulatory instruments within the set of innovation policy instruments, because they are still not broadly taken into account and implemented. Significantly, the recent focus on demand-oriented innovation policies (see Chapter 12 by Edler in this book) produces a further push for regulation as an instrument for innovation policy.

FUTURE CHALLENGES FOR REGULATORY POLICY SHAPING INNOVATION

The different roles of regulation for shaping innovation approaches ranging from a rather mainstream economic approach over the innovation systems approach to the differentiated roles of regulations in the innovation phases have shown various challenges, problems and shortcomings.

The variety of regulations has different and often ambivalent impacts on innovation. There is a significant lack of awareness regarding the issue within the regulatory bodies (Blind et al. 2004). The various interest groups and stakeholders involved in the regulatory process are also often not aware of the opportunities innovations can have for their own interests. Furthermore, the implementation of regulations is crucial for the incentives of companies to innovate.

Based on these general insights, I suggest four sets of proposals for more innovation-friendly regulatory policies:

- Strengthen the focus on innovation in regulatory policy. Regulatory bodies responsible, e.g., for the protection of competition, health and safety or the environment, should give more consideration to the opportunities of innovation in general for achieving their traditional goals. Also the major stakeholders in regulation, except industry, have to check systematically the positive influence of innovation on their organisations' objectives.
- Approaches to increase the quality of the regulatory framework regarding innovation. Regulatory bodies have to react more proactively to trends in science and technology relevant for their regulatory framework by intensifying the contact to the science and technology community, implementing 'regulatory foresight' exercises, and observing ongoing standardisation processes. The coordination and the division of work between standardisation and regulation activities have to become more efficient. The extent of self-regulation has to take into account the trade-off between the gains in flexibility and the loss in legitimacy and acceptance among stakeholders. Regulatory bodies have to focus on those types of regulations or shape regulations in a way which maximises the positive and minimises the negative impacts for innovation. The impact assessment approach of the European Commission already takes innovation into account but needs to be specified and to be accompanied by methodologically advanced impact assessment tools. The performance criteria of regulatory bodies have also to integrate indicators measuring the promotion of new products and services in balance with their other objectives.
- Coordination of policies of regulatory bodies to foster innovation. Since innovation is a complex process and the different types of regulation interact, the promotion of innovation by regulatory policies requires a comprehensive approach, co-ordinating or even integrating the regulatory policies of all the regulatory bodies, e.g. it is not sufficient to set a favourable framework for research, it is also necessary to stabilise the demand for innovative products and services.
- Improved implementation of regulations to foster innovation. The implementation of regulations has to be harmonised in order to reduce the risk and the costs of companies introducing innovations. Approval times have to be reduced, since they are very negative for the expected return of investment in longlasting and expensive R&D resulting in innovative products and services. The transition of regulatory bodies into service providers for the general public, but also for companies, represents a promising strategy also to promote their general support for the introduction of new products and services.

ACKNOWLEDGEMENTS

I would like to thank Philip Shapira for helpful comments on this chapter. The basic research was conducted within the project 'New Products and Services: Regulations shaping New Markets' funded by DG Enterprise of the European Commission. All omissions and mistakes are my own.

REFERENCES

Averch, H. and Johnson, L. (1962), 'Behavior of the firm under regulatory constraint', *American Economic Review*, 52 (5), 1052-1069.

Bassanini, A. and Ernst, E. (2002), 'Labour market institutions, product market regulation, and innovation: cross country evidence, ECO/WKP(2002)2', Paris: OECD Publishing.

Baumol, W., Panzar, J. and Willig, R. (1982), *Contestable Markets and the Theory of Industry Structure*, New York, USA and San Diego, USA: Harcourt Brace Jovanovich.

Besen, S. and Raskind, L. (1991), 'An introduction to the law and economics of intellectual property', *Journal of Economic Perspectives*, 5 (1), 3-27.

Bijker, W., Hughes, T. and Pinch, T. (1993), *The Social Construction of Technological Systems: New Directions in the Sociology and History of Technology*, Cambridge, MA, USA and London, UK: MIT Press.

Blind, K. (ed.) (2007), 'Innovation and regulation: editorial', special issue of *The International Journal of Public Policy*, 2 (1-2), 1-4.

Blind, K. (2004), *The Economics of Standards: Theory, Evidence, Policy*, Cheltenham, UK and Northampton, MA, USA: Edward Elgar.

Blind, K. and Gauch, S. (2008), 'Research and standardisation in nanotechnology: evidence from Germany', *Journal of Technology Transfer*, forthcoming.

Blind, K. and Jungmittag, A. (2008), 'The impact of patents and standards on macroeconomic growth, a panel approach covering four countries and 12 sectors: the empirical economics of standards', *Journal of Productivity Analysis*, forthcoming.

Blind, K. and Jungmittag, A. (2005), 'Trade and the impact of innovations and standards: the case of Germany and the UK', *Applied Economics*, 37 (12), 1385-1398.

Blind, K., Bührlen, B., Kotz, C., Menrad, K. and Walz, R. (2004), 'New products and services: analysis of regulations shaping new markets', European Commission DG Enterprise, Luxembourg.

Blind, K., Edler, J., Frietsch, R. and Schmoch, U. (2006), 'Motives to patent: empirical evidence from Germany', *Research Policy*, 35 (5), 655-672.

Brousseau, E. (1998), 'The link between regulation and innovation: some preliminary remarks', in F. Leone and J. Hemmelskamp (eds), 'The Impact of EU-Regulation on Innovation of European Industry', report EUR 18111 EN, The European Commission Joint Research Centre, Institute for Prospective Technological Studies, Seville, pp. 93-120.

Carlsson, B., Jacobsson, S., Holmen, M. and Rickne, A. (2002), 'Innovation systems: analytical and methodological issues', *Research Policy*, **31** (2), 233-245.

David, P. (1987), 'Some new standards for the economics of standardisation in the information age', in P. Dasgupta and P. Stoneman (eds), *Economic Policy and Technological Performance*, Cambridge, UK: Cambridge University Press, pp. 206-234.

David, P. and Steinmueller, W. (1994), 'Economics of compatibility standards and competition in telecommunications networks', *Information Economics and Policy*, **6** (3-4), 217-241.

David, P. and Greenstein, S. (1990), 'The economics of compatibility standards: an introduction to recent research', *Economic Innovation and New Technology*, **1** (1), 3-41.

Day, K. and Frisvold, G. (1993), 'Medical research and genetic resources management: the case of Taxol', *Contemporary Policy Issues*, **11**, pp. 1-11.

Hall, B. and Ziedonis, R. (2001), 'The patent paradox revisited: an empirical study of patenting in the U.S. semiconductor industry, 1979-1995', *The Rand Journal of Economics*, **32** (1), 101-128.

Hekkert, M., Suurs, R., Negro, S. and Smits, R. (2007), 'Functions of innovation systems: a new approach for analysing technological change', *Technological Forecasting and Social Change*, **74** (4), 413-432.

Jaffe, A. and Palmer, K. (1997), 'Environmental regulation and innovation: a panel data study', *Review of Economics and Statistics*, **79** (4), 610-619.

Jaffe, A., Peterson, S., Portney, P. and Stavins, R. (1995), 'Environmental regulation and the competitiveness of U.S. manufacturing: what does the evidence tell us?', *Journal of Economic Literature*, **33** (1), 132-165.

Johnson, A. (2001), 'Functions in innovation system approaches', paper for DRUID's Nelson-Winter Conference, Aalborg, Denmark.

Kemp, R. (1998), 'Environmental regulation and innovation: key issues and questions for research', in F. Leone and J. Hemmelskamp (eds), 'The Impact of EU-Regulation on Innovation of European Industry', report EUR 18111 EN, The European Commission Joint Research Centre, Institute for Prospective Technological Studies, Seville, pp. 12-39.

Mowery, D., Nelson, R., Sampat, B. and Ziedonis, A. (2001), 'The growth of patenting and licensing by U.S. universities: an assessment of the effects of the Bayh-Dole act of 1980', *Research Policy*, **30** (1), 99-119.

Nicolas, F. and Repussard, J. (1988), 'Common standards for enterprises', Office for Official Publications of the EC, Luxembourg.

Nicoletti, G., Scarpetta, S. and Boylaud, O. (2000), 'Summary of indicators of product market regulations with an extension to employment protection legislation', Economics Department, working papers No. 226, Paris: OECD Publishing.

OECD and Eurostat (2005), *Oslo Manual: Guidelines for Collecting and Interpreting Innovation Data*, Paris: OECD Publishing.

OECD (1999), *Regulatory Reform in the United States*, Paris: OECD Publishing.

OECD (1997), *The OECD Report on Regulatory Reform*, Vol. I, Sectoral Studies, Paris: OECD Publishing.

Porter, M. and van der Linde, C. (1995), 'Toward a new conception of the environment-competitiveness relationship', *Journal of Economic Perspectives*, **9** (4), 97-118.

Schmoch, U. (2003), 'Service marks as novel innovation indicator', *Research Evaluation*, **12** (3), 149-156.

Shapiro, C. (2001), 'Navigating the patent thicket: cross licenses, patent pools, and standard setting', in A. Jaffe, S. Stern and J. Lerner (eds), *Innovation Policy and the Economy*, Cambridge, MA, USA and London, UK: MIT Press, pp. 119-150.

Stiglitz, J. (1975), 'Incentives, risk, and information: notes towards a theory of hierarchy', *Bell Journal of Economics*, **6** (2), 552-579.

Swann, G., Temple, P. and Shurmer, M. (1996), 'Standards and trade performance: the UK experience', *Economic Journal*, **106** (438), 1297-1313.

Swann, P. (2005), 'Do standards enable or constrain innovation?', *The Empirical Economics of Standards*, London: Department of Trade and Industry, 76-120.

Swann, P. (2000), *The Economics of Standardization: Final Report for Standards and Technical Regulations, Directorate Department of Trade and Industry*, Manchester: University of Manchester.

Tassey, G. (2000), 'Standardization in technology-based markets', *Research Policy*, **29** (4-5), 587-602.

Teece, D. and Coleman, M. (1998), 'The meaning of monopoly: antitrust analysis in high-technology industries', *The Antitrust Bulletin*, **43** (3-4), 801-857.

Viscusi, K. and Moore, M. (1993), 'Product liability, research and development, and innovation', *Journal of Political Economy*, **101** (1), 161-184.

Walz, R. (2007), 'The role of regulation for sustainable infrastructure innovations: the case of wind energy', *International Journal of Public Policy*, **2** (1-2), 57-88.

Worldbank (2003), 'Starting a business: procedures to start a business'.

11. Innovation, Defence and Security

Jordi Molas-Gallart

INTRODUCTION

There has long been a relationship between civil and military innovation. Many technologies used in warfare have non-military applications, and often skills and capabilities developed elsewhere have been adapted for military use. Over the years, however, this relationship has changed and has also been perceived in different ways. The observer of military innovation during the second half of the twentieth century would have identified a clearly bounded military innovation system, where technology variation and selection dynamics were defined by procurement and research policies, which had in turn developed out of a perception that technology was a key factor in providing military security and superiority. Yet, the separation between military and civilian innovation systems has not always been so clear-cut. This chapter explores this evolving relationship, focusing on the experience of the US and some Western European countries since the Second World War and the systemic shift that is occurring as we enter the twenty-first century. I will argue that the sharp distinction between military and civilian innovation systems that characterised most of the twentieth century is starting to fall apart. Mutually interdependent changes in technology, the political and security context, and innovation and procurement policies are leading to a new, more complex and integrated, innovation system.

The integration of military and civilian innovation within a single innovation system is not new. Historically, inventors, artisans and scientists have worked to improve military technologies while, at the same time, being engaged in other pursuits. Innovations in weaponry were often developed by technologists and philosophers who were not themselves military men (O'Connell 1989). This 'duality' was not perceived as problematic. Further, modern society has been deeply influenced by technologies and forms of organization developed initially for military applications. With the industrial revolution social technologies that enhanced and supported hierarchical organizations and central control were transferred from military applications to civilian contexts. For instance, military officers oversaw the construction of the US railroad network in the mid-nineteenth century (O'Connell Jr. 1985), and forms of organization developed by members of the US Army Ordnance Department after 1815 diffused to other areas of the economy and provided the basis of what was to be known as the 'American System of Manufactures' (Smith 1987a).

By the nineteenth century, military technology was drawing on established scientific capabilities, while military requirements triggered innovations that were to play an important role in industrial development. Arguably, what became different in modern industrial societies was the intensity of the relationship between military activity and technological innovation, the effort invested in supporting scientific and technological activities for military purposes, and the impact of such efforts on the nature of military conflict. The slaughter of the First World War was the first major scale warning of the destructive potential of science and industrial technologies when applied to military purposes. From then on, the activities of increasingly specialized arms manufacturers and the concern about the destructive power of new weapons led to a stream of critical analysis,[1] and the search of policies that could control the emergence and diffusion of new, increasingly devastating, armaments.

Yet, the systematic application of scientific and technical knowledge to weapons development had just started. The Second World War led to the biggest scientific effort ever seen by humanity to develop means of destruction. The results were astounding and have their clearest exponent in the Manhattan Project and the development of the nuclear bomb.[2] This experience prompted a new approach to defence research policy and military procurement that came to characterize the industrial and technological development of the post-War period.

THE INNOVATION SYSTEM IN THE DEFENCE INDUSTRIES AFTER THE SECOND WORLD WAR

Following the end of the Second World War, in the often quoted *Science – The Endless Frontier* (Bush 1945), Vannevar Bush argued not only for more but also 'more adequate' military research. At first sight it may seem surprising that after the monumental efforts invested in military research during the war, and the overall decline in US military expenditure that followed its end, the most important document of the time defining a rationale for public investment in research should argue that more, not less, defence research was needed. The lesson derived from the War had been clear: large scientific and technological efforts and their results (including the atomic bomb, radar and decryption devices) had been central to Allied success, but the margin of victory had been tiny. As Bush (1945) wrote:

> The bitter and dangerous battle against the U-boat was a battle of scientific techniques – and our margin of success was dangerously small. The new eyes which radar has supplied can sometimes be blinded by new scientific developments.

1 See for instance the work of Noel-Baker (1936).

2 A classic account of the Manhattan Project can be found in Rhodes (1986).

Bush proposed that, in addition to the research work carried out by the armed forces and focusing on development tasks, a civilian-controlled organization funded directly by Congress would be charged with initiating military research programmes. The quest to manage and conduct military research out of the control of the military harks back to the experience of the Manhattan project. And yet, for its attempt to free scientific research from the strictures of military command practices, the military innovation system that developed in the subsequent years was unmistakably 'top-down' and 'closed' (Edwards 1997). Paul Edwards has argued that this 'closed' system is characterized by a clearly defined network of specialized labs (including but not limited to university laboratories), in close relationship with the Armed Forces, and specialized industries, developing, producing and deploying centralized command and control systems as favoured in military applications, and accompanied by a 'discourse' (language and worldview) emphasizing the virtues of centralization and control. Closure does not naturally mean a complete isolation from its surrounding environment: although system agents established a dense network of relationships among themselves and more tenuous links with their surrounding environment, they were, as we will see later, subjected to external influences. I will first describe the system that developed during the Cold War period, stressing the mutual interdependence between the policy rationale developed by academics and analysts, the 'selection environments' implemented through procurement and standardization policies, the military technologies that they developed, and the research and production infrastructure that generated such technologies. I will then describe how this system is changing in the twenty-first century.

The Policy Rationale

The development of the military innovation system that characterized industrialized Western economies during the Cold War was supported by a discourse developed mainly by leading US academics and scientists. Working from different academic backgrounds and environments, they developed a policy rationale that underpinned the expansion of a new military innovation system after World War II. Some of the physicists and scientists active in the military R&D effort contributed to the policy debate, providing a set of policy justifications in the form of military strategies, accompanied by active lobbying. Especially well-known is the activity of Edward Teller, one of the scientists in the Manhattan Project, later co-founder and director of the Lawrence Livermore National Laboratory (one of the Department of Energy 'nuclear weapons labs') and better known as the 'father of the hydrogen bomb'. From the 1930s Teller worked both in government research establishments (Los Alamos, Lawrence Livermore) and universities (University of Chicago, George Washington University, University of California at Berkeley). Teller combined his scientific and research management work, with a very intense advocacy activity (in favour of the development and deployment of nuclear weapons, opposing the nuclear test ban, convincing President Reagan to launch the 'Strategic Defense Initiative', and even planning the use of hydrogen and nuclear bombs for large civil engineering projects). That his autobiography was titled *A Twentieth-Century Journey in Science and Politics* accurately reflects this dual activity. Of course, Teller was not the only scientist involved in policy advocacy; the clash with some of Manhattan Project colleagues, and in particular Robert Oppenheimer, was a personal as well as a political one. Many of the scientists that had been involved in the Manhattan Project did later oppose what they perceived as Teller's militaristic tendencies and the development and deployment of an increasing number of ever more destructive nuclear weapons.

Economists, operational analysts and civilian strategists also contributed to the development of policy theories related to nuclear and military strategy, as well as military procurement policies (including research and innovation). In the US, the RAND Corporation played a crucial role in providing a bridge between applied social research and policy development and implementation. RAND was set up in 1946 by the US Air Force, and soon became an independent research organization with an expanding portfolio of clients and initiatives. During the 1950s and 1960s the RAND Corporation was the producer of many influential studies carried out for the US Armed Forces and the Department of Defence. Most of the best-known defence analysts of the period worked at RAND. Some like Bernard Brodie started their careers in universities and then moved to RAND, others like Herman Kahn carried out most of their work there, while still others like Thomas Schelling spent visiting periods or contributed to RAND projects while working mainly from universities. We can distinguish three main fields in which RAND's work was influential:

1. The development of theories of nuclear deterrence and of foundations of military strategy, including the application of games theory to strategic problems. Bernard Brodie set the basis for nuclear deterrence theory in 1949, before joining the RAND Corporation in 1951 where he continued work on this and was mentor to a cohort of analysts including Herman Kahn. Kahn famously applied games theory and systems analysis to nuclear strategy, reaching the notorious conclusion that a thermonuclear war could be won. Thomas Schelling, a Nobel laureate, developed most of his career as a Professor of Economics at Harvard, but worked for a year at RAND before producing one of his best-known books, *The Strategy of Conflict* (1960).
2. The development of systems analysis and their application to operational planning and procurement management.
3. The application of insights from economics to the development and management of defence policies including, but not limited to, defence procurement. The RAND Corporation generated some of the seminal research that gave rise to the field of 'defence economics', particularly Hitch and McKean's book *The Economics of Defense in the Nuclear Age* (1960).

Obviously academics also made contributions to the policy process outside RAND. In the field of defence procurement two Harvard academics, Peck and Scherer, carried out an extensive study of procurement practices, which identified many of the problems defence procurement still faces today: *The Weapons Acquisition Process* (Peck & Scherer 1962). The study identified the impact of the difficulties stemming from the technological uncertainty associated into large projects that require state-of-the-art technologies based on recent research. The complexity of large projects combining thousands of components and technologies still under development, was already leading to delays in production and development and substantial cost overruns (often doubling or trebling initial cost estimates). In response to these problems project management tools were developed in collaboration with defence customers, firms and research groups.[3]

The abundant contribution of scientists and theoreticians to policy development continued over the following decades, although not with the intensity that had characterized the early 1960s. For instance, Jacques S. Gansler moved from industry, to consultancy, to the Pentagon, and concluded his career in academia, publishing books and journal articles which had a considerable impact on policy debates during the late 1980s and 1990s. Indeed, throughout this period many scientists moved between the defence industry, defence-related think tanks, academia and government. The degree to which their contributions had an influence on US policy and practice, or rather provided a justification or a convenient 'discourse' for political decisions, is a matter for a debate whose complexity cannot be properly addressed here. Yet, their authors belonged to the military research and production system, they had access to confidential data, their reports were also often confidential, and their own personal mobility often provided a channel for the implementation of their recommendations.

POLICY IMPLEMENTATION: DEFENCE PROCUREMENT AND THE DEFENCE RESEARCH SYSTEM

The large funds that continued to be invested in defence research and procurement after World War II were channelled through defence ministries, the armed forces, and defence agencies. The way in which the policy rationale interacted with other factors and agents to constitute a closed innovation system is evident in the US. The independent civilian organization carrying out defence research as envisaged by Vannevar Bush did not quite materialize, although the inheritors of the Manhattan

3 For instance, PERT was developed in the late 1950s to be applied to the Polaris nuclear submarine project and was the result of a collaboration between the government agency in charge of managing the project (the Special Projects Office), a consultancy (Booz Allen Hamilton) and Lockheed Corporation. PERT was rapidly adopted in defence production and other sectors, and spawned a copious literature covering adaptations, improvements and guidance. It has been estimated that by 1964 more than 1000 books on PERT had been written (Morris 1994).

Project, the Department of Energy 'nuclear weapons labs', maintained a certain degree of independence being managed by firms or universities and reporting to the DoE, and not to the Department of Defense. But there are reasons why, even in the case of the independently managed weapons' labs, the research system that evolved was far from open and flexible. The nature of the specialized technologies under development made the free circulation of results and personnel typical of other scientific fields, inadequate here. Access to the facilities was carefully controlled, as they were the results the research laboratories reached. Even the research funded by defence ministries at universities was carried out under conditions far away from the celebrated world of peer review and publicly supported scientific commons that allegedly characterizes academic research.[4]

The interaction between the policy rationale that postulated that scientific research and technological development were crucial for national security and the economic and socio-political realities led to a substantial concentration of research resources in defence or defence-related defence agencies. In the US, for instance, throughout the decades after the Second World War, defence has consistently accounted for almost two-thirds of all public expenditure in R&D. In France and the UK the figure revolved around 50% until recently. Under these conditions the organizational culture of military organizations was to influence the organizational conditions under which military innovations took place. Many defence research laboratories evolved into large bureaucratic and centralized establishments.[5]

4 For the extent to which fundamental research carried out at universities during the second half of the twenteenth Century is still equated to the generation of scientific commons see, for instance, Dosi, Llerena et al. (2005). On a view that sees the military support for academic research in the US as degrading American science and limiting intellectual freedom see Leslie (1993).

5 There is an important exception to this trait: the US Advanced Research Project Agency (ARPA, later Defense Advanced Research Projects Agency – DARPA). DARPA was set up in 1958 as a response to the 'technological surprise' created by the Soviet successful launch of Sputnik. DARPA is a Defence Agency not tied to any operational role, whose objective is to supply 'technological options' for the Department of Defense (Defense Advanced Research Projects Agency 2005). DARPA's managers are employed for a maximum period of six years and the Agency has special contracting authority that frees it from having to apply Federal Regulations. The result is a flexible funding agency, constantly changing its personnel and operating with a high degree of freedom. Although it is naturally closely connected to the DoD, its flexibility allows it to fund projects outside the traditional military R&D networks. The result can, at times, be that open forms of innovation activity are being funded by the DoD, the most famous example being the development of ARPANET, the precursor of the Internet. However, DARPA remains an oddity in the world of defence research. Despite much talk about the creation of an European DARPA, the model has not even come close to replication in Europe. Further, in the US it accounts for a very moderate share of the Department of Defense R&D budget (around 4.5% in 2007).

Specialized defence companies and their divisions were also characterized by large bureaucratic and management structures, set up to deal with the exceptional administrative requirements imposed by the military customers (see below). In addition to the administrative practices, the sensitive nature of the technologies being developed fostered the segregation of military research and production facilities.

Within this system, state customer organizations played a key role. Not only did they control funding, but they established technical and procedural rules through standards and specifications, and often they were directly involved in the development, design and even manufacturing of defence products. Customers in the defence procurement departments and agencies maintained a dominant technological role, setting up what in evolutionary language would be called the 'selection environment', restricting diversity through standards and specifications, and promoting variation through the constant funding of new systems.[6] In the countries at the forefront of military technology, customers defined the performance characteristics of the systems to be developed, fully-funded their development, and through a complex system of specifications and standards controlled even minor technical details of the weapons they required. Government specifications, standards and regulations determined in detail the technical characteristics and, in addition, the procurement, production and maintenance processes of all weapons systems. They provided system stability.

The procurement processes that resulted were often unique and complicated. Military producers soon specialized in supplying this peculiar consumer and built and expanded whole administration systems to deal with the processes imposed by their customers in defence ministries and agencies. Additionally, governments controlled large research infrastructures in the form of defence research establishments, and consequently had substantial 'in-house' technological capabilities with which to support their extensive responsibilities in arms development and production. Further, the military was responsible not only for the operation but also for the maintenance of all large military systems. Weapons maintenance and upgrades were conducted by the Armed Forces themselves or by government-controlled agencies. Finally, in many countries government control extended to the ownership of the main military producers. When governments did not directly own defence-related firms, they had in place regulatory regimes to check their ownership

6 The tension between stability and change is one of the constants in the co-evolutionary literature. Authors have distinguished between variety and diversity generation. While variety refers to changes within a category ('a selection of possibilities that share a common attribute space'), diversity 'concerns a selection that spans different attribute spaces' (Allen 2001). From this perspective, the intervention of defence ministries and agencies prevented diversity through the establishment of strict and complex defence standardization systems (Molas-Gallart 2001a) but within these parameters systems evolved giving way to increasingly complex and technologically sophisticated equipment.

imposing, for instance, limitations on foreign control and ownership of defence suppliers.

In a way, the system of governance that ensued was relatively straightforward. Government agencies and departments ultimately made the choices that would determine the characteristics of the weapons to be deployed, while setting at the same time the regulatory framework within which the industry that made them operated. This form of 'customer-led innovation' was based on substantial in-house capacities providing the procurement agencies and military forces with an 'intelligent customer' capacity. Technologically, the typical outcome of this process was the generation of bespoke, specialized systems incorporating tailored sub-systems and components.

THE ROLE OF INDUSTRY

It would be wrong, however, to see the private industry as a passive receiver of the regulatory environment that its customers imposed. Shored up by substantial procurement programmes, and by R&D contracts that reduced the need to make their own investments in weapons development, the defence industries in countries like the US, the UK and France grew to become large industrial conglomerates. Producing relatively short runs of increasingly complex defence systems, with very high and growing fixed costs, the stimuli for industrial concentration were present from the very beginning. In most defence-related sectors industrial concentration through mergers and acquisitions progressively reduced the number of players in the market. In the US, industries like defence aerospace, which was populated by dozens of 'systems assemblers' in the 1940s, saw the number of prime contractors reduced to just a couple by the end of the century. As firms became larger their power vis-à-vis government agencies also grew.

Defence agencies and ministries developed their policies with the collaboration (formal and informal) of the large industrial conglomerates controlling the sector. This was only natural, for a policy to be successful those responsible for their implementation had to be brought close to the policy definition process. Yet, such close cooperation between defence firms and procurement agencies was soon portrayed as a case of dangerous collusion, with private industry enmeshed with their government customers to the detriment of the public interest. This is the concern that lies behind Eisenhower's warning about the growing power of the 'military-industrial complex', and which is reflected in the literature on 'iron triangles' in the defence sectors. Adams (1982), for instance, described defence as a 'special kind of business', in which industry had developed an 'intimacy' with its customers in Federal Government, inhibiting adequate cost control and therefore proper democratic accountability. The iron triangle was composed of industrialists, defence officials (including the military) and politicians in Congress and the Executive, with revolving doors allowing personnel 'transfers' among the three sides of the 'triangle', thus reinforcing its strength, and its segregation from external influence.

The concept of 'iron triangles' is not unique to the analysis of the defence sector, and it can be considered as part of a wider literature on policy networks (Jordan 1990). The type of policy environment that developed in the area of defence industrial policy can be described as a 'policy community' (Rhodes & Marsh 1992). Policy communities involve a small number of large groups, they are stable (it is difficult for new groups to enter a policy community), share a common understanding of policy problems, and operate within a specific policy domain (in our case defence). Policy communities limit policy participation; while group leaders are incorporated into the policy process and involved in policy development, policy formation remains close to others. The policy process becomes hierarchical.

The close linkage between industrialists, military users and other groups (such as defence ministry officials, parliamentarians, or defence scientists) involved in policy formation leads to groups that evolve in relative isolation and are sheltered from external influences, shaping policy definition and the nature of future technological development according to their own interests and worldview. From a practical point of view the communities may become so close as to blur the boundaries between private and public groups. More than three decades ago Galbraith had argued that in areas like defence procurement the distinction between the public and private spheres was at best blurred. Referring to the defence industries, he wrote:

> In the nature of the market, one organization or enterprise sells to another, and the boundaries between the two are sharp. This same delineation characterizes the private firm selling, say, powdered milk to the Department of Agriculture. But when planning replaces the market and identification and adaptation supplement pecuniary compensation, matters are very different. No sharp line separates government from the private firm; the line becomes very indistinct and even imaginary ... Each organization ... is an extension of the other. (Galbraith 1985)

The extent to which 'pecuniary compensation' became secondary in the defence sector and other 'planned industries' is obviously subject to debate. Yet, most academic analysis of the US and global defence industries published between the 1960s and 1980s agree in describing an industry that is so near to its customers that its behaviour no longer responds to the one to be expected from firms operating in free markets. Melman (1974) saw military firms as seeking to maximise costs and subsidies from State management. The operating procedures of defence firms were infecting the whole US economy, which according to Melman was turning into a 'permanent war economy'. There is a lot in Melman that resonates from previous studies carried out from less critical managerial perspectives. The fact that costs were not a major consideration for firms working for the US Department of Defense had already been pointed out in the seminal study by Peck and Scherer (1962) on the weapons acquisition process. They observed how prices in the defence industries were determined by cost + fee schemes and how, given the military's emphasis on performance, military firms would introduce products well before technological and market uncertainties had been solved (unlike in civilian markets). This is why military producers would lead commercial practice in the utilization of scientific resources and the introduction of new technologies but, underpinned by cost + fee contracts, would often produce very expensive products plagued by cost overruns. This is what, twelve years later, Melman (1974) interpreted as 'cost-maximizing' behaviour.

In part, the high costs stemmed from the administrative overheads needed to deal with the complex administrative processes imposed by the peculiar, extensive and exhaustive regulatory environment to which I have already referred. Melman estimated that administrative overheads in military products were two or three times those of civilian products. But more importantly, unbridled from costs concerns, the defence industries have tended to generate increasingly complex and more expensive products, 'gold plating' with refinements each new generation of arms systems. It is this practice that, when extended to the rest of the economy, proves inimical to economic competitiveness.

A few years later Mary Kaldor developed the implications of increased complexity and gold plating practices in her much quoted book *The Baroque Arsenal* (Kaldor 1982). Military systems were becoming 'baroque' in that they increasingly incorporated more functions, making them more complicated to operate and maintain, more expensive to procure and support, and more prone to breakdowns. She attributed this trend to the combination of the military's conservatism that kept the type of main weapons systems untouched, with the technological progress required to make these systems more 'capable' and, above all, justify new generations. In other words, the military was focusing on the large weapons systems around which their strategies, organization and even their internal promotions mechanisms revolved. They were seeking increasingly capable fighter aircraft (air force), battle tanks (army) and large aircraft carriers and frigates (navy), even when, Kaldor argued, precision-guided munitions and missiles were making them more vulnerable. To justify new generations of the systems, the argument continued, more capacity and therefore complexity and costs were built upon them. Paralleling the focus on large systems of the customer, the structure of producers revolved around prime contractors who were the firms responsible for assembling and delivering the large platform. The platform maker was the prime contractor and controlled the supply chain.

Throughout Peck and Scherer, Melman and Kaldor and many other authors in between, a characteristic of the defence industries that emerges clearly is how they differ from the practices prevalent in civilian markets. While civilian firms were oriented to the mass production of cost-efficient products that were easy to maintain and operate and cheap to purchase, defence industries focused on achieving the reverse: the production of increasingly costly (and therefore in smaller numbers), extremely complex and difficult to operate and maintain military systems. As commercial markets for high technology products were growing at a higher pace than for military systems, and the potential for growth in the civilian arena was clearly much bigger, those countries that had focused their technological development on the support of defence technologies were facing serious economic problems and were bound to lose competitiveness (Kaldor, Sharp et al. 1986). Reliance on military technology and production was a hindrance to future economic development and competitiveness. The closed systems of defence procurement agencies, military users, defence laboratories and specialized military manufacturers were ill-suited for conversion and diversification into commercial markets and were becoming a burden for the future.

These analyses were taking place at a time when countries with relatively high defence budgets like the UK and the US were facing serious domestic economic difficulties and countries with low budgets and small defence R&D expenditures, like Germany and above all Japan, were the toast of investors and economic analysts alike for their economic success. By the turn of the century the situation had turned around. Germany and Japan had been stagnant for over a decade and were having difficulties in extricating themselves from this situation, while the US economy grew strongly supported by the success of its ICT industries and other high-technology sectors. As the biggest spender in defence research and procurement was leading the world in virtually all technology-intensive sectors, the analytical tone changed.

Authors started to stress the military roots of many of the innovations that were core to our recent technological development (Hambling 2005) and European policy reports suggested the importance of increasing defence research not only to support European security but to close the funding gap with the US, which was associated with different rates of innovation and economic growth (Navarro 2000). In less than 20 years Melman and Kaldor's arguments had been turned on their head.

THE SYSTEM UNDER STRESS: THE ROOTS OF SYSTEMIC CHANGE

The situation described in the previous section displays the traits of a stable system. For instance, from a policy perspective, strong policy networks (policy communities, iron triangles, etc.) are aligned with policy stability (Rhodes & Marsh 1992). If relatively small groups of actors control policy formation by reinforcing each other, how can this system change? Public policy analysts have identified different change factors, including policy learning (Bennet & Howlett 1992), the emergence of new groups and ideas (Richardson 2000), and the influence of external events and alterations in the power patterns (Wilson 2000). From a co-evolutionary perspective these factors are understood to be mutually interdependent and are themselves influenced by policy changes through feedback mechanisms.

This section reviews how the changes that have affected the military system of innovation have fed on each other and are affecting both the ways in which the variety of military systems is generated and the processes by which military systems are selected. Stressing factors in the security and strategic environment have combined with changes in military technology and in policy design, to begin breaking up the boundaries that sheltered military research, and technological and industrial activities from other technological activities. This section summarily reviews the transformations in economic and technological conditions and in the strategic environment that started to emerge during the late twentieth century, and how they relate to changes in policy rationales and policy interventions. It is a change that will probably be slower than many anticipate, but it is likely to be profound as it occurs through the interplay of changing perceptions and theories about the role of military R&D, changes in the military innovation system and the way it relates with its broader environment, and the transformation in the security and defence policy landscape. The outcome is a move away from the closed system that characterized defence innovation in the decades after the Second World War, with new players and new technological dynamics being introduced into the system.

The Changing Relationship between Military and Civilian Technologies

The high expenditure in modern technologies and the unrestrained approach to cost considerations had placed the defence industries at the forefront of technological development. Defence agencies played a key role as 'first users' of new technologies, developing and acquiring goods at the early stages of their 'technological cycle' when, because of their high costs, civil markets could not absorb them (DeGrasse Jr. 1983). Defence agencies played a central role in the early development of the computing industry,[7] and of many other key technologies like sonar, radar, jet engines, transistors, numeric control machines, containerized transport, ball bearings, etc.[8]

It was not until the technologies matured and their costs fell that civilian markets started to adopt and adapt technologies that had first been developed for military environments. Yet, the growth potential of these new civilian markets was much larger and soon the level of R&D investment they could support dwarfed the research expenditure that even the US Department of Defense could afford.[9] From the late 1970s an emerging awareness progressively developed among analysts and policy makers that a shift of balance was taking place, with civilian technologies

7 See for instance, the comprehensive study by Flamm (1987).

8 There are many sources providing lists of 'military spin-offs' to civilian use. For an example see Smith (1987b).

9 For instance, in electronics the share of defence in the world market was already rapidly falling by the early 1980s). While in the early 1960s the only users of integrated circuits were government agencies (their applications were limited to military and space uses), by 1978 the US government accounted only for 10% of the domestic market, a percentage that kept falling rapidly in the following years and is, today, continuing to fall.

starting to lead in terms both of their performance and their cost, and that in key areas commercial investments in R&D were outstripping parallel military efforts.[10]

The new situation was described as one in which 'increasingly, in fields such as materials, electronics and aerospace, it is the civil sector which is leading the way and the military sector which is adopting and adapting their products' (Cuthbertson 1983). In other words, the sources of technology variation were moving outside the military system. Soon, even as defence budgets were skyrocketing under the Reagan administration, the perceived weakness of the US civilian industries started to be seen as a security risk because it endangered the superiority of US military technology (Manufacturing Studies Board 1987). Defence industrial strength would have to be built on the foundations of strength in civilian technology (US Congress Office of Technology Assessment 1989), and the military would have to learn not to control the development of the components and sub-systems it was using, otherwise it would end up buying expensive and sub-standard products.[11] Using evolutionary language, actors were learning that the technology selection dynamics would have to be adapted to the changes in the sources of technology variation.

Besides the market changes in sectors that had originally been developed with military funding, the late twentieth century saw the development of new industrial technologies originally targeted for civilian markets but with important military implications. Biotechnology and nanotechnologies are perhaps the two most relevant fields of current technological development with important defence and security military implications. Nanotechnology, for instance, has extensive military uses in areas like small distributed sensors, metal-free small arms and armed autonomous systems (Altman 2006). Developments in biotechnology can also have alarming outcomes related to the development of new pathogens that could be used in traditional warfare or in terrorist actions. As new sectors are becoming of interest to the military customer, defence agencies are looking for new suppliers to enter the, still singular, world of military procurement; yet, to entice them there is a need to simplify the over-complex procedures and overbearing regulations that characterize the market. Again, the selection dynamics have to be changed to adapt to the shifting sources of technology variation.

As the cost and performance improvements in civilian components and systems were rendering equivalent military alternatives obsolete and new civilian technologies gained military relevance, the closed nature of the defence system of innovation was put under pressure by a set of technological developments that it had, in part, help unleash. To be able to adopt and adapt technologies developed elsewhere, both the procurement processes dominant during the Cold War based on extensive military specifications, and the practices of isolation and secrecy that characterized defence

10 For an analysis of this process and its implications see Molas-Gallart and Walker (1992).

11 For instance, in the early 1990s it was argued that the specialist microprocessors the US Department of Defense was acquiring were ten times more expensive and two generations older than the equivalent commercial components (Weidenbaum 1992).

research establishments would have to be abandoned. But this, as we will see below, turned out to be a very difficult task.

SECURITY-DEFENCE CONVERGENCE

The second main stressing factor that came to disrupt the system of innovation built during the Cold War was the changing nature of the security threats that the defence agencies were facing, and the institutional implications of these changes. There are two main reasons for this change. First, the end of the Cold War meant that any possibility of a massive conventional conflict confronting the armies of the Warsaw Pact and NATO disappeared. Instead the types of conflict in which the armies of the NATO countries got involved during the 1990s responded to another type of hostilities, defined by Mary Kaldor as 'New Wars' (Kaldor 1999). New Wars have emerged due to or alongside with the dissolution of the power of the State, and were characterized by civil conflict and power fragmentation, with regional and local control passing to smaller armed groups (warlords) engaged in a variety of activities ranging from guerrilla warfare, to racketeering and arms trafficking. Violence is directed mainly towards local populations, and the role of foreign armed forces shifts from that of a traditional fighting army to having to conduct internal security operations. Often the armies have to act as police.

Second, a main security concern has emerged in the shape of so-called 'New Terrorism', a shift from 'old' forms of terrorism targeted to specific individuals and smaller groups of symbolic value. Instead New Terrorism is bent on inflicting the maximum amount of damage possible, often trying to kill as many civilians as possible in an indiscriminate way. The attacks of 9/11 were unprecedented for the scale of civilian casualties inflicted and the determination to cause such damage. Further, the attack that had caused the most casualties in the continental USA since the nineteenth century had been carried out using commercial aircraft. The determination to cause large civilian casualties and the availability of a whole range of new and old commercial technologies that could be applied to such an end, increased the concern about the possibility of terrorists using innovative tools, like radiological or biological means, to carry out massive attacks.

The shock of 9/11 and the heightened perception of new security threats have triggered a new policy agenda in which traditional military defence goals are conflating with national security objectives in what is increasing called 'defence and security policy'. There is therefore a direct linkage between this external set of events and the development of new policy rationales that are only just starting to be implemented. The coming together of defence and security agendas is a development the importance of which cannot be overstated. The system described in the first section of this chapter was characterized by a clear distinction, at least from an institutional perspective, between military defence against the threat posed by regular foreign armed forces, and internal domestic security. Different government departments dealt with the different policy environments and distinct policy networks evolved. From an economic perspective they had different procurement needs, and typically were supported by different supply networks. Further the budgets allocated to domestic security were only a fraction of those allocated to military defence.

Although an institutional separation continues to exist between the organizations primarily tasked with domestic security and those concerned with foreign defence, the distinction between both arenas of policy activity is blurring. The term 'security and defence' is seldom defined but has already found its way into a myriad of policy documents and points to the fusion of both areas of policy activity. The defence agencies have to deal with more diffused threats in their military interventions and often operate as security forces abroad. They also are naturally concerned by the implications of terrorist activities 'at home', which directly affect them. Meanwhile organizations like the newly created US Department of Homeland Security have been supplied with unprecedented levels of resources for a national security organization, and are starting to develop their own far-reaching research programmes, supplied by both firms new to these markets, as well as traditional defence suppliers. These developments are one of the factors breaking up the closed policy communities that had characterized the defence policy environment during the Cold War period. New and strengthened government and private players organized around the security agencies are breaking the boundaries of the old defence policy networks.

From a technological perspective, there has been a vast increase in the number of research fields falling under the scope of security and defence agencies. Many technologies that were developed by civilian organizations for civilian objectives are now perceived as having important security implications. I have already referred to the examples of biotechnology, and its potential use in bioterrorism. There is also growing concern about the potential vulnerability of physical infrastructures to terrorist attack. In both cases, the scientific communities that have started to work for the security agencies (analysing vulnerabilities and possible attack scenarios, developing defensive measures, etc.) are often newcomers to the world of 'defence and security' research programmes. Further, because of the perception that many areas of research, from biotechnology to civil engineering, have security implications, formal controls on the flow of knowledge and technology transfer are extending to more areas of research, as well as to teaching and industrial activities. As security concerns pervade all kinds of social activities, the sharp distinction between the 'military' and the 'civil' starts to fade away.

SHIFTING POLICY RATIONALES

The changes in the security context discussed above have led to attempts to reform defence innovation and industrial policies. The pressures to alter old approaches are strong, and have been conveyed through many government and consultancy reports, journal articles and books, mainly in the US but also in the UK and elsewhere in Europe.

In the industrial and technology front, it progressively became clear to military procurement experts that the systems in place for developing and manufacturing arms had to be overhauled. In the US a succession of 'blue ribbon' commissions had, over the decades, suggested avenues to reform procurement and thus tackle the spiralling growth in the cost of arms systems and the many inefficiencies identified in the procurement process. In the mid 1980s the debate intensified and reached the public after the highly publicized Pentagon procurement 'scandals' that accompanied the bonanza of Reagan's military build up. Both 'outside' critics and experts and analysts within the procurement system were putting forward solutions that called for a profound reform of the procurement process. For instance, Jacques Gansler, by then Executive VP of a large Washington consultancy firm and a faculty member at the John F. Kennedy School of Government at Harvard University, argued in two books (Gansler 1989; 1995) that 'the deeply entrenched practices [in weapons acquisition] that have developed over several decades must be changed' (Gansler 1989), that defence firms had to branch out into civilian production and that the US industry had to undergo a 'dramatic' transformation 'into a largely integrated (civil/military) structure' (Gansler 1995). Gansler had the occasion to apply its theories when, under the Clinton administration, he became Under-Secretary of Defense for Acquisition, Technology and Logistics, in charge of defining the Pentagon's defence procurement policy. Yet, the practices were indeed entrenched, and the combination of external pressures and a coherent 'policy theory' would prove insufficient to enact radical change in the short term. The reform attempts met with the inertia of consolidated practices and strong institutions that had evolved over decades; consequently, the implementation of new procurement approaches has been slow.

Nevertheless, new policy rationales in the procurement area are accompanied and mutually reinforced by the emergence of new military doctrines and approaches to military strategy. In the 1990s, military strategists and thinkers were considering the implications of the emergence of new Information Technologies on the conduct of military operations. The US analyst Martin Libicki (1994), for instance, argued that warfare would come to be dominated by a 'mesh' of inter-related small sensors and small intelligent projectiles. Surveillance and networking technologies would become crucial while large military platforms would be increasingly vulnerable to attacks from waves of small missiles. This, the argument continues, would eventually lead to the end of large platforms and the large organizations built around them, and the increased use by the military of mass-produced small weapons and systems, many of which will be of commercial origin or embody commercial technologies.

Similar visions of change propelled official Pentagon policy. The notions of a 'Revolution in Military Affairs' and of 'transformation' were pushed, among others, by officials in the US Department of Defense. In 1993, Andrew Marshall, then Director of the 'Office of Net Assessment', within the Office of the Secretary of Defense, considered the potential for 'military revolution' in an internal memo.

Soon a community of Pentagon officials, think tank analysts and academics developed the notion of the need for a Revolution in Military Affairs centred around the exploitation of Information and Communication Technologies in military operations, the focus on network-based (rather than platform-based) operations, and therefore the need for a profound change in the way the military procured and used military hardware (Ross 2007). These ideas were explicitly supported during President George W. Bush's first electoral campaign and became official Pentagon policy expressed through a surge of statements and internal documents. RMA is linked to the new capabilities that technological change can offer to the military. The view is that new ICTs provide the capacity to fight in a way that is completely different to the approaches that had characterized the twentieth century, and that, therefore, the doctrine, organizational structure, procurement policies and arms systems used will also need to be 'transformed'.

Changes in the policy discourse do not necessarily mean that policy practice and institutions will follow suit. Practices and organizations are likely to be affected but seldom in the ways that the proponents of change had envisaged, such is the institutional inertia of the existing organizations and the complex relationship between environmental conditions and the selection and implementation of new policy options. More than a decade after the RMA notions started to be developed in the US, there are many analysts who are deeply sceptical about the real impact that the new policies have had. Andrew Ross argues, for instance, that 'the U.S. military transformation enterprise is not (yet) as transformational as it is cracked up to be. It features not discontinuous but incremental change; not disruptive but sustaining innovation; not generation-skipping technological leaps ahead but small steps forward; not breakthroughs (radical or otherwise) but more of the same' (Ross 2007).

Ross continues to argue that the transformation agenda 'poses less of a challenge for the U.S. innovation system's defense research, development, and production capabilities than the rhetoric of transformation advocates might suggest' (Ross 2007). In and out of its own, policy rhetoric has little strength to change the institutional organizations and practices that have been set over several decades. A co-evolutionary perspective allows us, however, to focus on mutually interdependent factors that can bring about systemic changes albeit over longer periods of time. Although the revolution in military affairs has not taken place as a revolution, a process of change has been unleashed by the combined and interdependent changes in the technological and security environments, and the shifting policy proposals reviewed in this section.

The next section discusses the institutional response to the changing environment and policy rationales. First, I analyse the ambiguous redefinition of institutional roles and the emergence of new policy actors. Second, I provide two examples of specific policies that are being implemented to open the world of defence procurement and research to increased interaction with its broader 'civilian' environment: procurement reform, and the privatization and commercialization of government defence research establishments.

INSTITUTIONAL RESPONSES TO CHANGE: TOWARDS A NEW SYSTEM

Security: the Emergence of New Institutional Players and the Blurring of Military/Civilian Boundaries

The heightened concern over security threats has led to the emergence of new 'security' agencies and security R&D and procurement programmes channelled either through new departments like the US Department of Homeland Security (DHS) or through the redefinition of roles carried out by existing organizations like the European Commission. One of the indicators of the increasing importance of security agencies and agendas is the budgets they control. Of particular interest to innovation issues is their research activities. In the US, the Department of Homeland Security (DHS) R&D budget grew rapidly to reach a peak of almost $1.3 billion in 2006.[12] This figure is small if compared with the R&D budget of the Department of Defense (over $72 billion), but is nonetheless higher than the defence research budgets of most European countries.

The DHS has a broad remit, going well beyond defence against potential terrorist attacks. For instance, on a single day (8 May 2007) the home page for the Department of Homeland Security (http://www.dhs.gov/index.shtm) included a note on the federal assistance for the victims of a tornado in Kansas, a link to a speech by President Bush on immigration, and an announcement of $24 million in Federal grants to increase the security of non-profit organizations 'deemed high-risk for potential international terrorist attack'. Yet, this same breadth of activities blurs the distinction between the defence against politically motivated violent attacks, that were the traditional remit of the military and security agencies, the protection against natural disasters, and the control of social trends that may be deemed destabilising.

A related development is the entry of existing institutions into the field of security. An interesting example is provided by the way in which European Union institutions are acquiring security and defence roles. The European Commission's 7th Framework Programme (FP7) includes a 'Security Research Programme' with, again, vague objectives: to 'provide more security for citizens', and 'promote the innovativeness and competitiveness of security-related industries' (CORDIS 2006). There is no definition of what kind of 'security' are we talking about, neither is there any indication of what kind of firms and activities will be considered as forming part of the 'security-related industry'. In fact, the European Commission has long accepted that 'security is an evolving concept' (CORDIS 2006), and following the example of the DHS, the 'Security Research Programme' goes beyond the fight against terrorism to consider, for instance, measures to prevent and deal with natural disasters. FP7 defines the objectives of the Security Programme as follows:

12 In 2007 the DHS R&D budget was cut by about 10%.

> to develop the technologies and knowledge for building capabilities needed to ensure the security of citizens from threats such as acts of terrorism and (organised) crime, natural disasters and industrial accidents while respecting fundamental human rights including privacy; to ensure optimal and concerted use of available and evolving technologies to the benefit of civil European security; to stimulate the co-operation of providers and users for civil security solutions; to improve the competitiveness of the European security industry and to deliver mission-oriented results to reduce security gaps. (Council of the European Union 2007)

The document goes on to assure the reader that the 'Security theme will maintain and exclusive civil orientation'. Yet, later on, the same document states that '[r] ecognizing that Security research covers areas of **dual use** technology relevant to both civilian and military applications, coordination with the *European Defence Agency* (EDA), who will consult its Member States about national programmes, will ensure complementarity wherever necessary'. In other words, the programme has an exclusive civil orientation, but as it covers dual-use areas, it cannot be exclusively civilian!

There are reasons why the European Commission is, on the one hand, striving so hard to separate military from civilian activities, while recognizing that it is futile to attempt to make a strict separation. This contradiction can be traced, among other reasons, to the traditional opposition of some Member Countries to any measure that could imply any 'militarization' of the European Union. Military affairs were not to be addressed by the European Commission and for a long time their mere mention was almost taboo. The European Union was a civilian institution. This situation has slowly changed overtime, but political opposition to the change has remained strong throughout. First the Maastricht Treaty introduced the 'Common Foreign and Security Policy' (CFSP) as one of the 'legs' of the European Union, although in article J14 it called on the Western European Union (WEU), rather than the Commission, to elaborate and implement Union decisions with defence implications. This attempt to introduce security and defence issues within the scope of the EU, but to maintain them separate was also to be progressively abandoned. The EU Council decided in the 1999 Cologne meeting to transfer the WEU activities to the EU, and appointed a EU 'High Representative' for the CFSP explicitly focusing on the so-called 'Petersberg Tasks' (peacekeeping, humanitarian and rescue operations, but also the use of military force in crisis management and 'peacemaking' operations). Finally, the establishment by the EU of the European Defence Agency (EDA) in July 2004 explicitly included military responsibilities. EDA's main objective is to improve Europe's 'defence capabilities in the field of crisis management' and to sustain the European Security and Defence Policy. To achieve these goals the Agency was tasked with the promotion of increased European armaments cooperation and collaborative defence research.

By 2007 the institutional situation continued to be, at best, confusing. The WEU continues to exist formally although most of its activities and responsibilities have been transferred to other EU bodies. The EU and the EDA are the main actors in a complex web of shifting institutional responsibilities and ambiguously defined tasks, straddling military and security tasks. Their memberships are broadly overlapping but by no means identical. The political opposition by some countries and political groups to the allocation of military responsibilities and tasks to the European Union partly explains the confusion. Within this context, mixing vaguely defined 'security' goals with the defence against natural disasters and other clearly civilian objectives (industrial accidents) helps make the claim that the programme is civilian, but blurs even further the boundaries between military and civilian activities.

The conceptual confusion in the definition of security and its relationship with 'defence' and 'military activities' helps to rear a climate of policy ambiguity on which to shore up institutional formation. This allows, for instance, the European Commission to move towards funding research relevant for military purposes without making the move explicit. As the FP7 text recognizes, many research activities envisaged in the Security Research Programme are related to fields of activity germane to military R&D.[13] Accordingly, many leading military firms are likely to play a crucial role in the Security Research Programme, in the same way that they did in their predecessor: a pilot programme known as PASR.[14]

The DHS and EU examples show how new institutional arrangements are evolving to respond to the new security conditions that emerged with the diffusion of large-scale indiscriminate terrorist attacks, and the perception of a new type of security threat. Yet, these changes are not simply the result of a unidirectional cause-effect relationship in which policy reacts to changes in environmental conditions. The EU had dealt with 'security' issues during the whole of the 1990s, and the ambiguity in the concept of security and its relationship with military defence played an important role in allowing controversial policy steps. Further, changes in the technological and industrial context (see above) were questioning the notion that 'defence technologies' were clearly identifiable and distinct from 'civilian' ones. Also, the security-defence convergence discussed above, and the emergence of 'new wars' in which the lack of domestic security is intertwined with international military conflict provided a policy rationale for the new institutional architectures being developed. These trends were reinforced after 9/11.

There are therefore reasons in the security and technological environments that help explain the disorder in the definition of institutional tasks for the defence and

13 These include, for instance, 'security systems integration, interconnectivity and interoperability' and 'intelligent surveillance and border security'.

14 The European Commission launched the 'Preparatory Action for Security Research' (PASR) in 2004, following on the adoption by the European Council on December 2003 of a European security strategy: 'A secure Europe in a Better World'; a document providing a framework for European concerted activity in the field of security.

security agencies, but, as it has been shown for the European case, the ambiguity and confusion in the definition of institutional objectives has also played a political role. The European Union is trying to separate military from civilian-related activities while at the same time pursuing 'coordination' and *de facto* accepting that such a sharp distinction between, for instance, military and civilian research is not possible. This is politically expedient, but it is a type of expediency that feeds on technological changes and policy transitions. Meanwhile the distinction between security and military policies is further blurred and the 'closed' nature of 'military technology' falls apart. I now turn to another set of interdependent policy changes that is further helping to break the closed nature of the military innovation system.

DEFENCE PROCUREMENT REFORM

Opening the Military Market to Civilian Products and the Demise of 'Milspecs'

Defence agencies in several countries, but particularly in the UK and the US, have launched ambitious procurement reform initiatives to respond to the changes in the technological environment discussed above. In particular, as defence organizations increasingly perceived commercial technologies as offering more powerful, flexible and cheaper solutions than those that could be obtained from bespoke military products, they developed policies to allow an encourage the use of commercial components and sub-systems. Military procurement jargon distinguishes:

- Commercial-off-the-shelf technologies (COTS), products available in open commercial markets; and
- Non-developmental items (NDIs), non-COTS products whose development has, however, been funded privately.

The attempts to use both COTS and NDIs, however, generate multiple implementation problems. Initially, there is a need for regulatory change: the large number of military specifications (milspecs) and standards that were a formal requirement for most military purchases have to be eliminated whenever possible (Molas-Gallart 2001a), and other contractual procedures have to be simplified to encourage new suppliers. Yet, the process of change becomes more complex as these modifications require other adaptations as well. The use of COTS and NDIs calls for the adoption of life-cycle costing, long-term maintenance contracts and the introduction of 'obsolescence management' practices to be able to deal with the risk of adopting commercial components with short shelf lives, over which the customer holds no Intellectual Property rights, into complex systems with long life-cycles. Although, defence ministries welcome the savings brought about by shifting design and development responsibilities and their associated risks to suppliers, there is a potential for unexpected long-terms costs high enough to cancel the initial savings.

If the operator becomes dependent on a private supplier for the continuous delivery over the years of specific components and spare parts, there is a possibility that the supplier will try to exploit this situation by imposing higher prices. Even components initially widely available through commercial markets can rapidly become obsolete and then available only, if at all, from a diminishing base of specialist suppliers. As the buyer loses control over system design and configuration, the management of intellectual property rights over the life cycle of a system has become a bone of contention between suppliers and customers.

In short, defence procurement reform is not limited to a few changes in regulation and practices but involves a thorough change on the whole procurement system and culture. Not only do defence customers have to deal with a wider base of suppliers with whom they lack the close links of familiarity characteristic of their relationship with traditional military contractors, but they have to develop new attitudes, policies and systems to protect their interests as their control over product design fades away.

COMMERCIALIZING GOVERNMENT DEFENCE RESEARCH ESTABLISHMENTS

A further important policy initiative is the incipient commercialization and, in some cases, privatization of defence research laboratories. Traditionally government-owned defence research establishments had carried out their work fully focused on defence objectives and in relative isolation from related research communities in academia and elsewhere. The combination of reductions in defence research budgets in the early 1990s, with the technological trends described above impelled them to open up their activities to (1) exploit commercially in other markets the capacities they had accumulated and the technologies they had developed through their defence work, and (2) capture and exploit for defence (and security) purposes technologies developed elsewhere.

Although there are important differences across countries in the structure and management of defence laboratories, there is a general trend to open them up to new markets, widening their strategic functions, and increasing their commercial activities.[15] When this has happened, as in the transformation of the British Defence Evaluation and Research Agency into a private firm (QinetiQ) and a smaller 'retained' government defence facility (the Defence Science and Technology Laboratories), the new institutions have taken on a much expanded set of activities and are trying to penetrate new markets. QinetiQ has become very active in the security markets (including airport and infrastructure security, strong aircraft

15 I have analysed elsewhere the context and some of the problems triggered by commercialization of government defence laboratories in the UK (Molas-Gallart 2001b; Molas-Gallart and Tang 2006).

structures, and monitoring systems) and is penetrating other commercial markets through spin-offs, joint ventures and technology licensing arrangements, while maintaining its core activities in defence R&D. The outcome is, again, the opening up old defence-related institutions by involving them in a variety of new activities for a variety of new clients in security agencies and the broader commercial world.

In the US, the nuclear weapons laboratories were among the research establishments that had to adapt themselves to the end of the Cold War. Although a nuclear weapons mission remained for the laboratories, it was clear that new research missions would have to be found to sustain their existing research capacities (Crease 1993). The laboratories tried to diversify by entering co-operative research agreements with firms, sharing facilities with commercial users, providing technical assistance to SMEs, and generating technology-based spin-offs. Of the three nuclear weapons laboratories,[16] Lawrence Livermore was the one that went the furthest in terms of formally redefining its original mission: from a weapons laboratory focusing on defence work it adjusted its mission to cover any objective 'of national importance where we can make unique and valuable contributions' by solving 'complex scientific and technical problems' (Lawrence Livermore National Laboratory 1996).

It remains to be seen whether it will be possible in the long term for the old government defence research establishments to maintain their capacities by acquiring a diversified portfolio of activities, indistinctively receiving funding from both military and civilian customers. Whether their capabilities should be indistinctively applied to other civilian objectives has long been a topic of contention. Yet, whatever the differences across laboratories and their ultimate success in retaining existing research capabilities, what is important for our discussion here is that the change in strategy calls for new customers, new managerial practices and new approaches to the definition of research priorities. To be successful in a more commercialized and competitive research environment the laboratories will have to open up at least part of their facilities to collaboration and interaction with groups outside the relatively narrow world of the defence communities on whom they had initially focused all their efforts.

CONCLUSIONS

This chapter has focused on the models of defence research and innovation prevalent in the US and other Western European countries during the Cold War, and analysed the systemic change that is currently taking place. I have argued that the military innovation system is transitioning from a situation that I have characterized as 'closed', to an open system in which there are no longer clear boundaries separating military technologies from the rest. The emerging situation is that the closed systems

16 Sandia National Laboratories, Los Alamos National Laboratory and Lawrence Livermore National Laboratory.

of innovation and policy development that were characteristic of the Cold War system are opening up to new actors and, consequently, the boundaries between military production and the rest of the economy are increasingly difficult to draw. The sources of variation in the technologies used by the military have moved from being internal to the military innovation system, to be found increasingly in civilian environments. Consequently technology selection processes are being adapted. In line with the arguments developed in this book (Smits intro chapter), I have shown that this process cannot be presented as the result of autonomous developments in military technologies or even of transformations within the military innovation system. Instead the factors that explain the shift are systemic in the sense that they involve a complex interaction between a broad set of institutions and the conditions within which they develop. The institutions concerned are not only those directly related to military innovation (defence research organizations, agencies, procurement organizations, military firms, and others) but a broader, and enlarging, set of institutional actors (such as international organizations, security departments, and other commercial firms). The relevant policy environment also goes beyond the definition and implementation of military innovation policies, to involve security policies and shifts in the global security environment, in particular the emergence of global terrorist threats. In this sense, also in line with the arguments developed in this book, the process of technological change in the military and security arenas can be characterized as a multi-level, multi-actor affair:

- At the highest level, there have been important changes in the conditions within which the defence technology systems operate.[17] In particular, the terrorist attacks of 9/11, and the emergence of terrorist groups organized globally but often operating locally is leading to the redefinition of the roles of the institutions involved in defence and security tasks and broadening the set of research and technological activities that are relevant in a security context.
- Technological change outside the military industries has also triggered policy responses, both from a defence procurement angle and from the wider definition of military doctrine. The degree to which broader technological change can be considered as 'external' to the military innovation system will vary across technologies. In the most relevant case, however, the rapid development and improvement of commercial ICTs, many of the technologies involved can trace their origins back to military research and investments. Technological change cannot thus be considered as an autonomous, external variable.
- Scientists themselves were vital to the provision of the rationale that underpinned the closed military innovation system of the Cold War. Policy analysts, academic

17 The literature sometimes refers to this broader environment as 'landscape conditions'. Geels, for instance defines the 'socio-technical landscape' as a set of deep structural trends external to the regime' forming an external structure or context for the interaction among system actors (Geels 2005)

or linked to think tanks and other policy organizations, have also had an important effect in providing the rationale for defence procurement reform and for the definition of new military strategies. Again, however, these analysts do not operate in a vacuum; their work can be seen as responding to changes in the security and technological environment, and to the perceived deficiencies of existing military innovation systems.

- Although policy proposals and theories have been developed to respond to technological changes, their implementation has been slower than their proponents had hoped for. In other words, institutional change does not respond automatically to environmental conditions. Even apparently limited policy initiatives, like the elimination whenever possible of military specifications and standards, have far-reaching consequences on the nature and structure of military production (in this case the loss of control of the military client over product definition). These, in turn, require further policy measures (the need to develop long-term approaches to procurement) that eventually will lead to systemic change.

The transition between two systems of innovation in the military field provides a good example of the complexity of the relationships among technological change, policy theories, policy practice and environmental conditions. It is neither a smooth nor a rapid transition process. The institutional inertia of the old military organizations and agencies is built around a system of procurement practices and regulation that is tightly linked with current operational practices. The coming together of security and defence policies will give rise to new policy coalitions and institutional arrangements, but the new coalitions need time to gel and the new institutional roles are not yet settled given the current confusion about the distribution of tasks and the relationships among different organizations. Change is likely to take place over a period of decades rather than years.

REFERENCES

Adams, G. (1982), *The Politics of Defense Contracting: The Iron Triangle*, New Brunswick, USA and London, UK: Transaction Publishers.

Altman, J. (2006), *Military Nanotechnology: Potential Applications and Preventive Arms Control*, London, UK and New York, USA: Routledge.

Allen, P. (2001), 'A complex systems approach to learning, adaptive networks', *International Journal of Innovation Management*, 5 (2), 149-180.

Bennet, C. and Howlett, M. (1992), 'The lessons of learning: reconciling theories of policy learning and policy change', *Policy Sciences*, 25 (3), 275-294.

Bush, V. (1945), 'Science, the endless frontier', report to the President by Vannevar Bush, Director of the Office of Scientific Research and Development, U.S. Government Printing Office, Washington, DC.

CORDIS (2006), 'Verheugen makes the case for security research in FP7', Retrieved 5 May, 2006, http://cordis.europa.eu/fetch?CALLER=NEWS_SECURITY_ALL&ACTION=D&QM_EN_RCN_A=25244>.

Council of the European Union (2007), 'Corrigendum to council decision 2006/971/EC of 19 December 2006 concerning the specific programme "cooperation" implementing the Seventh Framework Programme of the European Community for research, technological development and demonstration activities (2007 to 2013)', Official Journal of the European Union, L54/30-L54/80.

Crease, R. (1993), 'The national labs and their future', *Forum for Applied Research and Public Policy*, **Winter**, 96-104.

Cuthbertson, I. (1983), Response of HMG to the questions posed in working paper 2 (Revision 2) of the UN Study on Military Research and Development, Arms Control and Disarmament Unit, Foreign and Commonwealth Office.

Defense Advanced Research Projects Agency (2005), 'DARPA bridging the gap: powered by ideas', DARPA.

DeGrasse Jr., R. (1983), *Military Expansion, Economic Decline: The Impact of Military Spending on U.S. Economic Performance*, Armonk: Sharpe.

Dosi, G., LLerena, P., et al. (2005), 'Science-technology-industry links and the "European Paradox": some notes on the dynamics of scientific and technological research in Europe', LEM Working Paper Series, Laboratory of Economics and Management, Sant'Anna School of Advanced Studies, Pisa.

Edwards, P. (1997), *The Closed World: Computers and the Politics of Discourse in Cold War America*, Cambridge, MA, USA and London, UK: MIT Press.

Flamm, K. (1987), *Targeting the Computer*, Washington, DC: Brookings Institution Press.

Galbraith, J. (1985), *The New Industrial State*, Boston: Houghton Mifflin.

Gansler, J. (1995), *Defense Conversion: Transforming the Arsenal of Democracy*, Cambridge, MA, USA and London, UK: MIT Press.

Gansler, J. (1989), *Affording Defense*, Cambridge, MA, USA and London, UK: MIT Press.

Geels, F. (2005), *Technological Transitions and System Innovations: A Co-Evolutionary and Socio-Technical Analysis*, Cheltenham, UK and Northampton, MA, USA: Edward Elgar.

Hambling, D. (2005), *Weapons Grade: How Modern Warfare Gave Birth to Our High-Tech World*, New York: Carrol & Graf Publishers.

Jordan, G. (1990), 'Sub-governments, policy communities and networks: refilling the old bottles?', *Journal of Theoretical Politics*, **2** (3), 319-338.

Kaldor, M. (1999), *New and Old Wars: Organized Violence in a Global Era*, Stanford: Stanford University Press.

Kaldor, M. (1982), *The Baroque Arsenal*, London: Andre Deutsch.

Kaldor, M., Sharp, M., et al. (1986), 'Industrial competitiveness and Britain's defence', *Lloyds Bank Review*, **October** (162), 31-49.

Lawrence Livermore National Laboratory (1996), 'Institutional plan', FY 1997-2002, Lawrence Livermore National Laboratory, University of California, Berkeley.

Leslie, S. (1993), *The Cold War and American Science: The Military-Industrial-Academic Complex at MIT and Stanford*, New York: Columbia University Press.

Libicki, M. (1994), 'The mesh and the net: speculations on armed conflict in a time of free silicon', Institute for National Strategic Studies, National Defense University, Washington, DC.

Manufacturing Studies Board (1987), *Manufacturing Technology: Cornerstone of a Renewed Defense Industrial Base*, Washington, DC: National Academy Press.

Melman, S. (1974), *The Permanent War Economy: American Capitalism in Decline*, New York: Simon & Schuster.

Molas-Gallart, J. (2001a), 'Government policies and complex product systems: the case of defence standards and procurement', *International Journal of Aerospace Management*, 1 (3), 268-280.

Molas-Gallart, J. (2001b), 'Government defence research establishments: the uncertain outcome of institutional change,' *Defence and Peace Economics*, **12** (5), 417-437.

Molas-Gallart, J. and Tang, P. (2006), 'Ownership matters: intellectual property, privatization and innovation', *Research Policy*, **35** (2), 200-212.

Molas-Gallart, J. and Walker, W. (1992), 'Military innovation's growing reliance on civil technology: a new source of dynamism and structural change', in W. Smit, J. Grin. and L. Voronkov (eds), *Military Technological Innovation and Stability in a Changing World*, Amsterdam: VU University Press.

Morris, P. (1994), *The Management of Projects*, London: Thomas Telford.

Navarro, A. (2000), 'The gap in defence research and technology between Europe and the United States', report submitted on behalf of the Technological and Aerospace Committee by Mr Arnau Navarro, Assembly of Western European Union, Brussels.

Noel-Baker, P. (1936), *The Private Manufacture of Armaments*, London: Gollancz.

O'Connell, R. (1989), *Of Arms and Men: A History of War, Weapons, and Aggression*, Oxford, UK and New York, USA: Oxford University Press.

O'Connell Jr., C. (1985), 'The corps of engineers and the rise of modern management, 1827-1856', in M. Smits (ed.), *Military Enterprise and Technological Change: Perspectives on the American Experience*, Cambride, MA, USA and London, UK: MIT Press, pp. 87-116.

Peck, M. and Scherer, F. (1962), *The Weapons Acquisition Process: An Economic Analysis*, Boston: Harvard Business School Press.

Rhodes, R. (1986), *The Making of the Atomic Bomb*, New York: Simon and Schuster.

Rhodes, R. and Marsh, D. (1992), 'New directions in the study of policy networks', *European Journal of Political Research*, **21** (1-2), 181-205.

Richardson, J. (2000), 'Government, interest groups and policy change', *Politcal Studies*, **48** (5), 1006-1025.

Ross, A. (2007), 'From bullets and bombs to bits and bytes: U.S. military transformation or business as usual?', PRIME-INNOMIL workshop on 'Re-Evaluating Defence R&D and Innovation Dynamics' PREST, University of Manchester, Manchester.
Smith, M. (1987a), 'Army ordnance and the "American system" of manufacturing, 1827-1856', in M. Smits (ed.), *Military Enterprise and Technological Change: Perspectives on the American Experience*, Cambridge, MA, USA and London, UK: MIT Press, pp. 87-116.
Smith, M. (1987b), *Military Enterprise and Technological Change: Perspectives on the American Experience*, Cambridge, MA, USA and London, UK: MIT Press.
US Congress Office of Technology Assessment (1989), 'Holding the edge: maintainig the defense technology base', U.S. Government Printing Office, Washington, DC.
Weidenbaum, M. (1992), *Small Wars, Big Defense: Paying for the Military After the Cold War*, Oxford, UK and New York, USA: Oxford University Press.
Wilson, C. (2000), 'Policy regimes and policy change', *Journal of Public Policy*, **29** (3), 247-274.

12. Demand-Based Innovation Policy

Jakob Edler

INTRODUCTION

Innovation policy has become tremendously differentiated.[1] In part, this is a consequence of a more systematic understanding of the innovation process as described in innovation system approaches (see Chaminade & Edquist in this book), which has led to a broad understanding of functions of innovation systems and related policies to support these functions (e.g. Bradke et. al. 2007; Hekkert et al. 2007). Strikingly, although the approaches of the national innovation system originate partly in the inclusion of potential users (Lundvall 1988; 1992), and although the literature on national innovation systems includes the users of new knowledge and the customers for innovations, the demand side has long been neglected. The triangle of policymakers, policy and innovation analysts and the business community (in their claims towards innovation policymaking) had long paid little attention to stimulating demand in innovation policy. In terms of policy, all three have, *grosso modo*, focused on supply side strategies and activities, where subsequently the differentiation of research and innovation policy has essentially taken place.

However, within the last few years, demand orientation has received growing attention again, mainly – but not exclusively – in the form of public procurement to spur demand. Efforts are being made, above all at the European level, to raise awareness among policymakers and the business community to the potential of public demand for innovations. More ambitiously, the potential of 'lead markets' for the innovativeness and competitiveness of Europe is being tested (European Council 2006). Policy analysts have played some role in shaping this new discourse and informing policy learning processes (see Kok et al. 2004; Edler et al. 2005; Wilkinson et al. 2005; Aho et al. 2006; Edler 2007a; Georghiou 2007). At least in Europe, the comeback of demand-based innovation policy tools is an interesting case of discursive policy development. Yet, it is not decided if the rhetoric in demand-based

1 See for example Bozeman (2000). For a good overview on innovation policy measures, see the European database TRENDCHART, which takes account of innovation policy measures in all EU member countries and a large set of non-European OECD countries (www.trendchart.cordis.lu); for research policy see the European data source ERAWATCH cordis.europa.eu/erawatch/.

innovation policy (DBIP) and initial policy design attempts will take off and result in a genuinely new focus for innovation policy, thus transforming our understanding of the policy functions within innovation systems. In this chapter, DBIP is defined as: a set of public measures to increase the demand for innovations, to improve the conditions for the uptake of innovations or to improve the articulation of demand in order to spur innovations and the diffusion of innovations.

At the current stage of 'potential transformation', the purpose of this chapter is threefold. First, it seeks to conceptualise demand-oriented innovation policy. It thus discusses the rationales for turning towards demand stimulation as an innovation policy means (in the first section) and then introduces a typology of demand-based innovation policy (section two).[2] Second, it illustrates some current governance practice (in the third section) in order to discuss key characteristics of the policy development process in DBIP. Third, a final section will highlight three dimensions of challenges of DBIP, providing a model of the complex relation within the policy triangle of policy practice, strategic intelligence and innovation practice.

This chapter does not engage in discussion on the relative value of pushing demand or supply through policy in terms of innovation effects.[3] It rather starts from the assumption that demand does play a role, both in inducing new needs into the R&D process and in shaping the conditions for marketing and diffusion, and discusses the role of measures that tackle the demand side of the market equation in the area of innovation policy.[4]

CONCEPTUALISATION OF DEMAND-ORIENTED INNOVATION POLICY

What is the reasoning behind demand orientation in public policy? A succinct answer to this question must differentiate two levels. First, the conceptual core, stemming from a certain understanding of market and system failures, which point towards a number of structural hindrances hampering the market introduction and the market diffusion on the demand side, as well as the transformation of needs into

2 This chapter will not deal with military demand and the latest developments in terms of security – defence and its meaning for innovation and economic development. For a treatment of the relation of defence and innovation see Molas-Gallart in this book. Further, while the articulation of demand is covered in the conceptual discussion and typology, this aspect of demand-based innovation policy has been dealt with elsewhere in much more detail (e.g. Smits 2002; Moors et al. 2003).

3 There still is a lack of empirical data on the level of importance of demand for technological innovation, a fact highlighted as early as 1979 by Mowery & Rosenberg.

4 The author is very grateful to the participants of two PRIME workshops for very valuable comments, especially to Philip Shapira, who within the editorial team had the thankless task of specifically commenting on this chapter. Of course the usual disclaimer applies.

market signals. Overcoming these hindrances with the help of public policy is the core of demand-side innovation policy.

The second level relates to the political target orientation. Demand concepts support a certain set of policy targets that are based in social needs. Demand-oriented concepts are as a rule applied to attain societal, sectoral policy goals (e.g. sustainability, mobility) or to support market creation induced by economic policy, for example in the form of lead markets.

Innovation Hindrances at the Demand Side[5]

Why should policy intervene in the market dynamics at all? Two major reasons stand out. First, policy may intervene where the demand for innovations is insufficient or has not even started, but where a technology or product has a high potential benefit (Geroski 1990). Second, the articulation of demand may be insufficient, as human and social needs are not automatically translated into clear market demands (Mowery & Rosenberg 1979), and it is within the realm of public policy to turn needs into articulated demands.

1. As for the first of the two arguments, innovations that are technologically speaking 'mature' for the market do not achieve a breakthrough or do not spread quickly enough. No positive dynamic via increased quantities, decreasing manufacturing costs, reduced prices and increased demand kicks in. The reasons for this are manifold:
   a. *High entry costs*: At the beginning of the market cycle products or services naturally have higher entry costs. The first users of innovations, so-called 'launching customers' or 'lead users' must therefore be prepared to assume a large part of the manufacturer's development and learning costs (technically speaking, for various reasons, they have a low price elasticity).
   b. *Lack of information or awareness*: Potential customers for innovations may not be adequately informed about the benefit, safety or reliability of innovations, thus early users bear a greater risk. State actors themselves – i.e. political decision makers and implementing agencies – then act as a risk-taker, contribute to a better risk assessment on the part of the consumers (information, labelling etc.) or create safety expectations in the performance by passing high safety and performance standards and other regulations on product information (for the latter see Blind in this book).
   c. *Learning and adjustment costs*: Innovations necessitate learning and adjustment costs, not only for their own direct use but also for the complementarities with other products. Early users have higher costs to bear here too, as they assume

5 Chaminade & Edquist in this book give a detailed account of market failures; here I concentrate on mentioning the most important ones in terms of the demand side, as these are important to understand the relevance of policy.

the learning costs of the later users. The more early users gain experience with new products or services, the less the costs of the changeover for new users later. In this case the state actors can also become active and lower the transaction costs of adapting to new products, either by timely use of an innovation and demonstrating its use, or by setting financial incentives.

d. *Lock in-effects and path dependencies*: A radical change in technologies is often hampered by so-called technological trajectories. Technologies have built up their own application context; they link in with specialised infrastructures and complementary products. Users of well established and contextualised technologies are 'locked in'. To break out of these trajectories is very costly and demanding and needs strong impulses and incentives.

e. *Lack of network effects*: Products whose consumer benefit rises with the number of users (e.g. network effects in telecommunication) are initially less attractive at market entry and run the risk that the network effects will not occur. Such network effects are especially frequent in the ICT area.

2. The second argument, the insufficient transformation of needs into market demands, has two variants. (a) An *inadequate interaction between demanders and manufacturers*: Demanders with needs exceeding the technological state-of-the-art are important stimuli for the producers to generate and implement ideas in innovations. This means users must communicate their needs clearly to the potential producers, or collaborate with them on innovative solutions, thus raising the probability that innovations will emerge for which (b) a *promising market* exists. To make such relationships between producers and users ('user-producer interaction': von Hippel 1986; Gregersen 1992; Moors et al. 2003) possible and promote them, or to enter systematically into such interactions with producers can therefore contribute towards stimulating marketable – and widely acceptable – innovations, and also to accelerating innovation diffusion. The potential customers must signal their needs to the manufacturers, and need to have adequate competences to formulate these needs.

More fundamentally, demands are market signals, and 'in order to become meaningful in the marketplace, needs must be translated into demands expressed in and mediated through the marketplace' (Mowery & Rosenberg 1979).[6] Often, societal preferences are not transmitted sufficiently into signals to the market place, as, for example, certain groups do not exercise enough influence to be heard, and better articulation is called for (Smits 2002). This might mean that in certain

6 Mowery and Rosenberg have stressed the distinction between 'need' and 'demand' strongly, and have defined the latter as a function of preferences and income, denoting a systematic relationship between price and quantities (ibid.).

technologies there is a push into the market that does not reflect risks and fears attached to this technology.[7]

The Political Objectives of Activating Demand

Next to the unleashing of innovation dynamics through overcoming insufficiencies in the marketplace, there is a political dimension, i.e. a range of political objectives pursued by demand-oriented policies.

Growth and competitiveness

The first level is the economic one: like every innovation-oriented policy, the demand-oriented policies also aim for economic competitiveness and growth. The basic idea is that the demand for innovations is sufficiently advanced and large to induce manufacturers to invest in innovations.

Demand stimulation to reach societal goals

Ministries and agencies often pursue societal goals with the support for market introduction and diffusion of certain innovations, and they do so in many different ways across the various policy areas. One out of many examples of this is demand subsidies for photovoltaic in the area of environmental policy. The innovation lever of measures designed to meet societal targets derives from the fact that in general new needs are formulated in order to attain social goals, and innovative solutions are called for. It is a central, basic assumption of this chapter that the necessary demand-pull is integrated in the sectoral policy goal via challenging public demand (Gregersen 1992; Edquist 1994).

This close connection, namely to understand demand-oriented policy as innovation policy to achieve certain policy goals such as sustainability, energy efficiency, or mobility, is still being insufficiently examined in the literature and applied in policy practice. Although this has been repeatedly called for,[8] the implementation as an innovation policy is largely missing. The cases of the UK and Sweden below will show that this explicit connection has been consistently formulated and implemented. This means that effective approaches to demand-oriented innovation policy are usually found where sectoral policy goals are pursued in which market stimulation can be meaningfully utilised.

7 One case in point is innovation in functional genomics that raises fears with potential users that are not reflected in the traditional supply oriented mode of innovation generation and related policy (Moors et al. 2003).

8 One very visible example is the German Sustainability Council (e.g. Hauff 2003).

Policy Support for Lead Markets: an Integrated Approach

The demand for innovations or the support of market introduction or diffusion can – in a strategic approach – also be applied as an economic policy instrument, in order to create so-called lead markets. The concept of lead markets[9] assumes that in a certain market new technologies or services are introduced and that the design of these technologies or services is generally accepted as 'dominant design' in the global markets. This concept describes important general conditions for the acceptance and diffusion of innovations in the market. As a political concept it represents – ideally – the combination of the economic policy and the social policy approach and the connection of supply-oriented and demand-oriented measures. The discussion about lead markets summarises thus essential characteristics of ambitious, systematic demand-oriented policies. Many of the characteristics of lead markets were already highlighted in essence by Porter in his pioneering work *Competitive Advantage of Nations* (1990). Here, besides the factor endowment, the industrial structure and the firm strategy (competitive situation), the domestic demand plays a crucial role in the innovation dynamic of countries.[10]

To raise the potential of lead markets, one must thus be aware of the characteristics of such markets and of the policy opportunities to shape markets accordingly (Porter 1990; Beise et al. 2003; Jacob & Jänicke 2003; Meyer-Krahmer 2004). These characteristics include: high per capita income and/or low price elasticity, high tendency to (quick) acceptance of innovations, critical mass of demand, specific, innovation-driving problem pressure (or high significance of clear political goals),[11] good framework conditions for rapid learning and adaptation processes or suppliers, pioneering regulations, adequate technological and productive competence in the entire valued added chain and supporting services.

---

9 For a discussion of lead markets, see Meyer-Krahmer (1999; 2004) Meyer-Krahmer and Reger (1997); Beise (2001); as well as Beise et al. (2003).

10 We do not imply that demand conditions are the predominant dimension; we do not intend to argue which dimension is more important for innovative dynamics in innovation systems. The Porter approach, as many other approaches dealing with the importance of demand conditions and user integration for innovation, is based on the perspectives of companies and it is its merit to have the demand brought back in the discussion of national competitive conditions. On the macro-level it has been analysed that supply side conditions still are of somewhat higher importance than demand conditions. For very early evidence see Pavitt (1971), cited in Mowery & Rosenberg (1979).

11 Beise et al. (2003) show that the probability to release innovation dynamics into the market increases when policy goals are set and pursued, that are forward-looking and ambitious on a worldwide scale. The call to support innovations which have worldwide market potential naturally raises the question whether the state can recognise such innovations and trends (Beise 2001), whether it can 'back the right horse'. The connection to global trends and to global problem pressure is a fundamental condition here.

Consequently, a lead market policy is most likely to succeed if it combines supply-side measures with an appropriate configuration of these demand-side characteristics, whereby – ideally – the societal goal and the economic goal reciprocally reinforce each other.

A TYPOLOGY OF DEMAND-ORIENTED MEASURES

After the discussion of the patterns of reasons, the instruments, which governments can utilise in demand-side-oriented policies, can be typified.

Public Procurement[12]

Leverage of public procurement

The most direct form to stimulate innovations by demand-oriented policy is through public procurement itself. In Germany, the public purse invested €260 billion in products and services in 2003,[13] more than 12% of GDP. In Europe before the enlargement (EU 15) this share is even higher, at 16% (Georghiou et al. 2003; Wilkinson et al. 2005). In certain sectors – construction, public health and energy in public buildings – this public purchasing power constitutes the lion's share of demand.

This direct demand has – consciously or unconsciously – impacts on innovation behaviour. As early as the 1970s, empirical studies compared R&D subsidies and public procurement contracts without direct R&D procurement. They concluded that over longer time periods, public procurement triggered off greater innovation impulses in more areas than R&D subsidies (Rothwell & Zegveld 1981). The quantitative and qualitative meaning of public demand leads Geroski even to conclude that procurement policy 'is a far more efficient instrument to use in stimulating innovation than any of a wide range of frequently used R&D subsidies' (1990).[14]

The explanations why the ministries and public agencies at various state levels as demander generates or can generate innovations are manifold and can only be summarised here. Some of these reasons apply also to potent private demanders; others are restricted to public institutions as demander (Geroski 1990; Dalpé et al. 1992; Dalpé 1994). As already mentioned above, public demand as organised through ministries or agencies is frequently very 'demanding', necessitating

12 For a much more detailed account of public procurement as a means of innovation policy see Edler & Georghiou (2007).

13 Source: BMWA (Federal Ministry for the Economy and Labour, now BMWI, Federal Ministry for the Economy and Technology).

14 Geroski is one of the very few who have made an attempt to compare innovation and economic effects of supply-side and demand-oriented policy approaches.

		Societal Need	Private Need
	State (ministries, agencies...)	Direct public procurement	X
End-user	State and private	Cooperative procurement	X
	Private	Catalytic, state-induced procurement	Private procurement

7406

Figure 12.1 Matrix of procurement policy

innovative solutions to fulfil public tasks for society. This applies in the military and non-military area. Dalpé has shown, especially in research-intensive fields, public administrations are often more demanding than industrial or consumer demand. They act more often as 'lead users' for new innovations than private demanders (Dalpé et al. 1992). Second, in connection with political tasks or even missions, ministries or agencies – for reasons of fulfilling their public tasks – are often more willing or able to pay the higher entry price at the beginning of the life cycle of innovations. Further public demand may lead rapidly to a critical mass, in particular by bundling the needs of various government bodies. Such public demand creates clear incentives for manufacturers and reduces their market risk. Finally, government demand for innovative products additionally sends strong signals to the private users; the diffusion impulses are sometimes much stronger than those triggered by purely private demand.

Types of public procurement

There are various ways to utilise public procurement to spur innovation. Government procurement can in general be organised to promote the realisation of innovation, e.g. when 'innovation' is one essential criterion in the tender and assessment of tender documents. Procurement is strategic if the demand for certain technologies, products or services is embedded into sectoral policy and if it deliberately stimulates a market. Again, strategic procurement is also, as a rule, associated with sectoral policy and therefore largely neither initiated nor co-ordinated by the ministries responsible for innovation.

Further, public procurement may also be linked to private use of the goods purchased (Figure 12.1). So-called cooperative procurement is given when government agencies buy jointly with private demanders and both utilise the innovations. We talk of catalytic procurement when state actors are involved in the procurement or even initiate it, but the purchased innovations are in the last instance used exclusively by the private end-user.[15]

15 This distinction goes back to the theoretically founded and empirically productive work on the innovation-inducing procurement system, which was presented by a European team of analysts at the end of the 1990s. See Edquist & Hommen (2000).

Promotion of Private Demand

While public procurement has attracted more attention in recent years and is high on the European policy agenda at least, the support of private demand to spur innovation explicitly is still undervalued. Several forms of support can be distinguished.

Direct financial promotion of private demand

One obvious public instrument to spur private demand for innovative products is financial support measures. In the literature on innovation, these measures play a subordinate role, and the financial influence exercised by private demand is not a prominent topic in the present discussion about the demand orientation in relationship with the Lisbon Strategy (see above). Two types of financial support can be distinguished. Under certain conditions private or industry demanders can receive a subsidy or can lighten their tax load (tax relief). The pre-conditions for these measures to be applied are derived in most cases from societal needs and sectoral policy targets, very frequently for instance in the area of sustainable development or energy efficiency. A prominent example is the '100,000 Roof Programme' in Germany, where a financial incentive was provided to private house owners who installed a photovoltaic system on their roofs. As a rule, such instruments are utilised to support the diffusion of innovations.

Awareness building/competence building/information

Besides the monetary influence on demand, there is a 'soft steering' process which is geared to the willingness and ability to accept, demand and apply innovations. The potential of such measures has been highlighted through an empirical survey of more than 1,000 enterprises and 125 associations asking about the conditions for accepting innovations in the marketplace (BDL 2003). The following four hindrances to purchase and apply innovations were mentioned:

- The uncertainty of customers about security and quality of innovations (75% of the respondents);
- A lack of awareness about innovations and their functionality (67%);
- The high costs of the innovation, respectively of the associated products and services (62%);
- Clients' lack of ability to use the innovation (52%).

By contrast, the rejection of technologies for political reasons (37%) and the generally negative attitude towards new technologies (46%) were less prominent obstacles. Demand competence is one of the 'key(s) to diffusion of innovations' (Gatignon & Robertson 1985). There are a number of diverse government measures that serve to improve information with the demanders and to dismantle cognitive as well as psychological barriers to demand for innovative products and services, which are summarised in Table 12.1.

A more fundamental demand orientation is the organisation of articulation of needs, preferences, ideas and fears as to new technologies and innovations in general. The starting point here is the abovementioned need to articulate societal need and translate it into demand in the market place. In terms both of lack of user – producer interaction and lack of articulation of demand in general, state actors have a role to play. It is the role of organising, brokering role various means of discursive means and strategic intelligence, such as constructive technology assessment or interactive/participatory technology assessment (Smits 2002; Moors et al. 2003).[16]

Regulations

In general, very little consideration is given to the relationship between innovation and regulation yet. Where systematic studies are being carried out, usually those regulations, which directly affect the manufacturers are discussed (so-called 'command and control' regulations; Blind et al. 2004; Blind in this book). Such regulations are as a rule the ones that influence production (ingredients/components, materials, processes etc.), performance (quality, compatibility) or consequences (health, safety, the environment) of products or services (e.g. recycling regulations, emission standards, etc.). All such regulations, however, directly affect the demand behaviour, for they reduce the information costs of the demanders and relieve them of appraising risks.[17]

Integrated approaches

A last category of measures are those which combine various instruments. Two different approaches can be distinguished here. There are strategic approaches, which are limited to the combination of different demand-side instruments, such as cooperative procurement, awareness building, demonstration projects and further education. The goal here is the acceleration of the discussion and establishment of new markets. The Swedish example discussed below of strategic demand for energy-efficient products is a typical example of such an integrated measure.

An even more comprehensive form is the approaches which combine demand-side and supply-side mechanisms. State actors design a policy for a selected technology, whereby the selection could be assisted by discursive strategic intelligence. The policy ensures not only the necessary factor endowment (R&D promotion), but also creates favourable demand conditions (quantity and quality), ideally aiming for dominant designs to be diffused into international markets (Lead Market concept, see above).

The demand-oriented approaches discussed here are summarised in Table 12.2.

16 For an overview of papers putting the construction and articulation of demand into focus, see Smits (2002).

17 For further details on regulations and their meaning for the uptake of innovation see Blind, in this book.

Table 12.1 Typology of demand-oriented measures

Instrument	Role of State	Method of Functioning
1. Public demand		
General procurement	Buy and use	State actors consider innovation in general procurement as main criterion (e.g. definition of needs, not products, in tenders)
Strategic procurement (technology-specific)	Buy and use	State actors specifically demand an already existing innovation in order to accelerate the market introduction and particularly the diffusion. This can include the targeted coordination of different government bodies and moderation with manufacturers.
		State actors stimulate deliberately the development and market introduction of innovations by formulating new, demanding needs. This can include the targeted coordination of different government bodies and moderation with manufacturers.
Co-operative procurement	Buy/use moderation	State actors are part of a group of demanders and organises the co-ordination of the procurement and the specification of needs. Special form: catalytic procurement: the state does not utilise the innovation itself, but organises only the private procurement
2. Support for private demand		
Direct support for private demand		
Demand subsidies	Co-financing	The purchase of innovative technologies by private or industrial demanders is directly subsidised
Tax incentives	Co-financing	Amortisation possibilities for certain innovative technologies
Indirect support for private demand: information and enabling (soft steering)		
Awareness building measures	Informing	State actors start information campaigns, advertises new solutions, conducts demonstration projects (or supports them) and tries to create confidence in certain innovations (in the general public, opinion leaders, certain target groups)
Voluntary labels or information campaigns	Supporting Informing	The state supports a coordinated private marketing activity which signals performance and safety features.
Training and further education	Enabling	The private consumers or industrial actors are made aware of innovative possibilities and simultaneously placed in a position to use them.
Articulation	Organising discourse	Societal groups, potential consumers are given voice in the market place, signals as to future preferences (and fears) are articulated and signalled to the marketplace.

EMPIRICAL ILLUSTRATIONS

The following section discusses a set of current or recent demand-oriented initiatives. This not only serves the purpose to signal the use of DBIP and to highlight a renewed interest in demand-based innovation policy. Rather, it also serves as a basis to understand governance challenges and governance principles when innovation policy is based on demand. These will subsequently be summarised in the last chapter in light of the trilateral 'dance' of policy practice, innovation practice and strategic intelligence.

Overview

A recent systematic overview of six countries (the UK, the USA, France, Sweden, Germany and the Netherlands) has shown that demand orientation is not yet an explicit and systemic element in national innovation policies (Edler 2007a). Despite the enormous market potential of public procurement and despite the overwhelming and obvious need for a change in consumption and production patterns to be supported by policy (e.g. towards goals such as sustainability and quality of life) there is no systematic usage of demand to spur innovation. To be sure, we find numbers of demand-oriented measures, such as tax breaks, demand subsidies etc., in all sectoral policies, but these measures are not integrated into innovation policy, and their usage is still somewhat limited (Edler et al. 2005). Similarly, measures of demand articulation appear to becoming more important, but in most countries these approaches are still rather limited.

In the last few years, the discussion about the role of the state in utilising demand for the production and diffusion of innovations appears to be developing a new dynamic at the European level. Various EU activities within the framework of the Lisbon strategy represent a catalyst to the international discussion.[18] It appears that public procurement, and more recently the Lead Market concept, is regarded as a means to unleash an upward spiral of demand, innovation activity and supply. In the Action Plan to implement the Lisbon Strategy, public procurement is explicitly named as an innovation policy instrument.[19] Further, there are efforts to systematically pave the way for Lead Markets within Europe activating demand and supply measures around certain technologies that promise to be supportive in attaining societal goals (European Commission 2006; European Council 2006). Subsequently, under the leadership of the Directorate General (DG) 'Enterprise' five

18 The EU Lisbon strategy aims to make Europe the most competitive region of the world and has led to a number of different catalogues of measures (http://europa.eu.int/comm/lisbon_strategy/index_de. html). One of the action lines is public procurement (http://europa.eu.int/comm/research/era/3pct/3pct-publicprocurement.html).

19 The new EU Directive on Procurement (2004/17/EC, 31.03.2004) foresees corresponding adaptations of the process.

Interservice Groups for Lead Markets were set up in early 2007 (eHealth, Sustainable construction, technical textiles, bio-based products and recycling). These aim at a broad, coordinated initiative combining the efforts of sectoral DGs with the expertise and competencies of DG Research and DG Enterprise – with demand measures playing a key role. Further, DG Enterprise has compiled a 'Handbook on Innovation Procurement' (European Commission 2007).

Within Europe, the first explicit approaches were made in the 1990s in some countries to utilise public procurement and to promote private procurement to create new markets and diffuse innovations. A study – again commissioned by the EU (Edquist et al. 2000) – provided a first systematic presentation of such cases. This study, however, also made clear that the various activities discussed – with the exception of Sweden (below) – in generally were not systematically designed and were not fed by an innovation-oriented procurement strategy. In addition, the measures were frequently restricted to direct public procurement, without realising the potentials of private demand.

However, also at the level of European countries, public demand has been rediscovered, at least as a potential tool for innovation, and again this rediscovery goes hand-in-hand with different kinds of analytical preparation. For example, the Irish government commissioned a study of the potentials of public procurement measures (Wilkinson et al. 2005). In the Netherlands, an internal group of government experts has defined the potential of public procurement for innovation policy. According to an action plan developed by the so-called 'Innovation Platform', the Dutch public procurers shall use at least 2.5% of their total procurement budget for properties or services not yet available to the market from 2006 on (special incentives). In addition, the professionalisation of public procurers plays a growing role within the political strategies. This includes the active diffusion of knowledge about the opportunities afforded by this legal frame amongst the diverse procuring agencies and a networking and actual training of public procurers, using examples of good practice. All this is organised by a Dutch project unit on public procurement and tendering (PIA) (Edler et al. 2005; Ebersberger 2007).

In Germany a so-called 'Impulse Circle Innovation Factor State' had been working on the possibilities of promoting innovation dynamics from the market place by means of a new adjustment of procurement in general, and strategic procurement measures in selected technology areas. This cycle has issued a booklet for strategic decision-makers and procurers alike in order to raise awareness across the administrations at all levels (Blind 2006). Within the Ministry of Energy, a 'New Deal' has been proposed (BMU 2006) that seeks to exploit the demand for eco-efficient technologies as a means for a strategy to combine sustainability and innovation at the same time (normative goal and innovation). The High-Tech Strategy of the country pledges to use public procurement to take up innovations in selected areas more systematically. Concrete steps to implement this dimension of the High-Tech Strategy still have to be taken (BMBF 2006). Before that, the ministry had tried to organise a future demand articulation with the so-called Futur process

(BMBF 2003; Edler et al. 2006). The Federal Ministry of Research and Education (BMBF) had organised a large-scale discourse in order to define 'Lead Visions' for the future that would then be translated into R&D programme priorities. This was one of the rare examples of a strictly demand-oriented foresight process, which, however, served to be overly challenging and did not fully result in the translation into research priorities. This again demonstrated how challenging it is to let demand articulation find its way into well-established R&D priority-setting mechanisms. Overall, all these attempts fall short of any structural adjustments, coordinations between ministries and a linking of policy approaches and related communities are not foreseen for the time being.

Compared to Europe and European countries, the US appears to have traditionally followed a very different path of demand-based innovation policy. At the Federal level there are no explicit strategies to integrate demand with innovation policy systematically and explicit innovation policy is largely focused on the supply side (Popper & Wagner 2002; CoC 2004; PCAST 2004). Nevertheless it is obvious that public demand has driven technological development to a large degree. First, defence – and recently security – procurement has had a key role the development for many technologies (GPS, Internet etc., see also James 2004). Second, there is a strong element of demand-orientation inbuilt into the mission-oriented research and technology policy approach in the US, and many policy measures across a variety of agencies in fact utilise innovation policy instruments (Breloh 2000). Large research programmes are defined and implemented in sectoral departments and agencies rather than in specialised research policy administrations as in most European countries. This links the mission of an agency, the societal goals it seeks to achieve, more directly to research priorities and support (Edler 2007b). In that sense the linking of (future) demand to the US policy in the area of innovation and research is systematically stronger than most European policies. However, this is again an indirect link, as – in general – it does not entail strong and systematic processes of demand articulation or stimulation of private demand, but rather derives policy priorities from intra-administrative priority setting.

Some Good Practice

A fairly successful example of a DBIP approach can be found in Sweden, where one of the largest systematic set of demand-based, targeted initiatives has been implemented to boost the production, improvement and diffusion of energy efficiency technologies. These initiatives aimed at transforming markets and combined for each of the various technologies a mix of various measures, mainly public procurement in combination with support of private demand (Neji 1998; Suvilehto & Överholm 1998).[20] The basic idea was that public procurement acts as a

20 Technologies covered were, for example, lighting systems, washing machines and dryers, heating systems, insulation systems etc.

catalyst for private procurement; the responsible agencies NUTEK and later STEM[21] evoked and bundled demand and organised large tender processes. In some cases, the agency simply organised private demand (catalytic procurement), in other cases public agencies procured the technologies for their own applications (co-operative procurement). The starting point of each programme was to acknowledge the need to transform markets along a more sustainable route. Subsequently a concrete technology with a high energy efficiency potential was defined through mobilising potential purchasers – public and private – and establishing a discourse with actors of the whole value and demand chain (including producers, wholesalers, craftsmen etc.). The agency complemented this complex, cooperative procurement activity with a set of marketing and support measures to raise awareness and to enable users, each tailored towards the peculiarities of each market (Suvilehto 1997; Suvilehto & Ofverholm 1998). In addition, for brand new innovations, a demand subsidy was granted in order to accelerate the process.

According to various evaluations this policy, by and large, achieved market transformation (in terms of types of products, criteria of purchasers) in many of the markets tackled (NUTEK 1994; Neji 1998; Interview NUTEK). In addition, the various awareness measures have resulted in a broader understanding of the meaning and availability of energy efficient technologies in Sweden.

In a more recent generation of such programmes in Sweden, the agency put more weight on the discourse, the demand articulation, through establishing user groups. These meet regularly, discuss opportunities and needs for technological improvement, and send signals to the producers and the state agency. The discourse in these groups may result in new and modified calls for tender that explicitly call for innovations. This discourse is often assisted by technology and market intelligence, investigating the potential for technological improvements as well as the readiness of Swedish producers to deliver and the scale of demand. All this discourse and intelligence then feeds the final decision to issue a call for tender. This activity is thus one form of organising articulation (Smits 2002), which relies mainly on resources within the specialised agency rather than activating external strategic intelligence. Moreover, the tenders in the STEM approach now generally contain a provision for technological innovation; the mere diffusion of existing efficient technologies is not aimed for. Again, depending on the degree of innovativeness, the winner of a tender may be further supported with a co-funding of demonstration and test installations. The first experiences with this approach have shown that not only one producer – the winner of the tender – but the whole sector is driven towards technological innovations as the competition is also technologically upgraded.[22]

---

21 The responsible agency was the former energy agency NUTEK, which is now responsible for industrial policy, regional development and enterprises; the new energy agency is STEM.

22 This is not yet based on sound evaluations, but on anecdotal evidence gathered in interviews with STEM.

A second example of good governmental practice in terms of cross-administrative discourse and mobilisation is the United Kingdom. In its innovation report 2003, demand orientation had become an explicit and integral part of innovation policy of the British Department for Trade and Industry (DTI 2004). The UK strategy was complex, it aimed to improve general procurement (efficiency and innovations) and strategic procurement (specifically targeted towards innovations), to integrate the local level (Office of the Deputy Prime Minister 2003), to consider the specific needs of SMEs, to spread across all ministries and seek to re-orient sectoral policies and procurement therein towards inducing more innovation. The overall aim was to mobilise 25% of public purchasing for innovation-oriented procurement. The guiding idea was that the improvement of public service goes hand-in-hand with innovations to be purchased through the public purse.

The DTI together with the Office of Government Commerce – the central agency organising public procurement at the national level – had started to implement the strategy according to the so-called Kelly Report, that provided a detailed, cross-governmental roadmap (DTI/OGC 2003). The OGC provided support with market studies and with activities to train decision-makers and procurers (internal strategic intelligence). State actors should become 'intelligent customers'. One further characteristic of the approach was horizontal coordination. Various interministerial working groups and overall coordination at the highest levels of ministries (DTI 2004) were established to define future needs. First pilot implementations have been launched, such as ProCure 21 in the health sector, a case study in Telecare and some broader activities in the area of sustainable procurement (co-ordinated by DEFRA[23]). The major characteristic of all the pilot applications has been that within them the administration tried to define public demand and industry capacities in an interactive and systematic way, mainly through heterogeneous working groups representing the most important stakeholders, and assisted by market studies. The overall success of the concept cannot, of course, be judged at this stage. Although the first phase of this strategy appeared promising, it remains to be seen whether the ministries will continue to co-ordinate themselves, and make the effort to decide on their mid- to long-term procurement strategies.[24]

23 Department for Environment, Food and Rural Affairs.

24 The author has conducted a number of interviews within the DTI and with DEFRA to assess the implementation of the strategy. However, the full implementation cannot be judged, especially as it is not obvious what the priorities of the responsible ministries are since the change from the Blair to the Brown administration in June 2007.

THE GOVERNANCE CHALLENGES OF DEMAND-ORIENTED POLICIES

In light of the conceptual variety opened up in the first and second section, and the illustration given in the third section, what systematic lessons do we derive for the understanding of the policymaking process and the implementation of DBIP? The major challenges of DBIP are on three levels: coordination, incentives and strategic intelligence, and all three are interrelated.

Coordination Challenge

First of all, policymaking in DBIP is more dispersed and needs more coordination than in traditional supply-oriented innovation policy. Concepts and instruments are designed and applied by a larger and more heterogeneous set of state actors at different policy levels.

The first dimension of heterogeneity is the hierarchical level and autonomy of public actors. We may differentiate the 'state' into decision maker at the political level on the one hand and operational policy designers and implementers in ministries and agencies on the other hand. The former may be national or regional ministries, European Commission (e.g. for regulation) as well as city councils at the local level. The latter are public actors that – within a given policy framework – design and implement policy on an operational basis. While this differentiation is true for most kinds of policies, demand-oriented policies face the challenge that both policy levels need a deep knowledge on the leverage, entry points and hindrances in terms of stimulating demand. Decisions must be taken close to public and private users and be informed about preferences, habits and aspirations, and thus the communication between the operational and the decision-making level is key for successful policymaking in DBIP (vertical coordination).

A second dimension to differentiate public actors for our purposes here is their specific remit. We distinguish between (1) sectoral policy actors, i.e. ministries and agencies, in the area of health, transport etc., whose activities may have a strong bearing on innovation activities; (2) public actors responsible for innovation policy and its implementation in the broadest sense; and finally (3) those that facilitate public activities within administration to organise demand operationally – a major example being procurement agencies and departments.

The coordination needs between these three types of public actors are challenging, and most often not sufficiently realised. A principal bottleneck to a large impact of demand-oriented innovation policy is the lack of awareness and readiness in sectoral ministries. Demand orientation has the highest leverage when combined with sectoral policy goals. Hence, there is a need to mobilise sectoral ministries and agencies for the innovation dimension. This calls for a combination of sectoral policy aims and innovation policy (and indeed research policy by means of supply support) and thus for inter-departmental coordination between innovation policy, sectoral policy and – in the case of public procurement – specialised procurement agencies at various levels, in short: sound horizontal policy coordination (see also Edler et al. 2003).

The ideas brought forward in some German approaches (Ministry of Energy with its New Deal, Ministry of Economics with its idea to trigger of public procurement for innovation) seem to suffer – at least until recently – from a lack of inter-ministerial coordination. Equally, the above mentioned Interservice Groups by the European Commission, while being a step in that direction, struggle with establishing a mutual understanding between the services. The concept of the UK strategy for public procurement and innovation focused on inter-ministerial and inter-agency coordination took care of the coordination requirements. Horizontal (inter-ministerial, inter-agency) and vertical working groups provided for coordination to build up converging expectations and foster mutual understanding and trust within government. This multi-dimensional coordination necessitates strong leadership and strong process management to keep up the momentum across governmental units. The starting point for strong coordination, however, appears to be an explicit vision that link different levels and sectors of government as well as signals the demand-based approach to society and markets – as was done in the UK (public procurement) and Sweden (stimulation of private demand).

Incentive Challenges

A major drawback to designing demand orientation as a tool for innovation policy has to do with wrong incentive structures. For example, incentives of innovation policy and sectoral policies do not naturally match. One obvious – and often neglected – challenge is related to the locus of value added on the side of the suppliers. The simple question is: Who benefits? Here, different policy administrations have different rationales and interests. Innovation policy is part of industrial policy, one way or the other, and the economic effect in industry is the prime criterion. Public authorities that implement demand measures, e.g. in procuring the latest technology for an agency, might be faced with a supplier structure that does not lead to added value in a region or country, but elsewhere. While the innovation would still be beneficial for a certain sectoral goal and for the procuring agency and its customers, the local (regional/national) producers would not benefit directly. The spatial distribution of effects on the supplier will thus always be an important element if demand-oriented policy is supposed to be an integral element of innovation policy.

There are means to ease this problem, and these have to do with dialogue and awareness. First, intensive and forward-looking user-producer interaction, with active participation of local, regional and national producers (depending on the policy level of a measure), can establish a co-construction of needs and innovation, increasing the likelihood of a match between future demand and its satisfaction. Such a discourse tends to have an in-built bias towards the producers near to the users.

Furthermore, the benefit for suppliers is not confined to selling a given product. Installation, maintenance and other accompanying services are upgraded and then realised directly for the user and thus evoke added value in local environments (local content), even if the initial innovation is produced and delivered elsewhere. Finally, an indirect effect can be realised as local producers are challenged through the increased technological competition, evoking an upscale of their own technological efforts and combining those with the advantages of being close to the users.

Public procurement for innovation poses further specific incentive challenges.[25] The incentive structures at the various levels of public procurement are adverse to introducing innovation into the procurement decisions. Public actors are largely risk averse, as they have to deliver their policies financed by the public purse and they need to justify their expenses and are asked for efficiency within their annual budget cycles. To procure innovation implies risk as innovations may fail and it often implies higher entry costs and additional switching costs (e.g. when leaving a technological trajectory). The net benefit of a procurement decision is often only realised in the longer term, and life cycle cost-benefit ratios are complex and hard to justify. Efficiency pressures within the procurement system may let the efficiency rationale prevail over consideration of long-term innovation.

Moreover, interaction with suppliers is called for, to monitor markets and to define the tender text. However, this has some legal limits, as no market actor is supposed to reap any advantage through this interaction. In addition to the need for expertise on technologies and markets (as with demand orientation in general) procurement policy also needs a high legal expertise on procurement regulations at the various levels of decision making.[26] All this thus puts an additional burden on public policymakers and implementers that is not matched by additional incentives.

Intelligence Challenge

All empirical examples of DBIP approaches highlight the need to build up the knowledge and informed discourse across administrations and between the various state actors and innovation practice (users, producers). Sophisticated intelligence must be built into the process at various levels, whereby policy practice and innovation practice are informed, enabled and connected through strategic intelligence. For DIBP, the challenges to build in strategic intelligence lie on three dimensions, whereby strategic intelligence takes on various roles: conceptual,

25 For a detailed discussion of these challenges see Wilkinson et al. (2005) and Edler & Georghiou (2007).

26 A considerable number of interviews with procurement officers and policy decisionmakers at various levels across Europe showed that one of the highest obstacles to procure innovations lies in the subjective insecurity as regards legal provisions, especially regarding the need for interaction with the market.

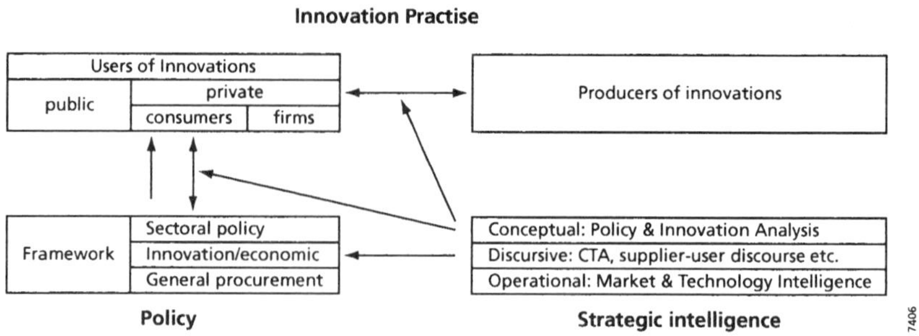

Figure 12.2 Policy practice, innovation practice and strategic intelligence in DBIP

discursive and operational.[27] A simple model of the triangular relationship between policy practice, innovation practice and strategic intelligence, delivering analyses and intellectual support illustrated these relationships in DBIP (Figure 12.2).

1. The basic function of strategic intelligence in DBIP is conceptual. The more strategic approach in the UK started with a set of conceptual reports on the needs and preconditions for demand-based (in this case public procurement based) innovation policy. All the European initiatives are based on enormous strategic intelligence and external advice. Both the so-called Kok Report revising the Lisbon Strategy (Kok et al. 2004) and the Aho report (Aho et al. 2005) on a more innovative Europe stress the importance of public procurement for innovation. Further, the Commission has called for various expert groups and studies to deal with the issue of (public) demand to spur innovation, be it in the context of a more general overview of promising policy measures (IPTS 2003; Georghiou et al. 2003), or in the more focused context of public procurement (Edler et al. 2005; Wilkinson et al. 2005). At the time of drafting this chapter, the DG Research is launching a study to be made on Lead Market potentials and ways of setting up Lead Market strategies combining demand goals and supply requirements. All these recent studies do not only signal new interest, but also the need to understand the ways and – more principally – means to mobilise demand for innovation in general and a sound understanding of the market or system failure that hinders an innovation taking place and being accepted in the market. If this mobilisation is to take place, conceptual knowledge is needed to frame this new – or renewed – policy. Such a conceptual strategic intelligence is the more important the more radical the new approach appears and the more it deviates from established rationales.
2. A further major function of strategic intelligence is discursive, i.e. to foster the organisation of demand articulation and bundling, supporting the definition of

27 In addition, within all of these functions effective policymaking needs – as all policymaking in fact – learning cycles to monitor appropriateness, gaps and effects at various levels (evaluative strategic intelligence).

societal goals and translating them into potential policy goals. One of the key challenges of demand-oriented measures that target a specific technology or market is the overwhelming complexity and thus the need for sound strategic intelligence and the formation of capabilities in administrations. First, how can it be ensured that the targeted innovation meets the preferences of users – and is responsive to future aspirations and fears? And why is support needed in the first place? Any decision for a targeted technology could miss the long-term preferences of users and could try to spur the private market but simply not succeed without, for example, permanent subsidies. This problem is less significant in public procurement, where public actors as users can define preferences and application of an innovation. Here, internal strategic intelligence units are often created that take care of awareness building, training and dicourse in administrations and also organise strategic discourse with user groups. As soon as measures target private users or aspire to take into account future needs, though, the match of user preferences and market responsiveness needs to be known in advance – or constructed. This means that for targeted measures ideally a systematic bottom-up approach is applied, where societal goals and the long-term expectations of users predefine a demand-oriented strategy. In doing so, demand-oriented policy is more responsive, i.e. it better reflects societal needs, meets long-term readiness to adopt innovations, and sends appropriate signals to the producers. To enhance responsiveness, strategic intelligence would thus have to include constructive technology assessment, user-driven foresight, user-producer interaction and the like (Smits 2002), as it is done within the more recent approaches of the Swedish agency STEM (see above).

3. Lastly, sophisticated strategic intelligence is needed on the operational level, in order to gain sufficient market knowledge (about the whole value and demand chains,[28] producer capabilities and readiness) and technological expertise. This is far from trivial. Policy-makers and administrators have to take concrete decisions in the market place, they have to know current and future technological options as well as the related cost-benefit ration, the technological interfaces and the switching costs. Even if an innovation has been defined as supporting societal goals and even if a system or market failure is obvious, it is still not trivial to define the right technology or type of product. While societal criteria for technologies (more efficient, more sustainable etc.) may help to single out a certain solution to support through demand-oriented measures, this technology may still be premature, and thus in the long term inefficient. If demand is triggered for a premature technology, prices may be far too high and – more problematic – the pressure on producers to invest in a further improvement of the technology might be eased, technological trajectories might be defined on a sub-optimal technological basis. Strategic intelligence is called for to provide

28 This takes into account that many important decisions in demand are not taken by the end user but by installation personnel, wholesalers and the like.

a sound assessment on the state-of-the-art of a technology as regards its market maturity.

It is obvious that the three functions of strategic intelligence are interrelated, and all three forms of strategic intelligence are important at all policy level. Conceptual strategic intelligence is a prerequisite for awareness building and underpins discourse and articulation, and operational intelligence is most effective when applied in a discursive manner.

CONCLUSION

The latest discussion on public procurement, mainly at the EU level, has brought demand considerations back into the discussion on innovation policy. The discussion in this chapter tried to give an overview of demand-oriented measures, provide some conceptualistic background and discuss major challenges. Given the variety of approaches, this had to be sketchy, abridged and could maybe not cut deep enough.

However, within the recent revival of activating public procurement for innovation, of fostering lead markets and using demand for a 'New Deal' in ecologically driven industrial policy and the like, one message is clear: there is a whole set of complex and interrelated preconditions for successful demand-based innovation policy. The chapter has shown that in order to realise the innovation potential through public demand or through the mobilisation of private demand, a high level of commitment and a consequent application of new forms of governance are called for. These must systematically include the application of strategic intelligence at different levels and in different forms. They must build in horizontal and vertical networking and coordination of ministries and agencies; they must find ways of strong interactions both with (potential) users and with suppliers; and they must provide for appropriate incentive structures within public administrations. They must further link demand with supply measures whenever necessary for a next generation of technologies. Above all, the selection of certain technologies needs a strong normative base (in terms of the underlying policy goal), and thus must be based on a strong articulation by potential users and potential producers and – as far as needed – strong awareness building. If this latter requirement is met, DBIP is neither a politically induced nor an illegitimate interference in the marketplace; rather it is the application of the foremost function of policy in the first place: making choices to fulfil societal goals. It appears that all too often these prerequisites are not met, and high ambitions in activating large-scale demand for innovations are shattered by lack of vision, commitment, coordination and information.

REFERENCES

Aho, E., et al. (2006), 'Creating an innovative Europe', report of the Independent Expert Group on R&D and Innovation appointed following the Hampton Court Summit. Rapporteur: Luke Georghiou, January 2006.

BDL (2003), 'The power of customers to drive innovation', report to the European Commission, Brussels.

Beise, M. (2001), *Lead Markets. Country Specific Success Factors of the Global Diffusion of Innovations*, Heidelberg and Berlin: Physica-Verlag.

Beise, M., et al. (2003), 'The emergence of lead markets for environmental innovations', FFU report, 02-2003, Berlin.

Blind, K. (2006), 'Regulation in nachfrageorientierte Innovationspolitik (Regulation in demand-oriented innovation policy)', in J. Edler (ed.), *Nachfrageorientierte Innovationspolitik: Benchmarking-Studie im Auftrag des Deutschen Bundestag*, (Demand-oriented Innovation Policy. Benchmarking Study on behalf of the German Parliament), Karlsruhe and Berlin: pp. 277-289.

Blind, K., et al. (2004), 'New products and services. Analysis of regulations shaping new markets', www.cordis.lu/innovation-policy/studies/gen_study11.htm.

BMBF (2006), 'The High Tech Strategy for Germany', Bonn and Berlin, www.bmbf.de/pub/bmbf_hts_en_kurz.pdf.

BMBF (2003), 'Futur. Der deutsche Forschungsdialog', Bonn. www.bmbf.de/pub/futur_eine_erste_bilanz.pdf.

BMU (Federal Ministry for the Environment, Nature Conservation and Nuclear Safety) (2006), 'Memorandum for a New Deal for the economy, environment and employment', Federal Ministry for the Environment, Nature Conservation and Nuclear Safety, Berlin.

Bozeman, B. (2000), 'Technology transfer and public policy: a review of research and theory', *Research Policy*, **29** (4-5), 627-655.

Bradke, H., Cremer, C., Dreher, C., et al. (2007), 'Developing an assessment framework to improve the efficiency of R&D and the market diffusion of energy technologies', report for the German Federal Ministry of Economics and Technology, Karlsruhe.

Breloh, P. (2000), 'Staatliche Förderung industrieller Innovation in der USA. Eine empirisch fundierte Analyse', Wiesbaden: Darmstadt Technischer Universität.

CoC (Council on Competitiveness) (2004), 'Innovate America. National innovation initiative report thriving in a world of challenge and Change', Washington, DC.

Dalpé, R. (1994), 'Effects of government procurement on industrial innovation', *Technology in Society*, **16** (1), 65-83.

Dalpé, R., DeBresson, C. and Ciaoping, H. (1992), 'The public sector as first user of innovations,' *Research Policy*, **21** (3), 251-263.

DTI (Department of Trade and Industry) (2004), 'Innovation report: competing in the global economy, the innovation challenge', December 2004, London.

DTI/OGC (Department of Trade and Industry/Office of Government Commerce) (2003), 'Increasing competition and improving long-term capacity planning in the government market place', December 2003, London.

Ebersberger, B. (2007), 'Nachfrageorientierte Innovationspolitik in den Niederlanden' (Demand-oriented inovation policy in the Netherlands), in J. Edler (ed.), *Nachfrageorientierte Innovationspolitik: Benchmarking-Studie im Auftrag des Deutschen Bundestag* (Demand-oriented Innovation Policy. Benchmarking Study on behalf of the German Parliament), Karlsruhe and Berlin: pp. 84-96.

Edler, J. (ed.) (2007a), *Nachfrageorientierte Innovationspolitik: Benchmarking-Studie im Auftrag des Deutschen Bundestag* (Demand-oriented Innovation Policy. Benchmarking Study on behalf of the German Parliament), Karlsruhe and Berlin.

Edler, J. (2007b), 'Nachfrageorientierte Innovationspolitik in den USA' (Demand-oriented inovation policy in the USA), in J. Edler (ed.) (2007a), *Nachfrageorientierte Innovationspolitik: Benchmarking-Studie im Auftrag des Deutschen Bundestag* (Demand-oriented Innovation Policy. Benchmarking Study on behalf of the German Parliament), Karlsruhe and Berlin: pp. 130-150.

Edler, J. and Georghiou, L. (2007), 'Public procurement and innovation: resurrecting the demand side', *Research Policy*, **36** (7), 949-963.

Edler, J., Hommen, L., Papadokou, M., et al. (2005), 'Innovation and public procurement. review of issues at stake', Study for the European Commission, final report.

Edler, J., Joly, P., Kuhlmann, S., et al. (2006), 'Understanding for a of strategic intelligence for research and innovation', The PRIME Forum Research Project, Final report, Karlsruhe.

Edler, J., Kuhlmann, S. and Smits, R. (2003), 'New governance for innovation. The need for horizontal and systematic policy coordination', *Fraunhofer ISI*, **31** S. (2/2003) (Fraunhofer ISI Discussion Papers Innovation System and Policy Analysis No. 2/2003), Online, www.isi.fhg.de/publ/downloads/isi03a04/new-governance.pdf (318 KB) (ISI-A-4-03), Karlsruhe.

Edquist, C. (1994), 'Technology policy: the interaction between governments and markets', in G. Aichholzer and G. Schienstock (eds), *Technology Policy. Towards and Integration of Social and Ecological Concerns*, Berlin, Germany and New York, USA: Walter de Gruyter; pp. 67-94.

Edquist, C. and Hommen, L. (2000), 'Public technology procurement and innovation theory.' in C. Edquist, et al. (eds), *Public Technology Procurement and Innovation*, Dordrecht, The Netherlands, Amsterdam, The Netherlands, Boston, USA and London, UK: Kluwer, pp. 5-70.

Edquist, C., Hommen, L. and Tsipuri, L. (2000), *Public Technology Procurement and Innovation*, Dordrecht, The Netherlands, Amsterdam, The Netherlands, Boston, USA and London, UK: Kluwer.

European Commission (2006), 'Putting knowledge into practice, a broad-based innovation strategy for the EU', COM (2006) 502 final, 13.9.2006, Brussels.

European Commission (2007), 'Guide on dealing with innovative solutions in public procurement. 10 elements of good practice', SEC (2007) 280, Brussels.

European Council (2006), 'Decision of establishing a competitiveness and innovation framework programme (2007-2013)', 26.09.2006, Brussels.

Gatignon, H. and Robertson, T.S. (1985), 'A propositional inventory for new diffusion research', *Journal of Consumer Research*, 11 (4), 849.

Georghiou, L. (2007), 'Demanding Innovation Lead markets, public procurement and innovation', NESTA Provocation 02, February, London. www.nesta.org.uk/assets/pdf/demandinginnovation_provocation_NESTA.pdf.

Georghiou, L., Amanatidou, E., Belitz, H., et al. (2003), 'Raising EU R&D intensity. Improving the effectiveness of public support mechanisms for private sector research and development, direct measures report of an ETAN expert group', 30.04.2003, Brussels.

Geroski, P. (1990), 'Procurement policy as a tool of industrial policy', *International Review of Applied Economics*, 4 (2), 182-198.

Gregersen, B. (1992), 'The public sector as a pacer in national systems of innovation', in B.-Å. Lundvall (ed.), *National Systems of Innovation. Towards a Theory of Innovation and Interactive Learning*, London, UK and New York, USA: Pinter, pp. 129-145.

Hauff, V. (2003), 'Nachhaltigkeit als Formel für politische Gestaltung', Rede auf dem Jahreskongress des Rates für nachhaltige Entwicklung, 1.10.2003.

Hekkert, M., Suurs, R., Negro, S., Kuhlmann, S. and Smits. R. (2007), 'Functions of innovation systems. A new approach for analysing technological change', *Technological Forecasting and Social Change*, 74 (4), 413-432.

Hippel, E. von (1986), 'Lead users, a source of novel product concepts', *Management Science*, 32 (7), 791-805.

IPTS (2003), 'Public procurement and R&D. A JRC/IPTS-ESTO fast track working paper', March 2003.

Jacob, K. and Jänicke, M. (2003), 'Leadmärkte für Umweltinnovationen', *Politische Ökologie*, 21 (August 2003), 19-21.

James, A. (2004): 'U.S. defence R&D spending: an analysis of the impacts', reporter's report for the EURAB Working Group ERA Scope and Vision. EURAB 04.011, Manchester. www.europa.eu.int/comm/research/eurab/pdf/recommendations10.pdf

Kok, W., et al. (2004), 'Facing the challenge. The Lisbon strategy for growth and employment', report from a High Level Group, November 2004, Luxembourg.

Lundvall, B.-Å. (ed.) (1992), *National Systems of Innovation. Towards a Theory of Innovation and Interactive Learning*, London, UK and New York, USA: Pinter.

Lundvall, B.-Å. (1988), 'Technological change and economic theory: innovation as an interactive process. From user-producer interaction to the national system of innovation', in G. Dosi, et al. (eds), *Technical Change and Economic Theory*, London, UK and New York, USA: Pinter, pp. 349-369.

Meyer-Krahmer, F. (2004), 'Vorreiter-märkte und innovation. Ein neuer ansatz der technologie und innovationspolitik', in F. Steinmeier and M. Machnig (eds), *Made in Germany 21. Innovationen für eine gerechte Zukunft*, Hamburg: Hoffman und Campe, pp. 95-110.

Meyer-Krahmer, F. (1999), 'Was bedeutet globalisierung für aufgaben und handlungsspielräume nationaler innovationspolitiken?' in K. Grimmer, et al. (eds), *Innovationspolitik in globalisierten Arenen. Neue Aufgaben für Forschung und Lehre*, Opladen: Leske und Budrich, pp. 43-74.

Meyer-Krahmer, F. and Reger, G. (1997), 'Konsequenzen veränderter industrieller F&E-strategien für die nationale forschungs- und technologiepolitik', in A. Gerybadze, F. Meyer-Krahmer and G. Reger (eds), *Globales Management von Forschung und Innovation*, Stuttgart: Schäffer-Poeschel Verlag, pp. 196-215.

Moors, E., Enzing, C., Giessen, A. van der and Smits, R. (2003), 'User-producer interactions in functional genomics innovations', *Innovation: Management, Policy & Practice*, 5 (2-3), 120-143.

Mowery, D. and Rosenberg, N. (1979), 'The influence of market demand upon innovation. A critical review of some recent empirical studies', *Research Policy*, **8** (2), pp. 102-153.

Neji, L. (1998), 'Evaluation of Swedish market transformation programmes. Lecture at the ACEE summer study on energy efficiency in buildings', Panel II.

NUTEK (1994), 'Effective market influence: an effect chain analysis of NUTEK's high frequency lightning campaign', report R 1994:70, Stockholm.

Office of the Deputy Prime Minister (2003), 'National procurement strategy for local government in England', Local Government Procurement Team, October 2003, London.

Pavitt, K. (1971), *The Conditions for Success in Technological Innovation*, Paris: OECD Publishing.

PCAST (Presidents' Council of Advisors on Science and Technology) (2004), 'Sustaining the nation's innovation ecosystem: maintaining the strength of our science and engineering capabilities', Washington, DC.

Popper, S. and Wagner, C. (2002), 'New foundations for growth: the U.S. innovation system. Today and tomorrow', MR-1338.0-OSTP, RAND, Washington, DC.

Porter, M. (1990), *The Competitive Advantage of Nations*, London, UK and New York, USA: Macmillan.

Rothwell, R. and Zegveld, W. (1981), 'Government regulations and innovation: industrial innovation and public policy', in R. Rothwell and W. Zegveld (eds), *Industrial Innovation and Public Policy*, London: Greenwood Press, pp. 116-147.

Smits, R. (2002), 'Innovation studies in the 21st century. Questions from a user's perspective', *Technological Forecasting and Social Change*, **69** (9), 861-883.

Suvilehto, H.-M. (1997), 'An evaluation of NUTEK's programme for more efficient use of energy', Final Report, National Board for Industrial and Technical Development, Stockholm.

Suvilehto, H.-M. and Överholm, E. (1998), 'Swedish procurement and market activities: different design solutions on different markets', presentation at the 1998 ACEEE Summer Study on energy efficiency in buildings.

Wilkinson, R., Georghiou, L. (ed.), Cave, J. (ed.), et al. (2005), 'Public procurement for research and innovation', report of an expert group, commissioned by the European Commission, Brussels.

13. Managing the Soft Side of Innovation: How do Practitioners, Researchers and Policymakers Deal with Service Innovation?

Pim den Hertog

INTRODUCTION[1]

Services and service innovation are ubiquitous. Some of these can be clearly observed in our role as end consumers. For example, new global positioning systems (GPS) are changing radically the ways in which we navigate through traffic and cities and form the basis for a whole new generation of location-based services. New retail models are appearing not only in the high streets but on the internet as well, leading to new service experiences. Digital marketplaces are transforming some consumers into traders. New business models – such as those exemplified by low-cost airlines or in retail banking – are changing our lives on a daily basis. In the business setting, innovative services are offered on almost all business processes including HR services, logistics, financial services and marketing. Electronic marketplaces have created both scale and niches to allow for new service offerings. Firms are confronted with all kind of innovative new services, which are wrapped around capital goods such as printer/copiers or advanced medical equipment. In fact these are the service activities where value is added at an increasing rate. Some manufacturers are even gradually transforming themselves into service providers or in some cases selling off their manufacturing activities completely (IBM is a well-known example here). In typical not-for-profit sectors such as education, safety and government we see new ways in which producers and final users interact, and new ways in which services are delivered and service concepts used. Think of one-stop shopping concepts or

1 This contribution is based on learning on the job when performing innovation and policy studies in service innovation. I would like to thank collaborators in these projects and especially Bart van Ark, Faiz Gallouj, Johan Hauknes, Jeremy Howells, Gjalt de Jong, Jari Kuusisto, Thomas Meiren, Ian Miles, Luis Rubalcaba, Jon Sundbo and Bruce Tether, to all of whom I am indebted. Many of the insights presented here are the result of joint research with these and numerous other persons.

clients having personal budgets for purchasing a customised set of services. Firms and organisations are increasingly unique mixtures of manufacturing-like and service-like activities. Both the bundling (service supermarkets) and unbundling (super-specialists) of these service functions can lead to innovative service offers.

Although one is tempted to see the above examples as recent phenomena, the development towards a service economy is a process that has been underway for some decades already. It is also clear that developing and providing innovative services involves both technological as well as non-technological innovation, knowledge and capabilities. How can it be explained, then, that when innovation scholars discuss innovation most of them discuss technological innovation? How can it be explained that when innovation and innovation management is discussed in the business community it is still mostly associated with R&D and technological innovation? How is it that statisticians measuring R&D and innovation still predominantly measure technological R&D and innovation and are taking such a long time not only to better cover service industries, but to develop innovation metrics more suited to service innovation? How can it be explained that in most countries the portfolio of R&D and innovation schemes is predominantly technology and manufacturing biased?

A possible explanation for what has been called the transformation to the soft side of innovation (den Hertog et al. 1997) is the co-evolvement of the domains of innovation practitioners, innovation scholars and statisticians and innovation policymakers on this issue. Previously, these three communities were long trapped in the same dominant paradigm that reduced innovation to mostly technological innovation. However, in the last few years a slow but steady process can be observed in which new views on the transformation to a service economy and its implications for state-of-the-art innovation practices, innovation research and statistics and appropriate (innovation) policy interventions are being developed. Service innovations were analysed more thoroughly, new theories and typologies developed and new statistical indicators introduced. Firms are increasing managing service innovation more explicitly. In some countries innovation policymakers have started to explore new, more services'-innovation-friendly R&D and innovation schemes. This has been neither a smooth process, nor a process that is finalised to now (rather, it is still building up momentum). It does, however, point at the relevance of the 'three dancers' analogy (discussed in the opening part of this book) in the service innovation domain.

In this chapter, I first discuss how service innovation can be conceptualised and defined. For this 'setting the scene' section, I use the 4D-model as developed by den Hertog and provide some stylised facts or lessons learned on service innovation. Subsequently, I address the question of why the services R&D and innovation theme was 'discovered' relatively late, presenting three different views on services R&D and innovation, i.e. the well-known assimilation, demarcation and synthesis or systemic approach to services R&D and innovation. How the research, firm and policy perspectives on service innovation have developed is explained. Then I take up the 'three dancers' metaphor as introduced in the first chapter of this book. It is shown how innovation practitioners (I), innovation research and theory, i.e. including the work of innovation statisticians (T) and public intervention strategies (P), have to a certain extent been trapped in the same technology-biased view on service innovation. Further, I point to developments that indicate that there is some interaction and mutual learning going on between the 'three dancers' and that we are at the brink of a paradigm shift regarding service innovation. Finally, in a concluding section, I present some suggestions for a future policy and research agenda for services innovation.

WHAT IS SERVICE INNOVATION ABOUT: A DEFINITION AND SOME STYLIZED OBSERVATIONS[2]

During the past decades, research on services and services innovation has mushroomed (Metcalfe & Miles 2000; Gallouj 2002; Tidd & Hull 2003). These studies have largely focused on the conceptualisation of service innovation (Gallouj & Weinstein 1997; Boden & Miles 2000; Howells 2001). Services are characterised as a specific economic activity that is distinctively different from goods-producing activities. This literature points at the intangible nature of most services, the overlap of the moment of production and consumption, the non-storability and low tradability of services, and the strong user-producer links. For a long period, the definitions of services were dominated by Pavitt's sectoral taxonomy of technological change that primarily characterised service industries as supplier-dominated sectors (cf. Soete & Miozzo 1989; Evangelista & Savona 1998; Djellal & Gallouj 2001). Likewise, the important theoretical contributions of Barras (1986; 1990) portrayed most service sectors as supplier-dominated and as being dependent on inputs from manufacturing before embarking on subsequent phases in the innovation process. Our point of departure aligns with the definition of Gadrey et al. (1995) who observed that

> to produce a service ... is to organise a solution to a problem (a treatment, an operation) which does not principally involve supplying a good. It is to place a bundle of capabilities and competences (human, technological, organisational) at the disposal of a client and to organise a solution, which may be given to a varying degrees of precision.

---

2 The subsection of the definition and the 4D-model is largely based on den Hertog & de Jong (2007) and earlier versions included in den Hertog (2000) and van Ark et al. (2003). The '10 stylized observations' is a slightly adapted version of the overview given earlier in van Ark et al. (2003) and den Hertog et al. (2003).

This observation suggests that technological, human as well as organisational capabilities are important for providing services. Additionally, the description allows making a difference between highly standardised service products or service formulas with quasi good characteristics such as fast food chains and more customised services such as consulting or legal services. The latter category usually requires a substantial part of tacit knowledge and interaction between the service provider and client.

Two further conclusions can be derived from the service innovation literature (den Hertog & de Jong 2007). Firstly, it has been increasingly recognised that many services deliver a – sometimes substantial – contribution to innovation processes. They are not merely passive recipients of others' innovations. Secondly, the emphasis on technological innovation has been moderated. Innovation in services or innovations in service functions represent multidimensional characteristics, which involve both technological and non-technological dimensions. The importance of non-technological elements has increasingly been acknowledged for service innovation. These achievements result in a better understanding of the peculiarities of services (Miles 1993), service management (Norman 1991; Quinn 1992; Løwendahl 2005), the role and importance of the interaction with clients (Kline & Rosenberg 1986) and the recombination of existing elements in new services (Henderson & Clark 1990). Manufacturing firms increasingly use innovation in service functions in order to differentiate their products. Vice versa, some business processes in service organisations increasingly resemble those in manufacturing.

Service innovation is seldom limited to a change in the characteristic of the service product itself. Innovation often coincides with new patterns of product distribution, client interaction, renewal of service delivery organisation and tapping the potential of new technological options, often involving information and communications technology (ICT). Figure 13.1 presents a conceptual model that maps the diversity of service innovations (den Hertog 2000). This model makes a distinction between a technological dimension of innovation on the one hand, and three non-technological dimensions of innovation on the other. The conceptual model represents four different dimensions of service innovation. The first dimension is a new service concept. Call centre services are an example of these 'conceptual innovations'. Companies that offer call centre services install, organise and recruit staff for their client's call centre. Manufactured products are typically highly tangible. This is often not the case with services. Service innovations may be embedded in a tangible product (such as automatic teller machines or ATMs) but the innovation self often is a new idea or concept (e.g. the concept of mono brand stores) of how to organise a solution to a problem.[3]

3 There is an ongoing debate on the 'newness' of many service innovations but this applies to 'new' manufacturing products as well and is part of this line of research. Services can be new to the providing firm, the regional, national or international market, or the client. The level of novelty in service innovations differs; although a particular service concept may already be familiar in other markets, it is key that it is novel in its application to a particular market.

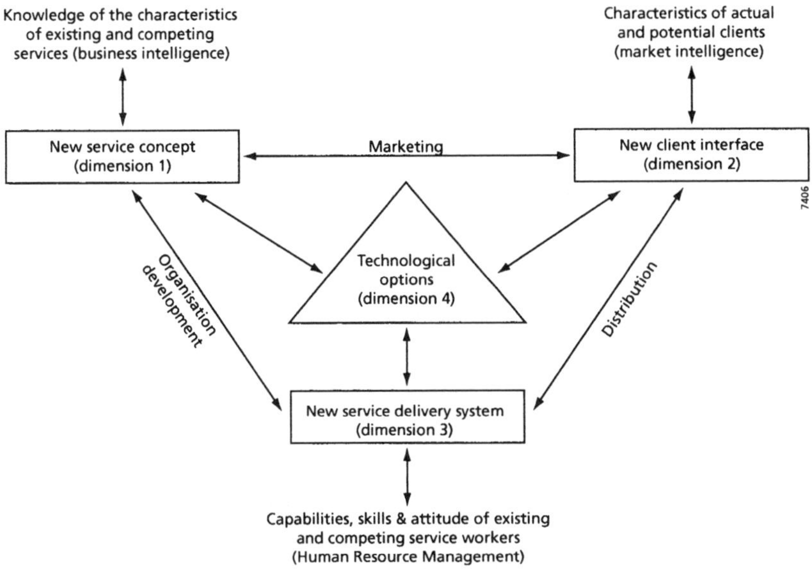

Figuure 13.1 Dimensions of service innovations

The second dimension is the client interface. The various generations of electronic banking (e.g. the use of mobile phones in retail banking) is an example where the 'client-interface dimension' is a dominant dimension. Electronic banking changed the way services are offered and how we interact with our banks. It further allowed for more self service as most of the data entry is now done by final consumers themselves. Clients are co-producing these (new) services. The interaction process between provider and client is an important source of innovation – more so when the business is itself offering support for innovation (which, for example, is the case in R&D or design services). The higher the degree of co-design or co-production of services, the more difficult it is to locate the prime source of the innovation.

In-house service delivery system and organisation is a third dimension discerned here. It refers to the organisational structure of the service company itself. For example appropriate management and organisational structures need to be in place to allow service workers to perform their job properly and to actually offer innovative services. New services may require new organisational structures, (inter) personal capabilities or team skills. The new type of service organisations that banks have introduced to offer typically more advice-intensive financial products and so increasingly replacing old-style banking counters where standardised services are offered is an example where the 'delivery system and organisation' perspective is dominant.

A fourth dimension is the cause and consequence of much debate in the service innovation literature, namely: technological opportunities. Service innovations are of course possible without technological innovations but are in practice highly intertwined. ICT can facilitate (but also constrain) service innovations. The use of convenience food and pre-cooked meals and ingredients combined with advanced kitchen appliances means that new types of services in the hospitality industry can be offered in an increasing variety of locations. The use of navigation software has changed the sorts of services that are on offer in, for example, the transport industry, but give rise to a whole new series of location-based ICT services as well.

These four dimensions may operate as separate innovation vectors of change. However, the linkages between the vectors may be even more significant. Often these cross-linkages are forged in practice by those responsible for marketing, organisation development and distribution. Launching a new service concept (for existing or new clients), for example, requires marketing expertise. In a similar vein, creating an adequate interface with clients, and adapting the service delivery system accordingly, requires knowledge of how services are produced, delivered and distributed. Hence, introducing and developing new services also requires considerable organisational knowledge.

To summarise, a particular service innovation may be characterised by one dominant feature related to one of the above mentioned dimensions. This feature will often prompt a set of changes in other dimensions, in order to bring about a successful innovation. Most service innovations will involve a combination of the four dimensions and their cross-linkages. The weight of the dimensions, as well as the interactions between them, will vary across service innovations, services and firms.

On the basis of the foregoing, service innovation can be defined as

> a new or considerably changed service concept, client interaction channel, service delivery system or technological concept that individually, but also in combination, leads to one or more (re)new(ed) service functions that are new to the firm, change the service/product offered on the market, and require structurally new technological, human or organisational capabilities of the service delivery organisation. (Van Ark et al. 2003)

This definition covers the notions of technological and non-technological innovation as well as the distinction between technological and non-technological aspects of innovation. This subtle though highly relevant distinction stresses the difference between the nature of the innovation itself and the characteristics of the innovation process and the organisations involved.

Having introduced a conceptual model and definition of service innovation, I provide some stylised observations below, which might help in developing a better understanding of what service innovation is about.[4] However, any generalisations should be made with care as variety within the overall services sector is probably as wide if not wider as the differences *vis-à-vis* manufacturing.

4 These observations are based on sectoral service innovation studies in which the 4D-model presented above was used as a mapping device.

1. In service innovation multidimensionality is the rule. Service innovations are most likely combinations of renewals in two or more of the four dimensions discerned above. The starting point for a sequence of innovations is difficult to pinpoint, but in principle any of the four dimensions is possible. The use of on-board computers in transport and logistics can be depicted as a pure technological innovation. However, it also changes the client interface (e.g. through advanced tracking and tracing services), requires changes in the service delivery system (truck drivers as front office workers, improvement of fleet management) and new service concepts can be realised (e.g. new sorts of contract logistics, synchronised production). Similarly, the use of advanced web-based knowledge management systems in technical engineering offer project teams the possibility of working virtually together and co-innovating with the client firms.
2. Existing dimensions are often renewed in other sectors through new combinations. The traditional view on innovation is biased towards mostly radical technological products and processes stemming from manufacturing. In services, however, a new combination of service elements that already exists on their own can (in a recombination) still emerge as an innovation.[5] 'Do it yourself' is a well-known concept in the home improvement trade, but applying it to financial services using electronic platforms is an important service innovation. Similarly, the possibility to lease a software package through an Application Service Provisioning (ASP) certainly is an innovation. Simple dichotomies like product and process or radical and incremental innovations do not suffice to understand service innovation.
3. The weight of each of the four dimensions in service innovation may change over time. Especially more substantial innovations need to be 'digested' and require adaptations in the other dimensions. Over time a shift in weight of each of the four dimensions is likely. In financial services a period of introducing new products is often followed by a period in which establishing new communication channels and co-ordinating these as well as restructuring the service delivery system are focussed upon. In retailing, substantial investments are being made in scanning registers. However, it takes years before this results in new service concepts being offered to individual clients.
4. Non-technological or organisational innovations in service innovation are important. The three non-technological dimensions mentioned in the 4D-model shape services innovation to a large extent. Service design is particularly important in services aimed at final consumers and is essentially about how a service is perceived.[6] If service functions are provided electronically, such as in

5 Earlier notions such as architectural innovation (Henderson & Clark 1990) and re-combinative innovation (Gallouj & Weinstein 1997) already pointed in this direction.

6 In an increasing number of cases the experience surrounding the actual purchase is more important than the actual purchase itself, often referred to as the experience economy (Pine & Gilmore 1999).

banking or in logistic services, the innovations regarding the shaping of client interfaces are relatively important. New service delivery systems seem to be less important as drivers of service innovations, but have greatly impacted on tasks and responsibilities of service workers. More employees have client contacts, ICT skills are mandatory in most service jobs and commercial/advisory skills are needed as well.

5. Although the role of ICT in facilitating service innovation is paramount, ICT is neither a sufficient nor a necessary condition. ICT adds intelligence to almost all service operations. The sheer fact that more information becomes available on, for example, the performance of service workers, the way users interact with a service provider or the status of service operations allows for further innovation in service functions. However, simply investing in ICT is not a sufficient condition for service innovation. ICT investments aimed at innovation need to be combined with substantial investments in new ways of working and new organisational structures to really materialise and result in better performance. Presenting ICT as a necessary condition for innovation to occur in services also goes too far. Especially conceptual innovations are quite often non-technological. Mono-brand stores, a new type of mortgage, the idea of adding light manufacturing to the wholesale and transport function are all examples of innovations that in principle are non-technological. Further it should be kept in mind that the sheer fact that ICT or another technology is important does not automatically imply that it is driving service innovation. For example, banking and retailing are huge service industries where demand rather than supply can steer technological development to an important extent.
6. Different firm strategies give rise to different service innovation strategies. Although there are dominant innovation patterns for innovation strategies at the level of the industry, the implementation of such strategies can still highly differ between individual firms. As in manufacturing there are trade-offs between specialisation versus one-stop shopping (e.g. in professional services like technical engineering), customisation versus standardisation (e.g. evident in logistics) and price strategies versus value added or quality strategies (which can be illustrated by the services offered by hotels and restaurants).
7. The rise of specialised service functions fosters service innovation. Starting from a functional rather than an industry approach, most firms in fact combine a mixture of various manufacturing and service functions. Over time more specialised service functions emerge and the room for innovation in these service functions increases, irrespective of whether they are insourced or outsourced (as both grow simultaneously). For example, part of the client interface can be left to a dedicated call centre, transport can be taken care of by a specialist logistic service provider, ICT systems are implemented and maintained by ICT specialists, the HRM function may be outsourced to HRM specialists and so on. Large service firms often combine the strengths of various functional specialists, a picture that is already more established in manufacturing (for example, in the

car manufacturing industry). As observed by Howells (2000) specialised services functions are needed as differentiators for both manufacturing and service products and in this context introduced the notion of 'service encapsulation'.

8. Innovations in service firms increasingly transcend borders of the individual firm and are mostly co-produced. Along the value chain the borders between these firms get blurred through outsourcing of service functions, through the use of networks of service professionals, and through mixed project teams in which client and contracting service firm co-produce solutions to problems. Technical engineering firms and ICT services mostly work jointly with clients to co-produce and sometimes co-innovate. A particular breed of service firms named knowledge-intensive business services do play a particular role in driving and facilitating innovation in client firms (see Miles et al. 1995; den Hertog 2000; Toivonen 2004). This blurring of boundaries is also reflected in the fact that some service firms have a reputation for a particular service function that is not directly seen as their core activity. For example, some competitive retailers have developed innovative logistic services, some banks are known for their innovative ICT capabilities and some ICT service companies have a reputation for their innovative approach to human resources management.
9. De- and re-regulation can trigger innovation in quite a number of service industries. In some service industries de- and re-regulation triggers the scope for innovation considerably through in fact changing the framework conditions. In retailing the liberalisation of closing hours of shops in combination with changes in the spatial planning regime has given rise to new types of shopping outlets at new locations. In financial services various legal changes and liberalisation of markers made it possible for banks and insurance firms and later other types of financial service firms to join forces and develop ingenious mix products.[7]
10. Understanding service R&D and innovation is hampered by the lack of adequate data. Measurement of services R&D and innovation is beset with problems,[8] basically as the proxy technological R&D – largely developed for measuring technological R&D in manufacturing – is used for measuring innovation. The current measurement issues regarding services R&D can be traced back to the services low-tech heritage (1), the ignorance as to some of the innovation specificities of service firms or invisibility or lower levels of formalization of R&D (2), and the resulting underreporting of R&D in services (3). The latter has to do with the first two points, but also with issues regarding definitions of R&D as used in surveys and, for example, R&D tax credit schemes and the

7 It is not by definition deregulation or liberalisation. Stiff environmental regulation, for instance, may trigger innovation as well as can be observed from technical engineering firms specialising on 'greening industry' and green financial products as offered by financial service providers.

8 This is discussed more widely a.o. in RENESER (2006) and PREST/TNO/Servilab/ARCS (2006).

ways in which in particular non-technological R&D is treated (RENESER).[9] So far, definitional and measurement problems are persistent, and obscure the contribution of services to innovation, productivity and economic growth.

WHY WAS THE SERVICES INNOVATION THEME 'IGNORED' FOR SUCH A LONG TIME?

Given the breadth and complexity of the service innovation theme as sketched in the preceding section and the observation that most advanced economies have turned into predominantly service economies, one might wonder why service innovation has not moved to the centre stage yet. This is a missed opportunity as service innovation offers a huge potential for raising competitiveness, productivity and ultimately economic growth and welfare. Why is the 'technological view' on innovation (Gallouj 2002) so predominant? Why is there as of yet no clear, widely accepted conceptual framework through which service innovation can be comprehensively analysed and its impact on economic output, employment and productivity growth adequately measured? How come that innovation scholars refer to the Schumpeterian definition of innovation,[10] but do not acknowledge this is a rather wide definition of innovation, which does not exclude non-technological or service innovation? Service innovation until recently seemed to be the domain of mostly a small group of innovation scholars and 'enlightened' practitioners, statisticians and policymakers. One of the root causes for this relative ignorance in our view is that there exist two basically different views on services R&D and innovation next to each other (assimilation versus the demarcation approach) and it is only recently that a synthesis between the two seems to emerge (synthesis or systemic approach). Below we first present these three basic views on service R&D and innovation and observe that the technological view on innovation was (and to an important degree still is) the dominant view or rather paradigm. Then we systematically explain how the research, firm and policy perspective on service R&D and innovation is developing.

9 This situation has improved in the last few years as in the most recent round of the Community Innovation Survey organisational innovation is on the radar of the statisticians (see e.g. Arundel et al. 2007) and attempts have been made to even formulate a composite service sector innovation index (for a critical review see Arundel 2006). The credibility of the latter is questioned and serious problems remain in measuring service innovation.

10 'New combination of inputs and the creation of new ideas leading to innovations in products (and services) and processes' (Schumpeter 1934; see Drejer (2004), for a Schumpeterian perspective on service-specific innovation concepts).

Approaches to services	Theory/practice/policy		
	Theory of services R&D and innovation	**Empirics & statistics on service R&D and innovation**	**Services R&D and innovation policy**
Assimilation approach	Technological R&D and innovations are the norm and services are simply less (technological) innovative. Look at the figures!	Technological R&D and Innovation is much lower in services, however variation between industries is large	No specific policy other than increasing service sensibility of existing generic R&D&I policies
Demarcation approach	Innovation in services is unique and specific service innovation theories are needed. The peculiarities approach to services	Major part of innovation in services is non-technological. Service innovation is mostly not managed explicitly	Specific/vertical service R&D and innovation policies are lacking and much needed
Synthesis/ systemic approach	Differentiating between manufacturing and services no longer works. Innovative service functions need to be part of wider innovation systems. A grand unified innovation theory is much needed	Service functions are ubiquitous (and intertwined) in services ánd manufacturing. They play a key role in creating innovative & differentiated products and services	Services need to be built in into systemic innovation policies paying attention to innovation policies and 'non-innovation policies'

7406

Figure 13.2 Three approaches to services R&D and innovation (den Hertog, 2006)

Three Different Views on Service R&D and Innovation[11]

Many innovation scholars, policy analysts, statisticians and policymakers have analysed, measured and reported on services R&D and innovation. The resulting different views on services R&D and innovation can be reduced to basically three different approaches as identified by Coombs and Miles (2000) and previously by Gallouj (1994; 2002). They came up with, focusing on services innovation, two main lines of analysis represented in the literature, and suggested a third approach as being possibly more productive. These three basic approaches – leaving out the perspective where service innovation is completely ignored, as we can safely say that not all researchers, practitioners and policymakers have passed this stage – are sketched out in Figure 13.2 and are summarised in Box 13.1.

Research Perspectives

The long and persistent technological view on innovation meant that service innovation was and to a certain extent still is neglected among innovation researchers, practitioners and policymakers. They were implicitly taught that innovation is about technological innovation in manufacturing firms. Even in the mid 1980s it was still the case that well-respected innovation scholars classified services predominantly as supplier dominated.[12] Most empirical studies on innovation took technological

11 A summary version of the argument given in RENESER (2006) and den Hertog et al. (2008).

12 The late Keith Pavitt in his 1984 article provided four archetypal innovation trajectories or patterns, i.e. 'supplier-dominated', 'scale-intensive', 'specialised suppliers' and 'science-based'.

Box 13.1 Three Approaches to Service R&D and Innovation

The *assimilation approach* originates from the idea that R&D and innovation are still about technological R&D and technological innovation, that some services industries may play a role here and that most existing policies need to be made more accessible for those service firms that perform technological R&D and innovation. Services can be included in existing R&D and innovation policy schemes.

The *demarcation approach* is based on the peculiarities of services, the predominance of types of innovation other than pure technological innovation and the different innovation processes or innovation styles of services R&D and innovation. As services R&D and innovation are conceptualised as rather different from the 'regular and well known' technological R&D and innovation, a plea is made for new ways of measuring and developing specific policy initiatives suited to services R&D and innovation. This would mean a development towards more specific or vertical R&D and innovation policies.

The *synthesis or systemic approach* starts from the idea that technology and non-technological innovation, manufacturing and services innovation typically support and need each other. Every firm has its own mixture of service and manufacturing functions. Most innovative firms realise they also have to differentiate themselves from the competition by adding extra service functionality to both manufactured goods and services. Taking this more systemic perspective would imply a broader perspective on services R&D and innovation and could for example result in proposals for all sorts of non-R&D and non-innovation policies that would benefit overall innovativeness and eventually competitiveness.

(*Source:* den Hertog et al., 2008)

innovation in manufacturing firms and industries as their point of departure, partly because these were described best in the available statistics. Innovation management handbooks discuss how to manage technological innovation[13] and typical case studies included were in manufacturing. Miles (2000) – one of the first service innovation researchers to systematically point at this systematic bias in innovation research and

Although originally aimed at describing technological trajectories at firm level, the Pavitt classification is mostly applied at the sectoral level leading to some generalisations (Hauknes 1996). Services have mostly been classified categorically as supplier dominated, suggesting that most service firms do not have an autonomous innovation function.

13 From the early 1960s a whole list of literature has emerged delving into the various components that shape innovative organisations and how the innovation process can be managed; see for an overview Tidd et al. (2005).

statistics – in this context spoke about the services sector as having a 'heritage of neglect' or the 'Cinderella sector'.

It is only since the 1980s and 1990s that the dominant technological view on innovation started to get questioned. Especially discussions on the knowledge economy from the mid-1990s onwards gradually broadened the focus from innovation in manufacturing and 'hard' technical knowledge and R&D, to innovation in services and other types of knowledge in both manufacturing and service industries. Also thanks to the increased use of ICT in services, the whole idea of service firms as being dependent on other non-service firms for their innovation and not being able to innovate autonomously started to be questioned and subsequently corrected. The first alternative (service) innovation typologies[14] and more empirical studies on service innovation started to appear. At that time most of these studies were predominantly pleas that asked for recognition of service innovation and explained over and over again how different service innovation is compared to manufacturing innovation. In fact we saw here the birth of the demarcation approach, mostly shaped by service innovation researchers. Their impact on innovation management at the micro-level was limited as the contributions of these scholars were conceptual in the first place. They did, however, advocate for dedicated service innovation statistics – and in some cases dedicated surveys among service innovators – and service innovation policies. There were some discussions and interaction with policymakers and statisticians (partly through service innovation studies sponsored by mostly strategic units at the policy level), but it proved hard to get the message of service innovation across and make structural changes to the way in which service innovation was managed in firms, covered in innovation statistics and in the actual design of innovation policy schemes. The fact that service innovation scholars have apparently not been persuasive enough is – in our view – due to a couple of serious flaws in service innovation research itself.

One of the main problems that service innovation research is struggling with is the fact that some of the most basic definitional problems have not been adequately tackled. It seems as if 'along the way' of a by now serious number of service innovation studies the definition of service innovation was broadened over and over again, making it increasingly hard to differentiate between good entrepreneurship and improvement of regular business processes and service innovation. This definitional issue seemed to be blurred further by the fact that some see service innovation as non-technological innovation (i.e. a discrete innovation still) such as organisational, service, design or transactional innovations, whereas to others service mostly concerns the non-technological aspects in the process of innovation (i.e. the process of innovation and the characteristics of the innovative firm). Drejer (2004) not only made a plea for a Schumpeterian perspective on service-specific innovation concepts, but in this context rightfully remarked that there is a 'problem of mingling activities that might lead to innovation with actual innovation'. As observed earlier

14 See e.g. Soete & Miozzo (1989) and den Hertog et al. (1998).

the consequence of this might be 'that the innovation concept becomes detached from the original meaning as an economically successful introduction of something new, thereby being a contrast to acting within the boundaries of routine systems' (Drejer 2004). This means that in defining service innovations there has to be a link again between the type of innovations and the economic performance of firms (den Hertog et al. 2005). It might well be that the considerable definitional problems have prevented the actual development of a body of theory concerning service innovation with hypotheses which could be systematically tested as this body is still rather bleak.

Another explanation is that the statistical apparatus is attuned to measuring technological innovation and – even if attempts to make R&D and innovation statistics more sensitive to service R&D and innovation are underway – this means that especially the policy discussion on service innovation is fed with very partial numbers where 'round pegs are put in square holes'. This is changing, but slowly. A few countries have experimented with measuring service innovation as part of their innovation surveys, e.g. Australia and the Netherlands. In the 3rd edition of the Oslo Manual on guidelines for collecting and interpreting innovation data (OECD 2005b) organisational and marketing innovation have a more prominent role in addition to the traditional emphasis on technological innovation. However, it takes years before new indicators are formulated, tested, reported and find their way to policy circles.

Firm Perspectives

A major reason why services innovation has not yet reached centre stage is that service innovation management in firms is underdeveloped. The perspective of the individual service firm or organisation reveals that service innovation is not widely perceived yet as an activity for which one needs to develop a strategy that needs funding and would benefit from some specific attention by management. Most senior management – whether educated and trained in innovation or not – still see innovation mostly as technological innovation or at best as ICT-based innovation. The results of a recent research project into the R&D needs of business-related service firms in which the R&D and innovation (RDI) efforts of 21 mostly large European service firms[15] in the United Kingdom, Germany, Spain and the Netherlands were analysed, confirm this (RENESER 2006). Although most of these generally speaking advanced service firms showed some form of structured attention to services, RDI was mostly less formalised, more distributed, less explicitly managed and funded compared to what you would find in a similar set of manufacturing firms. A dedicated long-term services R&D and innovation strategy (and hence management) at the level of the board of management was quite rare. Formalised methods for deriving services RDI were rather limited as well. The research also revealed that even among this quite advanced subset of service firms formalised, pure

15 Including four hybrid firms, i.e. manufacturing firms developing into providers of service solutions.

services R&D and innovation were the exception rather than the rule. However, in practice, important services R&D and innovation activities were hidden under labels such as business development, service improvement, or new service development, without being recognised as services RDI. Many were also hidden in client-specific solutions. In less than half of the cases, more formal methods for managing services RDI portfolios were used, and similarly at project level, less than half of the firms used more formalised (mostly rudimentary models) for service development.

An additional problem which might explain why service R&D and innovation is not so prominently on the radar of both innovation policymakers and innovation scholars is that service industries are mostly newer industries that are less well organised professionally or as a lobby. Their contacts with (innovation) policy-makers and academia are less well developed compared to representatives from manufacturing industries that may have longstanding links with policymakers and academia. They are therefore less capable of 'stealing' the governmental innovation agenda, benefit from academic research or set the agenda in a way that would be beneficial to innovative service firms.

Policy Perspectives[16]

Now that we have dealt with the service innovation researchers and service innovation practitioners we turn to the perspective of the innovation policymakers. How can it be explained that in most countries R&D and innovation policies are biased towards technology and manufacturing R&D and innovation? Why are innovation policies geared to facilitate service innovation more the exception than the rule? A combination of factors – linked to a similar bias in the standing innovation research and innovation practices in firms – can explain this. Most dominant here is the dominance of market failure argumentation within government and the hesitance to adopt innovation systems thinking including the notion of systemic failure when the rationale for (new) policies is concerned.

Among innovation policymakers, there is a received wisdom that services perform less technological R&D and innovation compared to manufacturing firms. The result is that all too often it is concluded that no specific attention to R&D and innovation in services need to be paid. This is at least remarkable as there are numerous service firms that do invest in technological R&D and innovation and these need at least be treated in a similar vein as their peers in manufacturing. Another argument used to open up to R&D and innovation policies that are more suited to service innovators is the argument that innovation in services typically takes place close to the market. This would imply that the case for intervention aimed at facilitating innovation is less obvious. Put differently: service innovation is not sufficiently fundamental to be supported and policymakers fear that intervention might distort competition. Further, it is believed that there

16 This subsection is largely based on RENESER (2006) and den Hertog et al. (2008).

is little scope for spillovers across firms as services innovation involves so much organisationally-specific development. Put differently: externalities from investments in non-technological R&D and innovation are less obvious than externalities from investments in technological R&D and innovation.[17] Typically, these are the sorts of arguments on the basis of which, for example, national ministries of finance judge the acceptability of R&D and innovation schemes.

It can be questioned whether the line of argument above is based on the right assumptions. Firstly, service firms and service industries are more active in technological R&D than is mostly anticipated. R&D and innovation in business-related services and in particular knowledge-intensive business services are performing substantial technological R&D, mostly considerably higher compared to the average for manufacturing firms (RENESER 2006). These services are more likely to engage in R&D than other service industries and than most firms in manufacturing industries. Standard market failure argumentation[18] is as relevant for these industries as it is to manufacturing firms and industries. Secondly, it seems to be forgotten that investing in technological R&D is just one of the ways through which firms become more innovative (and eventually more productive and competitive). There are simply innovations that do require relatively more investments in non-technological innovations such as new organisational or marketing concepts, new client interfaces, new type of delivery organisations or new smart combinations of service and product elements. These investments are more difficult to pinpoint and to assess, but they are as real as investments made in technological R&D. Further, as these intangible investments trigger innovations that eventually lead to economic growth, there is no fundamental reason to not facilitate these. However, a prerequisite is still that there should be externalities involved and these softer types of innovations do come at a cost and require serious investments, i.e. social benefits are higher than private benefits and the individual entrepreneur would have to invest substantially. Thirdly, it can be seriously questioned whether technological R&D and innovation and non-technological R&D and innovation can be treated separately. These two are in economic reality increasingly difficult to disentangle and treat separately as most innovations today are multidimensional, i.e. smart combinations of new or advanced technology in combination with new service elements or smart services enabled by innovative use of technology as explained

17 The classical argument for supporting private and collaborative (technological) R&D in firms is that through these externalities, social returns to investments made in R&D are higher compared to private returns for the firms making these investments. The resulting underinvestment in technological R&D and innovation is a market failure, which could be corrected through supporting private R&D efforts.

18 For an introduction into market failure argumentation and its applicability to service innovation see van Dijk (2002), van Ark et al. (2003) and RENESER (2006).

earlier in this chapter.[19] In our view we need to look beyond market failure argument and adopt a broader view on service innovation and the possible need or rationale for service innovation policies, i.e. possible systemic failures. Such an approach, which is still in its infancy, would start from the idea that innovation in service firms (rather service functions) suffers from an innovation environment or innovation system that is not well suited to cater for the needs of service innovators.[20] This link between firm and innovation systems is two-sided. It does not only look at how the innovation system impacts upon the service firm, but also takes into consideration how the service firm or functionality impacts on the overall innovation system; i.e., how do services R&D and innovation contribute to the overall quality of the innovation system?

All in all, the three perspectives as presented in this section illustrate that the dominant technological view on innovation is well anchored in research, statistics, practice and innovation policymaking. They show remarkable similarities indicating the various communities do influence and learn from each other. In a certain way all three communities involved are trapped in the same dominant view or paradigm and to a certain extent their interaction reinforces this dominant paradigm. Changing the dominant paradigm regarding service innovation has proved an uphill battle and nothing less than a paradigm switch. In the next section we point at some signs of change that involve all three communities discussed and point at some (second-order) learning processes that are underway.

SOME INDICATIONS OF A PARADIGM SHIFT AND MUTUAL LEARNING

In the last few years we have seen a slow, but steady process in which new views on the transformation to a service economy and its implications for state-of-the-art innovation practices, innovation research and statistics and (innovation) policy interventions were exchanged. This has been neither a smooth process, nor a process that is yet finalised, but it is a process where service innovation practitioners, researchers and policymakers do interact and learn from each other. Below, I illustrate this by outlining how the measurement of service innovation is developing, how mixed agenda-setting events contribute to mutual learning, and how service innovation policies and service innovation strategies by firms are developing.

---

19 At the level of firms and industries, the artificiality can be observed as well by manufacturing firms developing into hybrid firms realising a considerable part of their turnover in service activities and service industries developing into firms with a sometimes impressive technological capability.

20 Possible systemic failures – adopting the O'Doherty & Arnold (2003) typology – regarding service innovation have been extensively discussed in den Hertog (2006) and den Hertog, et al. (2008).

Developments in Measuring Service Innovation

As already mentioned, the measurement problems regarding service innovation are longstanding and complex. The discussion on how to measure service innovation is a typical occasion where the 'three dancers' interact. Here statisticians and empirical researchers are forced to be precise on what they define as service R&D and innovation. Researchers need to link up with mostly firms to learn how they manage, finance and report service innovations. Policymakers interact with these researchers – as well as with service innovators directly – to learn how service firms actually deal with services R&D and innovation. All three have a need for better measurement of service innovation. Over the years the discussion on questions to be included in R&D and innovation surveys, their actual wording, the definitions used and the interpretation of their results have proved highly instrumental in having an informed discussion on service innovation. In the 3rd edition of the OECD Oslo Manual with guidelines for collecting and interpreting innovation data (OECD 2005b) organisational and marketing innovation got a more prominent role in addition to the traditional emphasis on technological innovation. The Third Community Innovation Survey (CIS3) collected for the first time information about new or significantly improved products or processes and related activities in manufacturing and service industries during the period 1998-2000. Contrary to CIS2, CIS3 not only dealt with new and significantly improved products (goods and services) and processes, but also refers to other creative improvements that might have been undertaken by your enterprise. However, this concerns only one question on '[o]ther important strategic and organisational changes in your enterprise'.[21] The last CIS4 did cover service innovation more fully and has resulted in separate analyses of innovation in service industries and even the construction of a services innovation index (see Arundel et al. 2007).

Typically the sort of statistical pictures that derive from these exercises are used as an input for policy analyses and policy proposals. This is evident in many national policy documents and in the European Union for example in a recently published Commission Staff working document (European Commission 2007) which is part of the Commission's efforts to develop a policy framework aimed at better supporting innovation in services. It should be noted that in parallel to the

---

21 Enterprises were asked whether they are engaged in any of the following activities in 1998-2000:

- Strategy: implementation of new or significantly changed corporate strategies.
- Management: implementation of advanced management techniques within your enterprise.
- Organisation: implementation of new or significantly changed organizational structures.
- Marketing: changing significantly your enterprise's marketing concepts/strategies.
- Aesthetic change (or other subjective changes): significant changes in the aesthetic appearance or design or other subjective changes in at least one of your products.

To get additional insights, Statistics Netherlands added additional questions in their innovation survey 2000-2002 (see den Hertog et al. 2005).

quantitative line of research qualitative empirical work and especially case study work is building up (e.g. RENESER 2006; Eröcal 2005; Tidd & Hull 2003). The creation of such practical evidence is on the one hand much needed for researchers to mostly substantiate or build service innovation theories, and on the other hand needed to inform policymakers at a very practical level as to how service firms deal with R&D and innovation. Further, these real life stories are a way of fuelling the conversation between practitioners, service innovation researchers and innovation policymakers.

Mixed Agenda-Setting Events

A second set of events that contribute to the interaction between the three dancers are typically agenda-setting events where policymakers look for interaction with academia and industry before defining policy frameworks and innovation policies. A signal that the 'dance floor' is open are the numerous conferences, workshops, consultation documents and the like that have been held in the last few years worldwide. The OECD, already a think tank where researchers, statisticians and policymakers exchange views, for example in 2005 launched its policy framework for facilitating service innovation (OECD 2005a) and later a major research effort in the area of knowledge-intensive service activities which also asks policymakers to consider their efforts to both facilitate and take benefit from these dynamic service activities adopting an innovation systems approach (OECD 2006). In a similar vein the European Commission has been active in putting the services innovation theme on the research and policy agenda. It has not only launched numerous policy studies on service innovation and its measurements (e.g. Howells et al 2004; PREST/TNO/Servilab/ARCS 2006; RENESER 2006), but also through for example a Policy Trend-chart workshop on innovation in services creates a venue where policymakers, researchers and industrialists mix, exchange views and work on drawing up recommendations for a services innovation policy agenda (Cunningham 2007). The Commission has also established a mixed expert group on service innovation. Its report (Expert group on service innovation 2007) was subsequently used in the working document that outlines the services innovation policy framework (European Commission 2007). At the national level we see in numerous countries examples of studies – mostly combined with (series of) workshops – where evidence of innovation scholars on the importance of service innovation and its impacts are feeding into policy discussions on the policy responses needed for furthering service innovation. A recent example is a set of studies presented by the Department of Trade and Industry in the UK (DTI 2007).

A somewhat different example of an agenda-setting development is worth noting here as well. Although in general service industries seem to be less well organised in terms of industry associations and less well represented in informal networks in which industrialists and policymakers meet, there are occasions where industry with the help of academic scholars try to influence public policies, including the need for building a research and education infrastructure for service innovation. IBM has been particularly active in recent years – and has joined forces in at least the US with a number of other major service firms – and is pushing hard to create nothing less than services sciences (short for Services Sciences, Management and Engineering, SSME). This has in the United States led to the establishment in 2007 of the Services Research and Innovation Initiative (http://thesrii.org/initiative.asp). The efforts of especially IBM and some other major firms to put service innovation on the map have spread to numerous countries and typically have led to national trajectories where mostly practitioners, researchers and policymakers interact. In July 2007 for example the Cambridge Service Science, Management and Engineering Symposium was held. The resulting Green Paper (IfM and IBM 2007) is targeted at the business, policy and education and science community and in fact spurs them to act upon the services innovation challenge and explicitly asks for responses by the various stakeholders before the document is published as a White Paper. This is an example of the 'three dancers' interacting on the topic of service innovation.

Service Innovation Policies and Firm Strategies

Service innovation has developed into one of the topics that typically features high on innovation policy research agendas at both the national and international level (EU, OECD). It is increasingly acknowledged that in advanced economies most room for productivity growth stems from services and service innovation. However, does the interaction and mutual learning between the 'three dancers' that is observed for the stage of agenda-setting also apply to the day-to-day practice of service innovation policies of policymakers and service innovation strategies of firms? Here admittedly, evidence is more anecdotal as service innovation policies are in most countries at best being conceptualised into service innovation frameworks and there are only a few countries that have translated these in concrete measures and schemes which are then systematically monitored and evaluated. In the EU the recently finalised Innovation Policy Project in Services or IPPS project (TEKES 2007) provides a nice overview of the service sensibility of the innovation policies in nine countries and two regions. Two examples of policy schemes feature as probably most advanced (see Box 13.2). Interestingly, both not only aim at supporting the development of service innovation in firms, but also at supporting academic research and at furthering the exchange between academia and industry. Remarkably, both are 50% publicly financed and 50% financed by industry. This seems an indicator that mutual learning between academia and practitioners and policymakers is built in, in the design of the actual schemes. This fits into a way of innovation policymaking where the development and actual design of new innovation schemes and customised policy mixes for specific target groups are defined in dialogue between industry, policymakers and academic researchers – the latter, amongst others, through providing strategic intelligence, policy analysis and performing monitoring and evaluation studies.

Box 13.2 Two state-of-the-art service innovation schemes[a]

Finnish Serve – Innovative Services Technology Programme 2006-2010
Tekes has launched Serve programme that seeks to facilitate service development in the targeted industries and it also promotes service-related academic research. Serve is a five-year technology programme that aims to boost the development of innovative service concepts and new service business models. It runs from 2006-2010 and the total budget is about EUR 100 million, of which half is public funding and other half comes from the participating businesses. Serve programme also provides Finnish businesses and research organisations links to national and international networks through seminars and industry-specific forums. Further on, it offers tools for product management and IP issues. Serve programme facilitates the development of innovative service concepts that can be reproduced or replicated. In addition it offers support for service R&D projects where some technology or systematic method is being applied. (TEKES, 2007, p. 14)

German R&D Programme Innovation With Services 2006-2010
The Federal Ministry of Education and Research has launched the R&D Programme 'Innovation with Services' that aims to help increase Germany's innovation capabilities and boost its competitive strength in the service sector. The R&D programme runs from 2006 to 2010; and has a total public budget of 70 Mio EUR. About the same amount of funds will be contributed by the participating businesses. The Innovation With Services programme seeks to enable businesses to draw on application-oriented research on a long-term basis and to test, introduce and implement research findings in actual operational practice. That means to stimulate integrated learning processes between research and industry, theory and practice. (TEKES, 2007, p. 17)

a The text description for the Serve and R&D Programme Innovation With Services 2006-2010 are direct quotes from their descriptions in TEKES (2007, p 14 and 17)

The 'three dancers' allegory seems to be less well visible in dedicated and more explicit service innovation strategies as formulated and practised by firms as these are scarce and hardly documented yet. As already mentioned, a series of 21 case studies among European mostly large service firms on the way they dealt with services R&D and innovation (RENESER 2006) revealed that R&D and innovation in these firms is in most cases less formalised, more widely distributed and less explicitly managed and funded compared to what is practised in (mostly large) manufacturing firms. A remarkable finding was that – notwithstanding the trend towards more open innovation models – there is considerable room to improve R&D collaboration between service firms and research organisations. In most cases cooperation on service R&D was poorly developed. Another remarkable result was that most large service firms were not well-connected to the R&D and innovation policy scene (apart from those that perform extensive technological R&D themselves). Possibly as a result most service firms judged that existing R&D schemes were of limited value to them and most of these firms even found it hard or unappealing to try and gain access to or participate in them. At the same time when performing the case studies a huge interest in cross-firm and cross-industry (lateral) learning was observed as how to manage service innovation. This seems to point in a direction where there is scope for lining up the dance floor for the communities of researchers, policymakers and practitioners to interact more closely and support mutual learning.

TOWARDS A SERVICES INNOVATION RESEARCH AGENDA AND POLICY MENU

In the preceding sections we have observed that practitioners, innovation scholars and statisticians, and innovation policymakers have for quite a long time been trapped in the same dominant paradigm which reduces innovation to mostly technological innovation. We also noticed, however, that it has been increasingly recognised that services are not merely passive recipients of others' innovations and actually contribute considerably to innovation and innovation processes. It was argued that innovation in services or rather innovations in service functions (which are well represented in typical manufacturing environments) are multidimensional and involve both technological and non-technological dimensions. Using the 4D-model it was illustrated that service innovation is seldom limited to a change in the characteristic of the service product itself, but coincides with new patterns of product distribution, client interaction and new service delivery systems. We then offered some explanation why the technological view on innovation was so dominant in research, statistics, practice and innovation policymaking for such a long time. Finally we pointed at some developments such as the ongoing work on the measurement of service innovation, typical agenda-setting events and the development towards more explicit service innovation policies in government and service innovation strategies in firms, which are indicative for interaction and mutual learning between the 'three dancers' on the 'dance floor'.

In my view, a further development in the direction of a systemic approach to service innovation would help in getting a better understanding of service innovation. Below I present some suggestions for future research. Then I discuss a menu approach to service innovation policies.

Future Research Agenda[22]

Although the services innovation theme is established by now, an impressive list with pressing research issues remain. Following the 'three dancers' metaphor, three sets of questions are presented here that refer to research issues related to the communities of, respectively, researchers and statisticians, practitioners and policymakers. Those need, of course, to be addressed in interaction between these communities as mutual learning and sharing of insights are needed to progress.

There is still a considerable lack of both coverage and detail in the way that R&D in services is defined and measured (both classical and more widely defined R&D). Challenges for statisticians and innovation researchers alike are, for example:

1. To provide more consistent definitions and typologies of service R&D and innovation sticking to the well established classical definition of (technological) R&D and developing in parallel an alternative track with (broader) definitions of service R&D that would include R&D and innovation categories which are more common to service firms;
2. To start experimenting with alternative indicators which start, for instance, from the knowledge intensity and characteristics of firms in terms of occupations, innovation intensity or strategic intent regarding innovation or measuring social science R&D;
3. To take into account the role of size in service R&D and innovation, for example by looking more specifically at how large R&D-performing service firms are dealt with in R&D and innovation statistics, and analysing the numerous smaller-scale R&D and innovation-intensive service firms that make up the tail end of the highly-skewed distribution of R&D efforts among service firms;
4. To analyse in detail how product-based manufacturing firms are changing into service firms and to assess how this impacts upon the shifting balance in the manufacturing-services dichotomy;
5. To analyse the link between service R&D, service quality, service productivity and their link to overall economic performance.

Issues that are typically derived from the community of practitioners and might inspire the firm-oriented research agenda include the following:

1. How to manage the change in role from a product-based manufacturing firm into a mixed manufacturing and service firm (hybrid firms).
2. How can a structured service R&D and innovation agenda be developed? Given that more and more service firms – especially the larger ones – are interested in formalising their R&D and innovation strategies, there is an increasing need to systematically develop these strategies.

---

22 This sub-section is largely based on RENESER (2006).

3. What roles can technology or technologies play in making service firms more innovative, and how does this translate into new organisational and workplace design? There is a need to integrate service engineering, service quality and service innovation lines of work in order to better understand the roles of R&D, design and related activities in service innovation processes.
4. In what way do typical service firms, as well as more hybrid firms, use technology for service development?

	Assimilation policies	**Demarcati on policies**	**Systemic policies**
R&D policies	• Increase accessibility of existing R&D support schemes • Inclusion of services in technological foresight and road mapping exercises • Include services firms in policies aimed at improving industry sciences relationships	• Support public R&D in services • Introduce vertical R&D programmes aimed at service industries (logistics,trade,etc.) • Services IPR instruments • Create dedicated Co Einservices R&D at RTOs and HEIs • Increase role of the humanities in service innovation	• Under standand support role of R&D services (KIBS) in innovation systems • Support for servicesR&Dinand through hybrid firms • Integrated R&D programmes paying attention to technological and non-technological R&D and innovation
Innovation policies	• Increaseaccessibilityof existing Innovation support schemes • Innovation management trainingand practices more geared towards supporting all types ofinnovation in all industries • Mobility schemes no longer limited to QSEs	• Introduce courses on services innovation management • Awareness campaign on the importance of services innovation • Identify service innovationrole models (including innovation in public sectors)	• Innovation and business support systems also supports services innov. • Availability and use of specialized services/KIBS • Increase transparency in KIBS markets • Insight into and international competitiveness of key service functions • Cluster and networktype of policies that deliberatelyinclude services • Government procurement policies • Support role of users in innovation
Non-innovation policies	• Increase coverage of services in regularand R&D and innovation statistics	• Use deliberately policies such as trade, competition education and training policies for fostering R&D and innovation in services • Regulations that might trigger services innovations • Analyse offshoring in services	• Useregulation and standardization to support innovation • Financial and credit systems that acknowledges intangible assets • Enhancehighlevel service capabilities e.g. through education and training policies • Policies aimed at increasing entrepreneurship

Figure 13.3 Examples of assimilation, demarcation and systemic policies aimed at facilitating services R&D and innovation (an earlier version was included in den Hertog and Rubalcaba, 2006 and den Hertog et al. 2008)

5. Analysis is needed of the international dimension of service R&D. Given that increasingly complex manufacturing and service operations are being outsourced and offshored, an analysis as to what the future international division of (R&D) labour in service R&D would look like is highly relevant.
6. Changing roles of services in different innovation systems and networks. The idea of having service R&D and innovation policies, or systemic policies in which there is a sensitivity towards the 'soft side of innovation', is not very well established as yet.

This raises numerous issues, which are in the first place essential for the community of policy-makers dealing with service innovation. Some of these issues are given below:

1. Market failures and their relevance to service R&D and innovation. A detailed, practical analysis of (especially) spillovers in a number of different service industries would be needed to address this issue more fully and, related to this, IPR and services.
2. How to more directly link service firms with public R&D? What options are available to involve service firms more directly in formulating research agendas?
3. What do effective innovation and technology diffusion policies that can overcome barriers to innovation and technological change in the services sector look like? This includes the issue of policy mixes and development of what has been labelled systemic instruments to better address services R&D and innovation.
4. How can 'non-innovation policies', which are not designed to support R&D and innovation in the first place, be used most effectively to facilitate service R&D and innovation? Apart from regulation, policy areas that spring to mind as having a direct effect on the scope for R&D and innovation in services are policies in the areas of education and research, labour markets, regional development, competition and entrepreneurship.

A Menu Approach to Service Innovation Policies[23]

Earlier, I referred to the three basic approaches to services, i.e., assimilation, demarcation and synthesis or systemic approach. Applied to policymaking, the assimilation approach could be summarised as a 'no regret' scenario as it does not imply a specific services R&D and innovation policy. It mainly is about increasing the service sensibility of existing generic R&D and innovation schemes. The demarcation approach as applied to furthering service innovation asks for services-specific R&D and innovation policies. The heterogeneity of the service sector at

23 This subsection is a summary of an argument included in den Hertog & Rubalcaba (2006) and den Hertog et al. (2008).

large and the peculiarities of some specific service markets in practice require specific measures. In the synthesis or systemic approach services are perceived as a systemic dimension of economic activity and thus a potential focus of innovation practically anywhere. Policy initiatives originating from the systemic perspective have passed the services/manufacturing distinction, take a broad approach to innovation and see both R&D and innovation as well as 'non-innovation' policies as vehicles for supporting R&D and innovation. Services are not only understood as an economic sector, but also as a horizontal dimension of the overall innovation system that may boost overall innovation and competitiveness. A menu of some possible policy options in each of the three approaches is given in Figure 13.3. We do differentiate here between R&D and already wider innovation policies and 'non-innovation' policies which can be said to impact upon the scope for innovation. The three approaches might suggest that policymakers would have to make a choice between the three columns whereas in practice policy options from all three may be combined. The key issue is that there is not one single solution, as for every individual innovation system, a (temporary) mix of instruments and policies will have to be developed to suit that particular innovation system. This not only depends on economic structure, history and institutional set-up, but increasingly so on the willingness of the 'three dancers' to jointly address research, management and policy issues and including the willingness to experiment and invest in mutual learning. Therefore, put up the services innovation dance floor and let's dance.

REFERENCES

Ark, B. van, Broersma, L. and den Hertog, P. (2003), 'Service innovation, performance and policy: a review', Research Series (6), Ministry of Economic Affairs, June 2003, The Hague.

Arundel, A. (2006), 'Can we measure and compare innovation in services', Trendchart statistics report prepared for the Policy Trendchart Report on 'Innovation in services', held 19-20 June, 2006, Porvoo, Finland.

Arundel, A., Kanerva, M., Cruysen, A. van and Hollanders, H. (2007), 'Innovation statistics for the European service sector', UNU-MERIT, 10 May, 2007, Maastricht.

Barras, R (1990), 'Interactive innovation in financial and business services: the vanguard of the service revolution', *Research Policy*, **19** (3), 215-237.

Barras, R (1986), 'Towards a theory of innovation in services', *Research Policy*, **15** (4), 161-173.

Boden, M. and Miles, I. (eds) (2000), *Service and the Knowledge-Based Economy*, London: Continuum.

Coombs, R. and Miles, I. (2000), 'Innovation, measurement and services: the new problematique', in J. Metcalfe and I. Miles (eds), *Innovation Systems in the Service*

Economy, Dordrecht, The Netherlands, Amsterdam, The Netherlands, Boston, USA and London, UK: Kluwer, pp. 83-102.

Cunningham, P. (2007), 'Innovation in services', Thematic report within the framework of the Inno-policy Trendchart, DG Enterprise and Industry, Brussels.

Department of Trade and Industry (DTI) (2007), 'Innovation in services', DTI occasional paper no. 9, DTI, London.

Dijk, M. van (2002), 'Marktfalen bij innovaties in de dienstensector', CPB-Memorandum No. 50, Netherlands Bureau of Economic Policy Analysis (CPB), The Hague.

Djellal, F. and Gallouj, F. (2001), 'Innovation surveys for service industries: a review', in European Commission, Innovation and enterprise creation: statistics and indicators, Luxembourg: (EUR 17083), pp. 70-76.

Drejer, I. (2004), 'Identifying innovation in surveys of services: a Schumpeterian perspective', *Research Policy*, **33** (3), 551-562.

Eröcal, D. (2005), 'Case studies of successful companies and lessons for public policy', in OECD (ed.), *Enhancing the Performance of the Services Sector*, Paris: OECD Publishing, pp. 239-269.

European Commission (2007), 'Towards a European strategy in support of innovation in services: challenges and key issues for future actions', Commission Staff working document, (SEC)2007 no. 1059, Brussels.

Evangelista, R. and Savona, M. (1998), 'Patterns of innovation in services. The results of the Italian innovation survey', working paper presented at the VIII Annual RESER Conference, Berlin.

Expert Group on Innovation in Services (2007), 'Fostering innovation in services: final report', Brussels.

Gadrey, J., Gallouj, F. and Weinstein, O. (1995), 'New modes of innovation. How services benefit industry', *International Journal of Services Industry Management*, **6** (3), 4-16.

Gallouj, F. (2002) *Innovation in the Service Economy: the New Wealth of Nations*, Cheltenham, UK and Northhampton, MA, USA: Edward Elgar.

Gallouj, F. (1994), *Economie de l'Innovation dans les Services*, Paris: L'Harmattan.

Gallouj, F. and Weinstein, O. (1997), 'Innovation in services', *Research Policy*, **26** (4-5), 537-556.

Hauknes, J. (1996), 'Innovation in the service economy', STEP report 7/96, STEP, Oslo.

Henderson, R. and Clark, K. (1990), 'Architectural innovation: the reconfiguration of existing product technologies and the failure of established firms', *Administrative Science Quarterly*, **35** (1), 9-30.

Hertog, P. den (2006), 'Knowledge intensive business services and their role in innovation', presentation at the OECD workshop on Knowledge Intensive Service Activities, 22 March 2006, Sydney.

Hertog, P. den (2000), 'Knowledge intensive business services as co-producers of innovation', *International Journal of Innovation Management*, **4** (4), 491-528.

Hertog, P. den and Jong, G. de (2007), 'Randstad's business model of innovation: results from an exploratory study in the temporary staffing industry', *Innovation: Management, Policy & Practice*, **9** (3-4), 351-364.

Hertog, P. den and Rubalcaba, L. (2006), 'Is there a rationale for services R&D and innovation policies?', RENESER working document prepared for DG Internal Market and Services as part of the Framework Study on the Research and Development Needs for Services (RENESER), Topical paper no. 1, Dialogic/Servilab, Utrecht and Madrid.

Hertog, P. den, Rubalcaba, L. and Segers, J. (2008), 'Is there a rationale for services R&D and innovation policies?', *International Journal of Services Technology and Management*, **9** (3-4), 334-354.

Hertog, P. den, Poot, T. and Meinen, G. (2005), 'Towards a better measurement of the soft side of innovation. First results of an experiment aimed at measuring non-technological innovation using an adapted innovation survey in the Netherlands', in J. Sundbo, A. Gallina, G. Serin and J. Davis (eds), *Contemporary Management of Innovation*, Basingstoke: Palgrave Macmillan, pp. 181-202.

Hertog, P. den, Ark, B. van and Broersma, L. (2003), 'On the soft side of innovation: services innovation and its policy implications', *De Economist*, **151** (4), 433-452.

Hertog, P. den, Bilderbeek, R., Marklund, G. and Miles, I. (1998), 'Services in innovation: Knowledge Intensive Business Services (KIBS) as co-producers of innovation', SI4S synthesis paper no. 3, published by STEP, Oslo.

Hertog, P. den, Bilderbeek, R. and Maltha, S. (1997), 'Intangibles: the soft side of innovation', *Futures*, **29** (1), 33-45.

Howells, J (2001), 'The nature of innovation in services', in OECD (ed.), *Innovation and Productivity in Services*, Paris: OECD Publishing, pp. 55-79.

Howells, J. (2000), 'The nature of innovation in services', report presented to the OECD 'Innovation and Productivity in Services Workshop' 31 October-3 November 2000, Sydney.

Howells, J., Tether, B. and Uyarra, E. (2004), 'Innovation in services: issues and trends', Final Report to DG Enterprise, Brussels.

IfM and IBM (2007), *Succeeding through Service Innovation: A Discussion Paper*, Cambridge, UK: University of Cambridge Institute of Manufacturing.

Kline, S. and Rosenberg, N. (1986), 'An overview of innovation', in R. Landau and N. Rosenberg (eds), *The Positive Sum Strategy*, Washington, DC: National Academy Press, pp. 275-305.

Løwendahl, B. (2005), *Professional Service Firms*, Copenhagen: Copenhagen Business School Press.

Metcalfe, S. and Miles, I. (eds) (2000), *Innovation Systems in the Service Economy: Measurement and Case Study Analysis*, Dordrecht, The Netherlands, Amsterdam, The Netherlands, Boston, USA and London, UK: Kluwer.

Miles, I. (2000), 'Services innovation: coming of age in the knowledge-based economy', *International Journal of Innovation Management*, **4** (4), 371-389.

Miles, I. (1993), 'Services in the new industrial economy', *Futures*, **25** (6), 653-672.

Miles, I., Kastrinos, N., Flanagan, K., et al. (1995), 'Knowledge intensive business services: their roles as users, carriers and sources of innovation', report to the EC DG-XIII Sprint EIMS Programme, PREST/TNO, Manchester and Apeldoorn.

Norman, R. (1991), *Service Management: Strategy and Leadership in Service Business*, London: John Wiley.

OECD (2006), *Innovation and Knowledge-Intensive Service Activities*, Paris: OECD Publishing.

OECD (2005a), *Enhancing the Performance of the Services Sector*, Paris, France: OECD Publishing.

OECD (2005b), *The Measurement of Scientific and Technological Activities: Oslo Manual. Guidelines for Collecting and Interpreting Innovation Data*, Paris: OECD Publishing.

O'Doherty, D. and Arnold, E. (2003), 'Understanding innovation: the need for systemic approach', *The IPTS Report*, 71, February 2003.

Pavitt, K. (1984), 'Patterns of technical change: towards a taxanomy and a theory', *Research Policy*, **13** (6), 343-373.

Pine, J. and Gilmore, J. (1999), *The Experience Economy*, Boston: Harvard Business School Press.

PREST/TNO/Servilab/ARCS (2006), 'The future of RD in services: implications for EU research and innovation policy', report commissioned by the Science and Technology Foresight Unit of DG Research of the European Commission, EUR 21959, Brussels.

Quinn, J.B. (1992), *Intelligent Enterprise*, New York: The Free Press.

RENESER (2006), 'Research and development needs of business related service firms', final report to DG Internal Market and Services, commissioned by DG Internal Market and Services of the European Commission, Dialogic/Fraunhofer/PREST/Servilab, Utrecht, Stuttgart, Manchester and Madrid.

Schumpeter, J. (1934), *The Theory of Economic Development. An Inquiry into Profits, Capital, Credit, Interest, and the Business Cycle*, Cambridge, MA, USA: Harvard University Press.

Soete, L. and Miozzo, M. (1989), 'Trade and development in services: a technological perspective', working paper 89-031, MERIT, University of Maastricht, Maastricht.

TEKES (2007), 'Innovation policy project in services – IPPS 2006-2007', policy blueprint report, TEKES, Helsinki.

Tidd, J., Bessant, J. and Pavitt, K. (2005), *Managing Innovation: Integrating Technological, Market and Organizational Change*, London: John Wiley.

Tidd, J. and Hull, F. (2003), *Service Innovation. Organizational Responses to Technological Opportunities and Market Imperatives. Series on Technology Management*, London: Imperial College Press.

Toivonen, M. (2004), 'Expertise as business. Long-term development and future prospects of knowledge-intensive business services (KIBS)', PhD thesis, Espoo: Helsinki University of Technology.

14. The Evolution of Innovation Paradigms and their Influence on Research, Technological Development and Innovation Policy Instruments

Patries Boekholt

INTRODUCTION

This chapter describes the historical evolution of public policies to enhance science, technology and innovation. Attention is given to the design and implementation of policy instruments in successive periods as the theoretical and empirical understanding of the innovation process has evolved. In line with the general argument of this book, instruments are analysed in conjunction with the gradual development of different analytical concepts of the innovation process and the evolving understanding of bottlenecks that constrain private research investment in science and technology.

The development of public policies is strongly interlinked with progress in thinking about innovation processes and the interconnection between science and innovation. The theoretical models used to understand how new knowledge is created and translated into innovations and conversely how innovations have spurred the creation of new knowledge have shown considerable developments over the past few decades. These models have affected policy makers and the way they have designed policy instruments to stimulate innovation. This chapter focuses on how governments and not-for-profit intermediaries have responded to the challenge of innovation. Thus, this chapter will not describe in detail the evolution in the thinking on the innovation process and the interaction between science and innovation, which is covered in earlier chapters, but rather describes how these developments translated into public science and technology policy instruments.

Four types of policy domains discussed in the literature are closely related: science policy, research (and development) policy, technology policy and innovation policy. Each of these policies addresses a specific part of the 'knowledge chain', although there are no clearly defined boundaries between them and there are shifts over time in views as to where these the overlap occurs.

According to the UK's Royal Society definition 'science policy deals with policy for science (covering issues such as funding of university research, intellectual property rights for scientific research and education policy) and public policy issues with a scientific aspect'.[1] Thus, it tends to deal with public sector science system and its wider context. The broader term research (and development) policy also includes the activities of private sector research performers, particularly when development activities are included. Technology policy is a specific aspect of research policy, which is restricted to policies aiming at the enhancement of technological improvement, often excluding organisational and other non-technological changes. And finally, innovation policy refers to policies stimulating the translation of knowledge into new commercial applications.

Each of these different concepts can be matched to specific aspects of policies that exist in today's industrialised countries. Which of these terms are used is rather arbitrary in the policy debate and partly depends on national cultures. Whereas some countries make an explicit distinction between research and innovation policy (governed by different parts of the system) others have a much more integrated approach to these policies. In policy practice, the demarcation of policy responsibilities – for science, research and innovation – varies from country to country. European countries typically have an Education and Science Ministry dealing with the university system and basic science while the Ministry for Industry deals with the part of technology and innovation policy that involves the business sector.

This chapter will use the generic label Research, Technological Development and Innovation (RTDI) policy, unless there is a specific focus on one type of these policies.

Underlying this chapter is the evolution of the concepts of innovation and innovation policy over time from the linear model to a more systemic and even 'holistic' model of innovation policy. Many authors have described this development in detail (see for instance Lundvall 1992; Smits & Kuhlmann 2004). This chapter is aimed to link these conceptual developments to their use in innovation policy, helping to reach an increasing 'sophistication' of policy instruments. A policy instrument can be defined as a government measure or programme that aims to change the behaviour and actions of the actors involved in the whole process from generating new ideas into innovative market introductions and solutions. The first section describes the period before innovation policy as such existed, where science and industry policy existed as separate policy domains. The following section discusses how the concept of innovation changed from a linear model to a more complex chain linked model. This inspired new types of policy instruments that were less supply-side driven. The third section describes the emergence of innovation systems thinking and the effect it had on innovation policy. The fourth section

1 On the question What is Science Policy? The Royal Society gives this definition at: http://royalsociety.org/page.asp?id=1168 (29 December 2007).

discusses the most recent trends where the challenge for the 'horizontalisation' of innovation policy asks for better coordination between different policy domains. Finally, the last section draws conclusions on the basis of this chapter, particularly concerning the interaction between concepts of innovation, concepts of innovation policy and the development of specific instruments.

THE CO-EXISTENCE OF SCIENCE AND INDUSTRY POLICY IN THE LINEAR MODEL

In the three decades after World War II an integrated RTDI policy did not exist. On the one hand, science policy was responsible for shaping and funding the (public) research system. On the other hand, there was industrial policy that focused on promoting industrial sectors.

In this period the notion of innovation derived from the linear innovation model: new knowledge was created through basic research in universities and public laboratories. Subsequently these ideas were passed on to the more applied research establishments, again either within industrial research and development labs or public applied research organisations. Those elements of knowledge that could be usefully transferred for commercial purposes were developed by engineers and handed over to the manufacturing departments who would subsequently rely on the marketing and sales persons to bring the resulting innovation on the market. Many innovation analysts have described this development from the linear model to more sophisticated models of innovation and it suffices to refer to the abundant pool of literature on this matter (Utterback 1974; Mowery & Rosenberg 1979; Nelson & Winter 1977; Dosi 1982; Lundvall 1992).

Consequently, in the first generation of RTDI policies – at that time mostly science policy – the typical role of governments was to ensure that ample high quality research efforts were made at universities and public laboratories, which would ensure a significant pool of knowledge for inventions that could subsequently be transformed into commercial innovations. An important aspect of this science policy was the education of a sufficiently large pool of highly skilled people to conduct research and innovation activities, either in the public or private sector.[2]

While today the investments in science and technology are more overtly linked to improving industrial competitiveness, the 1950 and 1960s cold war power relations spurred governments to invest in strategic research. This could be either related to defence issues or the management of critical resources (energy, information

2 It must be noted that although the linear model is depicted here as belonging to a past era, in policy practice many countries – and particularly countries that have only recently developed an RTDI policy still adhere to the linear model and invest mainly in their public research system. A second note is that in some domains (e.g. life sciences) a predominantly science push model still exists and operates with long technological life cycles with uncertain market opportunities.

technology, etc.). The manifesto for the new view of science was Vannevar Bush's 1945 report *Science: The Endless Frontier* (1945), which successfully argued the case for a US National Science Foundation and paved the way for the massive expansion of higher education and research since the war. Bush had spent the war doing a mixture of applied and use-inspired research to underpin weapons development. Yet his manifesto for post-war science was basic science and he argued that increasing science funding would automatically increase product and process innovation and therefore national competitiveness as well as military preparedness (Stokes 1997; Arnold 2007).

A number of countries also chose to set up a separate system of applied research, which in parallel with the basic research-oriented university system, would work more towards societal- and industry-oriented research. In the US an extensive Federal Laboratory system was set up in the immediate post war period (see Adams et al. 2003) focusing on defence, energy, aeronautics, computer science, treatment of disease and so on. The main channels of exchange between the laboratories and industry were contractor relationships, cooperative research agreements, use of laboratory facilities, movements of scientists and engineers and patent licensing. During the first decades no additional measures were found to be necessary to enable a better flow of knowledge between the Federal labs, the universities and industry. Adams et al. (2003) state that from 1940 to 1980 the laboratories engaged in internal research and supported research in firms and universities, but were not formally required to commercialise technology.

The German Fraunhofer Society, founded in 1949, and the Dutch TNO (The Netherlands Organisation for Applied Scientific Research) organisation founded in 1932 are further typical examples of where government, industry and scientists and engineers came together to establish an applied research system that could focus on specific technology areas and industrial sectors.

Industrial policy in those early post-World War II decades was not particularly focused on research and development, but more on creating good market conditions, and from the 1970s, after economic growth started to falter under pressure of global competition, on the restructuring of vulnerable industry sectors (coal and steel, shipbuilding, etc.). Rothwell and Zegveld (1985) and later Rothwell and Dodgson (1992) characterise this period as one where national industrial policy mostly had attention for large firms and agglomerations, also with the objective to support national 'flagship' companies. There were, however, great differences between Japan, the US and Western Europe. Whereas Japan had an aggressive industrial policy with long-term public and private sector strategies, the US focused on strategic sectors in line with its position as superpower, while Europe was more concerned with supporting traditional sectors. It is, however, a misconception that in contrast to Europe, the US did not choose strategic sectors, and choices were made bottom-up by the research actors and companies themselves. Federal government programmes and agencies invested in focused technology and sectoral domains throughout this period. The stories of Boston's Route 128 Corridor and California's Silicon Valley show the large influence of considerable funding from defence agencies in funding the key actors in that innovation system. Saxenian (1996) explains for instance how the Massachusetts Institute of Technology (MIT) became the nation's leading centre of research during the war, performing more military research than any other US university. MIT laboratories received one-third of the $330 million in defence research contracts during the 1940s and 1950s. Similar examples can be provided of the role of Stanford University and how its involvement in government research programmes encouraged companies such as Hewlett Packard and indirectly Intel. The emergence of high-tech industries in Massachusetts as well as those in the Silicon Valley area is strongly related to the strategic goals of US research and technology policy in those days.

Overall, RTDI policy instruments in the first decades after World War II clearly reflected the linear model where innovations would fall out of the knowledge production 'pipeline' starting with basic research, taken up by scientists and engineers conducting applied research either in public research laboratories or in private sector research laboratories. Thus government spending was mainly focused on:

1. Funding and co-funding of basic and applied research within existing institutions such as universities and laboratories;
2. Launching large research programmes in mission-oriented and strategic policy areas (defence, energy, aerospace, computers and so on).

Innovation policies in these early generations were supply-side driven: they focused on the R&D inputs into the economy and society but paid little attention to the importance of the demand for technology and innovation and the interactions between the different actors involved in the entire 'knowledge chain'.

THE CHAIN-LINKED INNOVATION MODEL: SECOND GENERATION INNOVATION POLICIES

Changes in the Understanding of the Innovation Process

During the late 1960s up to the mid 1980s, the first criticisms of the linear model emerged in the innovation literature. It started with the critique of the 'technology push' pipeline model and the insight that there was also something of a 'market pull' aspect in the origins of new ideas. Authors such as Schmookler (1966) and Utterback (1974) introduced the 'market pull' concepts and for many years scholars debated which of the two models was dominant (see Mowery & Rosenberg 1979). Rothwell and Zegveld (1985) argued in the early and mid 1980s that the relative importance of the two directions might vary considerably during different phases of the technology and industry life cycles. Kline and Rosenberg (1986) were among the first to move away from this 'one- or two-directional thinking' and introduced the chain-linked model, based on an interactive model of the innovation process, with multiple feedback loops between research, the 'corpus of scientific and technological knowledge' and the central chain of innovation within individual firms to develop ideas into products for the market. This meant that at several critical points in the knowledge production chain, interaction between the public and private R&D communities would be essential. This inspired the more sophisticated design of policy instruments to encourage a closer interaction between two important elements of what later became known as the innovation system.

In the 1990s the concept of the user-producer interactions in the innovation process also appeared in the innovation debate (von Hippel 1998). Von Hippel's argument mainly focused on the influence of consumers and early adopters on the innovation process. He identified 'lead' users who have needs that foreshadow general demand in the market place. However, although this debate found great response in the innovation analyst community, it did not have a great influence on the design of specific policy measures. Some experimentations were launched (for instance in the area of green procurement in Scandinavia), but the development of user-oriented policy instruments even today proves quite difficult.

More influential was the debate on the role of clusters and value chains (Porter 1990) and the interaction between firms in these value chains as a source of innovation. In combination with innovation systems thinking that surfaced around the same time, it supported the arguments for the support of networks and agglomerations. Cluster policy was mainly taken up by regional governments, as it proved difficult to have a more selective approach from the national perspective. In the decade when industrial policy had grown into a 'suspect' type of policy (e.g. too defensively oriented to saving 'national champions'), cluster policy got caught up in the policy debate between two factions in the economic and RTDI policy domains: one side arguing in favour of letting the market forces decide which R&D investments should be made, with the other faction arguing that in order to induce changes to the economic structure and to dynamise innovation systems, policies have to focus on existing and potential pockets of strength. This tension between these two opposing economic policy paradigms can still be seen today and strongly affects the debate on 'appropriate' policy instruments.

In Europe the inspiration for developing new policy instruments came mainly from looking 'with envy' at the developments in the United States and their drivers of innovation. In the course of the 1980s a number of characteristics of the US situation became apparent as factors behind its success:

- The rise of *new-technology-based small firms* (NTBFs), their fast growth and considerable market shares;
- The availability of risk-taking *seed and venture capital* that formed a basis for this fast growth of NTBFs;
- The innovation potential of a deep *pool of SMEs* who contribute significantly to economic growth;

- The role of a number of leading universities in *spinning-off* knowledge as a basis for the creation of yet more NTBFs.

While a number of these developments were mainly spurred by market developments in the US, in Europe the absence of these factors can be seen as the driver to launch policy instruments to 'create' similar conditions. The US in the meantime launched the Small Business Innovation Research (SBIR) programme in 1982 to support small innovative firms with grants to develop their inventions into marketable prototypes. This programme is still existing today and it has been recently 'copied' in various European countries such as Sweden, Ireland and the Netherlands.

Response in Innovation Policy

It took over two decades, however, before the 'innovation is not linear' message sunk in at the policy level and still today linear thinking is strongly present in many national and regional policy strategies and instruments.

The gradual understanding that the 'pipeline model' – where governments invest financial resources in public research and anticipate that inventions would 'automatically' be derived for industry to further develop and commercialise – no longer works well has led to a typical set of policy instruments designed to overcome these failures. However, the bulk of the existing policy instruments addressed directly to industry were – and today in many countries still are – subsidy and grant schemes that co-funded R&D projects in industry. The debate whether governments were justified to intervene in what was essentially a private sector activity had in the 1960s already settled around the concept of market failure (Nelson 1959; Arrow 1962) leading to underinvestment in research. Hauknes and Nordgren (1999) describe three basic factors that limit the attainment of a social optimum through private profit optimisation:

- Outcomes of knowledge-generating processes are uncertain.
- Knowledge is a (quasi-) public good, implying inappropriability.
- There are substantial indivisibilities in knowledge generation.

The direct state funding of private R&D activities became a mainstream type of policy instrument in many countries although the debate about the degree of 'market interference' is still taking place in policy today. Some countries chose to fund private R&D through soft loans and credits (e.g. Spain), others from a very early date used tax incentives to limit the financial risk of R&D activities (e.g. the Netherlands), many countries had subsidy programmes funding specific R&D projects (e.g. Finland, Germany) while other countries were more hesitant to provide direct support to firms (e.g. Switzerland and the UK).

Nevertheless, in the course of the 1980s and the early 1990s a set of typical measures and policy instruments were designed that fitted with the progress in understanding of the innovation processes. The main new types of instruments that were launched in industrialised countries and particularly in Europe were:

- Technology transfer mechanisms.
- Instruments to support the establishment of NTBFs and university spin-offs.
- Setting up of (regional) intermediary structures and programmes to help SMEs to access and absorb technologies.
- Instruments to secure finance for innovation (seed and venture capital).

Technology transfer measures

Particularly in the European context it became clear that both university research and public applied research were badly exploited, regardless that Europe was performing well in terms of producing high scientific quality. Few commercialisations came out of publicly funded research, and industry was hardly tapping into the pool of knowledge that was readily available. A few years later this was labelled as the 'European paradox' in the European Green Paper on Innovation (European Commission 1995). To repair this situation the first policy response was to stimulate technology transfer. It was clear that public research institutes did not have sufficient incentives to commercialise their work while the business sector was not sufficiently aware what was on offer in the universities and research centres.

Universities and laboratories were encouraged and supported to set up 'liaison offices' that could communicate to the outside world what expertise and research results were available at these many sites. Government funding was made available to set up these activities. Unfortunately most of these first generation technology transfer instruments were still very much inspired by the linear model. The activities of the liaison offices were mainly to try and sell the good ideas and expertise that was present in the research laboratories. Little effort was made to understand what type of knowledge was actually needed by the business sector and on which terms. At the same time the parts of the business sector which had built up relations with universities had direct contacts with the departments and professors they knew and had little interest in communicating through an intermediary organisation. Thus in Europe not many of these liaison offices were a success in their first generation.

Although the US was generally considered to be better equipped in terms of technology transfer from public research to private sector commercialisation, with the launch of the Bayh-Dole Act of 1980 the government gave an important impetus to technology transfer. According to Göktepe and Etzkowitz (2005), Bayh-Dole legitimised technology transfer and encouraged the growth of a profession by dramatically increasing the number of Technology Transfer Offices (TTOs) and the size of these offices. Mowery (2005) argues that this should be seen in the context of a long history university-industry interaction, unique for the US higher education system.

A typical policy instrument with its roots in this period was the creation of science parks. By creating small incubators and high-tech business parks, often in the vicinity of universities, the creation and growth of university spin-offs would be facilitated (see for instance Massey et al. 1992). Here again, many of the first generation science parks were not really successful as they used too simple linear model assumptions that good technological ideas would automatically be picked up by the market and become market successes. In addition, in order to stimulate entrepreneurship from university requires more than only the physical infrastructure. Many of the early generation science parks were mainly property development projects that had little to do with the coaching of young technology entrepreneurs towards a viable business model.

Absorptive capacity and the need for dissemination policies

A particular debate in innovation policy in the 1980s was finding the appropriate balance between policies to generate new knowledge (mostly intended to lead to radical innovations) and policies for the diffusion of new ideas (mostly leading to incremental innovations and adaptations). Everett M. Rogers' (1995) comprehensive book on 'Diffusion of Innovations' was first published in 1962. It was only with the third edition of the book in 1983 that the understanding of the adoption of innovations and the need to develop policies to help diffusion became accepted. Rogers noted that the number of publications on the subject had risen from 405 publications in 1962 to 3085 in 1983. Henry Ergas (1986) stated that 'the key problem for technology policies, therefore, is not that of generating new ideas, but rather facilitating their widespread use – and this is equally valid for emerging and mature industries'. He identified countries whose policy instruments were mainly oriented on major projects to achieve big breakthroughs (mission-oriented countries, e.g. France, the US, the UK) and those countries whose policies were mainly focused on upgrading the capacity of firms to respond to new technologies (diffusion-oriented countries, e.g. Germany, Switzerland, Sweden). Japan was seen as the exception doing both at the same time. This categorisation of countries reflects the debate in the innovation literature at that moment, and particularly the emergence of diffusions and technology transfer as important mechanisms to enhance the spillovers from what was still inherently an innovation system based on the linear model. A typical example that was often quoted a good practice in the 1980s was the German Steinbeis Foundation system already established in 1971 but developed as a regional Technology Transfer Centre in 1983. The aim of the centre was to give SMEs in the Baden-Württemberg region advice on technological developments, to bring them into contact with research centres and other companies. Following the region's economic structure, the centre was particularly strong in mechanical engineering where incremental rather than radical innovations were more important.

Wesley Cohen and Daniel Levinthal introduced the idea of 'absorptive capacity' to the R&D and innovation literature in a landmark 1990 article (Cohen & Levinthal 1990). They define it as 'the ability of a firm to recognise the value of new, external information, assimilate it, and apply it to commercial ends'. According to Cohen and Levinthal, it is impossible to predict what the 'right' level of investment in absorptive capacity is for an individual firm. This depends on individual circumstances. Equally, the concept of absorptive capacity has not been developed in such a way that it is readily amenable to comparative study. For the most part, the research and innovation literature treats 'absorptive capacity' as a black box. Purely from the literature, we can see five more or less distinct elements within the box (Arnold et al. 2004):

- Human capital, especially in the form of graduates, and especially scientists and engineers
- Ability to network with external sources of knowledge and other resources
- Organisation, routines and organisational processes
- Learning processes (cognition)
- Codification

The importance of 'absorptive capacity' as an underlying factor for innovation and the understanding that a great innovation potential lies in the SME population, translated into the following typical set of policy instruments launched in various countries:

- Innovation management tools and coaching schemes (e.g. the FRAM programme in Norway, the European MINT programme);
- SMEs-oriented 'mobility schemes' where technicians or graduate students conduct innovation projects within the firm (e.g. the UK Teaching Company scheme, the German *Innovationsassistenten* or Dutch 'Knowledge Carriers in SME' programmes;
- Networks of innovation agencies and technology centres, which have low threshold for SMEs, provide a range of free and commercial consultancy and refer companies to research centres and other expert organisations for technological support. Typical examples are the ANVAR (now OSEO) and CRITT networks in France, EOLAS (now) Enterprise Ireland, the Business Links in the UK and the Innovation Network (now Syntens) in the Netherlands. Most of these networks have transformed their missions and today apply a range of tools (innovation management, networking, technical support, training, feasibility grants, etc.) to help small businesses;
- Subsidy and grant schemes that would co-finance R&D work carried out by firms or on behalf of SMEs (e.g. collaborative research performed by applied research organisations).

Finance for innovation

Another explanation for the difference between the US and European performance was the perception that in the US an abundant source of risk capital is available by means of seed and venture capitalists. While in the US young entrepreneurs with risky ideas could find private investors to support their business ideas, European entrepreneurs were dependent on traditional risk-averse banks.

In most cases of technology-based early stage investments, the commercial prospects of the firm are very uncertain since the entrepreneur is dealing with new or underdeveloped markets. This complicates the assessment of market and growth potential and thus increases the financial risks. A large part of their assets are intangible, based on knowledge and skills, which does not provide a basis for collateral, similar to fixed capital investments. Furthermore start-up firms are often led by entrepreneurs with notable technological competencies, but a lack of managerial and commercial experience, which increases the burden of management support on account of the investor. There is often a strong emotional reluctance to involve external investors fearing a loss of control over the firm (Boekholt, p. 760 1996). In 1994 a survey of European schemes to support risk finance found four types of schemes:

1. Mechanisms to *reduce the loss* for investors. In this type of support schemes public bodies make an ex-post payment to the investor in case of failure of investment. Examples in Europe were equity guarantee schemes such as the German pilot scheme BJTU (Business Investment Capital for NTBFs) and the Dutch equity guarantee scheme for private investment companies (PPM which was already abolished in the mid 1990s). Another type of mechanism is 'innovation insurance', where the firm and the (public) funders can pay an insurance fee that covers the failure of a particular innovation project. Tax incentives are another possible mechanism. In the UK the Enterprise Investment Scheme had a facility to write off the losses of equity investments in unquoted companies under certain conditions.
2. Mechanisms to *increase the liquidity of and rewards for* investors: These include measures to facilitate exit mechanisms and to improve the functioning of secondary markets. There were several initiatives in Europe to develop (financial) instruments of these types. One example was attempts to establish a European Association of Security Dealers Automated Quotation system (EASDAQ), which would encourage financial deals in non-public firms.
3. Mechanisms to *reduce the current costs and/or scale* of investment for the investor. Here public bodies give an ex-ante financial support for the investor when investing in a specific type of business. This could be part of the investment sum (co-investment), or tax and interest relief for investment in technology-based firms and projects. Joint public-private Venture Capital Funds are the most common national and regional initiatives in this regard. Another type of public support was aimed at assisting the financier in the high management burden when selecting and counselling (young) technology-based firms.
4. Mechanisms to *attract new type of investors* in innovation financing. This last type of support aims to attract investors that have no familiarity with technology-based firms. These could be either very traditional institutional investors, or

wealthy private persons preferably with managerial expertise. In Denmark there is an initiative to set up business angels networks, where the Danish Technological Institute acts as a broker between individuals that want to invest in growth companies and entrepreneurs with a good innovation project but lack of finance.

More, recently an expert group for the European Commission (2003) developed a taxonomy of existing schemes related to finance of innovation showing a similar categorisation of current measures. Their inventory includes (1) public investment in venture funds instead of private investors (substitution), (2) public investment alongside private investors (co-investment, to increase the funds available), (3) overhead subsidies to private investment companies (to address the disproportionate costs of making very small early stage venture investments) and (4) refinancing or leverage of private investment on favourable terms (again to increase funds available).[3]

Nevertheless a point of debate remains whether the largest problem is a lack of available funds or a lack of well-developed business plans of young technology entrepreneurs. Therefore increasingly policy approaches to stimulate new technology-based firms tackles the issue from various sides: the coaching of young entrepreneurs, the provision of physical infrastructures (incubators and laboratories) as well as the linkage to public and private sources of funding. An example of such a packaged approach is the recently introduced TechnoPartner scheme in the Netherlands, which includes support for pre-seed and seed capital, support for technology transfer offices to improve their professionalism and support for the creation of a more entrepreneurial environment in universities.

THE INNOVATION SYSTEMS APPROACH

Innovation Systems as the Leading Concept in Policy

The innovation systems approach, as described in the introductory chapter of this book was much inspired by the work of Freeman (1987) and later elaborated by Lundvall (1992). The multiple national innovation systems (NIS) studies in the 1990s elaborated the concept and described the specific features of national and regional systems (Edquist 1997; Cooke et al. 2000). With the emergence of the NIS concept, additional rationales for policy intervention were provided and alongside the argument of 'market failure' came the arguments based on systems failures (see Smith 1997).

---

3 European Commission, Raising EU R&D Intensity, Risk Capital Measures, Report by an Independent Expert Group, 2003

The OECD report 'Dynamising National Innovation Systems' (2002) states that implementing the NIS approach reinforces the need for innovation policy to focus on complex interactions between innovators and their partners.

It took a good decade or so for the innovation system concepts to be accepted in policy practice and this at first in only a small number of countries (e.g. Sweden, the Netherlands). Sweden labelled the name of its agency VINNOVA as The Swedish Governmental Agency for Innovation Systems in 2005. In the first years of the century the Netherlands used the Innovation Systems model in leading policy papers as a model to analyse the strengths and weaknesses.

While it inspired the creation of some types of policy instruments, the concept of innovation systems was mainly used by policy makers as an analytical framework to decide what policy instruments are necessary to tackle specific 'system failures'. The reflection of the innovations system concept in policy instruments can be typically found in the many 'bridging' type policy instruments that aim to 'repair' gaps between two or more components of the system. The development of instruments for better industry-science relations is an example of a typical systems approach. At the same time the innovation system concept is typically a European concept that is not always used so explicitly in non-European countries. As in the 1980s, alongside institutional funding for public research, the emphasis in innovation policy was still on subsidy and grant schemes for industry R&D projects. Gradually in the course of the 1990s in many countries the emphasis shifted from grant schemes for individual firms towards collaborative grant schemes where some type of industry-academia interaction had to take place. Current knowledge about innovation systems as well as the functioning of clusters will have contributed to this development. Countries such as the UK and the Netherlands abolished single firm subsidy schemes all together.

In the last few years, however, these subsidy schemes have been slowly replaced or complemented by fiscal incentives for R&D (European Commission 2003).[4] Today many European countries have introduced or are experimenting with different forms of fiscal incentives that give companies the possibility to deduct R&D costs from their tax bill. The precise design of these schemes depends on the national fiscal and accountancy regimes.

The Policy Instruments in the Innovation Systems Approach

The innovation systems approach reinforced the further development of 'bridging instruments' – i.e., instruments that brought together actors from different components of the innovation system, in particular from public science and industry. As discussed above, such bridging instruments are not completely new: cooperative research programmes have a long history (e.g. the CRADAs in the US launched in 1980 or the SME collaborative research system AIF system in Germany established

4 European Commission, Raising EU R&D Intensity, Fiscal Measures, Report by an Independent Expert group, 2003.

in 1954). However, some new features of the instruments were greater interactive organisational and funding models including public-private partnerships, centres of excellence, and competence centres.

'Competence centres' are a comparatively new form of university-industry research alliance that do both near fundamental but also more applied, problem-oriented research. Their long-term nature and the comparatively high rates of subsidy involved allow them to have a structuring effect on sub-systems of innovation, educating and generating communities of research practice between research-performing institutions and industry and generating common, use-oriented research agendas with potential to have significant positive socio-economic effects (Arnold et al. 2004). The pioneers in this type of arrangements were again to be found in the US with the NSF-sponsored Engineering Research Centres programme that started the competence centres movement. However, the programme's establishment in 1985 predates the emergence of the national innovation systems literature.

By the turn of the century most countries had established a wide package of measures to stimulate innovation. No one model is prevalent but many instruments and measures exist in parallel. Particularly in the more mature countries the issue of transparency and simplification of innovation policy have become high on the political agenda, as there is a tendency to create more and more measures, while old ones are rarely abolished. Figure 14.1 provides a simple taxonomy of the types of policy instruments that have been put in place in advanced industrialised countries. In a national innovation systems perception, no single instrument can make the difference between performing well and badly. In addition, the co-existence of multiple instruments influence each other and could provide on the one extreme amplification of each other's positive effect or on the other extreme counteract and interfere with each other. Figure 14.1 distinguishes four constituents from a national innovation system perspective: by membership of the private or public sector and by their role as knowledge users or knowledge creators. All these constituents need to interact well if the system is to function smoothly. Therefore the bottom part of Figure 14.1 shows policies that are important for knowledge creators both in the private and public sector, while the top half addresses knowledge users, including parts of the private sector that need to improve their absorptive capacity and step up their innovation activities. The intermediary type of 'bridging' policies in the middle cross-sections of the table are specifically designed to link the outer quadrants together encouraging inter-linkages between these different actors.

Demand-Led Policies

Demand-led policies have seen several 'swings' in policy attention. As mentioned above, the insights in user-producer relations did not have an immediate effect on innovation policy development. In a broader sense demand-led policies through the involvement of stakeholders (i.e., the users of RTDI programmes) in the

	Public Sector		Private Sector
Knowledge Users	**Reinforcement Policies for Public Sector** Knowledge Users • Support for science/technology educational institutions • Support for government 'knowledge workers'	**Bridging Initiatives between Public and Private Sector Knowledge Users** • Public sector bodies providing technical information services • Public sector bodies providing metrology services • Public sector bodies providing patent and licence information • Public sector bodies providing consultancy advice to firms	**Reinforcement Policies for Private Sector Knowledge Users** • Mostly policies belonging to other policy areas (business development, training, regulatoryframework, etc.) • Subsidy for technology feasibility studies-technology audits • Subsidies for hiring technicians in firms
	Bridging Initiatives between Public Sector Knowledge Users and Knowledge Creators • ICT network infrastructuresupport • Training by research	**Bridging Initiatives between Public and Private Sector Knowledge Users and Creators** • Systemic STI policies	**Bridging Initiatives between Private Sector Knowledge Users and Creators** • Subsidised technology transfer and adoption schemes • Technology brokerage schemes • Support for the growth of private sector 'intermediaries' • Awareness campaigns • Technology demonstration initiatives
Knowledge Creators	**Reinforcement Policies for Public Sector Knowledge Creators** • Support for research equipment infrastructure in universities • Support for government labs • Support for basic science • Support for generic collaborative research between universities • Support for networks of research excellence	**Bridging Initiatives between Public and Private Sector Knowledge Creators** • Support for targeted collaborative research between universities and industry • R&D services provided by Government labs to industry • R&D services provided by universities to industry • R&D partner search initiatives • Technology transfer infrastructure • University/industry personnel exchange schemes forR&D staff • Spin-off promotion policies • University Liaison Offices • Targeting R&D to areas with commercial potential • Hiring subsidies for scientists	**Reinforcement Policies for Private Sector Knowledge Creators** • Subsidised market-oriented R&D • Product development assistance • R&D tax incentives • Innovation credits • Favourable IPR regimes • Inward investment schemes for 'R&D-intensive' industries

Source: STRATA-ETAN Expert Group, EU DG Research, 2002

Figure 14.1 A simple taxonomy of science, technology and innovation policies.

programming of policies through 'platforms', foresight and road mapping exercises is gaining support in some countries, while still very contested in others.

Today a specific type of demand-led policy is gaining in political momentum: the use of public procurement for innovation. The 2006 Lisbon Progress Report (European Commission 2006a) highlighted the need for 'using public procurement more as leverage to promote innovative private sector solutions'. The influential Aho report (Aho Group 2006) confirmed the need for action in this field calling upon governments to 'use public procurement to drive demand for innovative goods, while at the same time improving the level of public services'.

The interest in using public procurement as a mechanism to raise investments in innovation and research is growing in Europe and often reference is made to US examples of innovation-oriented public procurement (such as SBIR). However, practice has illustrated that good public procurement is difficult, and even more so when it concerns the procurement of innovations (see Edler et al. 2005). Only a few countries, particularly in the UK and Sweden, have developed a coherent set of public procurement policies with the aim of stimulating innovative solutions. Other countries are today developing strategies and approaches as well, but the gap between innovation policy makers and public procurers is still considerable.

Policy Learning and the Use of 'Strategic Policy Intelligence'

The majority of RTDI policy instruments have the R&D performers (researchers, research organisations, firms) as their primary target groups. In the course of the evolution of RTDI policy there has been a growing need to develop policies and tools to improve policy learning. One important influence has been the emergence of New Public Management. Accountability, transparency and the desire to minimise distortions arising from government policies while maximising their leverage are driving the trend towards greater evaluation (OECD 2000). These are elements of a new public management culture, which operates following clear policy objectives and targets for achieving those objectives. This type of culture is strongly embedded in English-speaking and Scandinavian countries but is strongly emerging also in countries such as the Netherlands and in the European Commission. Many of the evaluation tools have an *ex-post* character and aim to learn from past and ongoing policy implementation. Increasingly tools for prospective analysis are applied by policy makers and their stakeholders in order to identify long-term trends and thus to guide decision-making. These are not necessarily linked to the innovation systems approach as tools such as the Delphi Exercises, Foresight Studies Technology Assessment have already been used as policy planning tools since the 1970s and 1980s. However the 1990s did show a huge surge of these tools being used in all industrialised countries. Table 14.1 gives an overview of the types of strategic intelligence, which are needed across the different phases of policymaking. This has some resemblance to what Smits and Kuhlmann (2004) refer to as the rise of 'systemic instruments' in innovation policy. They identify the need for instruments to manage the interfaces in NIS, to (de-)construct and organise innovation systems, to provide platforms for learning and experimenting and infrastructures for strategic intelligence and to stimulate demand articulation, strategy and vision development.

Policy evaluation is an important element of policy learning. While RTDI evaluation tools and other policy planning tools have developed considerably over the past decade, the *European Innovation Progress Report* (European Commission 2006b, p 74) distinguishes considerable differences between European countries in terms of their use of these practices and their evaluation culture. Thus in addition to the availability of strategic intelligence tools, embedding of a learning culture

Table 14.1 Overview of strategic intelligence needed in different phases of policy making

Phase in cycle	Functions	Strategic intelligence
Policy design	• Assess future trends in science, technologies, sectors and markets and/or facilitate this foresight with stakeholders • Analyse bottlenecks in the national system of innovation (with stakeholders) • Translate political agendas into policies • Coordinate with other policy domains • Define the policy mix and portfolios (priorities) • Decide to terminate existing policies • Decide on new policy measures to tackle bottlenecks or grasp opportunities • Allocate resources to policy programmes(s) • Define programme rationale	Foresight studies, trend studies, roadmaps Benchmark studies, SWOT analysis, sector & cluster studies Systemic evaluations, Evaluations of RTDI programmes, Evaluations of S&T institutions Processes for stakeholder involvement
Programme design	• Elaborate new programme ideas into a new measure • Set rules of the game and funding modalities • Seek EU approval for new measures • Define monitoring and evaluation strategy for new measure	Problem analysis Needs analysis Stakeholder analysis Solution analysis Processes for stakeholder involvement
Programme management	• Manage and allocate programme resources • Provide reporting on programme • Inform stakeholders and (potential) participants of programme • Organise assessment of proposals • Execute programme evaluations	Programme management information systems -Evaluation practices and toolkits User interfaces
Programme administration & delivery	• Monitor project resources and progress • Interact with programme users on appropriateness programmes • Provide reporting on projects	Project monitoring information system User surveys

Source: Adapted from Boekholt et al. (2004).

throughout the public sector is a prerequisite for improving the effectiveness of policy instruments.

The Increasing Role of Internationalisation in RTDI Policy

From the late 1990s, globalisation and the internationalisation of R&D emerged as a major challenge in many industrialised countries. Concerns from the perspective of policy makers included the threat of the relocation of private sector R&D activities and the fear of brain drain (of talented researchers and highly skilled workers) to countries with a higher level of research excellence (e.g. US East and West coast areas) and/or cheaper and more abundant human resources for R&D (e.g. India and China). In addition given that multinational corporations are globally more mobile to search for the best research competences, a concern is whether the national innovation system is sufficiently attractive to be a magnet for foreign R&D investments, either through the location of private R&D facilities or through collaboration with national research centres. Few countries have a fully systematic R&D internationalisation strategy. The most developed is Germany, which has promoted a range of R&D internationalisation policy instruments.

A review of internationalisation policies in eight countries found three types of policies (Edler and Boekholt 2000):

1. Instruments to enhance national attractiveness towards foreign researchers and foreign private investment (e.g. incentives to attract researchers and investment subsidies for R&D-intensive firms).
2. Instruments to facilitate the absorption of knowledge created globally (e.g. scientific attachés monitoring technological developments abroad).
3. Complex approaches to encourage exchange and interaction in both directions and integrate multiple actors, within one country and/or between different countries.

In Europe, a wide set of instruments for transnational collaboration/coordination is applied including: multi-lateral research initiatives (often based on inter-governmental agreements) where membership varies from all EU and associated countries to a smaller selection of countries (e.g. COST or EUREKA, as well as intergovernmental organisations such as CERN, EMBL); bilateral research agreements between two countries; joint research funds (e.g. Nordic Council, European Science Foundation); European Commission programmes and particularly the Framework Programmes (FPs), the Competitiveness Programme (CIP), Structural Funds and actions such as ERA-NETs and Article 169 initiatives. As part of the European Research Area, the European Commission aims to encourage countries to coordinate national research policies and to open up national research programmes for foreign partners. So far this has not progressed very far and national objectives remain the dominant driver for policy makers. Nevertheless, in Europe the common agreement of the Lisbon goals and the Barcelona target of spending 3 percent of Gross Domestic Product (GDP) on R&D, (of which two-thirds should come from the private sector) have brought RTDI policy more centrally into transnational debates and has spurred an increased interest in cross-border policy learning. In addition, there is a realisation that critical societal issues such as climate change, future energy supply and the ageing population increasingly call for policy interventions that go beyond the national borders.

TOWARDS A MORE HOLISTIC APPROACH: INNOVATION POLICY ACROSS SOCIETAL AND POLICY DOMAINS

Today RTDI policy formulation and implementation is mostly in the hands of ministries for education and science and/or ministries for industry and the agencies that are attached to them such as research councils and innovation agencies.

Nevertheless, other policy domains such as health, environment, energy and transport have vested interests in commissioning, managing and even conducting research and development activities as well, although innovation is not at the heart of their policy mission. A large share of publicly-funded research and development is related to societal issues such as health care, disease control, mitigation of pollution, climate change, energy conservation and so on. This type of research is often tackled outside the realm of the classical policy instruments in RTDI. Several innovation analysts (Lengrand et al. 2002; OECD 2005) have labelled the 'third generation' innovation policy as one where innovation is at the heart of all other policy domains in the knowledge-based society. If we consider this as the policy paradigm after the linear model, the chain-linked innovation and the innovation systems approaches, this could even be called the fourth generation of innovation policies.

The Importance of Governance of Research and Innovation Systems

A major obstacle to achieve this 'latest' generation innovation policy is the governance of the innovation system. Policy formulation, design and implementation are still very much compartmentalised. Separate ministries and agencies have the responsibility for parts of RTDI policy, but due to a lack of coordination there is no overall strategic vision on RTDI policy and synergies between adjacent policy domains (e.g. health research and the promotion of bio-technology) are not used. In addition, in aiming to achieve the Lisbon objectives, stakeholders from industry have made it clear that ensuring 'framework conditions' conducive to innovation is even more important than public R&D investments. Therefore the challenge of today's RTDI policy is to cross the bridges between different organisational and policy boundaries.

A number of external challenges such as globalisation and increased international competition, the growing complexity and multidisciplinarity of research, pressing societal challenges that need a coherent approach, have several implications for the practice of innovation and research governance.

First, societal actors and institutions outside the research community are likely to be increasingly involved in the governance of research and innovation. For agenda setting and prioritisation of actions, this implies that a variety of stakeholders need to be involved. Given the wider perspective more strategic intelligence will be necessary to come to decisions and priorities (as discussed in 'Policy learning and the use of "strategic policy intelligence"' (section 3)).

Second, governance mechanisms are needed that handle the systemic nature of research and innovation and the need for policies to be coherent and co-ordinated across institutional boundaries. This implies horizontal co-ordination along three lines:

1. The co-ordination and attuning of different societal and economic goals of research and innovation. In policy terms this means co-ordination and attuning between research and innovation policies for stimulating industrial growth, for the better use of information technology, environmental preservation, a healthy population, good quality food, and so on;
2. The integration of knowledge creation (mostly basic research) and the use of knowledge for innovation. In an innovation system this could involve bringing together those actors that focus on different roles in the knowledge production chain. In policy terms this involves the integration of science, research and innovation policy;
3. The combination of knowledge from different science disciplines to tackle interdisciplinary research needs (e.g. bio-technology) and overarching societal problems that need such an interdisciplinary approach (e.g. climate change).

Third, governance mechanisms will be needed that can handle change in the innovation and research funding and performance system. Overlaps and boundary ambiguities accompany change, and it is not necessarily clear that eliminating these will improve system performance. Current debates about the funding models for universities are an example where governments struggle with how to handle the necessary modernisation of the university system.

Fourth, governance mechanisms should allow effective implementation of the actions decided upon. The most 'optimal' division of labour between ministries who tend to be involved in the formulation of overall policies and sometimes design of instruments and agencies involved in the implementation of policies is being debated in many countries.

Finally, some of the failures to which innovation systems are prone are the result of failures in governance. Governance systems need, therefore, to be sufficiently reflexive to allow such failures to be identified and acted upon. Thus, even if government authorities could formulate and design a 'perfect' set of policies, its effectiveness depends very much on the governance model in place to implement and use these policies.

Developing a Coherent Policy Mix

In line with the thinking of a holistic and coherent policy coordination is today's discussion about identifying the appropriate set of policies. The previous section gave an overview of the manifold of RTDI policy instruments that are in use today. Each of these instruments has different objectives, target groups, financial implications and management requirements. While in the 1990s there was still the idea that there are 'best practice' policy instruments which need to be disseminated, today the awareness that the success and failure of an instrument is very much dependent on the context (and governance) in which it is used. The debate amongst policy makers has shifted from finding the best policy instruments, to developing a portfolio of instruments that together, in combination, have the best possible positive effect on RTDI investments and usage. This is what is know as the 'Policy Mix' debate that has become a critical aspect of the 'Open Method of Co-ordination' amongst European policy makers. Again this is very much a European debate and one where Europe is probably ahead of, for instance, the US, where coordination between policy echelons (federal versus state, between policy domains) has not been addressed.

A policy mix can be defined as the combination of policy instruments, which interact to influence the quantity and quality of R&D investments in public and private sectors. Thus policy mixes are looking for holistic solutions to addressing RTDI issues and particularly to increasing (public and private) investments in R&D. These policies could fall within the remit of RTDI policy. Increasingly, however, policy makers are aware that other policy domains such as competition policy, fiscal policy, education and macro-economic policies affect the effectiveness of core R&D policies. The national systems of innovation approach can serve as an analytical tool to identify which parts and nodes of the system are weak and what policies should address these weak links in the system. In the words of the European Expert Group on Policy Mixes, "holistic policies" involve the construction of policy portfolios of considerable breadth aimed in the main at rectifying wholesale innovation system deficiencies that, if successful, would stimulate private sector investment in R&D as a natural consequence of a healthy and vibrant innovation system' (European Commission 2003[5]).

Again, the main obstacle to achieve this is that the national research and innovation governance systems are not adequately structured to develop these types of cross-cutting policy strategies, despite various attempts to coordinate policies through high level Research and Technology Policy Councils such as exist for instance in Finland, Austria and the Netherlands. In practice countries that try to tackle their "Policy Mix" issues tend to focus on the available package of RTDI measures and the degree of synergy, overlap and possible gaps. A holistic policy approach would also demand for interactive processes to involve the many different stakeholder groups in society that are beneficiaries of or dependents on RTDI policy measures. A coherent package of instruments is likely to ask for choices for certain societal issues, segments of the economy or scientific and technological domains. A purely top-down approach by representatives of the responsible government bodies will fail to grasp the specific needs and opportunities of the different constituents and

5 European Commission, Raising EU R&D Intensity, Improving the Effectiveness of the Mix of Public Support Mechanisms for Private Sector Research and Development, Report by an Independent Expert Group, Brussels, 2003.

Table 14.2 Overview of mainstream RTDI policy measures used in most industrialised countries

Main category	Types of instruments
Generic R&D policy (Public sector)	Discretionary institutional funding for R&D activities (e.g. block funding to universities and research centres) Competitive R&D project grants Support for R&D infrastructures Selective support for centres of excellence Structural reform of public research sector
Generic R&D policy (Private sector)	Discretionary institutional funding for R&D (e.g. block funding to firms) Competitive R&D project grants Competitive R&D loans R&D-friendly procurement
Thematic/sectoral policy	Selective (collaborative) R&D support schemes for existing high-tech sectors/technology domains Selective R&D support schemes for new high-tech sector Selective R&D support schemes for low- to medium-tech sectors Selective R&D support schemes for specific societal issue
Linkage policy	Collaborative R&D programmes without thematic focus Technology platforms and similar 'stakeholder programming' actions Cluster policies and regional growth pole policies National support for science parks and other co-location schemes Support for other university-industry linkage mechanisms (e.g. university liaison offices) Support schemes for spin-offs
IPR Policy	Reform of IPR regulations Support schemes to help secure IPR
Financial and fiscal policies	Risk capital for R&D measures Loan and equity guarantees for R&D investments R&D tax relief measures for companies
Human resources policies	Efforts to make ST&E more attractive for pupils, students and graduates Subsidies for hiring S&T personnel Tax incentives for (foreign) ST&E personnel Immigration policies for ST&E personnel

stakeholders in the innovation system. A completely bottom-up process of defining needs will most likely disperse efforts among many activities and instruments. Some of the strategic intelligence tools (e.g. foresights, roadmaps) described in the third section ('The innovation systems approach') could provide support, but here lies an open field for developing better multi-actor agenda-setting tools.

Today the 'full package' of possible policy instruments that can be used to foster innovation and particularly to raise the R&D investments is shown in Table 14.2. The choice of instruments to use in a specific country or region depends on the particular challenges and institutional context of a country. For instance in countries where the level of corporate taxation is not very high and the level of fiscal 'discipline' not very high, a tax incentive for R&D is likely to have limited effectiveness.

An important subject of debate in many countries and regions is finding the right balance between generic policies (e.g. policies that do not specifically target any sector, technology or societal domain) and thematic policies that have a clear focus. There are arguments in favour of and against each type of instrument as listed in Table 14.3. While some countries have decided to have mainly generic policies (e.g. the United Kingdom) others have chosen for mainly thematic policies (e.g. Finland, Germany), most countries have a combination of both types of measures. The innovation policy literature has extensively discussed whether governments should select specific technology domains or sectors or whether instruments should let existing market parties decide in which technologies they invest (see Autio et al. 2003; Gassler et al. 2006).

CONCLUSIONS

This chapter aimed to show the evolution in the concepts of innovation and its effects on the thinking about RTDI policy and particularly the instruments designed by policy makers to change behaviour and activities in the public and private research communities.

The overview showed that in the 1950s to 1970s there were two separate systems (the science policy system and the industrial policy system) and that each had different dynamics. The concepts of technology policy and later innovation policy emerged only in the mid 1970s to late 1980s. The first generation policy instruments clearly showed the characteristics of 'linear model thinking' and focused for instance on public research funding, e.g. bringing inventions and technologies from the shelves of the laboratories to the manufacturing plants of industry to be transformed into innovations. While in the 1980s the thinking was on the interactive chain link models and the many feedback loops between science, strategic research, applied research and the market, this only gradually led to more sophisticated bridging and networking policies. The instruments aimed at stimulating private sector R&D evolved from mere R&D subsidy instruments to large R&D performers, to one understanding the importance of SME innovation (and absorptive capacity) and the potential of new-technology-based firms. The innovation systems approach, gaining momentum among innovation analysts in the early 1990s, was only starting to receive acceptance as a useful analytical model around the turn of the present century and at first only in a small number of countries. Today it is more widely accepted, but at the same time its practical use as a guidance for how a good set of policy instruments can be designed is still limited. In recent years the debate has shifted towards the definition of 'holistic' and 'horizontal' policy approaches where different policy domains coordinate the strategic approach to tackle specific issues. In practice

Table 14.3 Arguments in favour of and against thematic policies

Arguments in favour	Arguments against
• More suitable for stimulating structural changes than the generic approach • More suitable in emerging technological fields where players are less well organised • Mobilises a larger critical mass of support, so the effect of intervention on specific field is larger • More efficient bundling of scarce government resources • Can stimulate R&D investment towards societal issues which cannot be resolved by market forces • Alignment with international trends and international programmes (e.g. FP7 EUREKA, ESA, etc.) • Can help generate technological leadership in a certain niche	• Works mostly on basis of picking winners and governments are not always good at selecting 'winners' • Markets are best at selecting most promising R&D investments • Risks of 'betting on the wrong horse' • Risk of lock-in in existing positions or industries that are no longer vital • Creates market distortions • Specialisation makes national economies more prone to external shocks and swings in the market • Scarce government resources are not sufficient to bring about structural changes • Vested interests in government and industry could interfere with a rational process • Risk of hampering the diversification process

Source: Boekholt et al. (2007).

this proves quite difficult to achieve, mainly due to the resilience of governance structures that reflect more compartmentalised division of labour in RTDI policies. In addition, the complexity of formulating and designing policies in a systematic portfolio goes beyond the competence and capacity of today's policymakers and would require a high-level systemic view, involving broader policy domains. It would also require more system level evaluations instead of multiple single programme evaluations. The innovation policy analysts for their part have also only just started to grasp how the interactions work between different types of policy mechanisms and between RTDI and non-RTDI policies.

Although we can sketch broad developments in the thinking on innovation processes and innovation policies, there is no clear one-to-one mapping of these shifting paradigms. For instance examples of systemic and chain-linked policies existed *avant la lettre*. Nevertheless the mainstream evolution of policy making and design of instruments followed new insights with a certain time lag. One can easily say that up to the start of this century RTDI policy making was about a decade behind what was 'accepted' conceptual thinking in the realm of the innovation analysts. With the improvement of 'strategic intelligence', the communication between the innovation analysts and the innovation policymakers through various networks and discussion platforms intensified, but also with the relative 'maturity' of the innovation analysis field, the time and conceptual 'gap' between innovation analysis and policymakers seems to have diminished. Nevertheless, particularly for RTDI policy newcomers (e.g. the new EU member states, regions entering the RTDI domain), the threat of falling for the linear model trap still appears to be present. By establishing cross-border learning communities at least they have ample opportunities to have much faster learning curves than the majority of industrialised countries had.

REFERENCES

Adams, J., Eric, P. and Jeffrey, L. (2003), 'The influence of federal laboratory R&D on industrial research', Rensselaer working papers in economics, Number 0301.

Aho Group (2006), *Creating an Innovative Europe* (20 January 2006), http://ec.europa.eu/invest-in-research/action/2006_ahogroup_en.htm.

Arnold E. (2007), 'Governing the knowledge infrastructure in an innovation systems world', paper for the Inter-American Development Bank, Washington, DC.

Arnold, E., Allinson, R., Muscio, A. and Sowden, P. (2004), *Making Best Use of Technological Knowledge: A Study of the Absorptive Capacity of Irish SMEs*, Brighton: Technopolis.

Arrow, K. (1962), 'Economic welfare and the allocation of resources for invention', in R. Nelson (ed.), *The Rate and Direction of Inventive Activity*, Princeton: Princeton University Press.

Autio, E., Kanninen, S. and Wicksteed, B. (2003), 'Targeted technology programmes: a conceptual evaluation', report 2003/13, TEKES, Helsinki.

Boekholt, P. (1996), 'Financing innovation in the post subsidy era: public support mechanisms to mobilise finance for innovation', *International Journal of Technology Management*, **12** (7-8), 760-786.

Boekholt, P., Arnold, E. and Heide, M. de (2007), 'The use and effectiveness of programmatic policies, some examples and evidence from around the world', Ministry of Economic Affairs Research Series, The Hague.

Boekholt, P., Arnold, E. and Deuten, J. (2004), 'The role of innovation policy agencies in the Finnish, Swedish and Flemish innovation systems', report for the Dutch Innovation Platform, Technopolis, Amsterdam.

Bush, V. (1945), *Science: The Endless Frontier*, A Report to the President by Vannevar Bush, Director of the Office of Scientific Research and Development, Washington, DC: United States Government Printing Office, July.

Cohen, W. and Levinthal, D. (1990) 'Absorptive capacity: a new perspective on learning and innovation', *Administrative Science Quarterly*, **35** (1), 128-152.

Cooke, P., Boekholt, P. and Tödtling, F. (2000), *The Governance of Innovation in Europe: Regional Perspectives on Global Competitiveness*, London, UK and New York, USA: Pinter.

Dosi, G. (1982), 'Technological paradigms and technological trajectories: a suggested interpretation of the determinants and directions of technical change', *Research Policy*, 11 (3), 147-162.

Edler J., et al. (2005), 'Innovation and public procurement: review of issues at stake', Final Report, Fraunhofer Institute Systems and Innovation Research, Karlsruhe.

Edler, J. and Boekholt, P. (2000), 'Internationalisierungsstrategien in der Wissenschafts- und Forschungspolitik: Best Practices im Internationales Vergleich', Study for BMBF, ISI-Fraunhofer and Technopolis, Karlsruhe.

Edquist, C. (ed.) (1997), *Systems of Innovation: Technologies, Institutions and Organizations*, London, UK and New York, USA: Routledge.

Ergas, H. (1986), 'Does technology policy matter?', CEPS papers No. 29, Centre for European Studies, Brussels.

European Commission (2006a), *Lisbon Progress Report.*

European Commission (2006b), *European Innovation Progress Report, Trend Chart 2006*, European Commission, DG Enterprise.

European Commission (2003), *Raising EU R&D Intensity*, Report by an Independent Expert Group, Brussels: European Commission.

European Commission, DG Research (2002), 'Benchmarking national research policies: the impact of RTD on competitiveness and employment (IRCE)', A STRATA-ETAN Expert Group Report, Brussels.

European Commission (1995), *Green Paper on Innovation*, Brussels.

Freeman, C. (1987), *Technology Policy and Economic Performance: Lessons from Japan*, London, UK and New York, USA: Pinter.

Gassler H., Wolfgang, P. and Rammer, C. (2006), 'Schwerpunktsetzungen in der Forschungs- und Technologiepolitik – eine Analyse der Paradigmenwechsel seit 1945', *Österreichische Zeitschrift für Politikwissenschaft (ÖZP)*, 35 (1), 7-23.

Göktepe, D. and Etzkowitz, H. (2005), 'Towards an assisted linear model of innovation: an exploratory study of technology transfer offices in the USA', Conference paper presented at the Triple Helix 5 Conference, Paper A183, 18-21, May 2005, Turin.

Hauknes, J. and Nordgren, L. (1999), 'Economic rationales of government involvement in innovation and the supply of innovation related services', STEP report R-08, Oslo.

Hippel, E. von (1998), *The Sources of Innovation*, Oxford, UK and New York, USA: Oxford University Press.

Kline, S. and Rosenberg, N. (1986), 'An overview of innovation', in R. Landau and N. Rosenberg (eds), *The Positive Sum Strategy, Harnessing Technology for Economic Growth*, Washington, DC: National Academy Press, pp. 275-355.

Lengrand & Associés, PREST, ANRT (2002), 'Innovation tomorrow, innovation policy and the regulatory framework: making innovation an integral part of the broader structural agenda', report for the European Commission, DG Enterprise, October.

Lundvall, B.-Å. (1992), *National Systems of Innovation: Towards a Theory of Innovation and Interactive Learning*, London, UK and New York, USA: Pinter.

Massey, D., Quintas, P. and Wield, D. (1992), *High-Tech Fantasies: Science Parks in Society Science and Space*, London, UK and New York, USA: Routledge.

Mowery, D. (2005), 'The Bayh-Dole act and high-tech entrepreneurship in U.S. universities, egg, chicken or something else?' in G. Libecap, 'University entrepreneurship and technology transfer: process, design, an intellectual property', *Advances in the Study of Entrepreneurship, Innovation and Economic Growth*, **16**, 39-68.

Mowery, D. and Rosenberg, N. (1979), 'The influence of market demand upon innovation: a critical review of some recent studies', *Research Policy*, **8** (2), 102-153.

Nelson, R. (1959), 'The simple economics of basic scientific research', *Journal of Political Economy*, **67** (3), 297-306.

Nelson, R. and Winter, S. (1977), 'In search of useful theory of innovation', *Research Policy*, **6** (1), 36-76.

OECD (2002), *Science, Technology and Industry Outlook*, Paris: OECD.

OECD (2005), *The Governance of Innovation Systems: Synthesis Report*, Paris: OECD Publishers.

Porter, M. (1990), *The Competitive Advantage of Nations*, London, UK and New York, USA, Macmillan.

Rogers, E. (1995), *Diffusion of Innovations*, Vol. 4, New York: The Free Press.

Rothwell, R. and Dodgson, M. (1992), 'European technology policy evolution: convergence towards SMEs and regional technology transfer', *Technovation*, **12** (4), 223-238.

Rothwell, R. and Zegveld, W. (1985), *Reindustrialization and Technology*, New York: Armonk.

Saxenian, A. (1996), *Regional Advantage, Culture and Competition in Silicon Valley and Route 128*, Cambridge, MA, USA: Harvard University Press.

Schmookler, J. (1966), *Invention and Economic Growth*, Cambridge, MA, USA: Harvard University Press.

Smith, K. (1997), 'Economic infrastructures and innovation systems', in C. Edquist (ed.), *Systems of Innovation: Technologies, Institutions and Organizations*, London, UK and New York, USA: Routledge.

Smits, R. and Kuhlmann, S. (2004), 'The rise of systemic instruments', *International Journal of Foresight and Innovation Policy*, **1** (1-2), 4-32.

Stokes, D. (1997), *Pasteur's Quadrant, Basic Science and Technological Innovation*, Washington, DC: Brookings Institution Press.

Utterback, J. (1974), 'Innovation in industry and the diffusion of technology', *Science*, **183** (4125), 620-626.

IV. New dynamics in the innovation policy dance

15. Innovation and Inequality

Susan E. Cozzens

INTRODUCTION

Global inequality is a changing phenomenon molded by a variety of interlocking dynamic forces. Technological change is one of these. If we picture the world system as a cross-tabulation of nation states and technological capabilities, we see a core of advanced and advancing nations, a small set of countries rapidly developing their capabilities, and a large number of countries struggling to maintain or build (Sagasti 2004). These groups correspond roughly to the economic hierarchy of nations, in which only four countries have moved into the top group in the last five decades: South Korea, Taiwan, Singapore and Hong Kong (Milanovic 2005). It is no accident that these four are also the models constantly offered for technology-based economic development.

By focusing on economic development, innovation systems research has contributed to the debate on how to reduce inequality between nations. From its inception, national innovation systems analysis has probed the differences between affluent and aspiring nations (Nelson 1993). Recent volumes in this tradition have focused on developing world regions: Africa (Muchie, Gammeltoft et al. 2003), Latin America (Cassiolato, Lastres et al. 2003) and Asia (Lundvall & Intarakumnerd et al. 2006), and their ideas have been incorporated into strategy documents for the international community (Juma, Yee-Cheong 2005). The innovation systems tradition has also paid significant attention to inequalities between sub-national regions, from old core high-technology regions through new ones to those that hope to be (Cook 2001). When inequalities are geographically based and can be addressed with concepts of economic development, innovation systems research has a great deal to contribute.

Innovation systems research, however, has not given much attention to another dimension of development, namely, inequalities between individuals, households, or groups. As mentioned earlier in this volume, the lack of a perspective on the role of individual actors in innovation systems is one of the major flaws of the innovation systems approach. This chapter attempts to fill that gap. Econometric studies have shown that income inequality within countries hinders economic development (Deininger & Squire 1998), and ethnic fragmentation has also recently been demonstrated to do the same (Easterly & Levine 1997). The econometric studies reflect the human reality that inequalities place barriers in the path of using innovation for development and keep it from achieving everything it could. We cannot have effective theory, policy, or practice in this area without addressing these issues.

Like inequalities between countries, inequalities between individuals and households are a complex phenomenon with many dimensions.[1] This chapter groups them into three broad types.

- First, most of the conventional literature on inequality is about differences in income, wage, or wealth between individuals or households. Economists refer to these as *vertical* inequalities (Stewart 2001). In this chapter, I refer to the public interventions that reduce them as *egalitarian*.
- Second, the bottom end of the income scale is designated as *poverty*, and many efforts to reduce inequality focus on reducing it. Economic growth can eliminate absolute poverty, which the World Bank measures as income per capita of $1 or $2 per day.[2] I refer to public interventions that aim to reduce poverty or make life better for the poor as *pro-poor* policies or programs.
- Third, culturally defined groups often show different profiles in relation to income, poverty and other social goods. These differences carry the label *horizontal*. Horizontal inequalities may appear by gender, ethnicity and religion. This chapter refers to public interventions to reduce the differences between culturally defined groups in the distribution of things people value as *fairness* policies and programs.

Reducing vertical inequalities, eliminating poverty and achieving fairness across horizontally defined groups are three types of re-distributive objectives that innovation policy might adopt.

Measures of inequality in income are neutral representations of grueling human realities.[3] One in five infants in poor countries dies before the age of one, as compared with five in a thousand in affluent countries. There are almost no under-five children who are underweight in affluent countries, but rates of 25-50% are

1 Milanovic (2005) has recently attempted to construct an estimate of household income inequality at the global level. That is, ignoring national boundaries, how unequal is the distribution of income among the world's households? The global Gini coefficient – a common measure of inequality that ranges from 0 (complete equality) to 1 (complete inequality) – is about 0.65, or approximately in the range of the highest individual country measures, leading him to the intriguing conclusion that there is no world middle class. Because individual and household global inequality is so hard to grasp empirically, this chapter will confine its attention to inequalities within countries.

2 These are calculated in constant-dollar purchasing power parity terms.

3 All these figures come from UNDP (2006). Although I am illustrating differences between individuals, the data are reported by country because comparisons by individuals, households and groups at the global level are not readily available.

common in poor ones; and adult undernourishment is commonly 30%. A child born in Australia, Ireland, Japan, or Italy can expect to live 80 years or more, while one born in many central African countries could expect less than 40. People living in affluent countries can assume access to clean water and basic sanitation, but less than half of those in many poor countries can make the same assumption.

If the definition of innovation includes new and better ways of solving societal problems, then it could be used to tackle these human development challenges as well. Communities could be actively involved in trying out new solutions to their challenges. Governments could innovate to become more efficient in the provision of basic services like highways, telecommunications, clean water and electricity. Industries could produce new, life-enhancing products, through environmentally friendly processes, creating employment. All these efforts could contribute to eliminating poverty and distributing the benefits of technological change widely.

In practice, however, innovation systems are organized in ways that reproduce and even amplify inequalities between individuals, households and groups. Even growth-producing technological change brings both benefits and costs, for example, by creating and eliminating jobs and creating effective but expensive new medical options. Thus, while technological change seen from one angle is a source of progress, from another angle it creates the potential to increase gaps between rich and poor not just in wealth but also in well-being.

Can innovation policies help to re-invent innovation practice so that it helps to reduce these inequalities rather than reproducing them? That is the central question of this chapter. This first section of the chapter therefore describes four processes of cumulative advantage that link the current practice of innovation with individual and group inequalities, through assets, employment, products and diffusion. The four processes illustrate changes in policy that have first responded to and then contributed to innovation practices, resulting in changing relationships to the distribution of benefits and costs in the knowledge society.

The second part of the chapter outlines actions that private industry, civil society, research institutions and the public sector are taking to change innovation practice, re-inventing it in inequality-reducing ways and forming what could be the nucleus of another innovation system variant, the 'human development innovation system'. While the literature identifies industry, research institutions and government as key actors in innovation systems, the 'human development innovation system' would be more 'bottom-up', with priorities and new ideas coming from actors and communities that scarcely appear in the standard accounts of innovation systems. Innovation systems theory will itself need to take these developments into account if it is to be capable of analyzing the dynamics of these newly-configured systems. The growing attention of innovation scholars, policy makers and practitioners for user involvement in innovation processes seems a promising development in this context.

Christopher Freeman (2000) has observed that techno-economic paradigm shifts increase inequalities in part by rendering the redistributive mechanisms of the old paradigms ineffective. After such dramatic shifts, societies must create new redistributive processes if they are going to re-establish shared prosperity. Global society is currently in need of such new mechanisms, and innovation practice and policy could contribute.

DYNAMICS

Assets

Ownership of assets is a central and age-old facet of the process of cumulative advantage. Wealth is still remarkably concentrated in the contemporary world, with 2% of the world's population holding 51% of its assets, and citizens of the countries of North America, Europe and high-income Asia holding almost 90% of it (Davies, Sandstrom et al. 2006). But ownership has also traditionally been a policy option for redistribution. For example, land reform – breaking up large hereditary holdings and distributing land to those who farm it – was an essential step in modernization and improvement in agricultural productivity. Likewise, middle class prosperity in the United States rests in part on the widespread ownership of business assets through tax-subsidized investments in pension funds.

One particular form of ownership has co-evolved with the modern innovation system, namely, ownership of intellectual property (Granstrand 2006). Based on a principle first established as part of the US Constitution, patent and copyright laws are one of the oldest forms of innovation policy, designed specifically to provide incentives for invention. As corporations have become the homes for most inventors, intellectual property has also become an important set of assets for firms. While a new technology remains under the control of the firm that invented it, the firm can charge monopoly prices for its product or services. These high prices (or 'monopoly rents', as economists call them), are one incentive for firms to innovate in radical ways rather than simply incrementally improve their products or services (Duffy 2004; Pogge 2005). The search for monopoly rents has turned the patent system into a huge enterprise, fed by the research and development efforts of large firms and guarded by an army of patent attorneys.

The relationships between technological learning, innovation and intellectual property protection have been explored extensively in the national innovation systems literature. Kim (1997) portrays Korea, for example, as moving from a period of dependence of imported technology through reverse engineering (which is easier when intellectual property protection is weak) towards its current status as an innovator nation. The literature also notes that many of today's affluent countries achieved their early economic growth while ignoring or skirting intellectual property laws. The ever-tightening control of intellectual property through the World Trade Organization is thus changing the conditions for technological learning in today's developing economies, 'kicking away the ladder' that other countries climbed (Chang 2002; Drahos & Braithwaite 2002). The co-evolution of global economic patterns and intellectual property regimes is thus a key research topic in understanding inequalities between nations. But do the new regimes have any role in relation to individual and group inequalities?

Several current controversies illustrate the connection. One concerns rights to indigenous knowledge. Indigenous knowledge is 'local and tacit knowledge that is unique to a given culture or society' (Rao 2006). It may be the knowledge of aboriginal groups about the plants in their territory (Marinova & Raven 2006), knowledge of traditional medicines (Bermudez, Oliveira-Miranda et al. 2005), plant varieties developed by farmers (Borowiak 2004) or language, oral traditions and epistemologies (Tatsch 2004). Regardless of content, indigenous knowledge is owned in ways that are different from those that intellectual property systems establish. This situation has led to attempts to appropriate that knowledge under IPR systems, taking it out of collective ownership (Timmermans 2003). Conflicts over these issues reflect inequalities between ethnic groups and between powerful and less powerful social actors. The imposition of Northern concepts of ownership fails to respect the cultural diversity of knowledge owners and raises the distributive issue of fairness (UNDP 2004). This aspect of innovation policy thus becomes a choice point between more and less fair policies, and an active place for innovation in redistributive mechanisms for the knowledge economy.

The debate over access to essential medicines, a controversy that has focused on the availability of antiretroviral drugs to treat HIV/AIDS in poor countries, provides a second example of intellectual property rules and human development in potential conflict. 'The worldwide HIV/AIDS epidemic has generated intense criticism of pharmaceutical drug prices', write Calfee and Bate (2004). Drug prices are not generated entirely by development cost or investment (Keyhani, Diener-West et al. 2005). Intellectual property laws, along with recent world trade laws, significantly affect price and therefore who has access to which drugs (Loff 2003). Non-governmental organizations (NGOs) advocating human health rights led the pro-poor fight to activate provisions of the international law that allows production and use of lower-priced generic versions (Matthews 2004; Sell & Prakash 2004). But the controversy stimulated a wide debate on changes in incentive structures and patent provisions to make it more likely that innovation in medicines would serve those who need it most (De George 2005; Pogge 2005).

These two examples illustrate that revamped intellectual property rules need to be part of the new redistributive institutions that the knowledge society needs to develop. They also illustrate that those rules will need to be developed on a global rather than national basis, since the relevant innovation system in each case involved a mix of national and transnational actors. In the larger picture, owned innovations are undoubtedly only a subset of new approaches that are improving life at both the bottom and the top of the global income scale; so perhaps a pro-poor, fair and egalitarian innovation policy would pay much less attention in general to intellectual property and provide a broader set of incentives for benefits that come through many routes.

Employment

Innovation has huge effects on people at the middle and bottom of the income distribution through its relationships to employment. Process innovations are considered essential to efficiency, productivity and competitiveness, but they often achieve these goals by eliminating jobs. At the same time product innovations, which create something new to be sold, usually create jobs. The overall process of economic growth requires both dynamics: increasing efficiency of existing industries and expanding new business (and thus employment) opportunities. At the generic level, current innovation policies are almost always oriented towards increasing productivity, and government innovation policies are closely linked sometimes synonymous with – their industrial policies. Policy, industry and employment patterns thus co-evolve.

While innovation systems studies have paid careful and continuous attention to the expansion of new industries, they have paid almost no attention to the contraction of older ones, and very little to the kinds and numbers of jobs generated or lost in either process. For example, in telecommunications, the automation of directory assistance services has drastically reduced the number of people necessary to look up numbers and therefore reduced costs for telephone companies, while creating jobs for those who write voice-recognition software and search algorithms. The mobile phone industry has created jobs in design, production and sales, while eliminating jobs associated with pay telephones and reducing those devoted to land lines. More radical innovations can replace whole product lines or industries, as the photocopier replaced carbon paper and the personal computer replaced typewriters.

The competitiveness of firms, industries, regions and countries depends on such processes of technological change, but where there is competition there are of course winners and losers. The new jobs created may be located in different places than the old ones, undercutting some local economies while building others up. Urban-rural divides are thus likely to be exacerbated.

A classic theory of inequality, the Kuznets hypothesis (1955) claims that rising inequality is a temporary phenomenon associated with the shift of labor from low-productivity, low-income agriculture into higher-productivity, higher-income industrial employment. Inequality will increase during the process, as the gap between the two first widens, then falls when stable low post-industrial rural employment levels are reached. Since women today constitute a large proportion, some say a majority, of the world's agricultural workers (FAO 2002; Hornady 2006), this pattern – admittedly still a matter of controversy[4] – widens gender as well as urban-rural gaps.

4 Scholars have debated the Kuznets hypothesis and tested it with contemporary data, with inconclusive results on its validity today. For example, contrast Deininger and Squire (1997), with Galbraith, Conceicao et al. (2000).

Kuznets saw industrialization as a temporary transformation process, but there is no end in sight for current trends. If more productive industries survive, grow and produce new employment opportunities, and if less productive ones decline, then the phase of increasing inequality would seem to stretch indefinitely into the future.

It is also the case that contemporary technology-based economic development is highly selective geographically, not even characterizing all urban areas. This is true whether it takes place in agriculturally-based economies or service-based ones. Galbraith and Hale (2003) have graphically demonstrated the accumulation of wealth in just a few counties in the US during the IT-based boom in the late 1990s, for example. Likewise, technology-based developments in China and India are confined to a few locations, leading Shanghai, Beijing and Bangalore to international fame, while leaving the countryside in both countries largely untouched in opportunity or wealth (Wan & Zhang 2006). Growing industries generate other jobs in an area through the multiplier effect, while shrinking ones remove those jobs from the local economy.

New jobs may also call for different skills. Older workers displaced by technological change will not qualify for jobs requiring skills they never acquired. Weak educational systems may not prepare young people for the emerging employment opportunities and disadvantaged social groups, including women in many countries, often have fewer opportunities for formal education. Economists invoke 'skill-biased technological change' as a major explanation for growing wage gaps observed in some OECD countries (Greiner, Rubart et al. 2004; Piva, Santarelli et al. 2005). When demand for higher skills rises, according to this theory, higher skill workers receive premium wages. The premiums will show up as rising returns to education and a growing gap in income between skilled and less skilled workers (Lemieux 2006). Where educational opportunities are unequally distributed by region, gender, or ethnicity, these horizontal differences will make the gap even greater.

In addition to the skill dynamic, processes of gender and ethnic identification and closure (Weeden 2002) may lead to uneven distribution of the new employment opportunities opened by technological change even beyond issues related to educational preparation. Since the jobs are in attractive new areas, they are likely to fill early with members of advantaged groups. Women and members of disadvantaged minorities groups may find the resulting work environments difficult. Where network connections have been found to be important business assets (Saxenian, Motohama et al. 2002), informal connections among old-boys' networks or within certain ethnic groups may create uneven patterns as well.

In summary, then, innovation creates gradual pressure upwards in the skills levels required in growing parts of the economy. If everyone's educational level could keep up with the rising demand for skills, this essentially positive dynamic would not exacerbate inequality. But since education is itself highly unequally distributed, between old and young as well as between genders and ethnic groups, the introduction of new technology into the workplace tends to exacerbate inequalities in employment and wages. Vertical income inequality is largely a function of the kinds of employment opportunities available in a society, and horizontal inequalities are primarily a function of where the jobs are located, who is prepared for them and who is hired for them. Technological change is strongly connected to both patterns. Egalitarian policies are needed to address the first issue and fairness policies for the second. A discussion of targeted retraining programs and thoroughgoing educational reform go well beyond the scope of this chapter, but surely represent the kinds of conditions governments need to provide as a starting point for maintaining social cohesion in the face of technological change in the workplace.

Consumer Products and Services

Technological change is not limited to workplace technology, but also includes the availability of consumer goods and services. These in turn play important roles in individual and family well-being. The process of industrial innovation, which takes new ideas and develops them, determines what products are available to consumers, who they are designed for, what problems they are intended to solve, and how much they cost. Service industries make the equivalent determinations for what they sell. Although product development is often influenced by a variety of public policies and programs, including health, safety and environmental regulation, I will focus in this section on the role of projected markets in determining the availability of products and services at various prices. The primary reliance on markets to stimulate the development of new technologies is of course a policy choice.

Let us consider how market dynamics affect the availability of products to consumers variously placed in terms of vertical and horizontal inequalities. A market can be thought of as a kind of democracy in which each person holds as many votes as units of disposable income. Each person in the US, for example, has on average $43,740 votes per year for the technologies to be made available to them (that is, the US GDP per capita), and each person in India has on average $720.[5] Each of the one billion people living at the World Bank's 'misery' level has $1 per day, and each one living at the official US poverty level has $10,000 or so per year.[6]

Because of these vast differences in available resources, consumers at different income levels buy different things, with the very poor usually focusing their small resources on basic needs like food and housing and those with more income adding other kinds of goods and services (transportation, education, health care, etc.) to their shopping baskets (Goktolga, Bal et al. 2006; Yankelovich & Meer 2006).

5 World Bank data, http://web.worldbank.org/WBSITE/EXTERNAL/DATASTATISTICS, accessed April 1, 2007.

6 The poverty threshold in 2006 for an individual was $10,488 for those under 65 and $9,669 for those over 65. U.S. Census Bureau, http://www.census.gov/hhes/www/poverty/threshld/thresh06.html, accessed April 1, 2007.

At the top of the income scale, consumers purchase luxury goods like boats and international travel.

Firms developing consumer products and services can and do think about reaching all of these markets. Innovation policies urge them to match their capabilities and strategies to markets in order to achieve maximum business success; and the same policies encourage their partner institutions like universities and government laboratories to support them in doing so. Firms may choose to include in their portfolios some products or services aimed at affluent markets. Technology-intensive products like sports cars and personalized drugs have small markets but are sold at high enough prices to cover their specialized development and production costs. Firms may also develop products such as video cameras and microwave ovens for 'mass' markets, where efficient volume production makes products available to many consumers, with low unit prices but high total profits because of volume. Finally, they can develop products and services to sell to the poor, a strategy called 'bottom of the pyramid' marketing (Prahalad 2005), attractive because of the large numbers of poor who might be reached. Likewise, firms can develop products designed for specific culturally defined markets, for example, women or ethnic minorities.

In practice, however, the 'democracy' of the market will lead to much more overall new product development effort going into things that middle-class or more affluent consumers will buy. Likewise, there may be other biases in choosing which products to develop. For example, the feminist movement has criticized the pharmaceutical industry for focusing on developing contraceptive pills for women when the male equivalents would be simpler to invent and produce (Jutte 2005).

What markets are unlikely to produce are basic goods and services for the poor, even though society as a whole would benefit from maintaining minimum standards in areas like health, nutrition, housing and education. This situation is called 'market failure': private markets produce less than the socially-optimal amount because the incentive is private, not public, return. There are a variety of potential public interventions designed to overcome market failures, among which are several market-based mechanisms. One of these is public procurement – that is, the government buys the item on behalf of the public, thus stimulating private sector development. In these cases, policies and products co-evolve through government action. Some public procurement of technology happens at the high end of the cost and sophistication spectrum, with defense technologies as a prime example. Some are in the basic needs areas, like public health care systems, which purchase medicines and medical services and distribute them free. Pro-poor subsidies for purchases in basic needs areas, such as food stamps or housing allowances, also use the market to serve broader public goals.

I have already introduced one area where such a failure occurs, the development of drugs and vaccines for diseases associated with poverty. This area also illustrates the creative new thinking that is going into using pro-poor public procurement to go beyond simply purchasing goods to stimulating innovation. I will return to some of these possibilities in the 'Options' section of the chapter.

Diffusion

Diffusion of new technologies gets considerable attention in innovation policy as a way of spreading the benefits of new technologies – another co-evolutionary dynamic. For example, diffusion of new technologies in business, from agriculture through service and manufacturing, gets considerable attention in innovation policy. The United States Department of Commerce has a Manufacturing Extension Program that stimulates states to set up outreach efforts that facilitate firms, especially small ones, to adopt new technologies.[7] Programs aimed at SMEs (small and medium-sized enterprises) in other countries usually have a similar orientation. In addition, innovation policies include incentives for the diffusion of technologies with high social benefits, such as energy technologies. In Germany, for example, energy can be sold from alternative energy sources onto the main grid at premium prices. As a result of such policies, Germany has a much higher rate of production of solar energy than countries with much more sunshine, including the United States (Rogers 1962; Klaassen, Miketa et al. 2005).

However, the diffusion of new technologies is often uneven; some groups get more benefits than others. Likewise, there are often costs associated with new technologies, and these may also be unevenly distributed.

The classic study in diffusion of innovation is Everett Rogers' (1962) analysis of the Green Revolution, which set the framework for several generations of later studies. From the earliest stages of this work, the differential effects of introducing new technologies have been prominent in the analysis.

The Rogers model includes several key concepts, starting with the innovation itself. Working in 1962, he took the innovation process itself to be one-directional, delivering a fixed innovation which is either used or not used; later generations of researchers have taken the co-evolution of innovation and users more seriously. The Rogers model also includes a change agent. In the Green Revolution case, this was the extension worker, while in a market version of the model it could be the marketing and advertising departments of the firm along with retailers. The core of the Rogers model is the 'adopters', those who decide to use the technology. He divides these into three groups: early, middle and late adopters. The growth in number of users over time takes the shape of an S-shaped curve. A few people try the innovation early, taking some risks in doing so. If they like it, they spread the word to others, who join the population of users at increasing rates until the market for the innovation is saturated.

According to Rogers, the diffusion process can widen existing socioeconomic gaps for several reasons. First, innovators and early adopters are people who seek

7 http://www.mep.nist.gov/, accessed April 9, 2007.

out innovations and can afford the more expensive ones. For example, real estate agents were among the earliest users of car phones. Second, professional change agents concentrate their efforts on innovators and early adopters. In the case of the Green Revolution, the change agents were the extension workers, who naturally concentrated their efforts on the people who were easiest to convince about using new strains. In the case of private sector products, advertising will likewise be oriented towards those who are most likely to buy, not those who need the product most. Third, innovators and early adopters achieve windfall profits by adopting early. In the case of the Green Revolution, for example, the early adopters produced larger quantities while prices were still high. By the time the later adopters increased their yields, prices had dropped because of the larger supply. Thus the whole process, left to operate according to its own logic, will lead to increased advantage for those with greater initial resources (Rogers calls these the 'Ups').

Rogers and his colleagues, however, outline a number of diffusion strategies that can be used to intervene in this dynamic and make diffusion to disadvantaged groups (the 'Downs') more likely. If the problem is that Ups have greater access to information than Downs, one solution is to tailor communication strategies to the Downs. This can be done, for example, by tailoring the level of language in the message and the channels used. Change agents can also organize Downs into small groups to hear the message. Change agents with a redistributional task should in fact concentrate their communication efforts towards Downs, since Ups are already more likely to adopt.

If the problem is that Ups have greater access to innovation evaluation information from peers than Downs, according to Rogers and his colleagues, then solutions include identifying opinion leaders among the Downs to work with; choosing change agents from among the Downs; and forming groups among the Downs to reinforce innovation decision making. If the problem is that the Ups have greater slack resources for adopting innovation than Downs, the solutions could include recommending appropriate innovations for Downs; creating social organizations that pool Down resources; including Downs in setting priorities for innovation programs; and establishing special diffusion agencies to work only with Downs.

These recommendations are obviously oriented to situations in which the public sector is trying to increase adoption of a technology through a change agent. Despite the move towards more market-based implementation of policies, there are still plenty of examples of government-led processes. The manufacturing extension services listed earlier are one such example, and many more can be drawn from public health, including the diffusion of condom use to prevent AIDS; childhood vaccines; safe drinking water systems in villages; etc. Some of the Rogers principles are being applied as pro-poor policies options in these settings, such as programs that encourage strong community involvement in planning and implementation. In the section on Options, I will return to the possibilities for private firms developing bottom of the pyramid markets to apply the principles as well.

The work on diffusion of innovations, however, deals largely with the distribution of benefits. The distribution of the costs of technologies has received much less attention, although it is an aspect of diffusion. The exception is the literature on environmental justice, which has correlated the spatial distribution of hazardous wastes and other technology-generated pollutants to poor neighborhoods and disadvantaged ethnic groups in the United States (Bullard 1990), as well as tracking the off-shoring of hazardous wastes of affluent countries to less affluent ones (Bullard 2005). The Downs are less able to avoid or fight off such undesirable living conditions. Innovation policies are shrinking the size of this problem by encouraging overall reduction of environmental impacts through product life cycle analysis (Norris 2006), but this technique does not include an accompanying analysis of differential costs and benefits, and thus does not yet incorporate the principle of fairness called for by the environmental justice movement.

Summary

In short, traditional approaches to innovation policy rely on mechanisms that, unless designed specifically to do otherwise, tend to increase inequalities. Strengthening intellectual property regimes can turn public goods into private ones, narrowing the group that benefits from them. A tight focus in innovation policy on competitiveness through raising productivity can create the conditions for increasing rewards to already advantaged places and groups. Leaving most of the stimulus for and diffusion of new products, processes and services to the market is likely to leave the needs of the poor unmet. Finally, the traditional policies do not pay attention to differential distribution of the costs of technological change.

Indeed, the whole framing of innovation policies, in a conventional framework of assumptions partly[8] shared by both innovation systems theory and practice, leaves the connections to these larger social dynamics out of the picture. Furthermore, the focus on R&D in firms neglects the possibilities of distributed demand-driven innovation. In the next section, I turn to some different uses of innovation policies that allow more room for bottom-up, creative responses from more actors in the broader social system.

---

8 The notion of system failure as opposed to market failure as the rationale for innovation policies offers opportunities to solve some of the problems caused by the more neo-classical-based approach. An example could be the importance the innovation systems approach attaches to processes of demand articulation by which better opportunities for bottom-up-based policies are created (e.g. Smits et al. in this book).

OPTIONS

In the introduction to this chapter, I invoked Chris Freeman's image of new techno-economic paradigms needing new redistributive institutions. The industrial revolution created its own working class, which in turn organized into unions to demand their fair share of its benefits. Likewise, the industrial revolution was accompanied by centralized welfare systems run by national governments and by compulsory mass education. These redistributive institutions were particularly well suited to the geographically concentrated activities of industrial production.

What redistributive mechanisms are being generated in the current economic environment, characterized as it is by multinational firms who are not just trading but producing globally; shrinking rather than growing roles for national governments; and an expanded civil society highly connected globally through the new communication channels? I will argue in this section that some new and distinctive redistributive processes are emerging in this new techno-economic regime, under phrases such as 'public-private partnerships', 'corporate social responsibility', 'socially responsible investing', 'fair trade', 'bottom of the pyramid marketing' and 'social entrepreneurship'. These new mechanisms are addressing some but not all of the processes of cumulative advantage described in the last section. Traditional redistributive approaches co-exist with the new ones and have strengths that should not be neglected in the effort to reduce the inequalities that are standing in the way of human development.

Assets

Let us consider again the area of assets to see what redistributive dynamics are developing. On the one hand, we noted that intellectual property rules are strengthening on a global basis through the expanding reach of the World Trade Organization. This trend has the potential to undermine the arrangements that have made low-cost versions of technologies available to the poor, with the prominent example being medicines. The traditional redistributive mechanism in medicine is usually public health services. But as private health care insurance has grown up beside public services in many countries, the gap in quality and access grows as well. Furthermore, in this case, as in many generated in world economy today, if the problem is global, the solution cannot be local.

Indeed, public health services were not the mechanism through which the happy ending to the essential medicines story was negotiated. Instead, there were two lines of action. On the one hand, the international community of WTO members insisted on provisions in TRIPS that allowed nations with pharmaceutical capabilities to use them for the public good (Lai & Qiu 2003; Barton 2004; Atik & Lidgard 2006). So what global institutions took away, they in part gave back. And on the other, a coalition of non-governmental organizations (NGOs) negotiated lower prices with generic manufacturers and used these as leverage with the name brand companies

(Pogge 2005). The power of civil society played an important pro-poor role. These are themes we will see in other examples.

A traditional mechanism is being used in a creative way in the essential medicines story, however, through an institutional invention called 'advance purchase commitments' (Berndt & Hurvitz 2005; Towse & Kettler 2005). Under this mechanism, a set of donors (that could include both private philanthropies as well as governments) contracts with a drug company to develop a drug that is important for the poor and to sell it to them at a specific price. The company then bears the technical costs and risks of developing it, but with an assured market. The first such agreement has just been written. Italy, Canada, Norway, Russia and the United Kingdom, along with the Bill and Melinda Gates Foundation, have committed to purchase 'vaccines against pneumococcal disease (pneumonia, meningitis, and other killers), which annually claims the lives of up to one million children a year'.[9] This mechanism uses the old idea of public procurement, but in the mode of a 'public-private partnership' with traditional private sector incentives. By spreading the benefits of new medicines more broadly, the mechanism is playing an egalitarian role.

Before we leave the topic of assets, let us note an egalitarian trend that becomes important in the next section. The citizens of the affluent countries of Europe, North America and Asia now commonly own shares in companies, either local or multinational, either directly or through retirement funds. This fact presents the opportunity for them to play a decentralized but significant role in shaping the policies of those companies.

Products

The complex relationships among shareholders, consumers and corporate leaders form the backdrop for some emerging patterns in product availability and design. The traditional part of the pattern is government regulation. Both in the past and currently, national governments regulate product safety and environmental quality (of course some more stringently than others and some with much better enforcement than others). The diversity and assertiveness of national environmental regulations has in fact led to an effort among multinational firms to stay ahead of policymakers in their own environmental standards, leading to such mechanisms as ISO14000, a high international standard that firms try voluntarily to meet (Curkovic, Sroufe et al. 2005; Potoski & Prakash 2005; Raines & Prakash 2005; Brodhag & Taliere 2006; Prakash & Potoski 2006). Neither government regulations nor voluntary mechanisms, however, are particularly redistributive.

Perhaps the most interesting development of the last few decades in corporate governance has been the rise of consumer and shareholder movements to address safety and environmental issues around new technologies through direct action in the corporation itself, sometimes with goals that are redistributive (Haigh &

9 http://blogs.cgdev.org/globalhealth/2007/02/incentives_for.php, accessed April 9, 2007.

Hazelton 2004; Klein, Smith et al. 2004). These movements supplement formal regulation, which is itself a standard part of traditional innovation policy. Consumer boycotts on specific issues stretch back for decades, to the boycott of Nestle products over its sales of baby formula in the developing world, which was framed as a fairness issue (Gaskin 1981; Post 1985; Delamothe 1989). A more recent and dramatic example is the consumer movement in Europe against genetically-modified foods. European consumers perceived risks to them in the products and rejected them, in part on the grocery shelves and in part through press coverage generated by NGO actions (Levidow 1999; Bourgeois 2001). In both cases, affluent consumers use their consumer power to shape corporate product choices.

In the same vein, the socially responsible investment movement has organized the shareholder value of constituencies in the North. A socially-responsible investment portfolio meets a set of criteria that go beyond profitability. This mechanism directs resources towards the parts of the business community that meet the moral standards of investors (Woodward 2003; Guay, Doh et al. 2004).

These mechanisms, however, have primarily negative, not positive power. They can stop products that are on the way or being sold, but they cannot actively generate products that serve the needs of the poor. While government procurement could in principle do so, in practice it has not used the innovation route effectively. Likewise, consumer organizations could take an active, creative role, stimulating a debate on how consumers could benefit from new technologies, for instance in food and health care. But we do not know of examples in which they have taken up the challenge.

Instead, some interesting bottom-up methods in the private sector are getting increasing attention. The best-known variant is 'micro-finance', pioneered by the Grameen Bank under Muhammad Yunus, the 2006 winner of the Nobel Prize (Yunus 2006). Micro-financiers lend very small amounts of money to low-income people, often women, to start their own businesses using their existing skills and simple materials. These businesses tend to produce products and services that are needed in low-income communities. In some cases, these are technological inventions, as in the case of the KickStart pump (Fisher 2006). Sometimes they use technologies in creative ways that make their benefits available in low-income communities, as in the 'telephone ladies' who sell minutes rather than subscription plans on mobile phones.[10] These are pro-poor businesses, a whole new source of innovation and creativity.

---

10 Close cousins to micro-finance are 'social entrepreneurship' (Cook, Dodds et al. 2003; Christie & Honig 2006; Dees 2007) and 'bottom of the pyramid' marketing (Prahalad 2005), both ways of thinking about small, local businesses building economic strength through providing goods and services that low-income communities need.

Employment

Along with issues of product and environmental safety, the consumer and shareholder-based civil society movements have sometimes taken up labor and trade issues. Many socially-responsible investment screens include a set of good labor practices, and the 'Fair Trade' movement raises the question of whether producers are being paid reasonable amounts by middlemen and corporations that market their goods in the North (Fridell 2004; Mutersbaugh 2005; Hira & Ferrie 2006). Likewise, some rugs are now certified as free of the inputs of child labor.[11] These movements, however, do not appear to be capable of addressing the more structural issues associated with employment. While they can raise the expectation that producers will be paid a minimum amount, they cannot exert pressure for industrial production to produce a range of jobs including those at middle wage levels.

The structural issues could be raised in more direct labor negotiations – again, a traditional redistributive mechanism still playing an important role. For example, in the Nordic countries and in Japan, strong labor participation in technological change led to the transformation of jobs toward greater worker autonomy and higher skills, rather than replacement of workers with automated processes (Yamada, Yamada et al. 1997; Gregory 2003). This transformation turned into an advantage in product quality that helped firms in these regions in global competition. Similarly, when a community is trying to lure a firm to invest in production there, the traditional land and tax incentives could be accompanied by asking for commitments about the levels of jobs created and local suppliers. But since these mechanisms are place-based, it is likely that they will need to be replaced eventually in the new techno-economic regime.

Diffusion

One side of the diffusion issues raised above was the uneven distribution of the costs of technologies, especially the environmental costs. This issue is being addressed, through a pattern of interaction among civil society, government and industry that by this point in the discussion should be familiar. The issue of environmental justice was raised first in the United States by a coalition of religious and civil rights activists, specifically working from their culturally-defined political positions. They took the issue to the Environmental Protection Agency, and eventually the President signed an executive order on the matter – a fairness policy (Bullard, Mohai et al. 2007). Government agencies have initiated modest programs, mostly stimulating local action around the issues and providing technical support.[12] The movement, however, has gone worldwide, aligning environmental activists in many places to oppose the

11 http://www.rugmark.org/child_labor_problem.php?cid=3, accessed April 9, 2007.

12 http://www.niehs.nih.gov/translat/envjust/envjust.htm, accessed April 9, 2007

globalization of garbage.[13] In the meantime, stronger and stronger environmental regulations in Europe and the United States have stimulated firms to clean up their waste dumps and effluents, and to build environmental sustainability into the design of many new products (Gungor & Gupta 1999; Maxwell & van der Vorst 2003). The results are getting better for everyone, including the disadvantaged communities who are still exposed disproportionately, but to a smaller risk. From the viewpoint of innovation systems, the important point to note in these developments is the technology-stimulating effect of the community-government-industry partnership, which operates as a newly configured innovation system.

The other part of the diffusion problem is making the benefits of technology more widely available. Several of the new mechanisms already mentioned have contributed in this area: advance purchase commitments in the area of essential medicines; micro-enterprise in consumer goods and sometimes in technologies needed for small-scale production. Universities can maintain or refocus efforts on educating leaders for public innovation and employees that in firms and other organizations are better able to identify technological options and align them with the firms and societies interests, rather than simply supporting private sector goals and objectives. Public research institutions can be turned to pro-poor objectives, as they have been in South Africa under democracy.[14] But the pro-poor mission is constantly in tension with the increasing pressure for such public research institutions to subordinate their public agendas to those of private firms – encouraged ironically by the growth-oriented versions of innovation systems concepts. Again, the entry of community-based or other civil society organizations into innovation system discussions can change the balance in those discussions.

SUMMARY AND CONCLUSIONS

Inequalities are standing between technologies and human development, and the concepts and practices of innovation systems, aided by innovation policy, are often helping. To distribute their potential benefits, technologies have to jump over the barriers of monopoly pricing, skill-biased technological change, and lopsided market development and diffusion to reach most of the world's people. Innovation theories and policies that are oriented entirely towards economic growth distract attention from the potential of technologies for human development.

The second section of this chapter has illustrated, however, that an interesting new system of innovation may be emerging, with a distinctive coalition of actors seeking human development objectives through pro-poor, egalitarian and fair innovation processes. Micro-entrepreneurs, local communities and other bottom-up

---

13 See for example http://blog.foreignpolicy.com/node/3747, accessed April 9, 2007

14 http://www.csir.co.za/plsql/ptl0002/PTL0002_PGE038_ARTICLE?ARTICLE_NO=7040183, accessed April 9, 2007.

movements in civil society are often in the lead in this coalition, with strong messages coming over the Internet from activists in the South to align with the consumer and shareholder power of those in the North. This new source of global pressure helps the international community build human development goals into innovation-shaping institutions like TRIPS and helps multinational corporations take up the challenge of doing well by doing good.

This 'human development innovation system', like its national, regional and sectoral cousins, is characterized not just by its distinctive set of actors, but in particular by the relationships among them: civil society articulating the goals and drawing attention to the issues, national governments setting the rules of the game, and corporations being pushed into a leadership position in order to stay in business. Current innovation systems theory, focused as it is on private firms as entrepreneurs, will need significant revision to be able to understand how governments can provide supportive environments for this next generation of public innovators. As was said already, the growing attention to user involvement and demand articulation may be promising first steps in this direction.

The innovation policy that co-evolves with the new systems will thus look much like the traditional ones in terms of process: a multi-actor, consultative space that encourages and supports new ideas. The key difference is that this will be a space where Freeman's vision of new distributional mechanisms for a changed technological regime can be put into practice and where non-traditional actors like users and consumers may get better opportunities to exploit technological potential in a way that is beneficial to them.

REFERENCES

Atik, J. and Lidgard, H. (2006), 'Embracing price discrimination: trips and the suppression of parallel trade in pharmaceuticals', *University of Pennsylvania Journal of International Economic Law*, 27 (4), 1043-1076.

Barton, J. (2004), 'TRIPS and the global pharmaceutical market', *Health Affairs*, 23 (3), 146-154.

Bermudez, A., Oliveira-Miranda, M., et al. (2005), 'Ethnobotanical Brazil, research on medicinal plants: a review of its goals and current approaches', *Interciencia*, 30 (8), 453.

Berndt, E. and Hurvitz, J. (2005), 'Vaccine advance-purchase agreements for low-income countries: practical issues', *Health Affairs*, 24 (3), 653-665.

Borowiak, C. (2004), 'Farmers' rights: intellectual property regimes and the struggle over seeds', *Politics & Society*, 32 (4), 511-543.

Bourgeois, L. (2001), 'GM foods: advance or decline of food safety?', *Ocl-Oleagineux Corps Gras Lipides*, 8 (4), 278-286.

Brodhag, C. and Taliere, S. (2006), 'Sustainable development strategies: tools for policy coherence', *Natural Resources Forum*, 30 (2), 136-145.

Bullard, R. (2005), *The Quest for Environmental Justice: Human Rights and the Politics of Pollution*, Berkeley: Sierra Club Books.

Bullard, R. (1990), *Dumping in Dixie: Race, Class, and Environmental Quality*, Boulder: Westview Press.

Bullard, R., Mohai, P., et al. (2007), *Toxic Wastes and Race at Twenty: 1987-2007. Grassroots Struggles to Dismantle Environmental Racism in the United States*, New York: United Church of Christ.

Calfee, J. and Bate, R. (2004), 'Pharmaceuticals and the worldwide HIV epidemic: can a stakeholder model work?', *Journal of Public Policy & Marketing*, **23** (2), 140-152.

Cassiolato, J., Lastres, H., et al. (eds) (2003), *Systems of Innovation and Development: Evidence from Brazil*, Cheltenham, UK and Northampton, MA, USA: Edward Elgar.

Chang, H. (2002), *Kicking Away the Ladder: Development Strategy in Historical Perspective*, London: Anthem Press.

Christie, M. and Honig, B. (2006), 'Social entrepreneurship: new research findings', *Journal of World Business*, **41** (1), 1-5.

Cook, B., Dodds, C., et al. (2003), 'Social entrepreneurship: false premises and dangerous forebodings', *Australian Journal of Social Issues*, **38** (1), 57-72.

Cook, P. (2001), 'Regional innovation systems, clusters and the knowledge economy', *Industrial and Corporate Change*, **10** (4), 945-974.

Curkovic, S., Sroufe, R., et al. (2005), 'Identifying the factors which affect the decision to attain ISO 14000', *Energy*, **30** (8), 1387-1407.

Davies, J., Sandstrom, S., et al. (2006), 'The world distribution of household wealth', World Institute for Development Economics Research, Helsinki.

De George, R. (2005), 'Intellectual property and pharmaceutical drugs: an ethical analysis', *Business Ethics Quarterly*, **15** (4), 549-575.

Dees, J. (2007), 'Taking social entrepreneurship seriously', *Society*, **44** (3), 24-31.

Deininger, K. and Squire, L. (1998), 'New ways of looking at old issues: inequality and growth', *Journal of Development Economics*, **57** (2), 259-287.

Deininger, K. and Squire, L. (1997), 'Economic growth and income inequality: re-examining the links', *Finance and Development*, **34** (1), 38-41.

Delamothe, T. (1989), 'Nestle Boycott', *British Medical Journal*, **298** (6675), 702-702.

Drahos, P. and Braithwaite, J. (2002), *Information Feudalism: Who Owns the Knowledge Economy?*, London: Earthscan.

Duffy, J. (2004), 'Rethinking the prospect theory of patents', *University of Chicago Law Review*, **71** (2), 439-510.

Easterly, W. and Levine, R. (1997), 'Africa's growth tragedy: policies and ethnic divisions', *Quarterly Journal of Economics*, **112** (4), 1203-1250.

FAO (2002), 'Gender and law: women's rights in agriculture', FAO Legislative Study, Food and Agriculture Organization of the United Nations, Rome.

Fisher, M. (2006), 'Income is development: KickStart's pumps help Kenyan farmers transition to a cash economy', *Innovations: Technology, Governance, Globalization*, **1** (1), 9-30.

Freeman, C. (2000), 'Social inequality, technology, and economic growth', in S. Wyatt, F. Henwood, N. Miller and P. Senker (eds), *Technology and In/Equality: Questioning the Information Society*, London, UK and New York, USA: Routledge, pp. 149-171.

Fridell, G. (2004), 'The fair trade network in historical perspective', *Canadian Journal of Development Studies-Revue Canadienne D, Etudes du Developpement*, **25** (3), 411-428.

Galbraith, J. and Hale, T. (2003), 'Income distribution and the information technology bubble', UTIP working papers No. 27, University of Texas, Austin.

Galbraith, J., Conceicao, P., et al. (2000), 'Inequality and growth reconsidered once again: some new evidence from old data', UTIP working paper No. 17, University of Texas, Austin.

Gaskin, I. (1981), 'Nestle boycott', *Journal of Nurse-Midwifery*, **26** (1), 39-41.

Goktolga, Z., Bal, S., et al. (2006), 'Factors effecting primary choice of consumers in food purchasing: the Turkey case', *Food Control*, **17** (11), 884-889.

Granstrand, O. (2006), 'Innovation and intellectual property rights', in J. Fagerberg, D. Mowery and R. Nelson (eds), *The Oxford Handbook of Innovation*, Oxford, UK and New York, USA: Oxford University Press, pp. 266-290.

Gregory, J. (2003), 'Scandinavian approaches to participatory design', *International Journal of Engineering Education*, **19** (1), 62-74.

Greiner, A., Rubart, J., et al. (2004), 'Economic growth, skill-biased technical change and wage inequality: a model and estimations for the US and Europe', *Journal Of Macroeconomics*, **26** (4), 597-621.

Guay, T., Doh, J., et al. (2004), 'Non-governmental organizations, shareholder activism, and socially responsible investments: ethical, strategic, and governance implications', *Journal of Business Ethics*, **52** (1), 125-139.

Gungor, A. and Gupta, S. (1999), 'Issues in environmentally conscious manufacturing and product recovery: a survey', *Computers & Industrial Engineering*, **36** (4), 811-853.

Haigh, M. and Hazelton, J. (2004), 'Financial markets: a tool for social responsibility?', *Journal of Business Ethics*, **52** (1), 59-71.

Hira, A. and Ferrie, J. (2006), 'Fair trade: three key challenges for reaching the mainstream', *Journal of Business Ethics*, **63** (2), 107-118.

Hornady, M. (2006), 'CAFTA & Women', *Z Magazine Online*, **19** (2), 1-3.

Juma, C. and Yee-Cheong, L. (2005), *Innovation: Applying Knowledge in Development*, London: Earthscan.

Jutte, R. (2005), 'The longue duree of contraceptive methods', *History and Philosophy of the Life Sciences*, **27** (1), 71-79.

Keyhani, S., Diener-West, M., et al. (2005), 'Do drug prices reflect development time and government investment?' *Medical Care*, **43** (8), 753-762.

Kim, L. (1997), *Imitation to Innovation: The Dynamics of Korea's Technological Learning*, Boston: Harvard Business School Press.

Klaassen, G., Miketa, A., et al. (2005), 'The impact of R&D on innovation for wind energy in Denmark, Germany and the United Kingdom', *Ecological Economics*, 54 (2-3), 227-240.

Klein, J., Smith, N., et al. (2004), 'Why we boycott: consumer motivations for boycott participation', *Journal of Marketing*, **68** (3), 92-109.

Kuznets, S. (1955), 'Economic growth and income inequality', *American Economic Review*, **45**, 1-28.

Lai, E. and Qiu, L. (2003), 'The North's intellectual property rights standard for the South?', *Journal of International Economics*, **59** (1), 183-209.

Lemieux, T. (2006), 'Increasing residual wage inequality: composition effects, noisy data, or rising demand for skill?', *American Economic Review*, **96** (3), 461-498.

Levidow, L. (1999), 'Britain's biotechnology controversy: elusive science, contested expertise', *New Genetics and Society*, **18** (1), 47-64.

Loff, B. (2003), 'Debate that "This house believes the essential drug concept hinders the effective deployment of drugs in developing countries" – patents and access to essential drugs', *Transactions of the Royal Society of Tropical Medicine And Hygiene*, 97 (1), 6-9.

Lundvall, B.-Å., Intarakumnerd, P., et al. (eds) (2006), *Asia's Innovation Systems in Transition*, Cheltenham, UK and Northampton, MA, USA: Edward Elgar.

Marinova, D. and Raven, M. (2006), 'Indigenous knowledge and intellectual property: a sustainability agenda', *Journal of Economic Surveys*, **20** (4), 587-605.

Matthews, D. (2004), 'WTO decision on implementation of paragraph 6 of the Doha Declaration on the TRIPS Agreement and public health: a solution to the access to essential medicines problem?', *Journal of International Economic Law*, 7 (1), 73-107.

Maxwell, D. and Vorst, R. van der (2003), 'Developing sustainable products and services', *Journal of Cleaner Production*, **11** (8), 883-895.

Milanovic, B. (2005), *Worlds Apart: Measuring International and Global Inequality*, Princeton: Princeton University Press.

Muchie, M., Gammeltoft, P., et al. (eds) (2003), *Putting Africa First: The Making of African Innovation Systems*, Aalborg: Aalborg University Press.

Mutersbaugh, T. (2005), 'Just-in-space: certified rural products, labor of quality, and regulatory spaces', *Journal of Rural Studies*, **21** (4), 389-402.

Nelson, R. (ed.) (1993), *National Innovation Systems: A Comparative Analysis*, Oxford, UK and New York, USA: Oxford University Press.

Norris, G. (2006), 'Social impacts in product life cycles: towards life cycle attribute assessment', *International Journal of Life Cycle Assessment*, **11**, 97-104.

Piva, M., Santarelli, E., et al. (2005), 'The skill bias effect of technological and organisational change: evidence and policy implications', *Research Policy* 34 (2), 141-157.

Pogge, T. (2005), 'Human rights and global health: a research program', *Metaphilosophy*, **36** (1-2), 182-209.

Post, J. (1985), 'Assessing the Nestle boycott: corporate accountability and human-rights', *California Management Review*, **27** (2), 113-131.

Potoski, M. and Prakash, A. (2005), 'Covenants with weak swords: ISO 14001 and facilities' environmental performance', *Journal of Policy Analysis and Management*, **24** (4), 745-769.

Prahalad, C. (2005), *The Fortune at the Bottom of the Pyramid*, Upper Saddle River: Wharton School Publishing.

Prakash, A. and Potoski, M. (2006), 'Racing to the bottom? Trade, environmental governance, and ISO 14001', *American Journal of Political Science*, **50** (2), 350-364.

Raines, S. and Prakash, A. (2005), 'Leadership matters: policy entrepreneurship in corporate environmental policy making', *Administration & Society*, **37** (1), 3-22.

Rao, S. (2006), 'Indigenous knowledge organization: an Indian scenario', *International Journal of Information Management*, **26** (3), 224-233.

Rogers, E. (1962), *Diffusion of Innovations*, New York: The Free Press.

Sagasti, F. (2004), *Knowledge and Innovation for Development*, Cheltenham, UK and Northampton, MA, USA: Edward Elgar.

Saxenian, A., Motohama, Y., et al. (2002), *Local and Global Networks of Immigrant Professionals in Silicon Valley*, San Francisco: Public Policy Institute of California.

Sell, S. and Prakash, A. (2004), 'Using ideas strategically: the contest between business and NGO networks in intellectual property rights', *International Studies Quarterly*, **48** (1), 143-175.

Stewart, F. (2001), 'Horizontal inequalities: a neglected dimension of development', working paper of the Centre for Research on Inequality, Human Security and Ethnicity, University of Oxford, Queen Elizabeth House, Oxford.

Tatsch, S. (2004), 'Language revitalization in native North America: issues of intellectual property rights and intellectual sovereignty', *Collegium Antropologicum*, **28**, 257-262.

Timmermans, K. (2003), 'Intellectual property rights and traditional medicine: policy dilemmas at the interface', *Social Science & Medicine*, **57** (4), 745-756.

Towse, A. and Kettler, H. (2005), 'Advance price or purchase commitments to create markets for treatments for diseases of poverty: lessons from three policies', *Bulletin of the World Health Organization*, **83** (4), 301-307.

UNDP (2006), 'Beyond scarcity: power, poverty, and the global water crisis', United Nations Development Program, New York.

UNDP (2004), 'Cultural liberty in today's diverse world', Human Development Report, United Nations Development Program, New York.

Wan, G. and Zhang, X. (2006), 'Rising inequality in China', *Journal of Comparative Economics*, **34** (4), 651-653.

Weeden, K. (2002), 'Why do some occupations pay more than others? Social closure and earnings inequality in the United States', *American Journal of Sociology*, **108** (1), 55-101.

Woodward, I. (2003), 'Morals, markets and money: ethical, green and socially responsible investing', *Journal of Economic Psychology*, **24** (3), 418-421.

Yamada, T., Yamada, T., Liv, A.C. et al. (1997), 'Value-added productivity of manufacturing industries in Japan', *Applied Economics*, **29** (10), 1317-1324.

Yankelovich, D. and Meer, D. (2006), 'Rediscovering market segmentation', *Harvard Business Review*, **84** (2), 122.

Yunus, M. (2006), *Nobel Lecture: Poverty is a Threat to Peace*, Stockholm: Nobel Foundation.

16. The Role of Technology Assessment in Systemic Innovation Policy

Ruud Smits, Rutger van Merkerk, David H. Guston and Daniel Sarewitz

INTRODUCTION

This book conceives of innovation as a non-linear, multi-level, and multi-actor game with many interactions or feedbacks among those actors. We assume that innovation theory, practice, and intervention develop by interactions among the worlds of science, policy, and practice (see Chapter 17 by Smits, Kuhlmann and Teubel in this book). According to this view, we can no longer see innovation as a given thing – as an invention. Instead, innovation is a systemic process involving a heterogeneous set of actors who are inspired by both the potential that science and technology offer and by the context in which they have to function. These actors are involved in a complex decisionmaking process that leads to innovative activity.

In this chapter, we will examine an important consequence of this shared conception: the need of actors for information that enables them to engage in innovative activities in an adequate and effective way. We call this information Strategic Intelligence (SI), and actors involved in innovation require it to develop their visions, strategies, and plans of action. Apart from this 'instrumental' role, SI helps to reflect on the development, interaction and effectiveness of innovation theory, practice, and intervention. By this, SI provides an important input in the further development of these three concepts.

There are many types of strategic intelligence (Kuhlmann et al. 1999; Tübke et al. 2002). In this chapter we will focus on one particular strand: technology assessment (TA). TA emerged in the US in the late 1960s and developed along multiple paths. In this chapter, we focus only on TA in the public sector specifically related to knowledge-based innovation on one hand, and social implications on the other. This view necessarily excludes aspects of TA in the private sector, foresight, and other TA-like activities in the public sector, such as environmental impact assessment. This focus is of necessity somewhat arbitrary, but it is through public sector innovation policies that, in our view, the interactions between TA as a form of strategic intelligence and innovation theory, practice, and policy are most clearly visible. We will begin the chapter with a brief overview of the development of TA, in the United States and Europe, over the last four decades, and discuss some aspects of its institutional and intellectual diversity. The next section will elaborate our concept of SI and of TA as a type of SI. We will then develop a framework that allows for a more differentiated evaluation of the role of SI in policy-making. We will conclude with two detailed case studies of more contemporary TA: real-time technology assessment and constructive technology assessment, both focusing on emerging nanotechnologies. From the earlier history and the recent examples, we will explore the (possible) role of TA in systemic innovation polices and whether a 'dominant design' for TA is in place. Finally, we draw conclusions about improving the impact of TA as a major form of SI in systemic innovation theory, practice, and policy.

TECHNOLOGY ASSESSMENT

A Moving Target

TA originated in the US at the end of the 1960s as a reaction to many unexpected negative effects of new technologies and the problems encountered in 'making technology work' in the way users intended. These two factors were important, although partial, rationales for the proposal in US Congress to establish an early warning system for possible effects of new technologies. TA was initially positioned – to use the terminology of Smits and Leyten (1991) – mainly as a 'watchdog'. Founded as a scientific discipline that would systematically identify and evaluate science and technology for their consequences for social, cultural, political, economic, and environmental systems, TA has since become more of a 'tracker' – in which it is not so much an outcome of scientific analysis as an ongoing process of analysing developments in science, technology, and innovation, their consequences, and the discussions about them. In Table 16.1 these two types of TA are characterized further.

Table 16.1 Watchdog and tracker TA characterized

Watchdog[a]	Tracker
Technology (impact) analysis	Trans-disciplinary analysis
Project oriented	Process oriented
Actors not heavily involved	Actors involved
Problem definition often implicit	Emphasis on developing problem definition
Early warning	Support actors in innovation processes
Only research	Research and discussion
Centralized TA capacity	Pluriform, decentralized TA capacity

a Coates (1975) provides a definition that reflects the watchdog type of TA: TA is the systematic identification and evaluation of the potential secondary consequences (whether beneficial or detrimental) of technology in terms of its impact on social, cultural, political, economic, and environmental systems and processes. TA is intended to provide a neutral, factual input into the decision-making process.

Nowadays, at least in Europe, TA is primarily expected to supply information that enables people involved in decision-making about innovation to determine appropriate strategies. TA thus aims to support decision-making; contribute to the socialization of decision-making on science, technology, and innovation; and improve the social utilization of science, technology, and innovation. Many of the participatory approaches developed within TA attempt to give a say to citizens who formerly had little or no voice in decision-making on science, technology, and innovation, but nevertheless bring important perspectives to the discussion. Indeed, the participation of stakeholders and users has been appearing more prominently on the agenda of the development of TA (Smits & Weijers 1991; Decker and Lodikas 2004; Sclove 1995).

This development, principally over the last 15 years, has also included the formulation of constructive technology assessment (CTA) (Rip et al. 1995) and other initiatives that reinforce the process character of TA, involve users more effectively, and enhance the link with policy (Grin et al. 1997; Guston & Sarewitz 2002). Although much progress has been made, the integration of TA in innovation policies remains a problem. This conclusion could, however, be phrased the other way round: until now, policy-makers have paid but little attention to the lessons learned by TA about strategic intelligence in innovation.

In hindsight, this shift in TA can be conceived as an ongoing learning process driven by interactions between science, policy, and practice. From practice it became clear that the impact-oriented, R&D- and project-based type of TA with its heavy influences from the linear model of innovation was not very effective. The watchdog TA showed many traces of the rational but rather optimistic rational planning theories from the 1960s (Simon 1976; van Vught 1980) demonstrating a great trust in the potential of scientific research to 'know' the future. The TA community drew the conclusion that TA should not focus so much on R&D but on the innovation process as a whole and organize its work as a complementary process carried out in close interaction with all relevant actors. This approach was supported and influenced by developments in innovation theory. The TA community learned from the emergence of the innovation systems approach with its emphasis on process with many, interlinked heterogeneous actors (Freeman 1987; Lundvall 1992), research in the role of users (von Hippel 1988, 2005; Akrich 1995; Oudshoorn & Pinch 2003), and evolutionary approaches like the Social Construction of Technology (Bijker 1995), evolutionary economics (Nelson & Winter 1982; Dosi et al. 1988) and Technology Dynamics (van de Belt & Rip 1987).

The lack of impact on policy-making stimulated many changes in the conception and practices of TA and was one of the major driving forces behind the shift from watchdog to tracker. The impact of these changes on policy, however, has not been clear. We suggest two reasons for this persistent lack of apparent impact: first, policy discourse continued to be dominated by linear accounts of innovation, so policy-makers were not asking questions that TA could answer (Smits & Weijers 1991); and second, a persistent tension existed between the world of TA with its societal, sometimes technology-critical orientation, and the world of innovation policy with its heavy emphasis on economic goals. In the US, the absence of a centralized innovation policy complicates the issue even further. These factors over the years proved to be important barriers to the development of a fruitful interaction between TA and policy.

The hypothesis in this chapter is that the gap between TA and policy practice is beginning to narrow, due predominantly to recent developments in, and interactions between, research, policy, and practice, in particular the growing influence of evolutionary and systemic approaches to understanding innovation. To explore this hypothesis further we first will dive deeper in the development of TA in the US and Europe over the last 40 years.

Forty Years of TA in the United States

From 1974 to 1995, it was said that TA in the US was whatever the Office of Technology Assessment (OTA) happened to be doing. While OTA existed, that formulation avoided a number of complications, among them the sheer diversity of OTA's activities, the role of TA in the US innovation system, and the role of the US Congress as the monopsonistic consumer of OTA's work.

The general arc of OTA's history is well known: there was a significant but under-recognized and widely distributed TA capacity in the executive branch of the US federal government (Coates 1975). Congress, which had been grappling with its authority over S&T through the 1960s, began to consider its own role in TA (Stine 1986). Caught in a wider battle with President Nixon over the institutional bases of political power, Congress augmented its own status through a host of innovations, including OTA (Bimber 1996; Bimber & Guston 1997). OTA evolved from a troubled start-up to a respected collection of brokers and analysts of technical analysis. The Republicans, who gained the majority in the House of Representatives in 1995 for the first time in two generations, targeted OTA as both partisan and redundant and cancelled appropriations for the office.

This story, however, neglects at least four crucial issues. First, OTA had external critics all along, particularly those (e.g. Bereano 1997; Sclove 1995) who disparaged its technocratic orientation. Even Harvey Brooks, who helped craft OTA's chartering legislation, was by the mid-1980s arguing for a TA that was both more engaged with the public and with researchers. Second, OTA strayed far from the foresight capacity that the original legislation contemplated, and conceived of its work more as policy analysis than as TA in that original sense. Third, OTA had internal critics who helped it establish a modest but perceptive reflexive capacity and who, had the office not been de-funded, might have succeeded in reorienting it. Fourth, OTA's policy impact was not easily apparent, thus helping to create the environment in which congressional Republicans found themselves able to close the office (we will expand on this below).

Reconstituting TA in the US

Immediately after OTA's demise, the consensus (or hope) among many of its former analysts was that the Republican ascendancy would quickly wane and OTA could be revived. These analysts, together with other friends of the office, constituted the private, non-profit Institute for Technology Assessment as a holding pen of sorts. Intertwined with this effort was XOTA, a loosely knit but formally incorporated group of former OTA employees who lobbied for OTA's re-establishment. XOTA has a board of directors and other officers but has done little more than maintain a directory of OTA alumni and a list serve, and host an annual social function.

Granger Morgan and John Peha, both of Carnegie Mellon University, began a larger effort in 2001 with a conference on S&T policy advice to Congress, funded by the Heinz and MacArthur Foundations and supported by 18 professional and academic science policy groups. Well-coordinated with congressional staff and members, the meeting was ostensibly a bi-partisan activity, allied with a bill introduced in Congress to re-establish a TA capacity. But Republican involvement was largely limited to moderate, Eastern Republicans with long tenures on the committees that dealt with science and technology issues. Morgan and Peha (2003) published a book from the conference, and congressional advocates made a tiny incremental step – authorization for a pilot technology assessment by the GAO (then, the General Accounting Office; now the Government Accountability Office) on biometrics (GAO 2002; also see Fri et al. 2002). Most observers agree that TA for Congress will remain moribund at least until another partisan realignment, and thus some (e.g. Sarewitz 1996) have called for a more widely distributed TA capacity.

Other Tools

The focus on OTA neglects other efforts in the US that can reasonably be considered as TA, e.g., the forecasting and environmental impact assessments mentioned above. One approach with a significant impact is road mapping, especially as developed by SEMATECH, a public-private partnership aimed at restoring US pre-eminence in the semiconductor industry. SEMATECH pioneered the use of roadmaps to identify future goals and bottlenecks for semiconductor R&D, for example, what kinds of innovations were necessary, and when, to remain on the trajectory described by Moore's Law. A key feature of the roadmaps was their dynamic nature: They evolved through time both to integrate technological advances and to include increasing detail on benchmarks and obstacles. While the value of such roadmaps has not been fully assessed, some in the industry give them credit for catalyzing the renaissance of US semiconductor technologies. Public sponsors of research in the US, particularly the Department of Energy and more recently the National Institutes of Health, have begun crafting roadmaps as well, although it is not clear whether these achieve the methodological sophistication or the level of detail of SEMATECH's.

Variations on TA also found a home in 'futurism', a loosely defined practice aiming to illuminate potential technological and social trajectories to support present-day decision-making. In particular, a number of small consulting groups, started in the 1980s, support the strategic decision processes of private and public sector institutional clients dealing with rapid technological change. By seeking to inform decision-making based on scenarios of the future, these groups occupied the intellectual ground abandoned by OTA shortly after its inception. These groups also tended toward technological optimism, although such groups as the Club of Rome and the WorldWatch Institute could be construed as engaged in future-oriented TA with a more pessimistic view.

Persistent tensions surrounding knowledge-based innovation have led, more recently, to government support of research on the 'ethical, legal, and social implications' (ELSI) of major S&T initiatives. The first major ELSI program in the US was an add-on to the Human Genome Project. Its goal was to conduct ethics research on genomics, rather than to influence or support decision-making or policy. More recently, the US Congress passed legislation mandating a societal implications program for nanotechnology (Fisher & Mahajan 2006), specifying that the program should help improve the societal value of nanotechnology research by involving both the nanotechnology researchers and the public. The legislation neglected to specify particular mechanisms for such involvement (Bennett & Sarewitz 2006).

Research on the social dynamics of knowledge-based innovation suggest that the decentralized approaches encouraged by such legislation might be more feasible and desirable than a single, national-level effort. Given the magnitude, diversity, and scope of the US innovation system, approaches rooted in the organization of R&D institutions themselves make sense. Melding cutting-edge innovation with ELSI research to enable more reflexive institutions farther upstream in the innovation process could be a more effective way to encourage the anticipatory governance of emerging technologies. One such approach has been described as 'real-time technology assessment' (Guston & Sarewitz 2002). A new effort to test this integrated approach, described in more detail below, has recently begun at the Center for Nanotechnology in Society at Arizona State University.

To wind up, OTA started as a rather lame watchdog and over the years turned into an efficient policy analysis institute with some qualities of a tracker. Lack of visible impact and a well developed innovation policy, not to mention raw politics, in the end caused its demise. Attempts to resuscitate the agency have not been very successful. At the same time outside OTA the development of formal and informal TA went on, resulting in a number of promising approaches integrating different perspectives on new technologies and better linked to major actors involved in innovation processes.

Forty years of TA in Europe

As in the US, we can characterize the first wave of TA in Europe in the 1970s as the rise of the watchdog. Fear of the negative effects of technology acted as a major driving force. Other important aspects in the early days were links with environmental movements, a societal anti-technology attitude, and student revolts. In retrospect, we can conclude that in this period, the institutionalization of TA in Europe failed because of the lack of interest of politicians and a too-naïve concept of TA (Smits & Leyten 1991; Smits et al. 1995).

Unlike in the US, a second wave of TA in Europe broke in the 1980s, characterized by the emergence of the tracker. TA was viewed in a broader and more sophisticated way, not simply avoiding negative effects but pursuing a better integration of science and technology in society. The anti-technology attitude faded, and TA received the opportunity to play a role in innovation as seen from an economic as well as a broader societal perspective (health, environment, well being). The interest in TA grew in policy circles at the national and supranational levels as was demonstrated by two European TA congresses organized by the EC; in 1987 in Amsterdam (de Hoo et al. 1987), and in 1991 in Milan (Smits & Weijers 1991). In the same period, due to the heavy economic recession, innovation policy came on the political agenda. TA developed into a more policy-oriented instrument, geared to support actors involved in innovation processes. By this, the gap between the societal and economic approach of innovations slowly narrowed. An illustration of the impact of this narrowing gap is the TA on the impact of what in those days was called micro-electronics (ICT). In the Rathenau report (Adviesgroep Micro-Electronika 1980) not only the economic impact of ICT, but also the societal impact was addressed. The same accounted for a TA on genetic engineering (Brede DNA Commissie 1983). This development is reflected in the rise of a new concept of TA as 'consisting of analysis of technological developments and their consequences and discussions about them. The aim of TA is to provide information that will help the parties involved to determine their strategy and enables them to define new objects for TA-research' (Smits & Leyten 1991). In this period, also a number of TA organizations were established of which most of them were linked to national parliaments (Vig & Paschen 2000). However it should be noted that this 'second wave' went more smoothly by far. Long and difficult political debates as, for instance, in Germany often preceded the establishment of the various TA organizations. Moreover their mandates and resources often were very limited. Apart from political struggles one of the major reasons for this troublesome birth was related to the fact that in those days innovation policy in most European countries still carried many elements of the linear model as was demonstrated by its strongly supply-side (R&D) orientation (see also the next section). As a result, the impact of TA on policy still was rather meagre.

Continuing its distinction from the US, European TA experienced a third wave in the early 1990s that focused on the further development of the toolkit of the tracker dog and attempts to strengthen links with policy. Keywords here are participation, demand articulation, TA as a process. TA more and more was viewed as a source of Strategic Intelligence, supporting actors involved in innovation processes to better handle the interface between supply of and demand for technology. Despite these intentions, TA in Europe still was in a rather isolated and sub-critical position. The already mentioned 2nd European TA Congress, organized by the European Commission, in Milan in 1990 took stock of the state-of-the-art of TA in Europe. Although a number of TA institutions were in place, the world of TA remained fragmented and the institutions themselves were sub-critical and often fell short of required quality standards and expectations for impact on policy. These shortcomings, in combination with the ever-growing importance of innovation, induced a debate on how to improve the link between TA and innovation policy, including discussions on quality and methodology. From such discussions emerged experimentation with TA approaches like 'interactive TA', 'participatory TA' and 'constructive TA' to improve the quality, impact, and interactivity of TA. As a sign of further institutionalization, the European parliamentary TA (EPTA) organization consolidated and expanded its reach; EPTA nowadays involves 16 members and associates.

We have mainly focused on the development of TA, but over this period, innovation policies in most OECD countries were evolving, too. Whereas in the early 1980s innovation policies were still very much supply driven diffusion policies – in line with the linear model of innovation – the demand side (users), and even systemic elements of innovation began to receive more attention. The almost exclusive focus on new inventions faded away, and it became clear that meeting broader societal goals as part of an innovation policy made sense from an economic perspective too. As a consequence, more actors and perspectives were included in innovation policies. This development confronted policymakers – who meanwhile had adopted the systems of innovation approach as the basis of their policies but struggled with the implementation – with the problem of how to involve non-traditional actors, as for instance users, in an effective and efficient way in their policies. In the US, such policies included the Bayh-Dole and Stevenson-Wydler Acts, as well as manufacturing extension policies for pre-competitive technologies such as the Advanced Technology Program, the rise of state and regional S&T policies, and the coordination of national initiatives in climate change, information technology, genomics, and nanotechnology. In Europe the integration of TA in the European Commissions Framework Program (the R&D program of the EC), the EC-Targeted Socio-Economic Research program concentrating on TA-like research and striving after improving the link with policy, the recent establishment of the EC Technology Platforms and the establishment of the Institute for Prospective Technological Studies, an EC Joint Research Centre, intended to provide Strategic Intelligence on innovation to the EC, are important examples. On the national level, in the Netherlands, linking TA to 'big' science programs in the area of nanotechnology, climate change, carbon storage, and genomics, as well as large scale research and practices programs that should improve the links between science and application in areas as agriculture (see Chapter 17 by Smits, Kuhlmann and Teubal), water management and mobility are more recent examples of this trend. The core characteristic of all these activities is attention to the needs, wishes, and constraints of users and other societal actors (including policymakers) in an early stage of the development of a new technology. Also the UK Foresight Program and the German Futur Program (see Chapter 17 by Smits, Kuhlmann and Teubal) are manifestations of this new type of TA. In the UK Foresight program panels of scientists, policymakers and industrialists, in a process combining research and debate, designed scenarios for the future development of some ten sectors of the British society ranging from transport to chemistry, when may be seen as a manifestation. The German Futur Program strived after starting a broad societal debate on the future of science and technology.

This combination of policy change and development of TA in the 1990s opened up opportunities for TA to better link with, and have influence on, innovation policies. But here too stubborn problems in the relation between TA and policy circles, as well as the still marginal position of TA until now, prevented a substantial impact of TA on policy (and other actors).

User Involvement

In innovation systems many heterogeneous actors play a role. From recent research it has become clear that users are of special importance here, as confirmed by the US and European TA experience. In the following we will focus on the role of TA and users in innovation processes to further illustrate the development and potential of TA and the resulting consequences for policy. Before we explore two more recent examples of TA developments that elaborate user involvement, we first dig a little deeper into the reasons why it is important to involve users.

As is clear from the above discussions, the selection, generation, introduction, and application of scientific and technological knowledge increasingly demands a two-way learning process. A fruitful interaction between producers and users benefits both parties (von Hippel 1988; Akrich 1995; Smits 2002). Smits and den Hertog (2007) articulate five reasons to justify such interaction. (1) Interaction provides for more effective articulation of social needs in the face of market failures and other limits of private initiatives. (2) Interaction can increase the competitive strength of enterprises by helping to equilibrate the innovative process and its products with public expectations. (3) Interaction can improve acceptance and social embedding of knowledge and technologies by tailoring innovations at an earlier stage. (4) Interaction can improve the learning capacity of society as a whole, but increasing the ability of users to articulate their needs and those of producers to become more open and responsive to them. And (5), interaction can enhance democracy by allowing for citizens to influence the course of science and technology.

Each of these reasons could be sufficient in itself to justify greater user involvement in innovation processes. All of them tend to increase the democratic and rational quality of decision-making because more perspectives are considered. These rationales also appeal to actors across multiple sectors, as private firms will find the second and third reasons critical to the bottom line, and governments will find a better articulation of needs and demands (the first), an enhanced learning capacity of society as a whole (the fourth), and greater citizen participation in innovation processes (the fifth) important for succeeding in achieving policy goals and in maintaining legitimacy.

In these interactions, expanded TA activities thus converge with, or take on the characteristics of, the broader category of Strategic Intelligence with which we started. In the following section, we return to SI and elaborate its role in systemic innovation policies.

STRATEGIC INTELLIGENCE AND SYSTEMIC INNOVATION POLICIES

In this section we discuss the nature of SI and the role it could play in systemic innovation policy. We start with recent insights from innovation studies and policy concepts based on these insights. We then turn to explaining the different dimensions and types of SI and conclude the section by further explicating TA as a specific strand of SI.

Trends in Innovation Policy

To put this discussion in a better perspective we first will go a little bit further in major developments in innovation policy in OECD countries.[1] In decades following World War II, most OECD countries had nothing in place that resembled an explicit innovation policy. Propelled by the economic recession, innovation policy started to develop in the late 1970s. It became clear that OECD economies could no longer compete on prices and wages, but had to shift to added value embodied in new or advanced products and services. First attempts to develop an innovation policy, inspired by the linear model, were strongly supply-oriented and dominated by financial instruments stimulating R&D. This policy was not very successful because firms did not incorporate R&D results. Roobeek (1990) pointed out that almost all OECD countries were facing the same type of problems.

The solution to this problem was sought in strengthening the intermediary infrastructure in order to be better able to ensure that new technologies also reached the firms. This resulted in the development of policy instruments stimulating

1 For more in-depth analysis of the development of innovation policies see Arnold et al. (2003); Edler et al. (2003); Kuhlmann (2003).

mobility of researchers from academia to private enterprise and in many technology transfer programs. Without doubt, these activities improved the utilization of new technologies, but in the early 1990s it became clear that policy could not be restricted to measures that only encouraged the production and diffusion of knowledge. A considerable mismatch between the needs of private firms and the knowledge produced was apparent all too often. A better interaction between the producers and the suppliers of knowledge was essential to be able to cope with this problem. This awareness was the start of the next phase in innovation policy: the user-oriented approach. In this phase it was not only the interface between the users and the producers that was improved, but also the supportive infrastructure was expanded by introducing new or improved forms of strategic intelligence, more advanced risk capital schemes, a high-level electronic infrastructure, changes in educational systems, and other conditions that facilitated innovation in networks and systems. Furthermore we see in this phase the addition of policy instruments to support companies at an organizational level in innovation processes (management advice and support) and instruments that enhance the interface between users and producers in both directions. The trend towards such instruments helped companies not only to absorb new technologies, but also to turn them into new and successful products and services. Bessant and Rush (1995) argued that innovation does not stop when a new technology is adopted by the potential user and that it is necessary to develop approaches that help to bridge the managerial gap. However, in the 1990s, firms were still very inwardly focused, or at best focused on bilateral relations, while financial instruments still heavily dominated the policy portfolio (Boekholt et al. 2001).

The developments as sketched in the foregoing imply that more and more actors were getting involved in the development and implementation of innovation policies. Through this evolution, the need for system-level thinking and intervention increased; we term this the systemic phase of innovation policy. Policymakers were discovering the concept of innovation systems (IS), which in turn was leading to a variety of richer visualizations, such as the rise of the so-called 'cluster approach' adopted by many OECD countries (OECD 2002). This great interest in the IS approach, however, does not imply that policymakers have been able to put the IS to work in their daily practice. In the context of this chapter, the most relevant problems attendant on a systems approach include the involvement of more relevant actors, broadening the scope of innovation policies, the empowerment of actors, organizing learning and experimenting, the set up of an integrated infrastructure for SI and the orchestration of instruments in order to realize goals defined on a systemic level and prevent unwanted interaction. In the following we will focus on one of these problems: the development of an SI infrastructure.

Strategic Intelligence: Theory and Policy

Strategic intelligence deals with the questions *who* needs *what* kind of information in order to let actors maximize their innovation efforts and how can this information be produced? The answers to these questions depend on which innovation model one adheres to. In a linear model, the technology is more or less given and only needs 'unveiling'. No strong distinction is made between invention and innovation, the problem definition is rather clear and simple from the beginning, the major audiences are policymakers and firms, and the most important goal is to inform actors on emerging technologies and their possible impacts. The production of SI thus proceeds along the lines of 'normal' science, the broad array of societal actors are hardly involved in this process, and the interface with them is limited to providing (some) input in the problem definition and to providing reports. Watchdog TA is in line with this perception of innovation processes.

In a systemic model of innovation policy, the nature and role of SI are far more complex. We may understand this complexity by looking at recent insights from innovation studies and characteristics of innovation policies based on these concepts, as summarized by Smits and den Hertog (2007). (1) Innovation is an interactive search process that takes place under uncertainty. (2) Innovation and technological change are endogenous processes and the result of co-evolution of technology and society. (3) Learning and creating learning environments are crucial to innovation. (4) Innovation and technological change are increasingly linked directly to scientific knowledge and involve a variety of transfer mechanisms. (5) In addition to knowledge creation, knowledge diffusion and knowledge utilization are crucial. (6) Innovation also demands knowledge and understanding of organizational innovation, services, and other 'soft' aspects of innovation.From this overview, and particularly from points 1, 3 and 6, it is clear that many aspects related to innovation processes ask for SI. SI may help actors involved to develop their visions and strategies and to turn these into concrete plans of action. Innovation policy concepts and instruments basing themselves on these theoretical insights also reflect this need. In the chapter by Smits, Kuhlmann and Teubal a number of so-called systemic instruments are proposed of which at least two explicitly refer to the need for learning and an adequate infrastructure for SI. This point may further be emphasized by referring to the plea for new types of governance in innovation polices, as stated in the so-called 'Karlsruhe Principles', formulated by Edler et al. (2003).

Dimensions of Strategic Intelligence

The foregoing not only points at the importance of SI for systemic innovation policies, but it also has consequences for the content and the nature of SI. With regard to content, it will be clear that the systemic approach emphasizes that although technology (the invention) may be given, the way in which technologies end up in our societies (the innovation) certainly is not. Moreover, it is often not clear which questions actors have to ask when confronted with a new technology. As a consequence, problem definition takes real effort, including identifying the broad and heterogeneous set of actors that should be involved and providing insight into their positions, goals, and interests. But a useful problem definition is only the first step in articulating the demands of the actors involved and in the remaining, complex, and highly interactive learning process. Hence, SI has to be conceived as a process that transcends the scientific arena to involve non-technical actors and non-technical information. In order to manage such a process, much attention has to be paid to (facilitation of) the various interfaces between actors with different backgrounds and expertises. SI also includes more than only providing knowledge, but also intervenes in, and partly shapes, this decision-making process. This makes the question of how SI is institutionalised in the innovation system very important. Questions like who will get involved, which questions will be addressed, and how will the results of SI impact upon decision-making, are influenced by the institutional position of SI.

We can thus characterize SI along the dimensions of process and content (Smits 2001) (Table 16.2).

Table 16.2 dimensions of SI.

Content	Process
• Tailor-made	• Articulation of demand
• Hard- and softside	• Mobilizing creativity
• Distributed character	• Elucidating 'tacit knowledge'
• Scale effects	• Assessment of the technological potential
• Facilitating learning	• Facilitating activities
• Mix between specific and generic	• Optimal link with decision-making
• Enlarging accessibility	• User involvement
• Insights from interaction	

TA as a Type of SI

We have mentioned that over the years multiple types of SI, each with their own goals and methodologies, have developed. Following Tübke et al. (2001), we can distinguish between five types of SI: Technology Forecasting, Technology Foresight, Technology Assessment, Evaluation, and Road Mapping. Each type is distinguished by its major task or goal, its field of application, and the kind of political or policy issues it addresses. Each type also has its own unique focus on different aspects of the process and content dimensions. Seen from the needs of innovation policy, one often has to combine these different types of SI.

TA distinguishes itself from other forms of SI by its task focus on support for decision-makers in national parliaments, and actors directly involved in innovation processes, as opposed to forecasting and road mapping, which concern concrete technological developments. It also distinguishes itself by addressing developing technologies in a problem-orientation, as opposed to foresight, which addresses a broader scope of early-warning functions, or evaluation, which addresses innovation policies retrospectively. TA also distinguishes itself through its concern for identifying options, its heavy emphasis on process (but not to the exclusion of content), and as a consequence, its extensive and intensive interaction with a wide variety of actors. These characteristics show much overlap with the tracker type of TA. For the remainder of this chapter, we will deal with TA as a prime example of SI.

IMPACT OF TA AND SYSTEMIC INNOVATION POLICY

A major issue regarding SI in general and TA in particular is impact: How should the impact of TA be conceived? And how can impact be measured?

If we consider TA, or for that matter SI itself, to be a kind of policy analysis, we might look to the scholarship on the impact of policy analysis for some initial guidance. The Anglo-American version of this literature, however, is long on attempts and short on findings, as it often concentrates too much on impact on 'the decision' rather than on other aspects of policymaking. Bimber's (1996) study of OTA, for example, demonstrated few instances of spontaneous acknowledgement of OTA's influence on a policy decision made by Congress or by individual members of Congress. Similarly, Bimber found relatively few mentions of OTA in the *Congressional Record* of debates on issues. OTA, less than other congressional agencies like the Congressional Budget Office, was not part of the vocabulary with which members of Congress justified their actions. There were, however, instances in which OTA's work was absolutely critical to a particular policy outcome. Indeed, to understand the hostility of congressional Republicans to OTA one must also understand, for example, OTA's impact on the debate over the Strategic Defense Initiative and other defense policy questions in the 1980s. Morgan and Peha (2003) document further concrete impacts of OTA as well.

Mostly, however, OTA was a facilitator of communication around an issue. OTA was usually one node in an issue network that contributed to congressional information on that issue (Whiteman 1986). In many issues, OTA was simply one among many nodes. In some, it was a critical node. Sometimes, for example, in its assessment of potential human genome projects (Cook-Deegan 1995), OTA played an important role in assembling the network. Even though OTA was not broadly participatory, it became expert in conducting stakeholder analysis and stakeholders' meetings, thus smoothing the policy waters until a time that Congress might decide to wade in. The impact of such efforts on problem definition and framing and on broader strategic intelligence are, however, hard to discern, and methods for identifying them, which would include detailed ethnographies of issues and information, are lacking.

Another reason why OTA's policy impacts may appear to be minimal is that the arena of US innovation policy is not usually geared to such input. There is no basic policy for science and technology, nor for innovation, that OTA could have helped construct. OTA's 1991 report, *Federally Funded Research: Decisions for a Decade*, is a detailed and sophisticated assessment – particularly when compared to post-OTA S&T policy documents like the 1998 report issued by the House Science Committee chairman Vernon Ehlers – but there was little legislative opportunity to apply such analysis. To talk about OTA's influence on innovation policy is to talk about its impact on little, rather than its little impact. In Europe too, policymakers often did not (or were not able to) formulate questions that TA could answer.

The previous OTA story shows that strategic intelligence, and TA in particular, needs a set of tools to assess its impact that can focus on learning across a broad swath of society. Such a learning focus is especially important for more participatory activities in which framing (set of relevant actors and problem definition) and learning are at least as important goals as immediate influence in any event (Guston 1999a). In the following we explore a framework to evaluate the impact of a TA. We first discuss that every TA has its own environment in which it operates, i.e. TAs are constrained by context and institutionalization. Second, assessing impact should follow the general dimensions of SI: content and process.

Environment of a TA: Context and Institutionalization

TAs are never carried out in isolation. As the TAMI (2004) project noted, the process and impact of TAs are dependent on their context and institutionalization. Context means the current situation of the technological field or problem at hand. Typical contextual questions include: How developed is the technology currently? Which actors are already involved and which only potentially? What governance arrangements have emerged around the technology to date?

Institutionalization is a less clear concept. Following TAMI, institutionalization involves five factors: problem selection and definition, the relation and involvement of important actors, the composition of the toolkit or choice of methods; the modes of project management and interaction; and the style and freedom of communication.

Both context and institutionalization can vary considerable from case to case and, moreover, can change during the course of a TA process, e.g., institutionalization is (partly) shaped through problem definition as existing TA institutions and resources constrain actors in their attempts to institutionalize a TA. Both also imply that impact must be addressed case-by-case, because the environment strongly determines which impacts are sought and what can be expected.

Assessing Impact: Content and Process

In assessing impact, we can talk about the role of a TA in influencing both the content and process of innovation policy. Again, the first of the pair is relatively straightforward. Typical content-related impacts will include improved understanding among decision-makers, stakeholders, and other publics of the TA issues involved, substantive analyses and recommendations regarding choices in innovation policy, and the manifestation of such understanding, analyses, and recommendations in new policy.

Typical procedural impacts include: identifying and empowering actors in such a way that they can play their roles effectively, e.g., by enabling actors to identify and meet their information needs; bridging the gaps of information and influence among actors;[2] and contributing to the acceptance and use of TA results in decision-making. Such procedural impacts not only contribute to the quality and usefulness in a more restricted way by facilitating the provision of useful content, but they also will enhance the quality of participation and empowerment in and impact of decision-making.

In the long run, the latter result, although less visible than the provision of (in particular instrumental) knowledge or content, often turns out to be a very important type of impact. One consequence of this process orientation is the need for continuous monitoring – to evaluate intermediate results, plan for any indicated redesign of the process, and appropriately attune expectations.

There are many concepts that can be used to fill in the content and process measures of impact. One perspective that fits particularly well with the vision of TA as SI is conceiving of impact as the ability of actors to engage in substantive, procedural, and reflexive learning (Guston 1999a). Learning is a necessary step prior to behavioral change, i.e., acting differently within the innovation process. Collective or social learning is of special interest, because a change in the so-called 'shared frame of reference' can occur. Such change means that actors position themselves differently when making up their mind about a particular issue related to the innovation process. An example would be when scientists make it a common practice to take societal and ethical concerns into account in their work.

Others include the instrumental and conceptual use of TA results (Caplan 1979), seven standards of utilization (Dunn 1980), anticipation, reflection, and learning (Schot & Rip 1997), and timing of receiving and using information (Caplan 1979). While we do not claim to know which conception of impact on content and process is the best – and it may be that different conceptions are appropriate for different contexts and institutionalizations of TA – we do claim that both the content and process dimensions are necessary components in every TA evaluation. TAs conceived as simple research projects will not be able to meet these conditions.

2 Geurts (1993) distinguishes between four types of gaps between: practical policy processes and science, different (scientific) disciplines, administrators, and citizens and between experts and laymen, between the producers and the users of knowledge. Bridging the gap, however, does not imply consensus. It only points at enabling a fruitful dialogue between actors with different frames of reference and interests.

In the next section we will explore how TA developed over the last 40 years from a 'research project'-like activity into a 'process-oriented' activity by focusing on two recent developments in TA – Real Time TA and Constructive TA.

TWO CASES OF RECENT TA APPROACHES: REAL TIME TA AND CONSTRUCTIVE TA

TA approaches in the US and in Europe are under discussion in this section: (1) Real Time TA activities at the Center for Nanotechnology in Society at Arizona State University, and (2) Constructive TA activities in the Dutch TA NanoNed program carried out in the Innovation Studies Group at Utrecht University. We start by giving the context, institutionalization, and set-ups of both approaches. Second, we discuss the impact along the dimensions of content and process. We do have to make two remarks; first, the RTTA example plays at a higher level than the CTA example. This means that the latter is much more detailed, but on the other hand also of smaller scope. Second, in the case of RTTA, we can only discuss intended impact as the program is still unfolding towards its full capacity and activities are just starting. For the CTA example, activities have been rounded up, and impacts are assessed and known.

Real-Time TA in the US

Real-time Technology Assessment (RTTA) has a lineage of multiple strands. Guston and Sarewitz (2002) conceive of it as: (1) an incremental alteration of CTA for local consumption in the US; (2) a response to the specific failings of the ELSI model; (3) a synthetic enterprise built on the analytic findings of science studies and innovation studies; and (4) an intellectual and social response to the challenges that knowledge-based innovation poses both to governance and to the social sciences. The test-bed for RTTA has become the Center for Nanotechnology in Society at Arizona State University, which was created in October 2005 under a $US 6.2 million grant from the National Science Foundation. As implemented at CNS-ASU, RTTA is somewhat evolved from its initial description in the literature, but its impulse, first, to conduct fundamental and applied social science and humanistic research alongside natural science and engineering research, and, second, to push attention to questions of the value of knowledge-based innovations further upstream, remains the same.CNS-ASU maintains four program activities within RTTA:[3]

1. Research and Innovation System Assessment (RISA), which provides surveillance of nano-scale research activities through a variety of (mostly quantitative)

3 Details on the research activities and outcomes of CNS-ASU can be found in its first annual report to the National Science Foundation, on-line at http://cns.asu.edu/about/sitevisit.htm.

methods. In particular, RISA activities at CNS-ASU map the ongoing research dynamics of nanotechnology through bibliometric and patent analysis, evaluate the public value of nanotechnology by understanding the relationship between promises on its behalf and emerging societal impacts, and assess the nanotechnology workforce needs of regions by conducting supply-and-demand analyses for regional nano-workforces.

2. Public Opinion and Values (POV), which gathers both broad-brush and fine-grained data about knowledge of and attitudes toward nanotechnology and its societal implications from both researchers and lay-citizens. In particular, POV activities poll a large sample of the US public about their knowledge of and attitudes toward nanotechnology in a longitudinal and comparative perspective, study in a quasi-experimental fashion the impact of media reports on how the public conceptualizes nanotechnology, and survey a large sample of nano-scale science and engineering researchers about their perspectives on nanotechnology, for comparison to both the public poll and other data drawn from scientists.
3. Deliberation and Participation (DP), which elicits input on technical and societal issues through deliberative and participatory exercises. In particular, DP activities create 'naïve' but technically validated nanotechnological scenarios for analysis and assessment by other CNS activities, propose responsible nanotechnological products through a trans-disciplinary undergraduate education module called 'InnovationSpace', critique the scenarios and other visions of nanotechnological futures with the tools of critical theory and critical analysis, and involve scores of lay-citizens in the assessment of nanotechnological issues through a National Citizens' Technology Forum.
4. Reflexivity Assessment and Evaluation (RAE), which scrutinizes the activities and impacts of CNS-ASU at the micro- and macro-level. In particular, RAE activities examine the changing identity, knowledge, and practice among nano-scale science and engineering researchers with whom we interact, attempting to identify changes attributable to our influence, and reflect on CNS-ASU through international and other comparisons. CNS-ASU also addresses two crosscutting areas of substantive inquiry: Freedom, Privacy, and Security, in which researchers are particularly concerned with how the design of nano-sensors may or may not build in important capacities for data acquisition and management; and Human Identity, Enhancement, and Biology, in which researchers are particularly concerned with the place of nanotechnology in a long history of questions about the relationship between human biology and novel technologies, e.g., nanotechnological enhancements.

Interaction between nano-scale science and engineering researchers on one hand, and CNS researchers on the other, is a crucial aspect of the activities. Interaction among the various elements of RTTA is another. To address the former challenge, we design CNS-ASU intentionally as a boundary organization (Guston 1999b), to provide concrete benefits to participating researchers as well as to facilitate social research and spur societal capacities for anticipatory governance. To address the latter challenge, we focus on the central role of activities like mapping the nanotechnology enterprise and generating technologically plausible scenarios of nanotechnological futures in informing our other research (RTTA thus incorporates aspects of evaluation and forecasting into TA work). We also interweave research, educational, and outreach activities across CNS to enhance interactivity and opportunities for substantive, procedural, and reflexive learning.

Through this complex of RTTA activity, CNS-ASU hopes to build a greater societal capacity to engage in anticipatory governance of nanotechnology: from increasing public understanding of and engagement in nanotechnology, to rendering nano-scale science and engineering researchers more reflexive about their own work and its consequences for society. Decision-makers are one among many target groups for dissemination of RTTA findings, but scientists and engineers themselves are the critical target group, as RTTA aspires to cultivate a more reflexive, more anticipatory, and more engaged scientific disposition. CNS-ASU can thus provide descriptive and analytic information for scholars and decision makers, but it can also potentially seed a cultural change in science that will make knowledge-based innovation and the people who do it more familiar with, and receptive to, governance.

New TA in the Netherlands: CTA, Bypassing the Collingridge Dilemma

As we saw above, participatory TA projects are well founded in the Dutch/EU style of doing TA. More recently we see that TA projects are also linked up with 'big science and technology' projects and sometimes even incorporated in them. In the case of nanotechnology in the Netherlands this is done through NanoNed, a nation wide consortium for nanotechnology. TA practices are embedded in and funded by this scientific endeavour. Projects are carried out at universities around the Netherlands. Here, we discuss one of these projects executed at the Innovation Studies Group at the Utrecht University. A reason for TA practitioners to choose this setup is a tendency among TA practitioners to 'be there' as early as possible, to 'get the chance to do it right from the start'. When these two points merge we end up with participatory TA approaches, for emerging technologies. Taking the well-known Collingridge dilemma (Collingridge 1980) into consideration, in early stage technologies chances for improvement are richest, but directions for improvement and further development hard to find; this project strives to bypass this dilemma.

Since nanotechnology is an awkward bundle of exotic technologies simply defined by scale, a choice for the subject still has to be made. Here, we choose Lab-on-a-chip for medical applications technology that, in contrast to most other nanotechnology applications, has a few applications on the market and is used in practice. Therefore, although still being an emerging technology, the relevant actors can more easily be identified, because some structure is in place. We find another issue of interest when we look at the technological field itself. In the case of Lab-on-a-chip technology, not completely, but to some extent accidentally, we found a field that 'longed' for CTA kind of activities (in terms of de Bruijn et al. (2002) there is a 'sense of urgency'). This was revealed by interviews with the science and business community.

Which actors to include in the CTA when applications are almost non-existent, and taking into account there is a clear lack of visibility outside the scientific and pioneering business arena? CTA strives to incorporate all relevant actors in the process. Looking at which actors will have a say in the development and use of the technology sooner or later gives anchor points to find the actors. Heuristics as the innovation chain, innovation systems, and social maps provide the necessary intellectual tools to do this. The actors were invited iteratively, meaning feedback from the actors was used in the course of the project. This assumes an important point of TA, which is that TA is an ongoing process where steering along-the-way is a necessity.

The approach consisted of the following three steps: (1) providing information to the participants, (2) constructing individual scenarios, and (3) dialogue workshops. Follow-up interviews are used as part of the data on which the impacts of the approach can be assessed. The project recognizes the interface between insiders and outsiders (21), where insiders work towards the realization of the technology and are committed to its success (e.g., science and business), and outsiders are selectors in the sense of having multiple options to solve their problem of which the technology under discussion is just one (e.g., professional users and ministries).

We thus managed to perform a CTA on an emerging technology. The impact that is aimed for was, among others, seen in terms of increased anticipation, reflection, and learning. This in turn creates awareness of how socio-technical dynamics work and broadening of perspectives of the involved actors. Both are necessary for actors to change their actions, the only way real influence on innovation processes can be expected.

Impact Assessment of Both Approaches

RTTA at CNS-ASU, as of this writing just one year old, is still too young to have concrete impacts that can be easily measured – especially given the difficulty of identifying such impacts. Yet there are some early indications of the kind of impacts that RTTA might have, particularly on the target audience of nanotechnology researchers.

In terms of impact on content, CNS-ASU has begun to educate through coursework modest numbers of undergraduates in the societal implications of nanotechnology and provide opportunities for the engineering, design, and business students in InnovationSpace to create new nanotechnological product concepts. At the graduate level, CNS-ASU has begun to 'cross-train' several nano-science doctoral students to enable them to include research on societal questions about their own nano-science in their dissertations. By the account of one laboratory director, engagement with CNS-ASU has resulted in the success of a large grant proposal, which was more socially aware and problem-oriented than the first version of the proposal, which had been rejected. We have also begun to generate substantive knowledge in a variety of areas, in particular an operationalizable, social science definition of nanotechnology (Porter et al. 2006).

At the procedural level, we have influenced Mike Roco (2006), the principal policy entrepreneur of the US National Nanotechnology Initiative, to adopt CNS-ASU's terminology of 'anticipatory governance' as a goal for addressing the societal implications of emerging nanotechnologies. We have also argued for a greater attention, in the area of human nanotechnological enhancement, to a frame that includes attention to democratic political values (Parsi et al. forthcoming) and to the role and particular needs of persons with disabilities (Wolbring 2006).

The impact of the CTA example is discussed now. An important effect in terms of content is that most actors realized the value of discussing views and applications with different actors early in the development of the technology, i.e. their thinking about the future of the technology, reflected in their work, changed. Some even said their assessment for cooperation is now broader. There is a strong wish to have more events that also include a wide variety of actors. This is an important learning point that, due to the collective effect, can also be seen as a small change in the frame of reference of the participants. At the same time, actual changes in action are marginal. It can be the case that this is a longer-term effect (as some participants indicated), although this remains to be seen. Having more than one workshop would probably also be beneficiary on this point, which can be taken as a lesson for other TAs.

The process dimension focuses on three questions: did we succeed in getting the 'right' user involvement, did we succeed in providing effective support to participate in the discussion, and did we succeed in facilitating the interface between insiders and outsiders? The involved actors originated from science, business (SMEs and large firms), professional users (general practitioners and hospital care), ministries, financial institutions, and health care insurers. Listening to the participants during the follow-up interviews, this actor involvement was just about right. The participants indicated they were content with the process and the support (making the scenarios which prepared them for the workshop) given to them. The workshop itself was stimulating and generally (besides some comments) found useful. Since the discussions were enlightened both by insiders as well as outsiders, the interface was successfully facilitated.

CONCLUSIONS

In this final section we want to highlight the implications of our foregoing discussion on three aspects of TA that are central to our concerns: its development; its impacts and its role as SI in innovation policy.

Development of TA

Did TA actually develop from a watchdog functon into a tracker function, or are we, over the years, more and more looking at TA through a different analytic lens? If, as we argue, the former is the case, can we go even further to conclude that a 'dominant TA design' has evolved?

An innovation systems perspective obviously provides an analytical lens for understanding TA that is distinct from the lens provided by the linear model. The history of TA practice, however, clearly shows real and substantial changes in the institutionalization and practice of TA itself. Whereas OTA carried out its TA in the relative isolation of its office and much TA in Europe in the early days was carried out by committees of experts (as for instance the Rathenau committee) and TA institutes were not very well linked to other actors in the innovation systems (as for instance most parliamentary TA organizations in their early days), TA has, since the mid 1990s at least, beome something rather different. TA nowadays is far more process oriented, more inclusive of relevant actors, and better linked to those circles that have a real say in the innovation process.

The explicit goals of TA have also changed from informing a limited set of actors on emerging technologies and their impact, towards identifying involved actors, articulating their information needs, and stimulating and feeding learning processes. TA methodology in the watchdog era hardly paid any attention to working with stakeholders to formulate problem definitions. Indeed, watchdog TA can be understood not as a process, but as an event charcterized by formal methods and techniques carried out by centralized TA institutions with rather simple interfaces with users and policy. Nowadays much attention in TA is paid to defining the problem in a carefully designed participatory process with a focus on learning. While watchdog TA was a multidisciplinary, scientific endeavor, TA now can be characterized as a transdisciplinary activity carried out in a distributed infrastructure using a mixture of formal and process-oriented techniques. The interfaces between the heterogeneous set of actors are carefully designed to serve multiple perspectives to the extent possible. All in all, we may conclude that the tracker drove away the watchdog not only in the perspective of the analyst but also in daily practice.

During this change a sometimes intense interaction between policy, theory and practice has been visible. New insights from innovation studies (innovation systems, evolutionary approaches, the role of users) influenced TA and policy-makers, and experiences with TA in practice served as empirical input for innovation researchers. The interaction between TA and policy, however, has remained problematic, with policymakers often unwilling or unable to absorb the results of tracker TAs. We speculate that this is because the insights of TA do not easily translate to actionable knowledge in the daily work of policy-makers.

Will TA continue to evolve? In innovation theory, innovations go through a process of variation and selection (Utterback 1994) resulting in what is called a 'dominant design'. Might such an account apply to TA? In answering this question we first have to realize that in this chapter we limit ourselves to a special type of TA: public sector, focusing on knowledge-based innovation and social implications. Furthermore it should be obvious that despite some general trends, captured in the notion of the tracker, many different TA approaches are still visible. To speak of a 'dominant design' thus would not acknowledge this richness. Nonetheless, we feel it reasonable to conclude that a class of TA with the characteristics of the tracker may be viewed of as a promising strand of Strategic Intelligence, and an improvement over the watchdog function.

Impact

How should we conceive of the idea of the impact of SI, and TA in particular, and how can such impact be measured?

In our overview of attempts to evaluate TAs we argued that impact should not be understood or assessed in terms of influence on particular decisions. For one thing, evaluating TAs in this way does not make much sense as long as policymakers do not ask the questions TA can answer. More importantly, innovation theory is not a theory of decisions but of systems and networks. Thus, regarding OTA it can be concluded that 'To talk about OTA's influence on innovation policy is to talk about its impact on little, rather than its little impact'. Understanding TA's impact thus in part awaits further development of innovation theory (researching implications for policy of the systems of innovation approach, (see Part III), and further development of innovation policies (implementing insights from innovation studies in policy, see Chapter 17 by Smits, Kuhlmann and Teubal). Our view is that TA impact must be understood as a multidimensional phenomenon (content and process) and that determining impact is impossible without consensus on a well defined and shared problem definition, specifying explicitly the type of impacts to be expected, as developed by relevant actors taking into account the context and institutional limits in which the TA should be performed. As a matter for future research, proper evaluation of TA can build on in-depth analysis on a case-by-case basis; past TA projects should be a rich source of information for developing a useful approach to impact assessment.

TA and Policy

In conceptualizing TA as an important strand of SI in systemic innovation policies, what are the barriers to fruitful use of TA in innovation policy, and how can TA more succesfully contribute to such policies in the future?

Our understanding of systemic innovation policy makes it clear that actors involved in innovation processes need SI that support them in developing visions and strategies and turning these strategies into action. We concluded that the characteristics of this type of SI match up well with important characteristics of tracker TA, particularly the support of actors in their learning processes, the commitment to process-based information supply, and the focus on a heterogeneous set of actors engaged in the innovation process rather than the output of R&D activities. Smits and den Hertog (2007) present a number of possibilities for how TA may aid systemic innovation policies (see Table 16.3), drawing in turn on the five systemic instruments as discussed in the chapter by Smits, Kuhlmann and Teubal in this book.

Yet we are still left with the challenge of barriers between TA and policymakers. Policymakers often do not use insights from TA, sometimes because they are not aware of the results, but more often because the results do not seem relevant to them. Despite great conceptual advances, innovation policy in most OECD countries is largely characterized by (Smits & Kuhlmann 2002):

- A high degree of departmentalization, sectoralization of the political administration, and low inter-departmental exchange and cooperation
- Heterogeneous, non-inter-linked arenas: often corporatist negotiation deadlocks (e.g. health-innovation-related policy in Germany)
- Failing attempts at restructuring responsibilities in government because of institutional inertia (e.g. Germany, the Netherlands, the UK, etc.)
- Dominance of 'linear model' of innovation in policy approaches (and of related economists as consultants) in many national authorities (e.g. ministries)
- Conceptualisation of 'innovation policy' as a very specific, narrow field focusing, for example, on the introduction of new technologies in SMEs, issues related to intellectual property regimes, or the development of venture capital.

Innovation policy itself, that is, still often fails to embrace the known complexities of innovation processes. A key issue of systemic innovation policy, therefore, is the involvement of a 'new breed' of innovation policymakers, employed under the rules of a more flexible staff policy (e.g. supporting job rotation with industry or non-governmental organizations), working in a reformed, systemically interlinked institutional setting, fostering experimentation and learning and starting from a nonlinear, more systemic perspective on innovation.

A further development of systemic instruments is necessary to help this new breed of policymakers play their role in an effective and efficient way. Innovation studies may be helpful in this context by taking stock of already existing systemic instruments and, based on this, conducting a thorough analysis of the functions such instruments can fulfill and the situations in which they can be applied. Innovation studies may provide additional help for policy makers by furthering the insight into the dynamics of innovation systems and in the links between learning at the micro level with characteristics of the overarching system. The already mentioned 'functions approach' (see Chapter 6 by Bergek et al. in this book) appears to be a promising avenue to realize this.

This brings us to the contribution TA researchers and practitioners may provide. Apart from the possible roles as already outlined in Table 16.3, TA researchers and

Table 16.3 The possible role of TA and users in a modern innovation policy (Smits & den Hertog, 2007)

Innovation system functions	Possible roles of TA and users
Managing interfaces	• Involve users in innovation initiatives in sectors/fields • Stimulate user-producer relations (per cluster) • Create a TA section in innovation research (*'TA-begleitforschung'*) • Assess intermediaries as 'brokers' between knowledge demand and knowledge supply
Building and organizing (innovation) systems	• Give users a role in innovation networks and systems with regard to newly emerging technologies. • Collect and concentrate systemic knowledge with regard to the steering of innovation systems • Conduct strategic TA studies in selected sectors/domains • Review the operation of existing innovation systems (e.g. clean-up and/or reorientation of the existing knowledge infrastructure)
Creating forums for learning and experimenting	• Involve users as co-developers in innovation experiments. • Develop innovation forums connected to social issues (safety, healthcare quality, administrative innovation) • Experiment with demand-driven innovation (e.g. in the steering of a part of the public knowledge infrastructure) • Experiment with public/private knowledge institutions (along the lines of the *Technologische Topinstituten* in the Netherlands) • Experiment with systemic innovation instruments • Conduct constructive TA studies • Experiment with strategic niche and transition management
Establishing an infrastructure for strategic intelligence	• Conduct awareness TA studies • Make use of synergies between exploration communities, TA, and evaluation, and create a central strategic intelligence clearing house • Invest in policy learning on the basis of strategic intelligence studies • Challenge TA researchers to come up with concrete proposals for innovation policy • Contribute to a (distributed) strategic intelligence infrastructure • Give users and/or potential users access to the strategic intelligence function
Stimulating demand articulation and development of strategy and vision	• Invest in forms of public participation such as public debates, consensus conferences, constructive TA, scenario workshops, and round-table conferences • Stimulate the parliamentary debate on issues involving science, technology, and innovation • Start a discussion on the structure of the national innovation system or parts of it (the arrangement of the knowledge infrastructure) • Develop nationwide innovation strategies in social domains

practitioners could also benefit from a greater integration of the insights provided by innovation studies. Until now, TA researchers and innovation scholars often operate in two different and separate environments, hardly referring to each other's work. The same is true for TA researchers and policymakers. TA researchers need to invest more in matching their problem definition, goals and methods with the needs, possibilities and limitations of policymakers. Institutional innovations that allow for greater cross-fertilization among innovation policymakers and TA practitioners may be a central part of any successful effort to enhance the value of TA for policymaking.

REFERENCES

Adviesgroep Micro-Elektronica (1980), 'Maatschappelijke gevolgen van de micro-elektronica', The Hague: Staatsuitgeverij.

Akrich, M. (1995), 'User representations: practices, methods and sociology', in A. Rip, T. Misa and J. Schot (eds), *Managing Technology in Society: The Approach of Constructive Technology Assessment*, London, UK and New York, USA: Pinter.

Arnold, E., et al. (2003), 'Research and innovation governance in eight countries', a Meta-analysis of Work Funded by EZ (Netherlands) and RCN (Norway), Technopolis, Brighton.

Belt, H. van den and Rip, A. (1987), 'The Nelson-Winter-Dosi model and synthetic dye chemistry', in W. Bijker, T. Hughes and T. Pinch (eds), *The Social Construction of Technological Systems: New Directions in the Sociology and History of Technology*, Cambridge, MA, USA and London, UK: MIT Press.

Bennett, I. and Sarewitz, D. (2006), 'Too Little, Too Late? Research Policies on the Societal Implications of Nanotechnology in the United States', *Science as Culture*, **15** (4), 309-325.

Bereano, P. (1997), 'Reflections of a participant-observer: the technocratic-democratic contradiction in the practice of technology assessment', *Technological Forecasting & Social Change*, **54** (2-3), 163-176.

Bessant, J. and Rush, H. (1995), 'Building bridges for innovation: the role of consultants in technology transfer', *Research Policy*, **24** (1), 97-114.

Bijker, W. (1995), *On Bikes, Bicycles and Bakelite*, Cambridge, MA, USA and London, UK: MIT Press.

Bimber, B. (1996), *The Politics of Expertise in Congress: The Rise and Fall of the Office of Technology Assessment*, Albany: State University of New York Press.

Bimber, B. and Guston, D. (1997), 'The end of OTA and the future of technology assessment', *Technological Forecasting & Social Change*, **54** (2), 125-130.

Boekholt, P., Lankhuizen, M., Arnold, E., et al. (2001), 'An international review of methods to measure relative effectiveness of technology policy instruments', final report, Technopolis, Brighton and Amsterdam.

Brede DNA Commissie (1983), 'Eindrapport van de commissie ter bestudering van de maatschappelijke en ethische aspecten van werkzaamheden met erfelijkheidsmateriaal', The Hague: Staatsuitgeverij.

Bruijn, H. de, Heuvelhof, E. ten and R. in 't Veld (2002), *Procesmanagement*, Vol. 2, Schoonhoven: Academic Services.

Caplan, N. (1979), 'The two-communities metaphor and knowledge utilization', *American Behavioral Scientist*, **22** (3), 459-470.

Coates, V. (1975), *Readings in Technology Assessment*, Washington, DC: George Washington University.

Collingridge, D. (1980), *The Social Control of Technology*, London, UK and New York, USA: Pinter.

Cook-Deegan, R. (1995), *The Gene Wars: Science, Politics, and the Human Genome*, New York: W.W. Norton & Company.

Decker, M. and Ladikas, M. (eds) (2004), *Bridges Between Science, Society and Policy; Technology Assessment: Methods and Impacts*, Berlin, Germany, Heidelberg, Germany and New York, USA: Springer-Verlag.

Dosi, G., et al. (1988), *Technical Change and Economic Theory*, London, UK and New York, USA: Pinter.

Dunn, W. (1980), 'The two-communities metaphor and models of knowledge use: an exploratory case survey', *Knowledge: Creation, Diffusion, Utilization*, 1 (4), 293-362.

Edler, J., Kuhlmann, S. and Smits, R. (2003), 'New governance for innovation: the need for horizontal and systemic policy co-ordination', report on a workshop held at the occasion of the 30th anniversary of the Fraunhofer Institute for Systems and Innovation Research (ISI), Karlsruhe, Germany, 14-15 November 2002, organised in collaboration with the Copernicus Institute, Dept. of Innovation Studies, Utrecht University, and the 'Six Countries Programme – the Innovation Policy Network (6CP)', ISI-FhG, Karlsruhe.

Fisher, E. and Mahajan, R. (2006), 'Nanotechnology legislation: contradictory intent? US federal legislation on integrating societal concerns into nanotechnology research and development', *Science and Public Policy*, **33** (1), 5-16.

Freeman, C. (1987), *Technology Policy and Economic Performance: Lessons from Japan*, London, UK and New York, USA: Pinter.

Fri, R., Morgan, M. and Stiles, Jr., W. (2002), *An External Evaluation of GAO's Assessment of Technologies for Border Control*, Washington, DC: GAO.

General Accounting Office (2002), 'Technology assessment: using biometrics for border security', GAO-03-174, Washington, DC.

Geurts, J. (1993), 'Omkijken naar de toekomst'. Lange termijn verkenningen in beleidsexercities' [Looking back to the future. Long-term foresight studies in policy exercises], address Tilburg University, Samson H.D., Tjeenk Willink.

Grin, J., Graaf, H. van de and Hoppe, R. (1997), *Technology Assessment Through Interaction: A Guide*, The Hague: Rathenau Instituut.

Guston, D. (1999a), 'Evaluating the first U.S. consensus conference: the impact of the citizens' panel on telecommunications and the future of democracy', *Science, Technology, & Human Values*, **24** (4), 451-482.

Guston, D. (1999b), 'Stabilizing the boundary between US politics and science: the role of the office of technology transfer as a boundary organization', *Social Studies of Science*, **29** (1), 87-111.

Guston, D. and Sarewitz, D. (2002), 'Real-time technology assessment', *Technology in Society*, **24** (1-2), 93-109.

Hippel, E. von (2005), *Democratizing Innovation*, Cambridge, MA, USA and London, UK: MIT Press.

Hippel, E. von (1988), *The Sources of Innovation*, Cambridge, MA, USA and London, UK: MIT Press.

Hoo, S. de, Smits, R. and Petrella, R. (eds) (1987), 'Technology assessment an opportunity for Europe. Proceedings and full papers of the 1st European Congress on technology assessment', published in the context of the European Congress on Technology Assessment (Amsterdam, 2-4 February 1987) by the Dutch Ministry of Education and Science in cooperation with the Forecasting and Assessment of Science and Technology Programme of the European Communities (FAST, EC-DGXII), The Hague and Brussels.

Kuhlmann, S. (2003), 'Evaluation of research and innovation policies: a discussion of trends with examples from Germany', *International Journal of Technology Management*, **26** (2-4), 131-149.

Kuhlmann, S., Boekholt, P., Georghiou, L., et al. (1999), 'Enhancing distributed intelligence in complex innovation systems', report published within the framework of the Targeted Socio-Economic Research Programme of the European Commission, ISI-FhG, Karlsruhe.

La Porte, T. (2000), 'Shaping the information revolution', in N. Vig and H. Paschen (eds), *Multivisioning the Future: Parliamentary Technology Assessment in Europe*, Albany: State University of New York.

Lundvall, B.-Å. (1992), *National Systems of Innovation: Towards a Theory of Innovation and Interactive Learning*, London, UK and New York, USA: Pinter.

Morgan, M. and Peha, J. (eds) (2003), *Science and Technology Advice for Congress*, Washington, DC: Resources for the Future Press.

Nelson, R. and Winter, S. (1982), *An Evolutionary Theory of Economic Change*, Cambridge, MA, USA: Belknap Press.

OECD (2002), *Dynamising National Innovation Systems*, Paris: OECD Publishing.

Oudshoorn, N. and Pinch, T. (eds) (2003), *How Users Matter: The Co-construction of Users and Technologies*, Cambridge, MA, USA and London, UK: MIT Press.

Parsi, J., Tosi, J. and Guston, D. (forthcoming), 'Anticipating the ethical and political challenges of human nanotechnologies', in F. Allhoff, P. Lin, J. Moor and J. Weckert (eds), *NanoEthics: Examining the Societal Impact of Nanotechnology*, Hoboken: John Wiley and Sons.

Porter, A., Youtie, J. and Shapira, P. (2006) 'Refining search terms for nanotechnology', white paper available at http://cns.asu.edu/cns-library/author.htm.

Rip, A., Misa, T. and Schot, J. (eds) (1995), *Managing Technology in Society: The Approach of Constructive Technology Assessment*, London, UK and New York, USA: Pinter.

Roco, M. (2006), *Keynote Address' Center on Nanotechnology and Society First Annual NanoPolicy Conference*, Washington, DC: National Press Club.

Roobeek, A.J.M. (1990), *Beyond the Technology Race. An Analysis of Technology Policy in Seven Industrial Countries*, Amsterdam, The Netherlands and New York, USA: Elsevier Science Publishers.

Sarewitz, D. (1996), *Frontiers of Illusion*, Philadelphia: Temple University Press.

Schot, J. and Rip, A. (1997), 'The past and the future of Technology Assessment', *Technology Forecasting and Social Change*, **54** (2), 251-268.

Sclove, R. (1995), *Democracy and Technology*, New York: Guilford Press.

Simon, H. (1976), *Administrative Behavior, a Study of Decision-Making Processes in Administrative Behavior*, Vol. 3, New York: The Free Press.

Smits, R. (2002), 'Innovation studies in the 21st century: questions from a user's perspective', *Technological Forecasting and Social Change*, **69** (9), 861-883.

Smits, R. (2001), 'The new role of strategic intelligence', in A. Tűbke, K. Ducatel, J. Gavigan and P. Moncada-Paternò-Castello (eds), 'Strategic Policy Intelligence: Current Trends, the State of Play and Perspectives', IPTS Technical Report Series, EUR 20137 EN, IPTS, Seville.

Smits, R. and Hertog, P. den (2007), 'Technology assessment and the management of technology in economy and society', *International Journal of Foresight and Innovation Policy*, **3** (1), 28-52.

Smits, R. and Kuhlmann, S. (2002), 'Strengthening interfaces in innovation systems: rationale, concepts and (new) instruments', report prepared on behalf of the EC-STRATA Workshop 'New challenges and new responses for S&T policies in Europe', 22-23 April 2002, Brussels.

Smits, R. and Leyten, A. (1991), *Technology Assessment: Waakhond of Speurhond* (Technology assessment: watchdog or tracker? Towards an integral technology policy), Amsterdam: Vrije Universiteit Amsterdam.

Smits, R. and Weijers, T. (1991), 'Technology assessment: watchdog or integrated part of technology policy?', keynote paper presented at the second European Congress on Technology Assessment, Milan.

Smits R., Leyten, A. and Hertog, P. den (1995), 'Technology assessment and technology policy in Europe: new concepts, new goals, new infrastructures', *Policy Sciences*, **28** (3), 272-299.

Stine, J. (1986), 'A history of science policy in the United States, 1940-1985', Science Policy Study Background Report No. 1. Task Force on Science Policy, Committee on Science and Technology, US House of Representatives, 99th Congress, 2nd session, US Government Printing Office, Washington, DC.

TAMI (2004), 'Technology assessment in Europe; between method and impact', EC-STRATA Program, Brussels.

Utterback, J. (1994), *Mastering the Dynamics of Innovation*, Boston: Harvard Business School Press.

Vig, N. and Paschen, H. (eds) (2000), *Parliaments and Technology: The Development of Technology Assessment in Europe*, Albany: State University of New York Press.

Vught, F. van (1980), *Toekomstonderzoek en Ruimtelijke Ordening*, Enschede: Technical University Twente.
Whiteman, D. (1986), *Communication in Congress: Members, Staff, and the Search for Information*, Lawrence: University Press of Kansas.
Wolbring, G. (2006), 'Scoping document on nanotechnology and disabled people for the Center on Nanotechnology in Society at Arizona State University', available at http://cns.asu.edu/cns-library/author.htm#W.

17. A System-Evolutionary Approach for Innovation Policy

Ruud Smits, Stefan Kuhlmann and Morris Teubal[1]

INTRODUCTION

A leading question addressed in this book is how to improve public policy, using insights from innovation studies and theory as well as from innovation practice and innovation policy. Innovation in this book is seen as a dynamic interactive process taking place in heterogeneous, more or less interconnected arenas with many heterogeneous actors, cutting across various levels of societal organisation. To explore this complex dynamic we adopt the innovation systems (IS) approach as a starting heuristic and assume – in line with the leitmotif of this book – that practice, policy and theory develop by mutual interactions. Three major conclusions from the foregoing chapters are central to the discussion in this chapter:

- Although innovation studies made advances over the last two decades, the prevailing IS approach has some major flaws. The concept is still disposing a too static character: there is a lack of adequate bridges between the system level and individual actors. Also the implications for advancing theory and for policy designs are still limited.
- Until now scholars in innovation studies did not pay sufficient attention to innovation policy as an object of research. Phrased differently, policy is considered as a trivial application of 'other' knowledge but not a field of knowledge in itself (Teubal 2002). This explains why researchers tend to not fully investigate the implications of the results of their work for policy.
- Partly as a consequence of this lack of theoretical interest (and partly as a result of inertia in policy making), actual policy making often follows the old linear mode or, while claiming to be designed from a systems perspective, applies a 'one-size-fits-all' approach (e.g. the fashionable 'network-building' funding policies

1 Joint work with Gil Avnimelech underlies the analysis of the first part of section four (Two Case Studies). See also Avnimelech and Teubal (2008b).

launched irrespective of knowledge fields, technology or market characteristics etc.; see Boekholt in this book).

As a consequence we rarely find strategic system-oriented innovation policies that build on innovation studies which suggest a revision of the institutional set up to achieve non-trivial changes in IS.

In this chapter we will focus primarily on innovation policy as an object of research in itself. We will concentrate on the question of how insights from innovation studies – from actual policies and from awareness of the interaction between the theory and practice of policy – may improve innovation policy in a practical and concrete way. Based on experiences with actual innovation policies and recent insights from innovation studies, including studies of innovation policies, we will focus on two major themes:

- At times IS are (or should become) subjects of dynamic structural change. Hereby the focus has to be on established systems in a steady state but also on processes of creative destruction and the formation of *Neue Kombinationen* (Schumpeter 1934). We have to understand better the implications of our systemic, co-evolutionary perspective on innovation processes, systems and policies for operational policies dealing with the steady state and for strategic policies facilitating structural change. We aim to improve the capability of these two types of policy at a conceptual level to better take into account the dynamic and systemic character of innovation processes and systems.
- Both policies deal with systems and are in need of policy instruments that take into account the systemic character of the policy object. Strategic policies dealing with radical changes in the institutional set-up need instruments that enable the development of new interfaces and arenas, a much broader process of consultation and the strategic intelligence to support this. Operational instruments have to find solutions for system failures (Woolthuis et al. 2005) that cannot be solved by one organisation or limited to the relation between two organisations. In this chapter we will propose so called systemic instruments, that are able to tackle the problems and challenges of IS in an orchestrated and more effective way than the current, often linear model-based innovation policy instruments (Smits & Kuhlmann 2004).

We want to stress that this chapter does not provide full solutions nor even does it pretend to provide an all encompassing analysis of the flaws and their possible solutions of innovation theory and policy as discussed in other chapters of the book. This chapter tries to take insights one step further by developing concepts and instruments, illustrated by practical cases, linking these insights to real-life policymaking.

In the first section we will summarise the debate on innovation systems (theory) and innovation policy while taking stock of the major insights and focusing on what we consider to be the main deficiencies of the prevailing approaches. In doing so we will give special attention to the IS approach and show that some of the insights from innovation theory still are not taken on board in this policy. In section two (Need for a Strategic Level of Policy), we will propose a systems-evolutionary innovation policy portfolio tackling two major perspectives: a strategic dimension of innovation policy designed to facilitate structural change of IS, as opposed to operational policies for IS in a steady state. In the third section (Need for Systemic Instruments), we will present the concept of systemic innovation policy instruments. Based on our analysis of major insights from innovation studies key requirements for systemic initiatives, both for strategic and for operational policy purposes, will be suggested. Section four (Two Case Studies), then will present two cases in which these proposed approaches are illustrated and further explained. The Israeli Venture Capital and High Tech Cluster case stresses the need for strategic policies to focus on the support of new, dynamic elements in the system. In the case of Dutch agricultural innovation policy systemic instruments have played a key role both in strategic as well as in operational dimensions. The last section winds up by drawing conclusions and presenting recommendations that may help policy makers and researchers to improve systemic innovation policies.

THE SYSTEM OF INNOVATION APPROACH IN AN EVOLUTIONARY PERSPECTIVE

We start by discussing major insights from innovation studies and will draw some conclusions on the implications of these insights that may help policy makers to develop a systemic policy approach that takes into account the co-evolutionary nature of innovation processes and systems. Hereafter we will go into some weaknesses of the IS approach that offer starting points to further the usefulness of innovation systems theory for policy.

Insights from Innovation Studies

Innovation studies have provided a number of insights that help to improve systemic policies. Below we introduce a selection of findings (building on Smits & den Hertog 2007) suggesting possible implications for policy (see Figure 17.1).

1. *Innovation is an interactive search and implementation process that takes place in uncertainty.* In innovation theory, an interactive innovation model has largely replaced the linear innovation model. The latter generally emphasises the central role of design, the feedback effects between the downstream and upstream phases of the earlier linear model, the overlapping of phases in the innovation process, and the numerous interactions between science, technology and the process of innovation in every phase of the process (Nelson & Winter 1977; Kline &

1 Innovation occurs in uncertainty.
2 Innovation is path-dependent.
3 Innovation co-evolves with society.
4 Innovation draws on varieties of actors.
5 Innovation builds on actors' learning capabilities.
6 Often, innovation is spurred by vivid interfaces with science and 'high technology'.
7 Often, innovation is driven by demand (including non- or low technological).
8 Innovation builds on 'soft factors' (e.g. innovation-related services).
9 Innovation is shaped by institutional contexts and public policy.

Figure 17.1 Nine lessons from innovation research (based on Smits & den Hertog 2007)

Rosenberg 1986; Imai et al. 1988; OECD 1992). Many innovations are the result of a group or systemic process rather than the product of an individual innovator, a fact that also gets strong emphasis in the IS approach.

Possible implications: With strong uncertainty, policy strategies favouring evolutionary variation are important, e.g. in the early phase of creation of a new industry. The same accounts for the early phase of economies where – prior to the identification of areas/industries/technologies providing the economy or region with a sustainable competitive advantage – business sector innovation is in its infancy. In terms of types of policies to be adopted, for contexts where business sector innovation is in its infancy, there may be strong reasons for adopting horizontal policies (Teubal 1997) for the support of R&D and innovation in firms.

2. *Innovation is path-dependent, or evolutionary and/or accumulative in nature.* Innovation is a search process directed by search rules or heuristics that largely remain implicit, though they indicate how certain problems should be tackled (Nelson & Winter 1977). In this context, some authors speak of 'technological paradigms' and more recently of 'technological regimes' (Rip & Kemp 1998) which provide the framework within which solutions to problems are sought. At the level of technologies or types of solutions, this can mean that certain search directions are not considered and further development of innovation tends to continue along the same lines (path dependency) and eventually even run the risk of 'lock-in' and irreversibility. This makes it much more difficult to switch to another (and possibly better) solution, merely because a routine has already been established.

Possible implications: Path dependency underlies the difficulty of targeting new areas, technologies and industries and the frequent failures in 'picking winners'. Still, targeting of new industries has succeeded in the past, e.g. in the Korean ICT sector and salmon industries in Chile; and it may be a very important tool which cannot be dispensed of, in particular when rapid strategic adaptations of IS ('regime shifts', 'system transitions', see e.g. Geels & Schot 2007) are required. This means that an appropriate 'evolutionary' targeting perspective should be developed implying a strong strategic process of selection by government (which, if successful, the system as a whole including the business sector will accept). Thereby it is imperative to assure that sufficient creation of new options and associated experimentation and learning ('variation') take place before a finalized targeting policy be implemented (Sercovich & Teubal 2007; Avnimelech & Teubal 2008b). This also means that new arenas of interaction and new mechanism of interaction and mutual adaptation should be objectives of pre-targeting policy making. Strategic Niche Management (Kemp et al. 1998) and various types of strategic intelligence (see Chapter 16 by Smits et al. in this book) as for instance Constructive Technology Assessments could be helpful in this context.

3. *Innovation and technological change are endogenous processes and the result of co-evolution of technology and society.* At present, the dominant view among innovation theorists (probably with the exception of many neo-classical economists) is that technology and innovation should not only be conceptualised as 'ready-made' artefacts ('cannonball view' of the impact of technology). Technology and innovation are endogenous, and therefore their development is not strictly deterministic or linear. In economic terms, they are usually the result of a complex process of interaction and development between demand and supply. In terms of more socio-cultural perspectives, they are the result of a co-evolution of technology and society, or of the 'technological culture' (Bijker 2001) in which we live.

   Possible implications: The centrality of co-evolutionary processes gives further weight to the perspective that focuses on sparking and sustaining cumulative processes of emergence (as one of the possible objectives of policy, see Avnimelech & Teubal 2008b). This – together with the implications of path dependence mentioned above, and the fact that cumulative processes must involve not only economic and technological factors but also societal features – implies that relevant and potentially successful policies could become very complex and difficult to undertake. Scholars have made a plea for 'economics and socio-politics of collective experimentation' (Felt et al. 2007), characterised by emerging or strategically created situations which allow to try things out and to learn from them. Here 'experimentation does not derive from promoting a particular technological promise, but from goals constructed around matters of concerns and that may be achieved at the collective level. Such goals will often be further articulated in the course of the experimentation' (Felt et al. 2007). Policy makers should be aware of this and strive to generate an additional set of strategic and operational skills and routines. A hypothetical example of the importance of societal factors with respect to venture capital policies is culture: it may very well be that a cultural shift towards innovative entrepreneurship may be required prior to the implementation of venture capital policies (this is one factor explaining the failure of German policies in this area in the early 1980s).

4. *Innovation as an activity is no longer the prerogative of a genius inventor working in 'splendid isolation', but is a systemic activity to which a variety of actors contribute.*

Some people have an almost romantic image of the inventor working in solitude in his or her lab or garage, coming up with inventions that lead to innovations. However, the great majority of innovations are achieved by groups of people. The terms 'multi-actor activity' and 'system activity' refer to this aspect. In addition to companies, many knowledge institutions, intermediary organisations and, in some cases, also the government (through various types of policy including science and innovation policy), play a role in these innovation processes. A special case of interaction is feedback between users and manufacturers (e.g. von Hippel 1988; Oudshoorn & Pinch 2003). The IS approach emphasises precisely this multi-actor character of many innovation processes.

Possible implications: There is a need for 'fora' facilitating exchange and debate of heterogeneous actors on innovation needs and potential (Edler et al. 2006). Public policy can act as a catalyst here, e.g. organise multi-actor foresight processes, or provide support for intermediary organisations enabling interaction between users and producers of innovative achievements.

5. *Learning and creating learning environments are crucial to innovation.* An important theme in innovation research is the way people arrive at innovation and the role that learning processes play in this. Distinctions between different types of learning (e.g. learning by searching, learning by doing, learning by using, and learning to learn (Rosenberg 1976; 1982)) and different types of knowledge (e.g. the basic distinctions between codified and tacit knowledge and between instrumental and conceptual knowledge (Caplan 1979)) are key here. This learning takes place at many different levels. Individuals must learn to develop in an increasingly complex knowledge society. Companies must learn to keep abreast with the latest insights in their branch of trade or industry and stand out from the competition, if possible. Policymakers must learn to organise economic and social learning processes as efficiently and effectively as possible and learn from earlier policy interventions (policy learning). As pointed out by Lundvall (1992) learning and experimentation are basic prerequisites for successful innovation.

Possible implications: While the promotion of learning should almost always be an explicit objective of innovation policies providing/creating conditions for learning deserves special attention of policy makers in the pre-selection phase. Learning is an inherent part of the variation process preceding policy induced strategic 'selection'. This learning process may have to encompass a large and varied set of agents including the government itself.[2] Creating room for experimenting and a strategic intelligence infrastructure providing innovating

2 An illustration in the context of venture capital and high-tech policies in Israel during the 1990s: there was considerable experimentation and learning by individuals, start-up companies, financial institutions and Government during the 15-20 years that preceded the spate of policy measures of the early 1990s. Much of this was collective in the sense of mutual adaptation of different groups of actors (the organisation, capabilities and strategies of high-tech start-ups, financial institutions and Government institutions). The outcome was a policy designed to

actors with the (tailor made) information on the possibilities and consequences of new technologies are important elements of policies that want to contribute to the conditions necessary for learning in pre-competitive stages of innovation processes.

6. *An important share of innovation and technological change are increasingly linked directly to scientific knowledge and involve a variety of transfer mechanisms.* A growing number of industries depend ever more directly on the progress of (fundamental) research for innovation. However, this applies not only to hard technology, but also to the soft aspects of innovation such as user behaviour, design of client interfaces and organisational change processes. The transfer mechanisms involved go beyond R&D collaboration and personal mobility. The nature and exact operation of the many knowledge transfer mechanisms are in practice varied and complex (see also OECD 2002).

   Possible implications: an insufficiently considered policy in this regard in Europe concerns the creation of a strong 'innovative entrepreneurship' module closely linked to and co-evolving with universities (both through research and training) and government labs. The core of such a module might be a high-tech start-up segment and a linked venture capital industry and/or market. Such a module would represent a new type of critical complementary interface to the usual links through technology transfer offices and supply and demand for new graduates in the engineering and other sciences. Such an interface may stimulate the development of new firms and stimulate – via the manifold linkages in innovation systems – less innovative firms to become more innovative.

7. *In addition to knowledge creation, knowledge diffusion and demand-driven knowledge utilisation are crucial.* For a long time, innovation has been associated with progress in the knowledge-intensive and high-tech industries. This has led to a preoccupation with radical innovation and high-tech activity in innovation theory, practice and policy. However in innovation both incremental and imitation innovations are involved (e.g. Hirsch-Kreinsen et al. 2005; Hirsch-Kreinsen 2008). This certainly accounts for the many, not so technology-intensive small and medium-sized enterprises. O'Doherty and Arnold (2003) note that these latter innovations which seek limited improvement of existing products and processes, and which are sometimes seen as 'diffusing' new technologies, are the numerically and economically dominant variety.

   Possible implications: Systemic innovation policy would seek to address also low-tech, incremental and imitation innovations, and policies would recognise

support a new concept of high-tech start-up company (which was considerably different from the old 'R&D companies'); by a new type of financial institution (venture capital organised as limited partnerships with an 'early phase' strategy of investment in high-tech start-ups); and involving a completely different role of Government (promoting emergence of a new market/industry through a targeted programme involving a one-shot government venture contribution of 100 M$). Both business experiments and policy learning were critical ingredients in the above (see below and Avnimelech & Teubal 2006).

the indirect relation with knowledge-intensive and high-tech areas of innovation and policy. A further implication is that diffusion policies, until now often trying to transfer new, often scientific and technological knowledge to SMEs (outside-in) have to be complemented with 'inside-out' policies that stimulate SMEs to articulate their need for – often also socio-economic – knowledge to support their innovative activities.

8. *Innovation also demands knowledge and understanding of 'soft factors' of innovation.* Over time the concept of innovation became restricted to technological innovation. Innovation studies and practice, however, showed that technology often only is a minor component in innovation – certainly in the service sector – and that many other, non-technical factors play a decisive role. This observation is even more relevant when we realise that about 95% of firms in Western economies are small or medium-sized. Many of these firms do not have their own laboratory and often only limited capabilities to acquire and absorb new technologies. Given also the rise of the BRIC countries as economic powers, innovation policies of Western countries will have to focus more on stimulating innovation in these not so technology-oriented firms, which constitute the backbone of our economies. In line with these insights, especially since the 1990s, more attention has been paid to innovation in services, organisational innovation or more broadly the soft – non-technological – side of innovation. Currently, many innovation policymakers are struggling with the question whether these require specific policies and/or how to incorporate these into their policy mixes aimed at furthering innovation.

   Possible implications: The foregoing is a further argument to get rid of the linear model with its heavy emphasis on the role of (new) knowledge and technology in innovation. Not only the capacity to produce new knowledge but the potential to exploit new and already existing knowledge should be made central. Improving the capacity of firms to develop innovation strategies as well as the competence to articulate, identify and acquire knowledge necessary to implement these strategies, is a crucial issue in this context. The forementioned shift from diffusion-based outside-in policies to policies that stimulate the capacity of firms to improve their own innovative potential (inside-out) is an important step in this direction. Stimulating changes in the educational system that aim to provide more people with entrepreneurial competences, the ability to link technological options with commercial opportunities and the ability to work in heterogeneous settings (open innovation) constitute a pivotal component of such a policy. Furthermore, policy should pay far more attention to organisational implications, stimulate the development of organisational innovation and point at the societal impact of innovations, the role of users and the importance of often neglected but possible important intangible assets like design.

9. *Institutions and policymaking matter.* In innovation the interaction of broadly defined institutions (or 'institutional set-up': regulation, rules, beliefs) and their

adaptability play a major role. Edquist (1997) distinguishes three functions of institutions: (a) reducing uncertainty by providing information; (b) managing conflicts and collaboration; and (c) providing stimuli. Moreover, a system-specific institutional set-up is quite likely, precisely because of differences in history and specific context. In an IS approach, a government can intervene in various capacities. It can act as a co-innovator, for instance as a launching customer or as an articulator of important societal problems in areas as health, the environment and safety, or by providing the necessary framework conditions. Not only market failure, but also system failure – an institutional set-up that does not give the right stimuli to encourage the innovation process – can serve as an argument for a more broadly defined public innovation policy.

Possible implications: The institutional and policymaking set-up itself should be considered as subject of strategy development and systemic policy designs in the light of innovation system dynamics. This would call for procedures of 'reflexive governance' (Voß 2007), taking into account that policy and its reliance on policy instruments is itself embedded in, and constituted by, broader ongoing changes (e.g. Voß et al. 2006). Reflexive governing strategies would not aim at full control, but interact with ongoing change with a view to modulate it – while recognising the ironies that are involved in doing so (Rip 2006). An important consequence of the foregoing is the need to develop competences, stimulate – and some times take part in – processes of creative destruction and the building of *Neue Kombination*. Creative destruction as far as government is concerned may include the abolishment of obsolete institutions, rules and regulations and organisations. The development of *Neue Kombination* may be stimulated by developing policies that stimulate the further development of new and promising sectors.

Weaknesses of the Innovation System Approach

At present, the IS approach is becoming the *de facto* standard in the world of innovation policy, even though its application can be, and is, very diverse and demonstrates – as we will see – severe shortcomings. Why is this model so attractive and which role does the IS approach play in current developments in the policy thinking (see also chapters of Chaminade & Edquist, and of Carlsson, Elg & Jacobsson in this book). The attraction of the IS approach has to do with several factors. In the first place, the IS concept seems to fit in with the idea of an economy as a comprehensive system of actors between which flow and feedback is possible and where interventions on one of the relations have consequences for other actors and relations in the system. The IS approach is a recognition of the fact that innovation (and polices to further innovation!) is no longer a 'single act', but the result of a complex interplay of sometimes very diverse actors, who arrive at innovation in interaction. Besides thinking in terms of market failure (which is strongly based on neo-classical economic theory), the failure of the broader innovation system is increasingly recognised as a basis for government intervention. A second factor that may explain the attractiveness of the IS approach is the finding that the IS approach offers both a flexible heuristic that is sufficiently malleable to be practicable for many different policymakers and an idiom that indicates a contemporary understanding of innovation (Sharif 2006). For instance, the IS approach recognises the roles played by different institutions in the innovation process, allows a broader interpretation of innovation as such (more than just technological innovation) and can be operationalised at different policy levels (supranational, national, regional/local, at the level of complex technological systems, clusters/sectors).

The third attractive aspect of the relevance for policymaking of the IS approach is its integrated character and the fact that this provides for an organisational framework (analytical, but also very practical) for innovation policy in practice. It offers a framework not only for visualising the main (perceived) problems surrounding innovation, but also for designing a mix of policy instruments to solve the problems once they have been identified.

Despite all these pros (see also Lundvall 2007), as already noted, the IS approach still shows considerable weaknesses. The major ones are:

- The IS approach in its traditional sense does not capture the dynamic aspects of innovation processes, systems and policies. The functions approach as discussed by Bergek, Jacobsson, Hekkert & Smith in this book is an attempt to deal with this deficiency, but still is in an earlier stage of development.
- An important element of the dynamic character of IS, the emergence of new markets and new industries is either ignored or not considered (see Blind in this book and Avnimelech & Teubal 2008b). This weakness is part of a broader weakness of the IS approach as far as policy is concerned.
- The consequences for both strategic and operational systemic policies including relevant institutions, agents and capabilities are hardly discussed.

In the following two sections we will focus on the third and – to a lesser degree – the second weakness. In the next section we will go into the need for strategic policies and how this policy type distinguishes itself from operational polices. In the third, we will introduce and explain the concept of systemic instruments. Hereafter, in section four two cases will further illustrate our major arguments.

NEED FOR A STRATEGIC LEVEL OF POLICY

With respect to system dynamics a major distinction has to be made between the steady or stationary state of a system on the one hand and a system in transition on the other. Creative destruction, e.g. the creation of new industries and markets, which are disruptive of existing arrangements, and the establishment of *Neue Kombinationen* are at the core of innovation processes and their impact, and therefore

State of innovation system	
Steady state	Structural change
Operational innovation policies • focus on implementation of existing set of policies • aiming at keeping system vivid and competitive • removing system failures	*Strategic* innovation policies • generating new set of policies • seeking better fit to new global or domestic context • facilitating creative destruction and the emergence of *Neue Kombinationen*

7406

Figure 17.2 Systems evolutionary approach for innovation policy

the basis for system transformation. Policies at these crossroads must incorporate a strategic element; they have to generate a new set of policies, seeking a better fit to a new global or domestic context. Strategic policies would go beyond the implementation of the existing operational policy portfolio designed to keep a given system profile vivid and competitive (see Figure 17.2).

A strategic level of policy (for simplicity we call this 'strategic policies') by necessity focuses on an extended policy process, one which begins with the setting of new strategic priorities and continues with efforts at articulating these in terms of a new set of policies. Their relevance is enhanced whenever strong changes in the global and/or domestic environments call for a new or adapted innovation system, one inconsistent with the existing set of policies. In contrast to the operational level of policy which focuses on the implementation of an existing set of policies, action at the strategic level aims at generating a new set of policies, one better fit to the new global and domestic context (Avnimelech & Teubal 2008b; Rosiello et al. 2008). In this context, policies should *grosso modo* be conceived as comprising three main components: incentives and incentives programmes; institutional and regulatory changes; and other actions (like technological forecasting, setting new priorities and creating new 'strategic' policy institutions).[3] The latter are critical when shifting from a dominant operational level of policy to a mode which includes both an operational and a strategic level.

Despite enhanced awareness of the need for analysing the emergence and transformation of IS there are still important gaps in existing research. In this section we will focus on one important aspect of this dynamic process[4]: the analysis of the emergence of new organisations or sub-systems such as new markets and new industries and associated emergent properties. The recent literature on cluster dynamics and venture capital have considered both the conditions that should be in place and the nature of the cumulative, feedback-driven processes which lead to the emergence of new, sometimes high-tech clusters and new venture capital industries

3 The examples of 'other policy actions' given above strongly relate to the systemic (policy) instruments analysed below (Section 4, Need for Systemic Instruments).

4 Here, we exceed Schumpeter's notion of Neue Kombinationen and include the long-term impact of creating and destroying sectors.

and markets (Bresnahan et al. 2001; Avnimelech & Teubal 2004a, b; 2006; 2008a, b). This literature and the associated Industry Life Cycle literature (e.g. Klepper 1996; 2001) should be duly incorporated into innovation system dynamics. Related to this is the recent emphasis on co-evolutionary processes requiring a truly dynamic perspective on changes in various sectors of society such as for instance the agro-sector, water management and the energy sector. Here the issue not only concerns co-evolution between the technology and the institutions supporting it (Nelson 1993), but other critical co-evolutionary processes involved in innovation system transformations, or 'regime transitions' (e.g. Elzen & Wieczorek 2005; Geels & Schot 2007).

Preliminary work (Avnimelech & Teubal 2008a, b among others) has shown that the implications of the strategic policy perspective can be momentous indeed both in connection with the inherent dynamic approach (with strategy linking the past with the future) and in connection with the objectives of policy. Foremost among the latter is the identification of emergence of new sectors, product classes, markets and clusters as legitimate objectives of policy to be activated, given favourable 'transition facilitating' or pre-emergence conditions, by the triggering and sustenance of cumulative processes with positive feedback. When such conditions are not present, policies might aim at generating them.

While existing research on agent interaction, agent capabilities and infrastructures is important and useful, it has not informed us sufficiently about the higher organisational level implications, e.g. concerning the emergence of new markets, industries and clusters. It should be noted that given the existing turbulent global environment it is very likely that these emergent entities and phenomena acquire a greater importance than in the past. The rise of the BRIC countries without any doubt will force Western economies to develop new sectors at a more rapid pace. In other words, while existing research significantly advanced beyond the policy objective underlying linear models – particularly beyond achieving R&D 'additionality' and even in the area of clusters – it still has to incorporate other emergent phenomena and associated dynamics centrally into its analysis. In fact, the under-emphasis of the objectives of creating new markets and industries – including new systems-evolutionary views of infant industry development – is particularly surprising. In Schumpeterian terminology: innovation scholars should pay more attention to the underlying dynamics of the creation of *Neue Kombination*, the necessary creative destruction and their potential impact in terms of new multi-agent structures of higher levels of organisation like new sectors, clusters, etc. (Avnimelech & Teubal 2008b).

Once emergence of new meso-level entities is recognised as one important aspect of policy, the dynamic analysis of creation of pre-emergence conditions and the facilitation of a transition from pre-emergence to emergence acquires greater significance. Related to this is the issue of timing of strategic policies: right timing may be important both for the creation of adequate domestic conditions as well as for conditions which, while critical, are exogenous to the focal IS. Moreover, this dynamic, strategic approach to policy also spills over to questions of design, since the system failures that policy must overcome relate first and foremost to the triggering and to the sustaining of cumulative processes of emergence and their overcoming may be intimately related to policy design.

Finally, existing systems of innovation approaches exaggerate the divide between 'positive' and 'normative' analysis or alternatively between the world of 'policy' on the one hand and the 'real world' on the other. Policy actors, institutions and capabilities are part and parcel of our description of an IS. Their interaction with other actors and the processes leading to new policies are integral parts of system dynamics. In fact, it is imperative to adopt a fully dynamic theory of innovation policy broadly defined, one involving appropriately linked phases and strong co-evolutionary processes with other components or elements of the system.

We conclude this section with two additional thoughts. First, contrary to past routines, policy aiming at innovation system transformation should aim at creating and supporting dynamic elements in the system, (creating *Neue Kombinationen* and supporting creative destruction) a fact requiring specific attention to emergence processes, the conditions for targeting new industries, markets and technologies. Furthermore a dynamic, systems-evolutionary perspective to the emergence of venture capital industries and markets and also of entrepreneurial high-tech clusters is needed. Second, present policies, policy routines and organizational capabilities are a far cry from what is required for the promotion of a dynamic start-up segment of companies and of the broader innovative entrepreneurial cluster into which such a segment would be embedded. The same accounts for the possibility of undertaking a continuous process of innovation-based, structural change-led economic growth.

NEED FOR SYSTEMIC INSTRUMENTS[5]

The main conclusion from the foregoing is that innovation is not an activity performed in isolation but one which involves a variety of actions within the system of which the innovating organisation or innovator forms part. In other words, innovation is a systemic activity (Guy & Nauwelaers 2003; O'Doherty & Arnold 2003). One of the consequences of this observation is that actors involved in innovation processes not only need instruments that focus on individual organisations (e.g. financial and managerial instruments) or on the relation between two organisations (e.g. diffusion- and mobility-oriented instruments), but also instruments that focus on the system level. We will call these 'systemic instruments', broadly defined as policy instruments aiming to influence coherent parts of the system in an orchestrated way. A second conclusion is that managing systems is not limited to steady state situations but also involves the management of structural changes. In this context since a number of years the central role of and the need for

5 Partly based on Smits & Kuhlmann (2004).

Table 17.1 Israel's high-tech cluster – selected structural elements (1969-2005)

Accumulated Figures for a period	1969-1976	1977-1984	1985-1992	1993-2000	2001-2005
Number of new high-tech firms creation	56	80	297	2,224	1,655
Israeli LP VC (PE) fundraised M$	0 (0)	0 (0)	~85 (~170)	7,155 (9,170)	3,533 (6,772)
Number of IPOs at US & EU (VC-backed)	1 (0)	13 (0)	19 (3)	133 (65)	24 (6)
Number of significant M&As (VC-backed)	0 (0)	0 (0)	2 (0)	91 (37)	53 (33)
Figure for an actual year	**1976**	**1984**	**1992**	**2000**	**2005**
# LP VC management companies (F-IB)	0 (0)	0 (1)	5 (2)	61 (30)	58 (32)
Share of ICT in manufacturing exports	NA	14%	28%	53%	46%
ICT manu. exports M$ (% of ICT sales)	NA	900 (51%)	2,660 (63%)	11,000 (88%)	11,100 (86%)
Software development exports M$	NA	5 (4%)	135 (23%)	2,600 (70%)	3,100 (81%)
R&D as % of GDP (OCS Grants M$)	2.2% (20)	2.4% (97)	2.6% (199)	4.5% (440)	4.6% (263)

* Some of the figures in the box are from non-official sources. ** F-IB means foreign investment banks.
*** Notice that while columns 3-6 represent 8 years, column 2 represent only 5 years.

Sources: CBS (2006), IAEI (2006), IVC (2006), and OCS – Office of Chief Science (2006).

systemic innovation policy instruments has been emphasised (Smith 2000; Edler et al. 2003; Guy & Nauwelaers 2003; O'Doherty & Arnold 2003). Today there exist already instruments that function at the system level:

- Instruments that aim at more than two heterogeneous actors within a system; in particular initiatives stimulating cooperation interactive learning
- Measures designed to reshape institutions in their context
- Instruments helping to identify and strategic priorities
- Forum approaches for information, debate, consensus-building.

Examples are non-product standards (Tassey 2000), foresight programs (Smits 2002b), integral approaches to revive sectors or stimulate the development of new sectors (see section four) and information campaigns of governments and branch organisations to raise awareness by the public and enterprises for the opportunities specific technologies are offering.

Based on an analysis of major trends in innovation theory Smits & Kuhlmann (2004) conclude that systemic instruments should accomplish a number of basic functions. Based on this concept we suggest considering the following functional dimensions of policy instruments:

- Re-shaping of innovation systems by construction and deconstruction of subsystems; building bridges and stimulating the debate, alignment and consensus (e.g. between science, industry and users of emerging technologies; Jacobsson and Johnson (2000) have shown this for innovative energy systems):
  - facilitating construction and deconstruction of subsystems, preventing of lock-in
  - supporting prime movers
  - ensuring that all relevant actors are involved
- Building cross-linking platforms and new spaces for learning and experimenting: learning by doing, learning by using and learning by interacting (Rosenberg 1982; Lundvall 1992);
- Stimulating demand articulation, strategy and vision development; e.g. building on the growing body of knowledge on the role of users in innovation processes (Oudshoorn & Pinch 2003; Moors, Enzing et al. 2003; Boon et al. 2008);
- Providing and exploiting an infrastructure for distributed strategic intelligence (building on technology assessment, foresight, evaluation, benchmarking etc.); building links between sources, improving accessibility for all relevant actors (clearing house) and stimulating the development of the capacity to produce strategic information tailored to the needs of the actors involved (Kuhlmann et al. 1999).

Clearly, these functions of systemic instruments are covering both operational and strategic policy missions, and they may play a role in innovation systems in a steady state as well as systems in transition. Figure 17.3 illustrates the relation of systemic instrument functions and operational or strategic policy missions: some systemic functions appear to be of more relevance for strategic than for operational purposes.

Given the co-evolutionary development of innovation policy, practice and theory, one would expect that traces from this type of policy instruments should be visible already in real life. In the fourth section, we will discuss two examples of systemic policies *avant la lettre*.

Having discussed the need for systemic instruments the question at stake is how far today's instrument portfolios meet the formulated requirements and how traditional instruments relate to systemic instruments. Although instrument portfolios of many countries went through a development heading in the direction of more systemic instruments[6], recent evaluations show that financial instruments still heavily dominate the innovation policy portfolio of many other countries. In their comparative analysis of the innovation policy instrument portfolios of nine

6 For an analysis of Dutch innovation policy illustrating this trend see Smits & Kuhlmann (2004).

State of innovation system

Policy approach		Steady state	Structural change
	Policy mission	*Operational* innovation policies • focus on implementation of existing set of policies (e.g. function 1 c) • aiming at keeping system vivid and competitive (e.g. 4) • removing system failures (e.g. 1)	*Strategic* innovation policies • generating new set of policies (e.g. function 1 a, b, c; 2, 3) • seeking better fit to a new global or domestic context (e.g. 3, 4) • facilitating creative destruction and the emergence of new combinations (e.g. 1, 2, 3 and 4)
	Functions of policy instruments	Functions of systemic instruments 1 Re-shaping of innovation systems a. facilitating construction and deconstruction of subsystems, preventing of lock-in b. supporting prime movers c. ensuring that all relevant actors are involved 2 Building cross-linking platforms and new spaces for learning and experimenting 3 Stimulating demand articulation, strategy and vision development 4 Providing and exploiting an infrastructure for distributed strategic intelligence (building on technology assessment, foresight, evaluation, benchmarking etc.)	

Figure 17.3 Systems evolutionary approach for systemic strategic innovation policy

countries, Boekholt et al. (2001) conclude: 'If we look at the current policy mix in the benchmark countries we see that supporting R&D in individual companies is still the major objective of innovation policies in many countries'.

While the rise of systemic instruments reflects structural changes in intervention strategies, this does not mean that the existing instruments have become superfluous. On the contrary, apart from adding new functionalities to the toolkit, systemic instruments often contribute to the effectiveness and efficiency and/or reshaping of the 'old' instruments. For instance, the impact of financial instruments aiming to further innovation focusing on stimulating public expenditure on R&D tends to become more intense when instruments that stimulate the diffusion of the results of that R&D are introduced. Similarly, the effectiveness of diffusion instruments will improve by instruments that aim to bridge what Bessant and Rush (1995) call the 'managerial gap'. It is quite feasible that systemic instruments have a positive impact on managerial instruments that support the development of innovation strategies because these instruments may help to obtain a better view of the context in which the innovating firm has to operate. Furthermore, systemic instruments may help to improve and even reshape existing instruments. The progress of Dutch innovation centres from organisations focused on the transfer of knowledge to consultants endeavouring to bridge the managerial gap is one example of such reshaping. The same accounts for the development of technology assessment from a purely scientific activity, attempting to predict all feasible positive and negative impacts of a new technology, towards a more policy-oriented instrument; an instrument that aims to provide the actors involved in innovation processes with specific, relevant information, and by doing so playing an important role in improving the interface between developers, users and regulatory actors (see Smits et al. in this book).

TWO CASE STUDIES

In the following we will illustrate the system-evolutionary innovation policies with two real-life cases: (1) the emergence of a vibrant venture capital industry in Israel since the 1990s; (2) the ongoing transition of the Dutch agricultural sector. Both cases dispose systemic policies on strategic as well as operational levels.

The Emergence of Venture Capital in Israel During the 1990s[7]

During the 1990s a new venture capital (VC) industry and market emerged in Israel, one oriented to the early phase finance and support of high-tech start-up companies. While the first US VC company was created in 1946 (American Research and Development Corporation, see Bylinsky 1976) a significant VC industry and market emerged in the US during the mid 1970s (Gompers 1994; Gompers & Lerner 1999; Avnimelech et al. 2005) in the wake of the ICT and integrated circuit revolution and the creation of NASDAQ in 1971. Its diffusion to Israel during the 1990s took place in the context of globalisation of the main capital market focusing on IPOs of young technology companies: NASDAQ (which was the main channel for Israeli VC exits). When strictly defined, Israel's VC industry during the 1990s became one of the largest VC industries in absolute terms (only second to the US) and the largest in relative terms (VC as percentage of GNP, see Avnimelech & Teubal 2006).

Accompanying the process of VC emergence was the transformation of Israel's high-tech industry – from a military-dominated industry to a sophisticated start-up-intensive high-tech cluster (see Figures 17.4 and 17.5). Several observers have suggested that this case probably represented the most successful instance of diffusion of the Silicon Valley model of high tech (and VC) beyond North America (Bresnahan et al. 2001; OECD 2003; Carmell & de Fontenay 2004).

Our perspective is useful for analyzing the impact of VC on an existing high-tech cluster or its possible contribution to the emergence of a new one. This issue has been largely ignored in the cluster literature, which seems to consider VC as one of many Marshallian 'inputs' in the cluster formation process. In contrast to this our analysis strongly suggests that a central aspect of the impact of VC on high tech is a VC/start-up co-evolution (Avnimelech & Teubal 2006). Through this effect, VC has been a central axis in the emergence of the start-up-intensive high-tech cluster in Israel. It also suggests that the absence of a VC industry with strong co-evolutionary effects with start-ups may be a significant factor in the non-emergence or limited diffusion of Silicon Valley model of high tech clusters beyond the US up to the 1990s.

The analysis of the Israeli case is cast in terms of an extended industry life cycle perspective where the central phase is the VC emergence phase (Phase 3, 1993-2000). Since the focus is to explain the 'causes' of emergence in Israel (and possible causes

7 For more details see Avnimelech & Teubal (2006; 2008a, b) and Breznitz (2008).

Background Conditions Phase	• Creation of high-tech industry and R&D/ innovation capabilities. • National concern for the financing of SME not necessarily high-tech start-ups. • No (or insignificant) formal VC activity; limited informal VC activity. ➔ The start-up-intensive high-tech cluster does not exist yet.
Pre-Emergence Phase Emergence Phase Early Emergence Late Emergence	• A technological revolution that assures a continued stream of new business opportunities for start-ups. • A significant increase in the supply of potential high-tech employees. • Growing acceptance of entrepreneurship and new mechanisms for supporting start-ups. • Increasing numbers of start-ups, creating excess demand for VC services • Growth of informal VC, e.g. angels & VC–related activities; and some formal VCs. • Experimentation (variation) & Learning by SU, VCs and policymakers. ➔ Some important qualitative aspects of the future entrepreneurial cluster are getting established e.g. patterns of specialisation (software and communication in Israel) and VC & SU organisational forms. Also a quantitative jump in the number of start-ups.
Emergence Phase Early Emergence Late-Emergence	• High rate of growth of VC & SU activity; large numbers of new VC & SU companies • Continuation of experimentation by VC & SU agents, enhanced selection • VC-SU co-evolution process: a cumulative growth process caused by positive feedbacks, collective learning, scale economics and network effects. ➔ Cumulative process of emergence of the start-up-intensive high-tech cluster, with a large component of quantitative growth. • Entry of less skilled VC & SU agents, excessive competition and overshooting

7406

Figure 17.4 Israel's high-tech cluster - selected structural elements (1969-2005)

of non-emergence in other countries, at least during the 1990s) we pay particular attention to the pre-conditions that, with the help of a Government targeted programme (the Yozma programme or simply Yozma), facilitated a successful outcome. These pre-conditions are expressed in terms of two phases – a background conditions phase (Phase 1, 1969-1984) and a pre-emergence phase (Phase 2, 1985-1992). Underlying the industry life cycle analysis there is also a dynamic analysis of innovation and technology policy with the focus on R&D grants to business firms during Phase 1 (a horizontal programme open to firms in any sector or technology) and Yozma's targeting of VC (and indirectly, the new entrepreneurial high-tech cluster), during Phase 3. A central piece of the normative analysis is to analyse the conditions under which policies and other events in Phase 1, through their effect on the emergence of Phase 2 conditions, created pre-emergence conditions for targeting venture capital.

Transition to a successful VC emergence process (Phase 3) involves two groups of conditions: first, those underpinning early Phase 3 'demand' for VC services; second, those underpinning rapid growth of VC supply.

An adequate 'demand' for VC services early during Phase 3 is a result of the appearance of a critical mass of start-ups during late Phase 2. The factors stimulating VC 'demand' (i.e. the pool of start-ups) are both 'internal' (domestic) and 'external' to the economy. For Israel, domestic factors included e.g. creation of a separately identifiable software industry during the 1980s; the restructuring of (and spin-offs from) large civilian-oriented companies and of defence-oriented industries; a cultural shift favouring technological entrepreneurship; identification of areas of sustainable competitive advantages within Israel's ICT sectors, and continued R&D

support for high-tech start-ups. 'External' factors stimulating 'demand' include the growth in global markets for ICT products, deregulation of communications markets; 'globalisation' of capital markets for technology companies (NASDAQ); a significant growth in global acquisition activities of MNEs; etc.

The 'supply side' conditions for a transition to the VC emergence phase includes domestic liberalisation of capital markets e.g. to create conditions for foreign flows of funds to the country, and other institutional adaptations e.g. of bankruptcy laws and of company or commercial law to make possible the foundation of Limited Partnerships, and of accountancy procedures and routines to make them SEC compatible. These would pave the way for a rapid Phase 3 'supply response' through inflows of funds from domestic financial & other sectors and private investors, from the global PE industry or other foreign sources, and from Government funds as part of a targeted VC programme.

The Role of Policy

Emergence as the objective of policy

Israel's Yozma programme successfully targeted the VC industry in Israel by sparking a cumulative process leading to the emergence of the industry during 1993-2000. Government money ($80M) seeded 10 hybrid funds ('Yozma Funds') and an additional $20M was directly managed by a government-owned fund ('Yozma Venture Fund'). Government's contribution leveraged an additional $150M mostly from reputable financial institutions and corporations from abroad and from Israel. This initial infusion of funds was invested in approximately 200 start-ups.

The policy process

The background to this programme was a set of new national priorities, which emerged in the late 1980s and early 1990s in the wake of changes in the external and internal environments of Israel. First, during the second half of the 1980s the Military industries laid-off hundreds of engineers and many start-up companies were created only to subsequently fail. Second, the massive immigration from the former Soviet Union during the early 1990s spurred the Government to search for means to employ the thousands of engineers that came to this country. Third, the grants to business sector R&D programme (the main policy of Phase 1 which in fact continues up to this day), which was the backbone to Israel's innovation policy since 1969, was increasingly perceived as being ineffective. The government concluded that the problem was not only lack of resources for the post R&D phases of the innovation process in companies but also lack of management and marketing capabilities. It spurred a strategic problem solving process, which led to the identification of VC and support of start-ups (rather than simply 'R&D additionality') as new national priorities. The outcomes were two VC-directed programmes – a failed precursor programme (Inbal) and the successful Yozma programme (implemented during 1993-7). Yozma's design played a crucial role in explaining its differential performance since both programmes had almost similar goals and their date of initiation differed by only one year with a five years overlap in implementation.

Sparking and sustaining emergence processes

The new infusion of VC in the aftermath of Yozma triggered a cumulative process with positive feedback in which more and more profitable VC activity 'today' spurred even more and more profitable VC activity 'tomorrow'. At the centre of this process was a VC/start-up co-evolution. Other dynamic processes were involved as well such as (1) entry of strategic investors in response to the early reputation earned from some excellent Yozma Funds' portfolio company exits during 1996-98; (2) this in turn extended the Israeli VC industry networks and added value abilities; (3) cluster effects enabling a wider set of VC/high-tech non-tradable services to be available locally (e.g. lawyers, financiers, accountants and consultants); (4) entry of MNCs into the cluster and continues stream of acquisitions by MNCs, which also extended the cluster's networks and capabilities; and (5) collective learning concerning the VC business, etc.

Importance of background and pre-emergence conditions

Within our framework the emergence or non-emergence of a new VC industry and/or market depends crucially on a set of pre-emergence, Phase 2, conditions. Examples are the demand for VC services as reflected in the pool of start-ups already operating prior to emergence, a process of liberalisation and de-regulation involving capital markets and foreign exchange, and a sufficiently strong process of variation and selection related to the organisation and strategy of both start-ups and VCs (undertaken by agents operating in the proto-VC industry/market). Other work on industry emergence or work on new industries has emphasized additional Phase 2 conditions, which could enhance the likelihood of new or infant industry emergence in general. Among these and depending on industry and context we have: high capability of early entrants and of proto-industry agents (Avnimelech & Teubal 2004c); a policy process and government policy capabilities which lead to the successful identification of a new set of relevant and appropriate strategic priorities (Avnimelech & Teubal 2005); and political legitimation required e.g. for a new regulatory and institutional environment. The framework can also be applied to explain non-emergence and the frequent failure of VC-directed policies. For example, the failure of the German VC industry to emerge in the first half of the 1980s (Fiedler & Hellmann 2001; Becker & Hellmann 2002) could be attributed to the first of the pre-emergence conditions mentioned above and to other factors. A deeper understanding of the typical pre-emergence conditions required for VC and of the processes leading to them could have induced policymakers to abstain from targeting VC directly and focus instead on policies to create favourable pre-emergence conditions (creation of a pool of start-ups in the German case).

The importance of programme design and timing

Once the strategic goal of promoting industry/market emergence is accepted, and given other favourable Phase 2 conditions, the major decision to be made is whether or not to implement a targeted VC policy. ITP programme design for VC emergence should consider the crucial links between (1) the structure of incentives and the extent by which a critical mass of capabilities is achieved; and (2) government's venture contribution and how this could influence both the structure of incentives and the mass of financial resources. The Israeli experience suggests that simple tax breaks to investors in start-ups (as has been common in Europe during the 1980s and 1990s) need not suffice. Rather, attracting world class players as limited partners would require strong incentives to the upside. This in turn requires a 'Government venture capital contribution' in order to seed private VC funds; and a 'buy option' to private investors in VC funds. Alternatively and more extensively, the scope and design of the programme should effectively deal with the various components of the relevant system failure. The Israeli case suggests that these relate to (a) entry of professional managers of VC companies; (b) participation of reputable/linked foreign partners; (c) achieving critical mass; (d) selection of a suitable form of organisation which supports the generation of capabilities; (e) spurring a collective learning process; and (f) accelerating appropriate 'selection' of VC organisation, strategy, etc. This is a formidable task: it complicates the structure of incentives; and it requires policy makers to take account of a wide range of other factors not directly linked to incentives (see previous subsection).

This means that in estimating the scope of government investments account must be made not only of the fixed costs of managing a government VC fund, but the specifics of capabilities – both internal and external – that have to be accumulated. Moreover the greater the scope of government investments, the greater the incentives to the upside that may be offered and, at least up to a certain point, the greater the expected entry/participation of skilled agents in the new industry. This is a crucial link that seems to have been missed in the literature. Also the point that leveraging 'public' venture investments to assemble a critical mass of private VC capabilities is the way to transform a potential private capital crowding-out effect into a strongly complementary public-private capital contribution.

Summing up this case illustrates a successful shifting from Phase 1 to Phase 3, involving a relative shift from 'neutral/horizontal' support of R&D in firms to strategic support of a new industry and market (venture capital), i.e. a systemic instrument. The successful shift was facilitated by enhanced activity at the strategic level of policy starting with 'search, research and discovery' on the part of the dominant innovation policy institution (the Office of the Chief Scientist) followed by identification of a new vision and strategy for an entrepreneurial high-tech cluster with large numbers of start-up firms and venture capital funds.

Strategic Systemic Policies in the Dutch Agro Sector

As a second case we discuss a strategic systemic innovation policy in the Dutch agro sector carried by INNONET and one of its followers, the TransForum programme. In doing so we will try to make clear: (1) structural changes in innovation systems need strategic systemic policies and instruments, and (2) systemic instruments also play a role in steady state policies.

The Dutch agro-sector was a great success story for many decades. Fuelled by a strategic policy – starting at the end of the 19th century – to stimulate a knowledge-intensive agro sector, the Netherlands managed – despite its limited size – to develop in the 1980s into the second exporter of agro products after the USA. The major characteristics of the Dutch agro-sector in that period are: focus on mass production, rather homogeneous markets, high, knowledge-based productivity growth and a supply-oriented knowledge infrastructure with excellent feedback from users. Systemic, long-term policies played an important role in the build up of this system. A complete new innovation system was created by the setting up of the Wageningen Agricultural University, the agricultural test stations and a system of information officers that spread the results of research to farmers and fed back their experiences. Platforms where farmers and researchers could meet to exchange the latest knowledge and experiences complemented the system. Also in the steady state systemic instruments like platforms for learning and experimenting, continuous efforts to manage interfaces between the university, test stations and farmers and a strategic intelligence infrastructure to help farmers to stay informed of the latest developments, played an important role in the day-to-day activities of the sector. In the 1980s this successful system ran against its barriers: price erosion caused by competitors with much better factor conditions (energy and labour costs) like Spain, Israel, Kenya and Ecuador and external effects such as pollution and epidemic diseases. The latter were not only detrimental from an economic perspective but also undermined the societal 'license to produce' of the agro-sector.

As a consequence awareness increased that structural changes in the sector were necessary. The Netherlands Council for Agricultural Research (NRLO), one of the four Advisory Councils on Research (ACR) in the Netherlands (and an important actor in the strategic intelligence infrastructure), took the lead in this process of structural change. Given its basic structure the NRLO, as all ACRs, is a reflection of the status quo in the system. The developments in the foregoing, however, urged for structural changes in the system. In the mid 1990s the NRLO shifted its focus from managing a steady state system to changing the structure of this system. A salient illustration was the change of its name to the Innovation Network Rural Areas and Agricultural Systems (INNONET). This 'new' ACR was established in the middle of the year 2000. Structural change is a big challenge for researchers, producers and users in the knowledge-driven Agro Innovation System (AIS). They have to cope with a very complex transition of a – because of its successful history very well entrenched – system with many strong incumbents. The focus of INNONET was on building networks supporting the development of a vision dealing with a structural change from a mass production and not so sustainable system to sustainable production of specialties. The question whether or not INNONET realised a strategic innovation policy resulting in structural changes in the agro system is still difficult to answer. In terms of the systemic functions, some results were reached regarding the management of interfaces and the development of a joint vision In terms of building and organising innovation systems results are less convincing and creative destruction of incumbent structures was hardly reached. Furthermore, the newly developed visions were rather abstract and did not (yet) translate in many concrete new initiatives. To wind up, although INNONET acted as an important prime mover of the debate on structural changes in the AIS, real changes in this system were hardly reached. The exclusive focus on research, a still too conservative approach, the lack of funds available for substantial investments in new (business) initiatives and – of course – the problem of creative destruction can be seen as the most important reasons.

In the early years of the new millennium INNONET was offered an opportunity to continue its activities at a more concrete level by the funds made available through the so called ICES-KIS programme. This programme aims to invest in strengthening the Dutch innovation system by improving the interface between the production of knowledge and the users of this knowledge. For the period 2004-2010, 1.6 billion Euro was available. INNONET was one of the major initiators of the TransForum programme, a 60 million programme focusing on the transition of the AIS. The programme started in 2004 from three basic assumptions:

1. Sustainable value propositions (*Neue Kombinationen*) focusing on sustainable specialties are central.
2. The AIS aims to establish connections between businesses, markets and knowledge systems.
3. Cooperation between parties that jointly define the unique Dutch positions, the so-called KOMBI parties (knowledge institutions, governmental authorities, societal organizations and the business community) is essential for the realisation of the perspective.

Whereas in the old AIS knowledge institutions played an initiating role, in the new system the knowledge infrastructure is guided far more by the needs of other actors in the system. This policy involves structural changes in the knowledge infrastructure (more demand-oriented), major changes in the way research funds are allocated, the interfaces between production and demand for knowledge, and asks for empowerment of actors to develop creative and sustainable business opportunities. Creative destruction of incumbent structures plays an important role in this transition. One of the changes that already took place was the – rather painful – reorganisation of the Wageningen University and its merger with applied research institutes into the Wageningen University Research centre.

In order to contribute to this transition, the TransForum programme developed three strongly interlinked components:

4. A scientific programme.
5. Innovative practice projects (IPPs).
6. Knowledge projects.

Scientific projects not only have to inspire IPPs but – even more – answer fundamental questions IPPs are confronted with. Examples comprise research in the management of innovation in international agro food chains, research in the visions on sustainable agriculture, the involvement of users in the start and further development of IPPs and marketing-oriented research focusing on how consumers may be stimulated to buy more sustainable products. Meanwhile 25 IPPs (new business propositions), are on their way ranging from 'greening supermarkets' to the 'development of transgenic research in apple resistance'. Knowledge projects concern the development of (new) elements of the innovation system that support the development and management of new business opportunities and stimulate the interaction between knowledge production and use. As such they are new structural elements of the AIS. Examples include the development of curricula for teaching purposes and the development of workshops and/or fora that facilitate the dialogue between the different KOMBI partners.

A first evaluation in terms of the systemic functions reveals the following. TransForum puts a heavy emphasis on the development of interfaces, more in particular between the KOMBI partners and on the development of new business opportunities with a strong link with scientific research. The programme contributes – bottom up – to the development of the building blocks of a new AIS. In doing so, two major problems became visible. First, developing more balanced interaction between practitioners and research appears to be a very stubborn problem. Different languages, (reward) cultures and goals make it difficult to develop fruitful interaction and alignment. Universities are in a difficult position in this game. Society is asking for a contribution of science to the solution of major societal problems. The internal dynamics of science, however, make the survival of university researchers and groups more then ever dependent on the verdict of their scientific peers. This is a crucial dilemma that can not be solved in the context of the Dutch agricultural research because it touches on the identity of all scientific research (cf. Gibbons et al. 1994).

A second major problem concerns the creative destruction of incumbent structures. To break down existing structures appears to be considerably more difficult in a successful sector with a long tradition and strong entrenchment in scientific, economic and political circles. Apart from increasing external pressure to change (e.g. a growing societal resistance against non-sustainable production), the only solution here are strategic policies including stimulating vision development, long-term investments in experimenting with new business opportunities, protection of these business opportunities until they are able to survive competition with the old options and adapting the AIS to these new ways of production (e.g. regulation, strategic intelligence infrastructure).

TransForum can be seen as a systemic policy instrument addressing all systemic functions in order to contribute to a major transition in the Dutch agro innovation system. INNONET strategically prepared the ground (and the minds of involved actors) for this programme. The experiences of INNONET and TransForum clearly demonstrate the need for a systemic approach. However, breaking through incumbent interests, securing long-term (financial) commitment and overcoming differences between the world of science and practice appears to be tough. Without a long-term, strategic, systemic and inherently long-term oriented innovation policy even the best systemic instruments will not be able to induce major structural changes in innovation systems. From the foregoing it also will have become clear that addressing systemic functions as 'management of interfaces', 'providing fora for learning and experimenting' and a 'strategic intelligence infrastructure' tuned to the needs of the KOMBI partners also are important in policies addressing the steady state of the new agro innovation system.

CONCLUSIONS

We started this chapter claiming that innovation policies should follow a systems perspective. It was concluded that, although many policymakers now have adopted this perspective, actual innovation policies are still largely inadequate in at least two respects: (1) they focus on the management of innovation systems in a steady state, not aiming at structural changes, innovation system transformation and processes of emergence of higher levels of organisation; and (2) policy approaches have not been complemented by instruments necessary to tackle systemic problems, instead they often stick to the traditional, often linear-model-based toolkit. We proposed two solutions for these problems: a distinction between operational (steady state) and strategic policies (dynamic state) and systemic policy instruments, both being part of emerging systems-evolutionary perspectives to innovation policy broadly defined. In this final section we will draw some conclusions regarding these two concepts and also point at some issues that deserve further attention from policy makers and innovation scholars.

Strategic Policies

A strategic level of policy (for simplicity we can call this 'strategic policies') by necessity focuses on an extended policy process, one which begins with the setting of new strategic priorities and continues with efforts at articulating these in terms of a new set of policies. Their relevance is enhanced whenever strong changes in the global and/or domestic environments call for a new or adapted innovation system, one inconsistent with the existing set of policies and with a national system of innovation, which is in a 'steady state'. In contrast to the operational level of policy which focuses on the implementation of an existing set of policies, action at the strategic level aims at generating a new set of policies, one better fit to the new global and domestic context, i.e. aimed at achieving the required changes in the economy and innovation system. Strategic policies should *grosso modo* be conceived as comprising three main components: incentives and related programmes, e.g. supporting R&D in firms; institutional and regulatory changes, e.g. new IP laws and regulations; and other actions such as technological forecasting, setting new priorities and, whenever relevant, a long-term innovation agenda and creating new 'strategic' policy institutions. The latter are critical when shifting from a dominant operational level of policy mode to a mode which includes both an operational and a strategic level.

Systemic Instruments

Systemic policies ask for instruments able to cope with the characteristics and dynamics of the management of systems. Until now linear-model-based instruments that are not well equipped to handle these characteristics and dynamics heavily dominate portfolios. For this we introduced the concept of systemic instruments able to intervene in an orchestrated way in coherent parts of the system to manage steady state systems and guide the transition of structural changes in the system. In particular these instruments should address four functions: reshaping of innovation systems, building platforms for learning and experimenting, stimulating demand articulation and vision development and providing a tailor-made strategic intelligence infrastructure. From our analysis it appears that there are strong indications that systemic policy instruments are the product of learning processes in which policy, theory and process heavily interact. In other words, they are the result of a co-evolution of innovation practice, theory and policy.

Still, many such instruments have not been analysed thoroughly; there are also hardly any systematic monitoring and evaluation procedures to facilitate this analysis. In consequence, there is a great deal of uncertainty as to the nature and the magnitude and the structural character of the impact of these instruments. Up to now, however, the potential to reinforce the efficiency and effectiveness of other instruments in the portfolio has not been exploited because policymakers' focus is still rather on linear-model-based instruments addressing individual organisations or relations between two organisations and not the portfolio as a whole.

Apart from the problems related to monitoring, evaluating and analysing the impact of these instruments, the most important problems with the further development of systemic instruments concern questions such as: how to creatively destroy systems, institutions and relations that no longer meet the new demands; how to organise effective learning and experimenting; how to ensure that systemic instruments achieve structural, long-lasting results; and how to provide actors with the right attitude and appropriate skills.

Agenda for Further Research

Though there are some early experiences with strategic policies and systemic instruments, as illustrated with our two cases, there still are a lot of open questions. In our view the most important ones include:

- Further development of a systems-evolutionary perspective to innovation policy broadly defined, in particular concerning the strategic level of policy and associated policy institutions.
- Analysis of co-evolutionary processes involved in systems of innovation transformation and of associated learning processes including those involving policy in its various dimensions: process, intervention and theory (including research into the impact of systemic instruments on these learning processes).
- In-depth analysis of a limited number of systemic instruments. The major issues that should be addressed in the analysis are: the contribution to the five functions, the realisation of intended (and unintended) impact, the impact on other instruments in the portfolio and the organisation of effective evaluation.
- In order to improve the insight into the use of systemic instruments a number of *Begleitforschung* projects (monitoring analysis) of major innovation projects with a strong systemic character, could be initiated. These projects could (1) provide innovation researchers with insights into the role of systemic instruments in complex and long-lasting innovation processes; and (2) offer actors involved in the innovation processes with concepts, information and instruments that might help them to realise their goals.

An essential condition for strategic innovation policies is the creation of a strategic level of policy including its organisational, capabilities and governance underpinnings. Policy must become more knowledge-based, drawing on Strategic Intelligence (see Chapter 16 in this book by Smits et al. on the Role of TA in Systemic Innovation Policy; Kuhlmann et al. 1999; Kuhlmann 2003) and strategic prioritisation (as a precondition for actual policies and changes in policies) should assume a central role within an expanded system-evolutionary policy process entailing the incorporation of an explicit strategic level.

REFERENCES

Avnimelech, G. and Teubal, M. (2008a), 'From direct support of business sector R&D/innovation to targeting venture capital/private equity: a catching-up innovation and technology policy life cycle perspective', *Economics of Innovation and New Technology*, 17 (1), 153-172.

Avnimelech, G. and Teubal, M. (2008b), 'Evolutionary targeting', *Journal of Evolutionary Economics*, 18 (2), 151-166.

Avnimelech, G. and Teubal, M. (2006), 'Creating venture capital industries which co-evolve with high tech clusters: insights from an extended industry life cycle perspective to the Israeli experience', *Research Policy*, 35 (10), 1477-1498.

Avnimelech, G. and Teubal, M. (2005), 'Evolutionary innovation and high tech policy: what can we learn from Israel's targeting of venture capital?', Science, Technology and Economy Program, Working Paper Series STE-WP-25-2005, Samuel Neaman Institute, Science, March 2005, Haifa.

Avnimelech, G., Kenney, M. and Teubal, M. (2005), 'The life cycle model for the creation of national venture capital industries: the US and Israeli experiences', in E. Giulani, R. Rabellotti and M. van Dijk (eds), *Clusters Facing Competition: The Importance of External Linkages*, Surrey: Ashgate Publishing.

Avnimelech, G. and Teubal, M. (2004a), 'Venture capital: start up co-evolution and the emergence and development of Israel's new high tech cluster – Part 1: macroeconomic & industry analysis', *Economics of Innovation and New Technology*, 13 (1), 33-60.

Avnimelech, G. and Teubal, M. (2004b), 'Strength of market forces and the successful emergence of Israel's venture capital industry: insights from a policy-led case of structural change', *Revue Economique*, 55 (6), 1265-1300.

Becker, R. and Hellmann, T. (2002), 'The genesis of venture capital: lessons from the German experience', working paper, Stanford University, Stanford.

Bessant, J. and Rush, H. (1995), 'Building bridges for innovation: the role of consultants in technology transfer', *Research Policy*, 24 (1), 97-114.

Bijker, W. (2001), 'Understanding technological culture through a constructivist view of science, technology, and society', in S. Cutcliffe and C. Mitcham (eds), *Visions of STS: Counterpoints in Science, Technology, and Society*, New York: State University of New York Press (Sunny Press), pp. 19-34.

Boon, W., Moors, E., Kuhlmann, S. and Smits, R. (2008), 'Demand articulation in intermediary organisations: the case of orphan drugs in the Netherlands', *Technological Forecasting and Social Change*, 75 (5), 644-671.

Bresnahan, T., Gambardella, A. and Saxenian, A. (2001), 'Old economy inputs for new economy outputs: cluster formation in the new Silicon Valleys', *Industrial Corporate Change*, 10 (4), 835-860.

Breznitz, D. (2008), *Innovation and the State: Political Choice and Strategies for Growth in Israel, Taiwan and Ireland*, New Haven, USA and London, UK: Yale University Press.

Bylinsky, G. (1976), *The Innovation Millionaires: How They Succeed*, New York: Charles Scribner's Sons.

Caplan, N. (1979), 'The two communities theory and knowledge utilization', *American Behavioral Scientist*, 22 (3), 459-470.

Carmell, A. and Fonteney, C. de (2004), 'Israel's silicon wadi', in T. Bresnahan and A. Gambardella (eds), *Building High Tech Clusters: Silicon Valley and Beyond*, Cambridge, UK: Cambridge University Press.

Edler, J., Joly, P.-B., Kuhlmann, S., et al. (2006), 'Understanding fora of strategic intelligence for research and innovation', The PRIME Forum Research Project (Fraunhofer ISI), Karlsruhe.

Edler, J., Kuhlmann, S. and Smits, R. (2003), 'New governance for innovation. The need for horizontal and systemic policy co-ordination', report on a workshop of the Six Countries Programme – the Innovation Policy Network (Fraunhofer ISI, working papers Innovation System and Policy Analysis, 2003/2; ISSN 1612-1430), Karlsruhe.

Edquist, C. (ed.) (1997), *Systems of Innovation. Technologies, Institutions and Organizations*, London, UK and New York, USA: Pinter.

Elzen, B. and Wieczorek, A. (2005), 'Transitions towards sustainability through system innovation', *Technological Forecasting and Social Change*, **72** (6), 651-661.

Felt, U., Wynne, B., Callon, M., et al. (2007), 'Taking European knowledge society seriously', report of the Expert Group on Science and Governance (European Commission), Brussels.

Fiedler, M. and Hellmann, T. (2001), 'Against all odds: the late but rapid development of the German venture capital industry', *Journal of Private Equity*, **4** (4), 31-45.

Geels, F. and Schot, J. (2007), 'Typology of sociotechnical transition pathways', *Research Policy*, **36** (3), 399-417.

Gibbons, M., Limoges, C., Nowotny, H., et al. (1994), *The New Production of Knowledge. The Dynamics of Science and Research in Contemporary Societies*, London: Sage Publishing.

Gompers, P. (1994), 'The rise and fall of venture capital', *Business and Economic History*, **23** (2), 1-26.

Gompers, P. and Lerner, J. (1999), *The Venture Capital Cycle*, Cambridge, MA, USA and London, UK: MIT Press.

Guy, K. and Nauwelaers, C. (2003), 'Benchmarking STI policies in Europe: in search of good practice', The IPTS Report, no. 71, IPTS, Seville.

Hippel, E. von (1988), *The Sources of Innovation*, Cambridge, MA, USA and London, UK: MIT Press.

Hirsch-Kreinsen, H. (2008), 'Low-tech innovations', *Industry and Innovation*, **15** (1), 19-43.

Hirsch-Kreinsen, H., Jacobson, D., Laestadius, S. and Smith, K. (2005), 'Low and medium technology industries in the knowledge economy: the analytical issues', in H. Hirsch-Kreinsen, D. Jacobson and S. Laestadius (eds), *Low and Medium Technology Industries in the Knowledge Economy*, Frankfurt: Peter Lang, pp. 11-30.

Imai, K., Nonaka, L. and Takeuchi, H. (1988), 'Managing the new product development process: how Japanese firms learn and unlearn', in K. Clark, R. Hayes and C. Lorenz (eds), *The Uneasy Alliance: Managing the Productivity-Technology Dilemma*, Boston: Harvard Business School Press, pp. 337-376.

Jacobsson, S. and Johnson, A. (2000), 'The diffusion of renewable energy technology: an analytical framework and key issues for research', *Energy Policy*, **28** (9), 625-640.

Kemp, R., Schot, J. and Hoogma, R. (1998), 'Regime shifts to sustainability through processes of niche formation: the approach of strategic niche management', *Technology Analysis and Strategic Management*, 10 (2), 175-195.

Klepper, S. (2001), 'Employee startups in high-tech industries', *Industrial and Corporate Change*, 10 (3), 639-674.

Klepper, S. (1996), 'Entry, exit, growth and innovation over the product life cycle', *American Economic Review*, **86** (3), 562-583.

Kline S. and Rosenberg, N. (1986), 'An overview of innovation', in R. Landau and N. Rosenberg (eds), *The Positive Sum Strategy: Harnessing Technology for Economic Growth*, Washington, DC: National Academy Press.

Kuhlmann, S. (2003), 'Evaluation as a source of strategic intelligence', in P. Shapira and S. Kuhlmann (eds), *Learning from Science and Technology Policy Evaluation: Experiences from the United States and Europe*, Cheltenham, UK and Northampton, MA, USA: Edward Elgar, pp. 352-379.

Kuhlmann, S., Boekholt, P., Georghiou, L., et al. (1999), 'Enhancing distributed intelligence in complex innovation systems', report published within the framework of the Targeted Socio-Economic Research Programme of the European Commission, ISI-FhG, Karlsruhe.

Lundvall, B.-Å. (2007), 'Post script: innovation system research. Where it came from and where it might go', mimeo.

Lundvall, B.-Å. (1992), *National Systems of Innovation: Towards a Theory of Innovation and Interactive Learning*, London, UK and New York, USA: Pinter.

Moors, E., Enzing, C., et al. (2003), 'User-producer interactions in functional genomics innovations', *Innovation: Management, Policy & Practice*, 5 (2-3), 120-143.

Nelson, R. (1993), *National Innovation Systems*, Oxford, UK and New York, USA: Oxford University Press.

Nelson, R. and Winter, S. (1977), 'In search of a useful theory of innovation', *Research Policy*, **6** (1), 36-76.

O'Doherty, D. and Arnold, K. (2003), 'Understanding innovation: the need for a systemic approach', The IPTS Report, no. 71, IPTS, Seville.

OECD (2003), 'Venture capital policy reviews', DSTI/DOC(2003), STI Working Paper 2003, Paris.

OECD (2002), *Benchmarking Science-Industry Relationships*, Paris: OECD Publishing.

OECD (1992), 'Technology and the economy. The key relationships', report of the Technology and Economy Programme, OECD, Paris.

Oudshoorn, N. and Pinch, T. (eds) (2003), *How Users Matter: The Co-Construction of Users and Technology*, Cambridge, MA, USA and London, UK: MIT Press.

Rip, A. (2006), 'A co-evolutionary approach to reflexive governance: and its ironies', in J.-P. Voß, D. Bauknecht and R. Kemp (eds), *Reflexive Governance for Sustainable Development*, Cheltenham, UK and Northampton, MA, USA: Edward Elgar, pp. 82-100.

Rip, A. and Kemp, R. (1998), 'Technological change', in S. Rayner and L. Malone (eds), *Human Choice and Climate Change*, Vol. 2, Washington, DC: Batelle Press, pp. 327-400

Rosenberg, N. (1982), *Inside the Black Box. Technology and Economics*, Cambridge UK: Cambridge University Press,.

Rosenberg, N. (1976), *Perspectives on Technology*, Cambridge, UK: Cambridge University Press.

Rosiello, A., Teubal, M. and Avnimelech, G. (2008), 'Towards the framing of venture capital policies: a systems-evolutionary perspective with particular reference to the UK/Scotland and Israeli experience', paper presented to the July, 2008 International Schumpeter Society Meeting, Rio de Janeiro.

Schumpeter, J. (1934), *The Theory of Economic Development*, Cambridge, MA, USA: Harvard University Press.

Sercovich, F. and Teubal, M. (2007), 'Innovation, technological capability and competitiveness: the policy issues in evolutionary perspective', typescript.

Sharif, N. (2006), 'Emergence and development of the national innovation systems concept', *Research Policy*, **35** (5), 745-766.

Smith, K. (2000), 'Innovation as a systemic phenomenon: rethinking the role of policy, enterprise and innovation', *Management Studies*, **1** (1), 73-102.

Smits, R. (2002b), 'The new role of strategic intelligence', in A. Tűbke, K. Ducatel, J. Gavigan and P. Moncada-Paternò-Castello (eds), 'Strategic Policy Intelligence: current trends, the state of play and perspectives', IPTS Technical Report Series, EUR 20137 EN, IPTS, Seville.

Smits, R. and Hertog, P. den (2007), 'Technology assessment and the management of technology in economy and society', *International Journal of Foresight and Innovation Policy*, **3** (1), 28-52.

Smits, R. and Kuhlmann, S. (2004), 'The rise of systemic instruments in innovation policy', *International Journal Foresight and Innovation Policy*, **1** (1-2), 4-32.

Tassey, G. (2000), 'Standardization in technology based markets', *Research Policy*, **29** (4), 587-602.

Teubal, M. (2002), 'What is a Systems of Innovation (SI) perspective to Innovation and Technology Policy (ITP) and how can we apply it to developing and industrializing economics?', *Journal of Evolutionary Economics*, **12** (1), 233-257.

Teubal, M. (1997), 'A catalytic and evolutionary approach to Horizontal Technology Policies (HTPs)', *Research Policy*, **25** (8), 1161-1188.

Voß, J.-P. (2007), 'Designs on governance. Development of policy instruments and dynamics in governance', PhD dissertation (ISBN 978-90-365-2585-5), Enschede and Berlin.

Voß, J.-P., Bauknecht, D. and Kemp, R. (eds) (2006), *Reflexive Governance for Sustainable Development*, Cheltenham, UK and Northampton, MA, USA: Edward Elgar.

Woolthuis, R. Klein, Lankhuizen, M. and Gilsing, V. (2005), 'A system failure framework for innovation policy design', *Technovation*, **25** (6), 609-619.

18. An Outlook on Innovation Policy, Theory and Practice

Philip Shapira, Ruud Smits and Stefan Kuhlmann

INTRODUCTION

Innovation – especially innovation aimed at turning scientific and technological potential into successful products, services and solutions – has risen up on the agenda of both developed and emerging economies over the past three decades. Structural changes in the world economy, including the growth of low-cost yet capable production locations in Asia and elsewhere, have forced advanced countries to shift increasingly away from cost-based competition to strategies built on added value. R&D and knowledge-based innovations are at the core of strategies to sustain economic leadership positions in an era of international markets and at the same time offer the promise of solutions to broader societal and environmental problems that, in turn, may stimulate further economic opportunities.

These transitions in corporate and national strategies have been accompanied by parallel shifts in approaches to the innovation process and to innovation policymaking. The linear pipeline of R&D leading through to the manufacture and marketing of new products has given way to models of innovation that draw on broader sources of knowledge and network linkages and which have a wider array of outcomes, including innovations in organizations and business models as well as products and process. Similarly, older policies which emphasized public R&D spending have been succeeded by systemic approaches to policy that seek to address the multiple factors, institutions and relationships that influence innovation performance. It has been recognized that R&D investments alone do not suffice to improve the innovative capacity of economies and societies. Policies also should strive to improve the conditions in the environment – the innovation system – in which firms and other actors are operating. Crucial to advancing this system is an iterative learning process between policy, theory and practice – what we called the 'three dancers' at the beginning of this book.

In this last chapter, we focus on changes in the relations between these three dancers and consider some of the consequences of these changes. The chapter is structured as follows. First, we review the major goals and opening points of the book and synthesize important trends in the wider world influencing the relation between policy, theory and practice. In the next section we then discuss some major observations about changes in the relation between the three dancers. These observations constitute the basis for conclusions and recommendations in the last section.

STARTING PROPOSITIONS REVISITED

The central theme of this book has been the interactions between policy, theory and practice of innovation. In so doing, we have targeted three major goals. First, to improve and update our understanding of the interactive relationships between innovation policy, theory and practice. Second, to search for options that can improve the added value, effectiveness and efficiency of policy and innovative practice. Third, to offer critical perspectives to inform ongoing discussions about the dynamics of innovation systems and related policies, as well as to generate questions that can guide future research.

In the first chapter of the book, a number of starting propositions were put forward. The following discussion reflects on these, taking into account insights gained from subsequent chapters.

Our first proposition was that innovation should not be perceived as a free-standing linear progression but rather as a complex multi-actor game taking place within a system of institutions, conventions and associations. Analyses of how innovation is influenced by the interplay of institutions and multiple actors have proven to be insightful in retrospective studies of the development and adoption of particular innovations. For example, Hargadon and Douglas (2001) detail how Thomas Edison's design and development of an innovative electric lighting system in New York at the end of the nineteenth century was influenced by a series of mediating economic, social and political institutions and organizations. Similarly, institutional and multi-actor perspectives have been engaged to understand the emergence of the military-industrial complex in the US and its impact on the development of new weapons systems, university innovation and the relationships between military and civilian innovation (Baack & Ray 1985; see also the chapter by Jordi Molas-Gallart in this book). Most importantly from the standpoint of this book, the innovation system heuristic has been increasingly adopted in recent years by policymakers and their advisors as an operational framework to rationalize and target government interventions (see for example Popper & Wagner 2002; OECD 2005; DIUS 2008).

Of course, analyzing the socio-economic and institutional relationships that influence innovation through the use of a system heuristic is not an entirely new idea. Lundvall (2007) reminds us that systemic socio-technical perspectives on innovation can be traced back to Friedrich List's 1841 notion of national systems of productions, as well as to Adam Smith's late eighteenth century observations about modes of learning, Karl Marx's mid nineteenth century writing about the relationships of technological progress, and Joseph Schumpeter's early-to-mid twentieth century insights into the dynamics of the innovation process. Yet, it is only since the late 1980s and 1990s that the concept of innovation systems has fully entered into innovation policy thinking and practice (as discussed by Chaminade & Edquist in this book). This development has been stimulated not simply by academic revival or advance, but also by real changes in underlying conditions for innovation. These include such factors as globalization, the rise of networks, the more open character of innovation, the complexities of relationships with public organizations and intermediaries, internal organizational changes, the growth of multi-disciplinary problem approaches, and greater recognition of the role of multiple market and institutional elements in influencing innovation – all of which have prompted private and public actors to re-conceptualize the innovation process in systematic terms.

As several of the chapters in this book have observed, innovation systems are not fixed and clearly delimited. As in all systems, boundaries are more or less porous and are moulded by the nature of innovation processes and their actors and relations, and by broader contexts including hard and soft institutions, markets and governance, and political, social, cultural and regional conditions. While there is naturally an ongoing search for patterns and commonalities across innovation systems, there is always a need to be conscious of the specificities of particular systems (especially when trying to transfer perceived 'best practices').

Ironically, even with the growing acceptance of multi-dimensional innovation systems approaches, the linear model still persists. Boekholt (in this book) points out that many innovation policies and instruments are still predicated on linear models of basic research followed by applied research leading to development, production and diffusion (Godin 2006) and that such models are retained and are active in the cognitive maps of the innovation process used by numerous policymakers (see also Boekholt et al. 2001). Why the linear concept endures is worthy of fresh research. Indeed, there is an argument that the linear model of innovation (albeit in a weaker form or in certain sectors such as pharmaceuticals) still has applicability (Balconi et al. 2008). Moreover, despite justified and important work in highlighting the contributions of organizational and non-formal (or hidden) modes of innovation (Damanpour 1991; NESTA 2007), it should not be concluded that science and technology no longer play important roles in innovation. Advances in – and application of – science and technology are the central sources of many innovations. Yet, even in these cases, the innovation systems approach indicates that science and technology may be necessary prerequisites while there are numerous other aspects that must be taken into account for an innovation to succeed. Moreover, the systemic approach stresses that that to become innovative, the capacity to exploit knowledge – from here and there, from now and then – is often as important as the ability to produce new knowledge. This observation has become arguably more relevant over the last decade as East Asian countries like China and India have rapidly developed their technological capacities by drawing, in large part, on knowledge transferred or acquired from Western countries. However, these countries are growing not only through knowledge acquisition but also through the related hand-in-hand building up of their own technological and innovative capabilities.

The second proposition of this book is that there are mutual interactions between practices, policies and theories of innovation. We do not want to state that these three domains co-evolve, since such a statement could provoke inappropriate comparisons with biological phenomena by suggesting that change in one sphere directly triggers change in another. Rather, we observe that changes in the relationships between the three domains are frequently mediated or bridged by learning processes. For example, in the 1980s, the realization by practitioners and policymakers that increased R&D investments were leading to disappointing results stimulated theoretical interest in and adoption of innovation systems approaches (OECD 1991). In the 1990s, the innovation systems framework provided the context for many countries and regions to develop innovation cluster policies, spurred by concepts developed by Porter (1990) and others (including Saxenian 1994; Rosenfeld 1997). In the current decade, innovation systems approaches are being augmented by a new corporate and policy interest in open and networked innovation strategies (Chesbrough 2005). In these examples, several different types of learning processes have been engaged to transfer insights from one domain to another. These include learning by doing and using, complemented by interactive exchange between policymakers and practitioners; learning by searching and comparison, where ideas are interactively communicated between innovation scholars and policymakers, with scholars in turn translating these experiences into new theories and insights.

One important conclusion from this is that the stimulation of conditions favourable to a variety of types of learning processes constitutes a crucial element of innovation policies. In this book, many examples of such learning processes have been discussed. Significantly, less attention has been paid to formal learning processes as they take place in our education systems. We accept that most learning and exchange about innovation processes occurs through informal means. Moreover, public policies for innovation typically focus on horizons that are shorter in term than educational policies and are administered by different agencies. There has been movement in recent years at the level of higher education to develop innovation missions in universities (Youtie & Shapira 2008). Explicit attention is also being paid at national and regional levels in several countries to the policy mix for innovation, to better align streams of policymaking including education and training, as well as other policy areas which influence innovation (Flanagan et al. 2008). Still, it is noticeable that the alignment of educational and innovation policies continues to be a difficult challenge, reflecting institutional rigidities and conflicting goals. This reminds us that structural divides can intervene in the relationships among innovation practice, policy and theory. Sometimes not all of the actors are in step, for example when administrative responsibilities for different aspects of innovation are inevitably divided among agencies for research, enterprise, universities, secondary education and finance, resulting in fragmented policymaking with limited coordination (Braun 2008; Edler & Kuhlmann 2008). Similarly, disciplinary divides and competing perspectives are common in academia, which can be hard for policymakers and practitioners to sort out. Even when there is shared will among multiple stakeholders, for example to better link educational policies with innovation objectives, system construction may itself limit the effectiveness of implementation. These barriers may be self-reinforcing and hence limit the likelihood of breakthroughs in innovation performance, as suggested by path dependency theorists (see discussions in Fuchs & Shapira 2005).

Building on this review of our major starting points, we now turn to trends in the wider environment of innovation systems and processes that are of particular importance for the relationships of the three dancers of theory, policy and practice.

TRENDS IN THE INNOVATION ENVIRONMENT

The development of innovation theory, practice and policy is intrinsically linked to and influenced by trends in their environment. Some of these trends have a more general character; others are more specific for one of the three dancers. Many of these trends have already been discussed in detail in the chapters of this book, including developments in systems of science and knowledge production (Martin), the influence of increased international competition (Smith), broader changes in the organization of firms (Dankbaar and Vissers), and evolutions in policy governance (Smits, van Merkerk, Guston and Sarewitz).

We anticipate that internationalization and globalization will continue to become one of the most dominant trends that impacts all three dancers (Archibugi et al. 1999; Narula 2003; Edler & Polt 2007). There has been a dramatic rise of Southeast Asian countries, particularly China and India, as internationally-competitive producers of new knowledge and sites for innovation (Lundvall et al. 2006; Zhou & Leydesdorff 2006; Altenburg et al. 2008). Brazil is among other fast developing countries likely to enlarge global roles in innovation in the coming period (Cassiolato et al. 2003), while Russia has immense potential to do so (Gianella & Tompson 2007). These nations are now participants in the international innovation policy race, and in the near future others will possibly join them. While these new dynamics refer to 'nations' it is also appropriate to consider what might be the future role of nation states in innovation systems. More than ever, technology and innovation related producer-user chains are working either 'local' (within clusters, districts and other agglomerations) or 'inter-local', including inter- and transnational networks: they are often less rooted within 'national' economies and institutions. Similarly, in many thematic innovation domains tacit knowledge is getting embodied and institutionalized (in people, rules, organizations) either 'locally' or 'inter-locally' (with – often web-based – collaboration and through growing international migration of knowledge workers). The locus of tacit knowledge is less and less 'national'. Globalizing firms and their research and development, universities and other research organizations have become international actors.

These ongoing globalization trends reinforce demands on OECD countries not only to enhance domestic innovation capabilities, but also to improve their capacities to access and exploit worldwide knowledge pools. Moreover, pressures are also increased in advanced economies to stimulate innovation strategies that can maintain an edge, for example approaches that develop close relationships with innovation users. As a consequence, new actors have to be involved in innovation processes, adding to the complexity of innovation processes. For example, innovation users often have distinct interests and knowledge as compared to traditional scientific and corporate innovation producers. New relations and related learning processes have to be constructed. This increased and intensified involvement of users can be perceived as another motor behind the trend towards more open innovation systems and processes (Oudshoorn & Pinch 2003; Boon et al. 2008). Another consequence of internationalization concerns the organization of the value chain. The international restructuring and relocation of R&D, production and marketing functions greatly impacts the organization of innovation processes in home countries (whether through the sending or receiving of these functions). This presents further challenges for policymakers and innovators, who increasingly need to recognize international dimensions as they formulate their own strategies. Improving insights about the internationalization of innovation and its consequences in turn is one of the major challenges for innovation scholars.

As we complete the writing of this chapter (in late 2008), a dramatic new twist in global economics appears – that of massive crisis and upheaval of financial systems. It remains to be seen whether this will generate an international economic recession that, while deep, can be turned around in one to three years, or a longer slump that will require up to a decade for recovery. Irrespective, there are many implications for innovation. On the one hand, with credit squeezed, demand down and uncertainty about the future, many companies (large as well as small) face retrenchment or collapse. In this environment, corporate R&D investment and projects for new products and services are greatly at risk. Increased government regulation is apt to affect innovation in the financial sector. Yet, from a longer-term view, as Joseph Schumpeter observed almost a century ago, innovators can take advantage from downturns in the business cycle, and upturns in growth can be aided at least in part by innovation. In November 2008, entries such as 'Innovation Loves a Crisis', 'Crisis Could be Right Time for Innovation' and 'Crisis, the Mother of Innovation' were being posted in entrepreneurship web blogs. Much of this is surely wishful thinking. Yet, as policymakers move through their first (and massive) tranche of emergency measures to shore up banks, address mortgage defaults and free-up credit, they can at some point be expected to turn to ways of regenerating growth and stimulating innovation. Again, there will be opportunities for innovation thinkers and practitioners to contribute to debates with policymakers about how best to do this.

Sustainability presents one of the most prominent opportunities for linking innovation not only to economic growth but also to environmental, social and developmental challenges. Attention to sustainability has indeed been a growing trend in the innovation domain (Meyer-Krahmer 1997; Elzen et al. 2004; Lehmann-Waffenschmidt 2007; Murphy 2007). We expect that policy demands and measures to address sustainability will have even greater influence on innovation in future years, just as innovators themselves focus on opportunities to address needs and demands for low-carbon, high-value products and services. In the US, the incoming Obama administration has pledged to invest $150 billion over ten years to support clean energy initiatives. To be effective, this initiative requires innovation across a spectrum of sectors, from automobile manufacturing and energy production to housing and recycling. Moreover, frameworks and incentives for sustainable innovation will be required. Other countries are implementing similar policies. However, as policies and programmes for sustainability become widespread, tensions will increase between proponents of a more sustainable society and those in favour of economic growth – indeed, the global economic crisis will amplify these strains. Here, linking sustainability with innovation could play an important role in bridging this gap. As perhaps the lack of attention to sustainability in this collection indicates, there are needs and opportunities for innovation scholars to more fully engage with this trend.

A more general but related trend concerns the need for sectoral transformation in major components of Western economies, and the time required to implement these changes. The challenges of heightened international competition and sustainable development generate both push and pull towards smarter and greener products and services. These processes of change involve fresh and more intense knowledge needs and new relations with different actors – which in turn can have profound impacts on the set-up of existing innovation systems (see den Hertog in this book). The following example illustrates the seriousness of these impacts. As was described in the chapter by Smits, Teubal and Kuhlmann, the Dutch agricultural sector has for many decades been successful through a knowledge-intensive but mass production approach. It is an approach with high environmental impacts. In the future, the sector is challenged to shift to more environmentally-friendly methods of production. This may only be possible by restructuring the Dutch agro innovation system from a supply-side oriented, top-down system producing for homogeneous markets to a bottom-up oriented system with close links to varied set of users and with fresh approaches to product, process and service innovation. In fact, the shaping of this new system, including the reshaping of policy instruments, institutions and actor roles and relations has been going on for some 10-15 years and is expected to go on for at least the next decade. This demonstrates the time and effort necessary to both break down and build up new innovation systems. It also puts in perspective the time horizons and resources that policymakers need to commit to see through major changes in sectoral innovation systems.

Trends impacting more specifically the innovation practitioner include the ongoing development of current technologies in such fields as computing, electronic networking and communications, and the emergence of new technologies like genomics or nanotechnology with a broad array of possible applications. Also to be addressed are challenges of increasing international competition, the drive to force firms to produce products and services with higher added value, changes in the nature of the research system and its relations with the outside world, market and regulatory uncertainty, and demands for sustainability. These trends have important consequences. Firms often have to develop more intense and stronger interfaces with a growing set of heterogeneous innovation system actors demonstrating new demands. An important consequence of the increased competition for 'added value' is that mature and small-mid-sized enterprises (SMEs) outside of glamorous high technology industries have to improve their innovative capacities (Hirsch-Kreinsen et al. 2005). Over the last decades many policies have been developed to stimulate innovation in SMEs (as Shapira notes earlier in this book). Those efforts which have tried to promote the take-up of innovation from outside (for example, through technology transfer) have not had great impact on most SMEs. In the future these firms will be forced to develop their innovative capacities from within. Employees with adequate qualifications and management capable of exploiting the knowledge and creative potential of employees are necessary preconditions. In order to increase the added value of their products and services firms have to develop new relations with knowledge institutions, knowledge-intensive business services or intermediaries in between in which they position themselves as demanding clients and no longer as passive receivers of knowledge.

Driven by the challenges of international competition and the rise of major societal problems like sustainability, policymakers experience great pressure to develop policies to contribute to an environment that facilitates firms to innovate and to take the lead in developing innovative solutions for societal problems. As Keith Smith (in this book) rightly observes, internationalization does not necessarily minimize the role of national/regional public innovation policy. He argues that innovation system elements such as those for knowledge infrastructure, risk management and education are 'largely independent of the constraints of globalization' and that they 'remain open to discretionary policy strategies'. However, many current strategies and policies are obsolete or ineffective. Policymakers are challenged to change their mindsets and develop new, more systemic policy concepts and instruments. The heritage of old mindsets as well as the far more complicated nature and long-term orientation of the systemic approach can cause many problems

Although the systemic approach has gained broad support in circles of innovation scholars, the consequences for policy are still rather unclear. One of the major reasons for this is that the approach has often been a static one. Research into the dynamics of innovation systems and in the emergence of new systems us still relatively recent. Also insights about various types of learning, a crucial element of innovation systems, have often not found their ways into policy implementation.

These trends put pressure on scholars to invest more in strengthening the systemic approach in such a way that these dynamics are taken into account, the role of various actors is better understood and ideas are developed for instruments that are better able to deal with the systemic character of innovation processes.

POSITIONED TO INFLUENCE CHANGE?

The concept of innovation as a systems activity is gaining ground among a broad range of actors. Various actors take the systemic character of innovation processes into account in their actions and policies. Reasons for this include the influence of conceptual and empirical research which, as clearly demonstrated, has broadly the systemic character of innovation processes (beginning with landmark works such as Freeman 1985; Lundvall 1992; Nelson 1993;,and continuing with subsequent research as discussed in several chapters in this book). This has been complemented by business-oriented research from authors like von Hippel (1988), Porter (1990) and Chesbrough (2005) who have provided influential models of innovation within systems of multiple producers, users, knowledge sources, and networks. Other push factors have included disappointment with the outcomes of more linear-based policies, financial, market, technological and strategic pressures which have stimulated firms to enter into more open types of innovation, and changes in the range and role of actors, intermediaries and institutions in the innovation process. These forces have all reinforced the point that innovation has a systemic character.

The observation that actors increasingly adopt a systems approach to innovation may seem obvious. However, although innovation always has been a systemic activity, for much of the post-World War II period, there was a perception that innovation was a straightforward supply-driven process that could be best influenced through public policy by investing in public R&D. This perception influenced not just policymaking but the allocation of public resources, tilting them to support the expansion of R&D in universities and national laboratories. This perception now seems to be waning. In part, this has been through the impact of innovation scholars – who were once marginal but who are now more engaged and drawn upon in deliberations by policymakers, research managers, industrialists and other stakeholders. We need to be cautious in over estimating impact. Many scholars are still bounded by conventional disciplinary boundaries and canons (for example, micro-economic theories of market operations are still often taught with knowledge and technology as exogenous variables). Yet, as demonstrated by the chapters in this book by multiple authors from different disciplinary backgrounds and countries, the embracement of systems approaches to innovation by scholars has become widespread.

The question thus arises: how well positioned is the scholarly innovation community to respond to emerging challenges in the innovation environment and system? Strength is found in the cumulative build-up of knowledge and the testing and validation of concepts about innovation systems that has occurred over the past two decades. Moreover, there is opportunity in that the increasingly complex and interlinked innovation problems that policymakers and practitioners face demands coming to grips with the systemic nature of innovation. However, there are weaknesses too. There are many different versions and perceptions of the theoretical concepts put forward under the systems of innovation umbrella. Furthermore, one should not be over optimistic about the role of scientific research in policymaking and practice. The gap between theory and practice can be a difficult one to bridge. In the field of innovation studies, there are even more reasons to take a critical stand on the relation between research and practice. Often theories are developed in close interaction with policymakers and other actors. By it is not surprising that these actors often wholeheartedly receive results of research. Often actors accept the results because they were part of their production and they are familiar to them instead of basing themselves on rigorous (scientific) evaluations. Moreover, these co-produced researches are rather context- and time-dependent. Attempts to transfer and apply them as 'best practices' in other contexts often fail.

OUTLOOK: HOW MIGHT THE DANCE PROCEED?

What now can be gleaned about how the dance between innovation theory, policy and practice might proceed? In this section, we explore issues and options related to each of these three domains and how they relate to the others. We make no claims to be exhaustive, rather focus on what we see are pressing and interesting issues for engagement.

Innovation Scholars and their Links with Policy and Practice

From the foregoing it is clear that innovation scholars (can) play an important role in the further development of innovation policy and practice. Innovation scholars are important as producers of the resources learning processes need and at the same time play a role in these learning processes. They can contribute more to these learning processes between the three dancers by focusing on the improvement of insights in the systemic character of innovation. In doing so they could focus on three major flaws of this approach: the still too static nature of the innovation systems approach, the underdeveloped insight in the role of actors at the micro level, how they are framed by the system and – in turn – impact on the system and the consequences for policy in terms of concepts, strategies and instruments. Attempts to tackle these problems already are visible (in several chapters of this book, including Smits, Teubal and Kuhlmann), but still a lot more has to be done before this approach provides results that may help policymakers and practitioners in their daily work. With this we do not want to suggest that all innovation scholars should focus on improving the innovation systems approach. What we want to stress is that scholars dealing with innovation should be aware of the fact that innovation is a multi-actor, multi-level game to be studied in an interdisciplinary context. Bearing this in mind, focused research on more isolated aspects of this system is of course necessary too.

The innovation systems approach is not (fully) in line with the role of a researcher as an objective outsider only. Until now, researchers dealing with innovation often take rather different stands regarding their involvement in real life innovation and policy practice. Sometimes the links between research and policy are too strong, preventing objective and sound research results. Sometimes, however, researchers do not want to get involved at all. For example, some scholars in the field of science, technology and society (STS) decline to work directly with policymakers or firms. STS started as a movement focused on the better embedment of science in society but over the years STS has become transformed into an academic discipline (Hackett et al. 2007). In its wish to get accepted by the scientific community, STS closely adhered to academic rules. This resulted in a distant, reflexive position. Over the last decade, however, it has become clear that STS cannot maintain this position and at the same time still take its original mission seriously. The need for contributing to and facilitating of learning in various forms sometimes makes a closer interaction with other actors necessary. To contribute to learning processes, researchers more often should take a deliberate and at the same time transparent role as a knowledge-intensive business service stimulating the transfer of relevant knowledge, if necessary translate this knowledge to make it useful for other actors and assist in articulating the demand for knowledge. In the chapter of Smits, van Merkerk, Guston and Sarewitz in this book, this information was called 'strategic intelligence'. The interaction with practitioners and policymakers that results from this role in turn will provide scientists with (often tacit) knowledge on the dynamics of innovation processes and the impact of policy interventions. Of course such a more active role may threaten the objectivity of the research. However, certainly in this multidisciplinary and problem-driven field, objectivity almost always is an unachievable goal. Instead of striving after absolute objectivity a more promising, and from a scientific point also acceptable, strategy would be to make the involvement of researchers as transparent as possible. Furthermore, to avoid tangling and disguise of interests, it will be necessary to reflect on the code of conduct of science and take care that this code of conduct is followed.

Adopting and shaping such a role will not be easy and will meet fierce resistance. The rules of the academic world reflected in the peer review system even force scientists away from engagement with practice. This of course is of extra importance for a young discipline such as innovation studies that has not yet acquired a broadly recognized position in the academic world. However, given the many attempts we see nowadays to stimulate science that is of high quality and at the same time relevant for society (see chapter by Martin in this book) and the examples presented in this book of the importance of mutual learning processes between science, policy and practice, adapting this new role might be considered one of the major challenges of innovation scholars. There is, meanwhile, encouraging evidence that explicit

'Science, Technology, Innovation, and Policy Studies' (STIPS) are turning into a self-standing field of academic reputation, education and research, with relative stable international networks, a number of accepted scientific journals, and regular conferences. An international community of STIPS has to facilitate complicated learning processes involving building bridges between worlds with rather different cultures, practices and reward systems. Also, STIPS would increasingly have to include both subject areas and colleagues from the 'non-OECD' world.

To stimulate this learning process it could be helpful to involve 'hybrid' actors. By this we mean people who work in more than one of the three domains, for instance scientists who become policymakers, consultants or entrepreneurs, entrepreneurs that became policymakers or policymakers turning back to academia. These hybrids experienced the differences between these worlds in their daily life and can provide valuable advice how to deal with them.

Policymakers and their Roles in the Innovation System

Policymakers – although they often advocate the systemic approach – still often rely on the linear model when developing policies and policy instruments. The lack of useful and practical insights from innovation scholars may be seen as an important reason for this. However, in changing the dance policymakers too have an important role to play. First, a more systemic approach as discussed in this book meets a couple of important blockades in the policy system. Systemic policies ask for horizontal policies cutting through well-defined and fiercely protected policy domains. A systemic policy approach requires – in principle – a global perspective, exceeding national borders and constituencies. Innovation actors *and* policy actors get embedded in complex international governance structures, driven by policy mixes and diverse actor strategies on multiple levels. Still, at the same time, national policymakers remain in charge of national budgets as (still) the major source of public investments in research and innovation; also national polity provides democratic legitimisation (legislation). Policymakers at all levels still frame their policies in a national systems perspective, firmly embedded in the political accountability of their national level. Existing international organizations are not yet sufficiently aware of upcoming common problems. Coordination mechanisms are still poorly defined, and truly global programmes are still odd exceptions; this holds in particular for non-OECD countries.

Also the short-term rhythm of our political systems provides problems given the long-term orientation of systemic policies. These are problems that cannot be solved by the other two dancers. A next problem policymakers have to solve is the problem of de-learning. The popularity of the linear model not only is caused by the lack of practical alternatives, it also reflects the tendency of bureaucratic organisations to segmentate interlinked fields of action in order to minimize the span of control and responsibility.

Even if one could suppress this mechanism, the shaping of systemic policies remains difficult. The first element concerns orchestration and long-term strategies. Strengthening the innovative capacity of economy and society asks for choices and a long-term orientation. Government could play a role here as an orchestrator stimulating, and if necessary organizing, the debate in (sectors of) our society on long-term innovative strategies. Broadly based and accepted innovation agendas may help to reduce the many uncertainties inherent to innovation and bundle forces necessary to survive in an ever more competitive world. The UK Foresight programme provides a rather successful example of such a guiding role of government and also countries like Finland, Japan and Taiwan recently developed such strategic initiatives. A second element relates to learning. The learning capacity is central to a high quality innovation system. As was demonstrated in the foregoing, the learning processes of and between the dancers are crucial for the further development of theory, policy and practice. In an innovation systems approach, learning refers less to the capacity to produce new knowledge and more to the capacity to select, acquire and handle knowledge and turn it into useful applications. In order to learn and develop visions, strategies and action plans actors need information on the options science and technology have to offer, what these options mean for their sector or domain, and need information on the consequences of the realization of these options. Often actors themselves will be able to acquire or produce the necessary information. However, some actors involved will not have the resources to do so and especially in the early stages of the development of scientific and technological options, actors are not aware of these options or find it too risky to invest in them. In such cases it is up to government to take care that there are institutions in place in the innovation system that provide the necessary information, i.e. 'strategic intelligence' (Smits et al. in this book). In order to stimulate innovation processes in an early stage and take care that all relevant actors have access to and are provided with the information they need to play their role, government should take the responsibility to develop an adequate strategic intelligence infrastructure. In many countries elements of such an infrastructure are already present, but given the growing importance of innovation for economic welfare and societal well-being, these infrastructures should be reinforced and better internationally interlinked. Understanding innovation systems and policies in a globalized context requires 'post-national' analytical approaches. In line with the metaphor of the 'three dancers', the various actors involved in innovation processes should play an important role in setting the agenda of this global 'strategic intelligence infrastructure'. So far there is no coherent conceptual and methodological framework addressing the multilevel and transnational nature of today's innovation systems, where the different spatial and organizational levels – from the regional, to national states, to the global – have distinct, but largely connected roles, and that take into account the strategies of key individual actors.

A second, more indirect aspect of the learning capacity of innovation systems concerns the educational system. This aspect becomes even more important because in order to survive on worldwide markets, no longer only a limited set of high-tech and big firms have to innovate. Future developments ask for a reinforcement of the innovative capacity of firms in a much broader range. During the last decades policies stimulated innovation in SMEs in a linear model mode focusing on technology transfer. Often these attempts to stimulate innovation from the 'outside-in' were not very successful. From a systemic perspective it makes more sense to change these policies to 'inside-out' policies. To phrase it differently, instead of trying to introduce innovative resources from the outside governments should focus on strengthening the innovative capacity of firms from the inside. Such a strategy asks for adequately educated employees. Employees, for instance, able to live and act in the world of science and technology and in the world of the market and are able to link these two. This asks for fundamental changes in our educational systems. Strange enough, however, until now innovation scholars have hardly addressed the educational system. Research in the requirements the knowledge society asks from the educational system is scarce. As a result our educational systems do not provide our labour markets with people that are adequately equipped to take part in the learning processes innovation requires. In the long run this might prove to be the major weakness of our innovation systems.

Practitioners – Exploring Opportunities for Greater Connectivity and Reflexivity

Innovation is often seen in gladiatorial combat terms. Innovators in companies and other organizations are expected to be champions of change and to work hard to overcome barriers. Thomas Edison's aphorism (1 percent inspiration, 99 percent perspiration) is an apt description. Failure is to be expected, and often successful innovators are heard to say that you need to fail one or more times to learn how to succeed. But is this 'school of hard knocks' the most efficient way of fostering innovation? Does it have to be the innovator versus the system, or are there ways to ensure that the system is more conducive to innovators? It could be argued that there are immutable laws of human nature (i.e. we tend to resist change) that mean that innovation is typically going to be an uphill struggle. Yet, at the same time, the way in which the system of innovation is constructed surely also provides rules and conditions behaviour, and this system is subject to change. To be sure, it can be hard to change established institutional routines. But, if we could change the system to be more favourable to the acceptance of innovation and to enhance desirable outcomes (for example, to meet goals of sustainable economies), how would we?

One direction is provided in the chapter by Smits, van Merkerk, Guston and Sarewitz where they raise the possibilities of forms of constructive technology assessment and real-time technology assessment – where the best possible uses of promising technologies are considered early on in the innovation cycle. Through the early engagement of users, consumers, community and environmental stakeholders, business, scientists and policymakers, deliberations can occur which can send signals and help channel actions. Coupling the results of such deliberations with supportive resource allocations and other incentives would increase the likelihood that desired societal goals as well as economic ones are addressed through collaborative rather than isolated or disjointed innovation processes. Similarly, as Edler discusses in his chapter on demand-driven innovation, greater use of demand mechanisms by policymakers could speed up the pioneering and take-up of desirable innovations. Success in moving along these collaborative innovation pathways is, of course, dependent on strengthening the links between policymakers and innovators, although there is probably much that can be done in fostering collaborative innovation through actions by private organizations (user-producer links) and intermediaries and advocacy groups (giving greater emphasis to the societal goals that innovation should address).

There may also be opportunities to stimulate greater reflexivity among those on the front-line of innovation, to learn from others – to avoid making old mistakes and to recover from new ones. Much of this learning is inevitably informal and perhaps unavoidably experiential. But, there are roles for educational communities and developmental organizations to engage more with the innovation process, through cross-training of young scientists (learning about management of innovation) and with young managers (learning about innovation and technology), facilitating exchanges and group innovation projects, and comparing practices across regions and innovation systems. Within companies and organizations, there may be new opportunities for managers to develop open innovation strategies and enable improved interfaces with other knowledge sources, resources and organizations, to facilitate the implementation of creative potential.

Other pathways are possible, and we recognize that many organizations around the world are indeed piloting new initiatives to foster new collaborative approaches to innovation which address a broad range of goals. The essential point is that while scholars, policymakers and practitioners each influence one another in the development of innovation concepts, policies and actions, there are also significant opportunities to collaboratively advance and improve the structure and operations of innovation systems themselves. We hope the arguments, ideas and cases discussed in this book will help in working through how innovation systems thinking, policy and practice can move forward over the next decade of change.

REFERENCES

Altenburg, T., Schmitz, H. and Stamm, A. (2008), 'Breakthrough? China's and India's transition from production to innovation', *World Development*, **36** (2), 325-344.

Archibugi, D., Howells, J. and Michie, J. (eds) (1999), *Innovation Policy in a Global Economy*, Cambridge, UK: Cambridge University Press.

Baack, B. and Ray, E. (1985), 'The political economy of the origins of the military-industrial complex in the United States', *Journal of Economic History*, 45 (2), 369-375.

Balconi, M., Brusoni, S. and Orsenigo, L. (2008), 'In defence of the linear model: an essay', Working Paper 216, CESPRI, Bocconi University, Milan.

Boekholt, P., Lankhuizen, M., Arnold, E., et al. (2001), 'An international review of methods to measure relative effectiveness of technology policy instruments', Technopolis, Brighton and Amsterdam.

Boon, W.P.C., Moors, E.H.M., Kuhlmann, S. and Smits, R.E.H.M. (2008), 'Demand articulation in intermediary organisations: the case of orphan drugs in the Netherlands', *Technological Forecasting and Social Change*, 75 (5), 644-671.

Braun, D. (2008), 'Organising the political coordination of knowledge and innovation policies', *Science and Public Policy*, 35 (4), 227-239.

Cassiolato, J.E., Lastres, H.M. and Maciel, M.L. (eds) (2003), *Systems of Innovation and Development: Evidence from Brazil*, Cheltenham, UK and Northampton, MA, USA: Edward Elgar.

Chesbrough, H.W. (2005), *Open Innovation: The New Imperative for Creating and Profiting from Technology*, Boston: Harvard University Business School Press.

Damanpour, F. (1991), 'Organizational innovation: a meta-analysis of effects of determinants and moderators', *Academy of Management Journal*, 34 (3), 555-590.

DIUS (2008), 'Innovation nation', Department for Innovation, Universities and Skills, London.

Edler, J. and Kuhlmann, S. (2008), 'Coordination within fragmentation. Governance in knowledge policy in the German federal system', *Science and Public Policy*, 35 (4), 265-276.

Edler, J. and Polt, W. (eds) (2008), 'Symposium: international industrial R&D – policy challenges', *Journal of Technology Transfer*, 33 (4), 331-336.

Elzen, B., Geels, F.W. and Green, K. (eds) (2004), *System Innovation and the Transition to Sustainability: Theory, Evidence and Policy*, Cheltenham, UK and Northampton, MA, USA: Edward Elgar.

Flanagan, K., Uyarra, E. and Laranja, M. (2008), 'Opening the policy black box: understanding innovation policy mixes in multi-level, multi-actor spaces', Working Paper, Manchester Institute of Innovation Research, University of Manchester, Manchester.

Fuchs, G. and Shapira, P. (eds) (2005), *Rethinking Regional Innovation and Change: Path Dependency or Regional Breakthrough?* Berlin, Germany, Heidelberg, Germany and New York, USA: Springer-Verlag.

Gianella, C. and Tompson, W. (2007), 'Stimulating innovation in Russia: the role of institutions and policies', Economics Department, Working Paper, ECO/WKP (2006) 67, Organisation for Economic Cooperation and Development, Paris.

Godin, B. (2006), 'The linear model of innovation: the historical construction of an analytical framework', *Science, Technology & Human Values*, 31 (6), 639-667.

Hackett, E.J., Amsterdamska, O., Lynch, M. and Wajcman, J. (eds) (2007), *The Handbook of Science and Technology Studies*, Cambridge, MA, USA and London, UK: MIT Press.

Hargadon, A.B. and Douglas, Y. (2001), 'When innovations meet institutions: Edison and the design of the electric light', *Administrative Science Quarterly*, **46**, 476-501.

Hippel, E. von (1988), *The Sources of Innovation*, Oxford, UK and New York, USA: Oxford University Press.

Hirsch-Kreinsen, H., Jacobson, D. and Laestadius, S. (eds) (2005), *Low-tech Innovation in the Knowledge Economy*, Frankfurt: Peter Lang.

Lehmann-Waffenschmidt, M. (ed.) (2007), *Innovations Towards Sustainability: Conditions and Consequences*, Heidelberg and Berlin: Physica-Verlag.

Lundvall, B.-A. (2007), 'Innovation system research. Where it came from and where it might go', Globalics Working Paper Series, No. 2007-01, The Global Network for Economics of Learning, Innovation, and Competence Building System. [http://www.globalics.org]

Lundvall, B.-A., Intarakumnerd, P. and Vang, J. (eds) (2006), *Asia's Innovation Systems in Transition*, Cheltenham, UK and Northampton, MA, USA: Edward Elgar.

Lundvall, B.-A. (ed.) (1992), *National Innovation Systems: Towards a Theory of Innovation and Interactive Learning*, London, UK and New York, USA: Pinter.

Meyer-Krahmer, F. (ed.) (1997), *Innovation and Sustainable Development: Lessons for Innovation Policies*, Heidelberg and Berlin: Physica-Verlag.

Murphy, J. (ed.) (2007), *Governing Technology for Sustainability*, London: Earthscan.

Narula, R. (2003), *Globalization and Technology: Interdependence, Innovation Systems and Industrial Policy*, Cambridge: Polity Press.

Nelson, R.R. (ed.) (1993), *National Innovation Systems: A Comparative Analysis*, Oxford, UK and New York, USA: Oxford University Press.

NESTA (2007), 'Hidden innovation', National Endowment for Science, Technology and the Arts, London.

OECD (2005), 'Governance of innovation systems', Volume 1, Synthesis Report, Organisation for Economic Cooperation and Development, Paris.

OECD (1991), 'Technology in a changing world: the technology-economy program', Organisation for Economic Development and Cooperation, Paris.

Oudshoorn, N.E.J. and Pinch. T.J. (eds) (2003), *How Users Matter. The Co-construction of Users and Technology*, Cambridge, MA, USA and London, UK: MIT Press.

Popper, S.W. and Wagner, C.S. (2002), *New Foundations for Growth: The U.S. Innovation System Today and Tomorrow*, Santa Monica: Rand Corp, CA.

Porter, M. (1990), *The Competitive Advantage of Nations*, New York: The Free Press.

Rosenfeld, S.A. (1997), 'Bringing business clusters into the mainstream of economic development', *European Planning Studies*, 5 (1), 3-23.

Saxenian, A. (1994), *Regional Advantage: Culture and Competition in Silicon Valley and Route 128*, Boston: Harvard University Business School Press.

Youtie, J. and Shapira, P. (2008), 'Building an innovation hub: a case study of the transformation of university roles in regional technological and economic development', *Research Policy*, 37 (8), 1188-1204.

Zhou P. and Leydesdorff, L. (2006), 'The emergence of China as a leading nation in science', *Research Policy*, 35 (1), 83-104.

Index